POSS-Based Polymers

POSS-Based Polymers

Special Issue Editor

Ignazio Blanco

MDPI • Basel • Beijing • Wuhan • Barcelona • Belgrade

MDPI

Special Issue Editor
Ignazio Blanco
University of Catania
Italy

Editorial Office
MDPI
St. Alban-Anlage 66
4052 Basel, Switzerland

This is a reprint of articles from the Special Issue published online in the open access journal *Polymers* (ISSN 2073-4360) from 2018 to 2019 (available at: https://www.mdpi.com/journal/polymers/special_issues/POSS_Based_Polym).

For citation purposes, cite each article independently as indicated on the article page online and as indicated below:

LastName, A.A.; LastName, B.B.; LastName, C.C. Article Title. *Journal Name* **Year**, *Article Number*, Page Range.

ISBN 978-3-03921-994-0 (Pbk)
ISBN 978-3-03921-995-7 (PDF)

Contents

About the Special Issue Editor

Ignazio Blanco is Full Professor of Chemical Foundations of Technologies at the Department of Civil Engineering and Architecture at the University of Catania. His research activities are focused on the following themes: synthesis and characterization of toughened thermoset blends; process technology of polymeric fiber-reinforced composites; heat capacities, fusion, and solid-to-solid transition of series of organic molecules; comparative kinetic studies of the thermal degradation of model polymers; comparative kinetic studies of the thermal degradation of nanocomposites; synthesis and characterization of aromatic copolymers; thermal characterization of films used in food packaging applications; synthesis and characterization of nanoparticles; POSS synthesis and characterization; and material characterization for rapid prototyping. To date, he has authored and/or co-authored 90 papers indexed on Scopus, 1 book chapter, and 80 proceedings for national and international conferences. The published papers have amassed 1550 citations corresponding to an h-index of 28.

polymers

MDPI

Editorial

POSS-Based Polymers

Joseph D. Lichtenhan [1], Krzysztof Pielichowski [2] and Ignazio Blanco [3,*

[1] Hybrid Plastics Inc., Hattiesburg, MS 39401, USA; lichtenhan@hybridplastics.com
[2] Department of Chemistry and Technology of Polymers, Cracow University of Technology,
   Ul. Warszawska 24, 31-155 Kraków, Poland; kpielich@pk.edu.pl
[3] Department of Civil Engineering and Architecture and INSTM UdR, University of Catania, V.le A. Doria 6,
   95125 Catania, Italy
* Correspondence: iblanco@unict.it

Received: 14 October 2019; Accepted: 21 October 2019; Published: 22 October 2019

The combination of functional polymers with inorganic nanostructured compounds has become a major area of research and technological development owing to the remarkable properties and multifunctionalities deriving from their nano and hybrid structures. In this context, polyhedral oligomeric silsesquioxanes (POSSs) have increasing importance and a dominant position with respect to the reinforcement of polymeric materials. Although POSSs were first described in 1946 by Scott, these materials, however, were not immediately successful if we consider that, starting from 1946 and up to 1995, we only find 85 manuscripts in the literature regarding POSSs. This means that less than two papers per year were published over 50 years. Since 1995, we observe an exponential growth of scientific manuscripts concerning POSSs. It has changed from an annual average of 20 manuscripts for the period 1995–2000 to an annual average of about 400 manuscripts, with an increase of 2800%. The introduction of POSSs inorganic nanostructures into polymers gives rise to polymer nanostructured materials (PNMs) with interesting mechanical and physical properties, thus representing a radical alternative to the traditional filled polymers or polymer compositions.

Polyhedral oligomeric silsesquioxanes, with Si vertices interconnected by –O– linkages, form three-dimensional nanometer size cage structures with substituents attached to silicon atoms. These substituents may contain reactive groups, such as hydroxyl or isocyanate. A combination of a rigid inorganic nanocore with organic vertex groups makes POSS molecules useful hybrid building blocks that can be chemically incorporated in the polymer matrix by copolymerization, grafting or reactive blending, or physically mixed by solvent casting or polymer processing by using, for example, the extrusion technique. Depending on the number and kind of functional groups attached to the POSS cage, silsesquioxane moieties can be chemically built into a polymer structure as a side group of the main chain or terminating end-group, as a main chain fragment, or as a macromolecular network knot.

Various routes of chemical decoration of POSS molecules with organic substituents offer new perspectives for synthesis of novel organic–inorganic hybrid materials with desirable—and often still unknown—properties. Interest in POSS-containing polymer composites and hybrid materials has been growing in the recent years as they show improved mechanical performance, thermo(oxidative) stability, and surface durability. Well-defined structure, non-toxicity, and enhanced biocompatibility, as well as biostability against oxidation, hydrolysis, and enzymatic attack under in vitro and in vivo conditions make functionalized silsesquioxanes desirable nanofillers in the biomedical field as implants and scaffolds. Other application areas include membranes, high-performance adhesives, flame retardants, aerogels, optical sensors, and shape-memory materials, to name a few. On the other hand, challenges with polyhedral oligomeric silsesquioxanes' tendency to agglomerate are still to be addressed during synthesis and engineering of nanostructured polymer-based composites made thereof.

This special issue, which consists of 18 articles, including two review articles, written by research groups of experts in the field, considers recent research on novel POSS-based polymeric materials.

Firstly, it was highlighted that, compared with the other most commonly used fillers, POSSs possess the advantage of being molecules, thus allowing the combination of their nano-sized cage structures, which have dimensions that are similar to those of most polymer segments, and production of a particular and exclusive chemical composition. These characteristics linked with their hybrid (inorganic–organic) nature allow researchers to modify POSS according to particular needs or original ideas, before incorporating them into polymers [1].

Liu et al. synthesized a novel organic–inorganic hybrid containing allyl benzoxazine and polyhedral oligomeric silsesquioxane (POSS) to be used for preparing epoxy resin composites in order to verify the effect of POSS on the thermal stability and flame retardancy of a prepared material. Improvement in thermal stability and flame retardancy was observed when the amount of POSS reached 10% or more, thus demonstrating that POSS nanoparticles can effectively protect the combustion of internal polymers [2].

Wei and co-workers, by using the graft-from method, prepared a series of heptaphenyl siloxane trisilanol/polyhedral oligomeric silsesquioxane (T_7-POSS) modified by polyols to be used for obtaining polyurethane nanocomposites at different POSS contents. The results showed that the polyol-terminated POSS particles overcame the nano-agglomeration effect and evenly dispersed in the polymeric matrix. The damping factor of resultant nanocomposites increased from 0.90 to 1.16, while the glass transition temperature decreased from 15.8 to 9.4 °C when POSS contents increased from 0 to 9.75 wt%. They attributed the improvement of the damping properties of the composites to the friction-related losses occurring in the interface region between the nanoparticles and the matrix [3].

POSS-derived Si@C anode material was prepared by Bai et al. with the copolymerization of octavinyl-polyhedral oligomeric silsesquioxane and styrene, for use in a lithium ion battery. The initial discharge capacity of the battery based on the as-obtained Si@C material Si reached 1500 mAh g^{-1}. After 550 charge–discharge cycles, a high capacity of 1430 mAh g^{-1} was maintained. By using combined XRD, XPS, and TEM analysis they showed the high potential of the novel electrode material and provided insight into the dynamic features of the material during battery cycling, which will be useful for the future design of high-performance electrode material [4].

Chen and collaborator used cationic octa-ammonium polyhedral oligomeric silsesquioxane (Oa-POSS) particles to improve the performance of traditional sodium alginate (SA) hydrogels that meet a greater application demand in the biomedical field. The characterization of the gels demonstrated that their properties depend on the content of both the uniformly dispersed Oa-POSS and poly(*N*-isopropyl acrylamide) (PNIPA) network directly. Furthermore, they observed that the gels with a hydrophilic PNIPA network exhibited better swelling ability and remarkable temperature responsiveness, and their volume phase transition temperature can be adjusted by altering the content of Oa-POSS. The deswelling rate of gels increases gradually with the increase of POSS content due to the hydrophobic Si–O skeleton of POSS [5].

By monitoring the polymerization kinetics by Photo-Differential Scanning Calorimetry (DSC) and photorheology, Marcinkowska et al. investigated the effect of monomethacryloxy-heptaisobutyl POSS (1M-POSS) on the process. They determined that at low concentrations the modifiers with a bulky substituent increase the molecular mobility by increasing the free volume fraction, which leads to an acceleration of the termination and slows the polymerization. At higher concentrations, they retard molecular motions due to the "anchor effect" that suppresses the termination, leading to acceleration of the polymerization. Thus, they considered the possibility of anchoring a monomer with a long substituent around the POSS cage, which further enhances propagation [6].

Song et al. synthesized a series of hybrid thermoplastic polyurethanes from bi-functional polyhedral oligomeric silsesquioxane and polycaprolactone using 1,6-hexamethylene diisocyanate as a coupling agent for testing their thermo-mechanical properties. They observed an increase in decomposition temperature and glass transition temperature when compared with pristine polyurethane, and an improvement of both storage modulus (G′) and loss modulus (G″), showing the possibility to further adjust these parameters by varying POSS content in the copolymer. In addition, the synthesized POSS polyurethanes demonstrated a remarkable effect in toughening commercial polyesters, indicating a simple yet useful strategy in developing high-performance polyester for advanced biomedical applications [7].

The modification of polyurethane foams with aminopropyl isobutyl-POSS (APIB-POSS) and aminoethylaminopropylisobutyl-POSS (AEAPIB-POSS) to enhance their mechanical and thermal properties was the object of the study of Członka and her collaborators. The results showed that the morphology of modified foams is significantly affected by the filler typology and content, which resulted in inhomogeneous, irregular, large-cell shapes, and further affected the physical and mechanical properties of resulting materials. The best results were obtained for the polyurethane modified with 0.5 wt% of APIB-POSS, showing greater compression strength, better flexural strength, and lower water absorption [8].

Aiming to bridge the gap between small-molecule surfactants and amphiphilic block copolymers Qian and coworkers designed and synthesized single-tailed giant surfactants carrying hydrophobic poly(ε-caprolactone) (PCL) as the tail and a hydrophilic cage-like polyhedral oligomeric silsesquioxane (POSS) nanoparticle as the head. To endow the POSS head with adjustable polarity and functionality, three kinds of hydrophilic groups, including hydroxyl groups, carboxylic acids, and amine groups, were installed to the periphery of the POSS molecule by a high-efficiency thiol-ene "click" reaction. The full characterization demonstrated that these giant surfactants can form nanospheres with different sizes in aqueous solution [9].

Li et al. synthesized, by a one-step grafting reaction, a hybrid flame retardant copolymer starting from methacryloisobutyl polyhedral oligomeric silsesquioxane (POSSMA), reactive glycidyl methacrylate (GMA), bis-9,10-dihydro-9-oxa-10-phosphaphenanthrene-10-oxide methacrylate (bisDOPOMA), and derivative functionalized graphene oxide (GO). They showed a remarkable enhancement of the composites' thermal properties by adding the graphene oxide hybrid flame retardant (GO-MD-MP). Furthermore, they observed an increase in the limiting oxygen index as well as the mechanical strength of the epoxy resin [10].

Wang et al. prepared blends of cyanate ester and phthalonitrile–polyhedral oligomeric silsesquioxane and studied their cure behavior by means of thermal and rheology experiments. The obtained copolymers showed high chemical reactivity, low viscosity, and good thermal stability. In addition, an increase in glass-transition temperature of the blends, compared to cyanate ester resin, was recorded, making them suitable for preparing carbon-fiber-reinforced composite materials via a winding process and a prepreg lay-up process with a molding technique [11].

The effect of preparation method, POSS content, and type, on the morphology, thermal, mechanical, and surface properties of poly(ε-caprolactone)/POSS derivatives nanocomposites (PCL/POSS) were studied by Cobos et al. Morphological analysis evidenced that amino-POSS with a longer alkyl chain exhibited a better degree of dispersion independent of preparation method, reducing the formation of POSS crystalline aggregates. They also showed how the incorporation of POSS derivatives into the PCL matrix improved thermal stability and enhanced the surface hydrophobicity of PCL [12].

Ueda and collaborators designed and prepared, by the casting method, dual-functionalized polyhedral oligomeric silsesquioxane (POSS) derivatives, which have seven fluorinated alkanes and a single acrylate ester on the silica cube, to be used as a filler for lowering the refractive index and improving the thermomechanical properties of poly(methyl methacrylate) (PMMA). They observed a large lowering of the refractive index. Moreover, the degradation temperatures and the storage moduli of the obtained films were greatly elevated by loading the POSS fillers [13].

Niemczyk and coworkers were engaged in the evaluation of a novel series of siloxane-silsesquioxane resins as possible flame retardants in polypropylene (PP) materials. Their results revealed that the functionalized resins formed a continuous ceramic layer on the material surface during its combustion, which improved both thermal stability and flame retardancy of the PP materials. This beneficial effect was observed especially when small amounts of siloxane-silsesquioxane resin were applied [14].

Li et al. prepared two different models of hybrid ionic liquids (ILs) based on polyhedral oligomeric silsesquioxanes (POSSs), showing excellent thermal stability and low glass transition temperatures.

They then focused their attention on the high sensitivity of these products for detecting nitroaromatic compounds, highlighting their great potential for the detection of explosives [15].

Lin and collaborators prepared and characterized mesoporous molecular sieves by using rice husk as a silicon source, thus also solving environmental pollution problems to avoid its burning as garbage, in addition to the desired field of application. Structural characterization showed evenly distributed and hexangular mesoporous structures. An increase in pore size was obtained, thus leading to an increase in ammoni—nitrogen adsorption capacity [16].

Finally, this special issue hosts two reviews. The first one, by Dudziec et al., highlights the significant number of papers on the design and development of POSS-based organic optoelectronic as well as photoluminescent (PL) materials. In view of the scientific literature abounding with numerous examples of their application (i.e., as Organic Light Emitting Diodes (OLEDs)), the aim of the review was to present efficient synthetic pathways leading to the formation of nanocomposite materials based on silsesquioxane systems that contain organic chromophores of a complex nature. A summary of stoichiometric and predominantly catalytic methods for these silsesquioxane-based systems to be applied in the construction of photoactive materials or their precursors was given [17].

The preparation of hybrid nanocomposite materials derived from polyhedral oligomeric silsesquioxane (POSS) nanoparticles and polyimide (PI) was the subject of the second review presented by Mohamed and Kuo. The two researchers discuss the various methods used to insert POSS nanoparticles into PI matrices, through covalent chemical bonding and physical blending, as well as the influence of the POSS units on the physical properties of PIs [18].

Conflicts of Interest: The authors declare no conflicts of interest.

References

1. Blanco, I. The Rediscovery of POSS: A Molecule Rather than a Filler. *Polymers* **2018**, *10*, 904. [CrossRef] [PubMed]
2. Liu, B.; Wang, H.; Guo, X.; Yang, R.; Li, X. Effects of an Organic-Inorganic Hybrid Containing Allyl Benzoxazine and POSS on Thermal Properties and Flame Retardancy of Epoxy Resin. *Polymers* **2019**, *11*, 770. [CrossRef] [PubMed]
3. Wei, W.; Zhang, Y.; Liu, M.; Zhang, Y.; Yin, Y.; Gutowski, W.S.; Deng, P.; Zheng, C. Improving the Damping Properties of Nanocomposites by Monodispersed Hybrid POSS Nanoparticles: Preparation and Mechanisms. *Polymers* **2019**, *11*, 647. [CrossRef] [PubMed]
4. Bai, Z.; Tu, W.; Zhu, J.; Li, J.; Deng, Z.; Li, D.; Tang, H. POSS-Derived Synthesis and Full Life Structural Analysis of Si@C as Anode Material in Lithium Ion Battery. *Polymers* **2019**, *11*, 576. [CrossRef] [PubMed]
5. Chen, Y.; Zhou, Y.; Liu, W.; Pi, H.; Zeng, G. POSS Hybrid Robust Biomass IPN Hydrogels with Temperature Responsiveness. *Polymers* **2019**, *11*, 524. [CrossRef] [PubMed]
6. Marcinkowska, A.; Przadka, D.; Dudziec, B.; Szczesniak, K.; Andrzejewska, E. Anchor Effect in Polymerization Kinetics: Case of Monofunctionalized POSS. *Polymers* **2019**, *11*, 515. [CrossRef] [PubMed]
7. Song, X.; Zhang, X.; Li, T.; Li, Z.; Chi, H. Mechanically Robust Hybrid POSS Thermoplastic Polyurethanes with Enhanced Surface Hydrophobicity. *Polymers* **2019**, *11*, 373. [CrossRef] [PubMed]
8. Członka, S.; Strąkowska, A.; Strzelec, K.; Adamus-Włodarczyk, A.; Kairytė, A.; Vaitkus, S. Composites of Rigid Polyurethane Foams Reinforced with POSS. *Polymers* **2019**, *11*, 336. [CrossRef] [PubMed]
9. Qian, Q.; Xu, J.; Zhang, M.; He, J.; Ni, P. Versatile Construction of Single-Tailed Giant Surfactants with Hydrophobic Poly(ε-caprolactone) Tail and Hydrophilic POSS Head. *Polymers* **2019**, *11*, 311. [CrossRef] [PubMed]
10. Li, M.; Zhang, H.; Wu, W.; Li, M.; Xu, Y.; Chen, G.; Dai, L. A Novel POSS-Based Copolymer Functionalized Graphene: An Effective Flame Retardant for Reducing the Flammability of Epoxy Resin. *Polymers* **2019**, *11*, 241. [CrossRef] [PubMed]
11. Li, X.; Zhou, F.; Zheng, T.; Wang, Z.; Zhou, H.; Chen, H.; Xiao, L.; Zhang, D.; Wang, G. Blends of Cyanate Ester and Phthalonitrile–Polyhedral Oligomeric Silsesquioxane Copolymers: Cure Behavior and Properties. *Polymers* **2019**, *11*, 54. [CrossRef] [PubMed]

12. Cobos, M.; Ramos, J.R.; Guzmán, D.J.; Fernández, M.D.; Fernández, M.J. PCL/POSS Nanocomposites: Effect of POSS Derivative and Preparation Method on Morphology and Properties. *Polymers* **2019**, *11*, 33. [CrossRef] [PubMed]

13. Ueda, K.; Tanaka, K.; Chujo, Y. Fluoroalkyl POSS with Dual Functional Groups as a Molecular Filler for Lowering Refractive Indices and Improving Thermomechanical Properties of PMMA. *Polymers* **2018**, *10*, 1332. [CrossRef] [PubMed]

14. Niemczyk, A.; Dziubek, K.; Sacher-Majewska, B.; Czaja, K.; Czech-Polak, J.; Oliwa, R.; Lenża, J.; Szołyga, M. Thermal Stability and Flame Retardancy of Polypropylene Composites Containing Siloxane-Silsesquioxane Resins. *Polymers* **2018**, *10*, 1019. [CrossRef] [PubMed]

15. Li, W.; Wang, D.; Han, D.; Sun, R.; Zhang, J.; Feng, S. New Polyhedral Oligomeric Silsesquioxanes-Based Fluorescent Ionic Liquids: Synthesis, Self-Assembly and Application in Sensors for Detecting Nitroaromatic Explosives. *Polymers* **2018**, *10*, 917. [CrossRef]

16. Lin, D.; Huang, Y.; Yang, Y.; Long, X.; Qin, W.; Chen, H.; Zhang, Q.; Wu, Z.; Li, S.; Wu, D.; et al. Preparation and Characterization of Highly Ordered Mercapto-Modified Bridged Silsesquioxane for Removing Ammonia-Nitrogen from Water. *Polymers* **2018**, *10*, 819. [CrossRef]

17. Dudziec, B.; Żak, P.; Marciniec, B. Synthetic Routes to Silsesquioxane-Based Systems as Photoactive Materials and Their Precursors. *Polymers* **2019**, *11*, 504. [CrossRef] [PubMed]

18. Mohamed, M.G.; Kuo, S.W. Functional Polyimide/Polyhedral Oligomeric Silsesquioxane Nanocomposites. *Polymers* **2019**, *11*, 26. [CrossRef] [PubMed]

polymers

MDPI

Article

The Rediscovery of POSS: A Molecule Rather than a Filler

Ignazio Blanco

Department of Civil Engineering and Architecture, University of Catania and UdR-Catania Consorzio INSTM, Viale Andrea Doria 6, 95125 Catania, Italy; iblanco@unict.it

Received: 31 July 2018; Accepted: 10 August 2018; Published: 11 August 2018

Abstract: The use of polyhedral oligomeric silsesquioxanes (POSSs) for making polymer composites has grown exponentially since the last few years of the 20th century. In comparison with the other most commonly used fillers, POSSs possess the advantage of being molecules. Thus, this allows us to combine their nano-sized cage structures, which have dimensions that are similar to those of most polymer segments and produce a particular and exclusive chemical composition. These characteristics linked with their hybrid (inorganic–organic) nature allow researchers to modify POSS according to particular needs or original ideas, before incorporating them into polymers. In this present study, we first start with a brief introduction about the reasons for the rediscovery of these nanoparticles over the last 25 years. Starting from the form of POSS that is most widely used in literature (octaisobutyl POSS), this present study aims to evaluate how the reduction of symmetry through the introduction of organic groups favors their dispersion in polystyrene matrix without compromising their solubility.

Keywords: polyhedral oligomeric silsesquioxanes; POSS; composites; thermal stability

1. Introduction

Undoubtedly, the materials that have characterized the last century, were the polymers, due to their ability to provide economic and structural benefits. Continuing in this direction and trying to improve mechanical, thermal and durability properties of polymers by adding various reinforcements to them, material experts are defining the composites as reference materials for the twenty-first century. In the design and assembly of a composite we have to fish in the sea of chemistry, that as we know contains organic and inorganic materials, each with different characteristic properties. Hence, the idea in developing hybrid compounds involves taking advantage of the best properties of each component and trying to decrease or eliminate their drawbacks. Based on this original idea, the prediction is that the defining material platform of the twenty-first century could very well be the hybrid, where two or more components are combined in a single material to give new and previously unattainable combinations of useful properties [1]. In this project, it must be taken into account that design of hybrid materials requires the use of building blocks, in particular those included in the nanotechnology field [1], such as dendritic polymers [2–4], carbon nanotubes [5–7], graphene [8–10], fullerene [11–13] and Polyhedral oligomeric silsesquioxanes (POSSs) [14–16]. The latter building block is the subject of this review.

Unlike the other building blocks, for which few traces are found in the literature before the 1980s, the earliest reports of chemistry relating to silsesquioxanes occurred at the end of the nineteenth century, which were produced by Buff and Wohler [17] and Ladenburg [18]. Just after the end of the second world war, the POSS molecule was first described as we know it today [19]. Considering the small amount of instrumental data available, it is truly surprising how Scott hypothesized not only the oligomer formula but also its structure, surmising that the poor solubility and the tendency to sublimate without fusion were due to its characteristic symmetry. About ten years had to pass before the cube-octameric structure of the molecule was established [20]. In order to appreciate the interest for this topic through the many years, one only has input the keyword "POSS" into the Scopus search engine. After their description as we know today, we only found 85 records in 1946 to 1995, which means less than 2 manuscripts per year in fifty years. Since 1995, we can observe an exponential growth of scientific works concerning POSS (Figure 1).

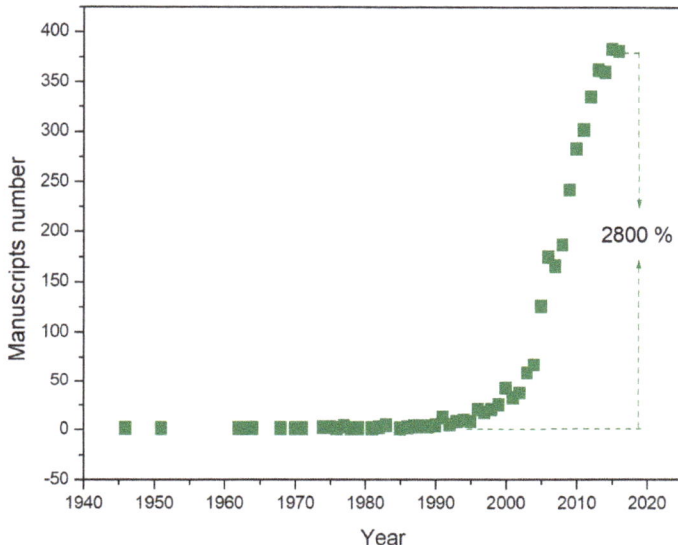

Figure 1. Time evolution of manuscripts with POSS as a topic, which were published on International Journals. Source Scopus.

It increased from an annual average of twenty manuscripts during the five-year period of 1995–2000 to currently reach an annual average of about four hundred manuscripts, with an increase of 2800% starting from 1990 and an increase of 1200% starting from 2000. It is worth noting that in the same period, there has been an overall increase in scientific publications but if we conduct the same operation by entering two other very common search terms, such as Clay or Natural fibers, it can be observed that these terms do not generate a number of manuscripts that is half the number of publications related to the POSS term.

The rediscovery of these materials is certainly due to the work of Feher who set up their synthesis with easily reproducible methodologies [21,22] and Lichtenhan, who understood the infinite potential of POSSs in being able to be mixed with polymers for making hybrid composites [23,24]. Lichtenhan begun to work with POSS at the *Air Force Research Laboratory* at the Edwards Air Force Base (California, USA), where he studied POSS-containing polymers as precursors to hybrid inorganic/organic materials. He subsequently founded the Hybrid Plastics company in 1998, which is actually recognized as the commercial leader for POSSs production. Thus, the most common POSSs are manufactured on a large scale nowadays so that researchers worldwide may purchase these nanomaterials at very competitive prices. Furthermore, the organic nature of the POSSs periphery, which we will analyze in detail later in the text, can be functionalized to generate hundreds of possible compounds. Both purchased and modified POSSs are extremely versatile and may be chemically bonded or physically blended into a companion material, resulting in a hybrid nanomaterial with the combined benefits of POSS and the companion material [1]. This is what we expect of composite technology at its best and it should not be confused with the simple combination of two materials for the sake of it that does not achieve any real design or provide an advantage in terms of material science.

2. Experimental

2.1. Nomenclature

The spatial structure of these molecules is not strictly cubic, as first described by Barry et al. [20], because they are actually spherical, hence providing them with the name of Polyhedral. This is because the silicon–oxygen cluster forms a polyhedron not a cube as the silicon–oxygen–silicon and oxygen–silicon–oxygen bonds do not form angles of 90°. The prefix oligo- is for the small number of silsesquioxane units present in the material (sometimes a specific prefix is used to indicate the exact number of repeat units, such as hexa-, octa- and deca, with octahedral silsesquioxane being the most commonly used). Sil- stands for silicon; -sesqui- is added because each Si atom is bound to an average of one and a half oxygens (-ox-); and finally, -ane is used because the Si atom is also bound to one hydrocarbon group. More broadly, silsesquioxanes can be classified into those with un-caged (random, ladder or partially caged) and caged structures [25]. These latter ones with caged structures, which have a general formula of $(RSiO_{1.5})_n$, where R is H or an organic group (alkyl, aryl or any of their derivatives), are also known as Polyhedral Oligomeric Silsesquioxanes. These chemicals are composed of a silicon and oxygen cage, which is externally completed by organic groups that are covalently bonded with the silicon atoms. The most common value of n is 8, thus generating a very highly symmetric structure. This is sometimes indicated in the literature with the symbol T_8, which has a diameter that is usually in the range of 1.5–3 nm (Figure 2).

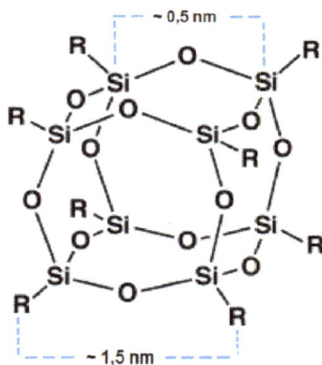

Figure 2. Molecular structure of the classical T8 Polyhedral Oligomeric Silsesquioxanes cage.

2.2. Materials

The various POSSs were prepared by a corner capping reaction of trisilanol with aryltrimethoxysilane or phenyltrimethoxysilane. Isobutyltrimethoxysilane, cyclopenthyltrimethoxysilane and phenyltrimethoxysilane were purchased from Aldrich Co. (St. Gallen, Switzerland). and used as received. Trimethoxysilane derivatives were prepared from the appropriate Grignard reagent and $Si(OCH_3)_4$ [26–29]. The different functionalized trisilanol molecules were prepared according to literature methods [30]. The various POSS/PS nanocomposites were obtained by the in situ polymerization of styrene in the presence of different quantities of POSSs. Styrene (Aldrich) was purified by passing it through an inhibitor removal column. We recrystallized 2,2-Azobis(isobutyronitrile) (AIBN) (98% Aldrich) twice from dry ethanol at temperatures lower than 40 °C and not in direct light. Toluene, which was the used solvent, was stirred over calcium hydride for 24 h and distilled in a nitrogen atmosphere.

Polymers **2018**, *10*, 904

3. Results and Discussion

3.1. Incorporation of POSS in Polymer Matrix

POSSs have shown interesting thermal, mechanical, optical and electrical properties [31–34]. Furthermore, due to their nano-dimensions, they have particularly proven to be the ideal candidates for incorporation into polymer matrices in order to improve their properties. Considering that the size of POSS molecules is usually in the range of 1–3 nm and that a POSS with the very common isobutyl periphery has a diameter of about 1.5 nm [35], these fall under the category of nanocomposites [36,37]. The inorganic silica-based core results in molecule rigidity, thermal stability and resistance to oxidation [38], thus proposing itself as one of the best fillers for polymeric matrices, especially considering its biocompatible nature. Nevertheless, it would be a venial mistake to consider the POSSs as simple fillers. POSS is a molecule rather than a simple filler particle and may be reactive or functionalized to the specific purpose. Thus, the modification of the external organic corona will increase (or decrease) the solubility of POSS in organic solvents [39] and its compatibility with polymers [40,41]. This compatibility and its subsequent capacity to undergo nanometric dispersion is driven by the nature of the organic groups attached to the silicon atoms, which thus strongly affects the properties of obtained nanocomposites [42]. From the available literature, it is also worth noting that when aliphatic groups are attached to the cage POSS, or when long arms are attached to the cage, the solubility in matrix increases, while the presence of short or rigid aromatic groups has the opposite effect [43]. On the other hand, the presence of aliphatic groups in the periphery worsen the thermal properties of the resultant nanocomposite, while the presence of aromatic groups has the opposite effect [44].

3.2. The Importance of POSSs Asymmetric Structure in Designing Composites

Taking advantage of this knowledge, POSSs with seven aliphatic groups and one phenyl group were prepared and used to reinforce Polystyrene (PS), which aimed to obtain nanocomposites with more thermal stability than the matrix. The corner capping reaction of trisilanol with trichlorosilane and/or triethoxysilane [45,46] was carried out to synthesize POSS, while POSS/PS nanocomposites were prepared by the in situ polymerization of styrene with different POSS amounts (3, 5 and 10 wt %) [47]. The PS nanocomposites with heterogeneous POSS showed an extraordinary increase in initial decomposition temperatures (+45 °C about in both oxidative and inert atmosphere) and in degradation activation energy (+40 kJ/mol about) [48,49]. This change was not only in respect to the former polystyrene but overall in respect to the nanocomposite obtained by synthesized styrene with octaisobutyl-POSS (oib-POSS). In particular, among the prepared phenyl heptaisobutyl-POSSs (ph, hib-POSS), the best performance was recorded for the nanocomposites with 5 wt % of POSS. Once the best POSS/PS ratio was established, we needed to explain the considerable increase in thermal stability found for ph, hib-POSS. Homogeneous POSS (with the all the same R groups) can be considered as a symmetric molecule, while heterogeneous POSS (seven same R groups and 1 different) can be considered as an asymmetric molecule. Calorimetric investigation also showed an increase in glass transition temperature (T_g), suggesting a preference in self-regulating phenomena for the symmetric POSS as opposed to the asymmetric ones (Figure 3). A good dispersion is obtained by changing one of the R groups to phenyl, thus allowing the association of the cage with the PS matrix. Such an association does not happen in the aggregation case where all R groups are the same.

Finally, SEM analyses showed the presence of aggregated POSS when oib-POSS is used to reinforce PS, which resulted in a better dispersion of nanoparticles when ph, hib-POSS was used [50,51].

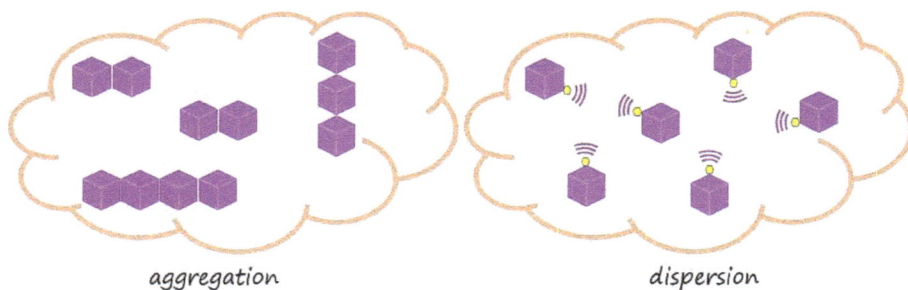

Figure 3. Aggregation vs. dispersion behavior of the different POSS molecules, octaisobutyl-POSS (**left**) and heptaisobutyl-POSSs (**right**), in PS matrix.

3.3. The Influence of POSSs External Groups' Rigidity on Solubility and Thermal Behaviour

The next step of the research was the synthesis of POSSs with seven phenyl and one alkyl group (ib, hph-POSS) in order to create the relative PS based nanocomposites (ib, hph-POSS/PS). In this case, a thermal investigation was carried out, which showed a dramatic increase in both $T_{5\%}$ (+116 °C with respect PS and +69 °C with respect to ph, hib-POSS) and E_a of degradation (+105 kJ/mol with respect to PS and +74 kJ/mol with respect to ph, hib-POSS/PS nanocomposite) values in an oxidative environment. The increasing trend albeit slightly less dramatic was also observed for the degradations performed in an inert atmosphere for both $T_{5\%}$ (+56 °C with respect to PS and +23 °C with respect to ph, hib-POSS/PS nanocomposites) and E_a of degradation (+49 kJ/mol with respect to PS). The greater reinforcement, which was due to the replacement of isobutyl with phenyl groups in the POSS corona, was confirmed by the calorimetric experiments carried out to measure the glass transition temperature of the prepared nanocomposites with polystyrene. We observed a T_g value that was increased by 28 °C compared to that of the oib-POSS/PS nanocomposite and increased by 12 °C compared to that of the ph, hib-POSS/PS nanocomposite. Thus, this supports the hypothesis that the synergistic effect from the introduction of more rigid (in respect to the aliphatic one) phenyl groups with maintenance of the asymmetric structure (by replacing 7/8 organic group at vertices of silicon cage) leads to an increase in the thermal stability of the resultant nanocomposites.

At this stage of the study, the only possibility to further improve the reinforcing effect while still keeping these two points (i.e., the presence of several phenyl groups on the periphery of the POSS molecules and their asymmetric structure) involves obtaining soluble POSSs without a large decrease in the initial decomposition temperature with respect to that of commercial octaphenyl POSS. This can be achieved through the replacement of only one hydrogen atom on one of the eight phenyl groups. Therefore, octaphenyl POSS with small substituents (fluorine and chlorine atoms, methyl and methoxy groups) in the place of one hydrogen atom on one of the eight phenyl groups was synthesized [40] and used to prepare the relative PS nanocomposites. The main concern of using these new slightly modified octaphenyl POSSs focused on their solubility in the PS matrix. These new POSSs were dissolved and the resulting spectroscopic analyses confirmed the structure and properties of the materials so that the subsequent characterization of the chemico-physical properties has been carried out. The $T_{5\%}$ values of all prepared nanocomposites in both oxidative and inert environments were much higher than those of neat PS and slightly higher than those of ib, hph-POSS/PS nanocomposites (+124–128 °C in air and +69–73 °C in nitrogen with respect to PS; +8–12 °C in air and +13–15 °C in nitrogen with respect to ib, hph-POSS/PS nanocomposites). The T_g value remained practically the same as that observed for the ib, hph-POSS/PS but this value was higher than that of PS.

3.4. The Use of Dumbbell-Shaped POSSs in Reinforcing Polymer Matrix

In recent years, a particular group of POSSs, which are namely "dumbbell-shaped" POSSs formed by an organic group covalently attached to more than one silicon cage, have attracted particular interest. The change in organic bridge leads to a change in geometry, length, rigidity and functionality of obtained POSS, which allows us to subsequently tune its bulk properties [52–54]. Polydispersity increased with filler concentration while the d spacing was influenced by phase selectivity and domain–filler compatibility [55].

Since we have studied polymer and copolymer systems for sulfonates membranes for several years [56,57], the next step could be the incorporation of POSS into these matrices. Therefore, keeping in mind that the compatibility (i.e., the presence, of groups in the molecular filler that are of the same nature as those present in the polymer chain) between the matrix and filler in these systems is of great importance [58], we do not want to limit the introduction of a simple T_8 cage into PS. For this reason, we synthesized dumbbell-shaped POSSs with a simple and functionalized aromatic bridge using the well consolidated corner capping reaction (Scheme 1) [59].

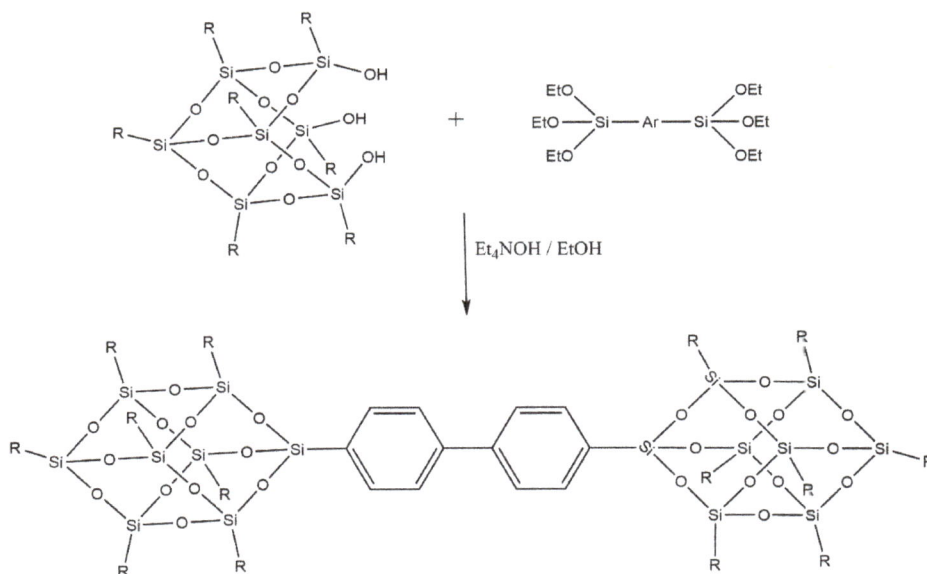

Scheme 1. Corner capping reaction of heptaisobutyl-tricycloheptasiloxane trisilanol with triethoxysilyl derivatives for the synthesis of dumbbell-shaped POSSs with an aromatic bridge.

Once synthesized and thermally characterized, a comparison was made with the unbridged systems, dumbbell shaped POSS showed a better thermal stability ($T_{5\%}$ values ranging from 310 to 410 °C in nitrogen and 290 to 380 °C about in air for the Dumbbell Shaped POSSs vs $T_{5\%}$ values ranging from 285 to 365 °C in nitrogen and 290 to 380 °C in air for the single cage hepta isobutyl POSS) [46,59]. When inserted in the polystyrene matrix, spectroscopically investigations revealed a poorer dispersion [44] which was attributed to the symmetric structure of the dumbbell shaped molecules that facilitate a POSS aggregation phenomenon (Figure 4). This behaviour led to a maintenance of the thermal stability just seen with the PS nanocomposites reinforced with the single cage hepta isobutyl POSS ($T_{5\%}$ values of about 290 °C in nitrogen and 260 °C in air) [36].

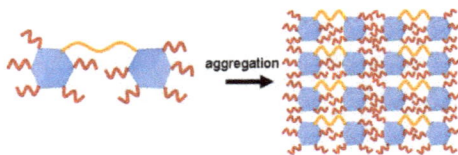

Figure 4. Aggregation phenomena hypothesis for Dumbbell-Shaped POSSs with an aromatic bridge when inserted in PS matrix.

In order to improve the dispersion of these new very thermally stable nanoreinforcements in the polystyrene matrix, the next design step was to synthesize a series of dumbbell-shaped POSSs with an aliphatic bridge instead of the aromatic one [60]. The result was that the presence in the molecular filler of jointed chains allowed the sufficiently free movement of silicon cages, which increased as a function of the bridge chain length, resulting in a better dispersion of bridged POSSs in the matrix [14].

In terms of the chemico-physical behavior, we found an increase in the resistance to the thermal degradation with respect to the PS reinforced with the POSS bearing an aromatic bridge, which was measured by the initial decomposition temperature and increased as a function of the alkyl bridge length (Table 1).

Table 1. Temperatures at 5% mass loss ($T_{5\%}$) for the various Dumbbell-Shaped POSS/PS nanocomposites with different alkyl bridge length among the POSS cages in static air atmosphere and in flowing nitrogen at 10 °C min^{-1}. Values from reference [61].

Nanocomposites	Nitrogen	Air
	$T_{5\%}/°C$	$T_{5\%}/°C$
T (CH$_2$)$_2$ T/PS	284	259
T (CH$_2$)$_4$ T/PS	287	263
T (CH$_2$)$_6$ T/PS	288	266
T (CH$_2$)$_8$ T/PS	289	273
T (CH$_2$)$_{10}$ T/PS	293	282

4. Conclusions

Concerning the synthesis of differently characterized POSSs and their incorporation into the polystyrene matrix, it is possible to draw a series of conclusions that can be useful for the designing of nanostructured polymers based on polyhedral oligomeric silsesquioxanes.

By taking for granted that the phenyl groups confer greater thermal stability to the resulting composites, which is supported by the literature [62], this present study confirmed that according to a few studies [63,64], it is preferable to disperse asymmetric POSS molecules in the polymer matrix in order to obtain a nanocomposite with good thermal resistance properties.

Therefore, in the designing of these materials we must ask whether the priority should be the dispersion in the matrix, and then the possibility to opt for the asymmetric un-bridged POSSs, or a higher compatibility with the polymer and then the possibility to opt for the symmetric, but with a bridge that can be functionalized, dumbbell shaped POSSs. An alternative route was proposed, which involved the addition of an aliphatic bridge, which allows the freedom of movement of silicon cages. Due to the spatial blockage, this cannot be achieved with an aromatic bridge.

Funding: Ignazio Blanco is grateful to the MIUR for the grant "Fund for basic research activities", and to the Department of Civil Engineering and Architecture of the University of Catania for supporting the project MATErials LIfe foreCAst (MATELICA).

Conflicts of Interest: The author declare no conflict of interest.

References

1. Hartmann-Thompson, C. *Applications of Polyhedral Oligomeric Silsesquioxanes*; Springer: New York, NY, USA, 2011; ISBN 978-90-481-3786-2. [CrossRef]

2. Mezzenga, R.; Boogh, L.; Månson, J.-A.E. A review of dendritic hyperbranched polymer as modifiers in epoxy composites. *Compos. Sci. Technol.* **2001**, *61*, 787–795. [CrossRef]

3. Díaz, I.; García, B.; Alonso, B.; Casado, C.M.; Morán, M.; Losada, J.; Pérez-Pariente, J. Ferrocenyl dendrimers incorporated into mesoporous silica: New hybrid redox-active materials. *Chem. Mater.* **2003**, *15*, 1073–1079. [CrossRef]

4. Liu, H.; Guo, J.; Jin, L.; Yang, W.; Wang, C. Fabrication and functionalization of dendritic poly(amidoamine)-immobilized magnetic polymer composite microspheres. *J. Phys. Chem. B* **2008**, *112*, 3315–3321. [CrossRef] [PubMed]

5. Punetha, V.D.; Rana, S.; Yoo, H.J.; Chaurasia, A.; McLeskey, J.T., Jr.; Ramasamy, M.S.; Sahoo, N.G.; Cho, J.W. Functionalization of carbon nanomaterials for advanced polymer nanocomposites: A comparison study between CNT and graphene. *Prog. Polym. Sci.* **2017**, *67*, 1–47. [CrossRef]

6. Vorobyeva, E.A.; Chechenin, N.G.; Makarenko, I.V.; Kepman, A.V. Heat Propagation in Anisotropic Heterogeneous Polymer-CNT Composites. *J. Compos. Sci.* **2017**, *1*, 6. [CrossRef]

7. Bertolino, V.; Cavallaro, G.; Milioto, S.; Parisi, F.; Lazzara, G. Thermal Properties of Multilayer Nanocomposites Based on Halloysite Nanotubes and Biopolymers. *J. Compos. Sci.* **2018**, *2*, 41. [CrossRef]

8. Kuilla, T.; Bhadra, S.; Yao, D.; Kim, N.H.; Bose, S.; Lee, J.H. Recent advances in graphene based polymer composites. *Progr. Polym. Sci.* **2010**, *35*, 1350–1375. [CrossRef]

9. Wang, H.; Hao, Q.; Yang, X.; Lu, L.; Wang, X. A nanostructured graphene/polyaniline hybrid material for supercapacitors. *Nanoscale* **2010**, *2*, 2164–2170. [CrossRef] [PubMed]

10. Shahil, K.M.F.; Balandin, A.A. Graphene-multilayer graphene nanocomposites as highly efficient thermal interface materials. *Nano Lett.* **2012**, *12*, 861–867. [CrossRef] [PubMed]

11. Wang, C.; Guo, Z.-X.; Fu, S.; Wu, W.; Zhu, D. Polymers containing fullerene or carbon nanotube structures. *Prog. Polym. Sci.* **2004**, *29*, 1079–1141. [CrossRef]

12. Tenne, R.; Enyashin, A.N. Inorganic Fullerene-Like Nanoparticles and Inorganic Nanotubes. *Inorganics* **2014**, *2*, 649–651. [CrossRef]

13. Zhao, D.; Ning, J.; Wu, D.; Zuo, M. Enhanced Thermoelectric Performance of Cu2SnSe3-Based Composites Incorporated with Nano-Fullerene. *Materials* **2016**, *9*, 629. [CrossRef] [PubMed]

14. Blanco, I.; Bottino, F.A.; Cicala, G.; Latteri, A.; Recca, A. A kinetic study of the thermal and thermal oxidative degradations of new bridged POSS/PS nanocomposites. *Polym. Degrad. Stab.* **2013**, *98*, 2564–2570. [CrossRef]

15. Carraro, M.; Gross, S. Hybrid Materials Based on the Embedding of Organically Modified Transition Metal Oxoclusters or Polyoxometalates into Polymers for Functional Applications: A Review. *Materials* **2014**, *7*, 3956–3989. [CrossRef] [PubMed]

16. Wang, K.; Pasbakhsh, P.; De Silva, R.T.; Goh, K.L. A Comparative Analysis of the Reinforcing Efficiency of Silsesquioxane Nanoparticles versus Apatite Nanoparticles in Chitosan Biocomposite Fibres. *J. Compos. Sci.* **2017**, *1*, 9. [CrossRef]

17. Buff, H.; Wohler, F. Ueber neue Verbindungen des Siliciums. *Liebigs Ann. Chem.* **1857**, *104*, 94–109. [CrossRef]

18. Ladenburg, A. Ueber aromatische Verbindungen, welche Silicium enthalten. *Ber. Dtsch. Chem. Ges.* **1873**, *6*, 379–381. [CrossRef]

19. Scott, D.W. Thermal Rearrangement of Branched-Chain Methylpolysiloxanes. *J. Am. Chem. Soc.* **1946**, *68*, 356–358. [CrossRef]

20. Barry, A.J.; Daudt, W.H.; Domicone, J.J.; Gilkey, J.W. Crystalline Organosilsesquioxanes. *J. Am. Chem. Soc.* **1955**, *77*, 4248–4252. [CrossRef]

21. Feher, F.J. Polyhedral Oligometallasilsesquioxanes (POMSS) as Models for Silica-Supported Transiton-Metal Catalysts: Synthesis and Characterization of $(C_5Me_5)Zr[(Si_7O_{12})(c-C_6H_{11})_7]$. *J. Am. Chem. Soc.* **1986**, *108*, 3850–3852. [CrossRef]

22. Feher, F.J.; Budzichowski, T.A. Syntheses of highly-functionalized polyhedral oligosilsesquioxanes. *J. Organomet. Chem.* **1989**, *379*, 33–40. [CrossRef]

23. Lichtenhan, J.D. Polyhedral Oligomeric Silsesquioxanes: Building Blocks for Silsesquioxane-Based Polymers and Hybrid Materials. *Comments Inorg. Chem.* **1995**, *17*, 115–130. [CrossRef]

24. Haddad, T.S.; Lichtenhan, J.D. Hybrid organic-inorganic thermoplastics: Styryl-based polyhedral oligomeric silsesquioxane polymers. *Macromolecules* **1996**, *29*, 7302–7304. [CrossRef]

25. Blanco, I.; Abate, L.; Bottino, F.A. Influence of n-alkyl substituents on the thermal behaviour of Polyhedral Oligomeric Silsesquioxanes (POSSs) with different cage's periphery. *Thermochim. Acta* **2016**, *623*, 50–57. [CrossRef]

26. Lee, J.Y.; Fu, G.C. Room-temperature Hiyama cross-couplings of arylsilanes with alkyl bromides and iodides. *J. Am. Chem. Soc.* **2003**, *125*, 5616–5617. [CrossRef] [PubMed]

27. Murata, M.; Ishikura, M.; Nagata, M.; Watanabe, S.; Masuda, Y. Rhodium (I)-catalyzed silylation of aryl halides with triethoxysilane: Practical synthetic route to aryltriethoxysilanes. *Org. Lett.* **2002**, *4*, 1843–1845. [CrossRef] [PubMed]

28. Weber, W.P. *Silicon Reagents for Organic Synthesis*; Springer: Berlin/Heidelberg, Germany, 1983; ISBN 978-3-642-68663-4. [CrossRef]

29. Manoso, A.S.; Ahn, C.; Soheili, A.; Handy, C.J.; Correia, R.; Seganish, W.M.; Deshong, P. Improved synthesis of aryltrialkoxysilanes via treatment of aryl grignard or lithium reagents with tetraalkyl orthosilicates. *J. Org. Chem.* **2004**, *69*, 8305–8314. [CrossRef] [PubMed]

30. Lichtenhan, J.D.; Schwab, J.J.; Reinerth, W.; Carr, M.J.; An, Y.Z.; Feher, F.J.; Terroba, R.; Liu, Q. Process for the Formation of Polyhedral Oligomeric Silsesquioxanes. U.S. Patent 6,972,312 B1, 6 December 2005.

31. Dou, Q.; Karim, A.A.; Loh, X.J. Modification of Thermal and Mechanical Properties of PEG-PPG-PEG Copolymer (F127) with MA-POSS. *Polymers* **2016**, *8*, 341. [CrossRef]

32. Mohamed, M.G.; Jheng, Y.-R.; Yeh, S.-L.; Chen, T.; Kuo, S.-W. Unusual Emission of Polystyrene-Based Alternating Copolymers Incorporating Aminobutyl Maleimide Fluorophore-Containing Polyhedral Oligomeric Silsesquioxane Nanoparticles. *Polymers* **2017**, *9*, 103. [CrossRef]

33. Huang, C.-W.; Jeng, S.-C. Polyhedral Oligomeric Silsesquioxane Films for Liquid Crystal Alignment. *Coll. Interfaces* **2018**, *2*, 9. [CrossRef]

34. Li, X.; Yu, B.; Zhang, D.; Lei, J.; Nan, Z. Cure Behavior and Thermomechanical Properties of Phthalonitrile–Polyhedral Oligomeric Silsesquioxane Copolymers. *Polymers* **2017**, *9*, 334. [CrossRef]

35. De Armitt, C.; Wheeler, P. POSS keeps high temperature plastics flowing. *Plast. Addit. Compd.* **2008**, *10*, 36–39. [CrossRef]

36. Blanco, I.; Bottino, F.A.; Cicala, G.; Latteri, A.; Recca, A. Synthesis and characterization of differently substituted phenyl hepta isobutyl-polyhedral oligomeric silsesquioxane/polystyrene nanocomposites. *Polym. Compos.* **2014**, *35*, 151–157. [CrossRef]

37. Kim, K.; Alam, T.M.; Lichtenhan, J.D.; Otaigbe, J.U. Synthesis and characterization of novel phosphate glass matrix nanocomposites containing polyhedral oligomeric silsesquioxane with improved properties. *J. Non-Cryst. Solids* **2017**, *463*, 189–202. [CrossRef]

38. Michałowski, S.; Hebda, E.; Pielichowski, K. Thermal stability and flammability of polyurethane foams chemically reinforced with POSS. *J. Therm. Anal. Calorim.* **2017**, *130*, 155–163. [CrossRef]

39. Blanco, I.; Abate, L.; Bottino, F.A. Mono substituted octaphenyl POSSs: The effects of substituents on thermal properties and solubility. *Thermochim. Acta* **2017**, *655*, 117–123. [CrossRef]

40. Li, S.; Simon, G.P.; Matisons, J.G. The effect of incorporation of POSS units on polymer blend compatibility. *J. Appl. Polym. Sci.* **2010**, *115*, 1153–1159. [CrossRef]

41. Blanco, I.; Bottino, F.A. The influence of the nature of POSSs cage's periphery on the thermal stability of a series of new bridged POSS/PS nanocomposites. *Polym. Degrad. Stab.* **2015**, *121*, 180–186. [CrossRef]

42. Ueda, K.; Tanaka, K.; Chujo, Y. Synthesis of POSS Derivatives Having Dual Types of Alkyl Substituents and Their Application as a Molecular Filler for Low-Refractive and Highly Durable Materials. *Bull. Chem. Soc. Jpn.* **2017**, *90*, 205–209. [CrossRef]

43. Fina, A.; Tabuani, D.; Carniato, F.; Frache, A.; Boccaleri, E.; Camino, G. Polyhedral oligomeric silsesquioxanes (POSS) thermal degradation. *Thermochim. Acta* **2006**, *440*, 36–42. [CrossRef]

44. Blanco, I.; Abate, L.; Bottino, F.A.; Cicala, G.; Latteri, A. Dumbbell-shaped polyhedral oligomeric silsesquioxanes/polystyrene nanocomposites: The influence of the bridge rigidity on the resistance to thermal degradation. *J. Compos. Mater.* **2015**, *49*, 2509–2517. [CrossRef]

45. Blanco, I.; Abate, L.; Bottino, F.A.; Bottino, P.; Chiacchio, M.A. Thermal degradation of differently substituted cyclopentyl polyhedral oligomeric silsesquioxane (CP-POSS) nanoparticles. *J. Therm. Anal. Calorim.* **2012**, *107*, 1083–1091. [CrossRef]

46. Blanco, I.; Abate, L.; Bottino, F.A.; Bottino, P. Hepta isobutyl polyhedral oligomeric silsesquioxanes (hib-POSS) A thermal degradation study. *J. Therm. Anal. Calorim.* **2012**, *108*, 807–815. [CrossRef]
47. Blanco, I.; Abate, L.; Bottino, F.A.; Bottino, P. Thermal degradation of hepta cyclopentyl, mono phenyl-polyhedral oligomeric silsesquioxane (hcp-POSS)/polystyrene (PS) nanocomposites. *Polym. Degrad. Stab.* **2012**, *97*, 849–855. [CrossRef]
48. Blanco, I.; Abate, L.; Antonelli, M.L.; Bottino, F.A.; Bottino, P. Phenyl hepta cyclopentyl–polyhedral oligomeric silsesquioxane (ph,hcp-POSS)/Polystyrene (PS) nanocomposites: The influence of substituents in the phenyl group on the thermal stability. *eXPRESS Polym. Lett.* **2012**, *6*, 997–1006. [CrossRef]
49. Blanco, I.; Bottino, F.A. Thermal Study on Phenyl, Hepta Isobutyl-Polyhedral Oligomeric Silsesquioxane/Polystyrene Nanocomposites. *Polym. Compos.* **2013**, *34*, 225–232. [CrossRef]
50. Blanco, I.; Bottino, F.A.; Bottino, P. Influence of symmetry/asymmetry of the nanoparticles structure on the thermal stability of polyhedral oligomeric silsesquioxane/polystyrene nanocomposites. *Polym. Compos.* **2012**, *33*, 1903–1910. [CrossRef]
51. Blanco, I.; Bottino, F.A.; Cicala, G.; Latteria, A.; Recca, A. Synthesis and thermal characterization of mono alkyl hepta phenyl POSS/PS nanocomposites. *Polym. Degrad. Stab.* **2016**, *134*, 322–327. [CrossRef]
52. Araki, H.; Naka, K. Syntheses of Dumbbell-Shaped Trifluoropropyl-Substituted POSS Derivatives Linked by Simple Aliphatic Chains and Their Optical Transparent Thermoplastic Films. *Macromolecules* **2011**, *44*, 6039–6045. [CrossRef]
53. Araki, H.; Naka, K. Syntheses and Properties of Star- and Dumbbell-Shaped POSS Derivatives Containing Isobutyl Groups. *Polym. J.* **2012**, *44*, 340–346. [CrossRef]
54. Araki, H.; Naka, K. Syntheses and Properties of Dumbbell-Shaped POSS Derivatives Linked by Luminescent π-Conjugated Units. *J. Polym. Sci. Part A Polym. Chem.* **2012**, *50*, 4170–4181. [CrossRef]
55. Spoljaric, S.; Genovese, A.; Shanks, R.A. Novel elastomer-dumbbell functionalized POSS composites: Thermomechanical and Morphological Properties. *J. Appl. Polym. Sci.* **2012**, *123*, 585–600. [CrossRef]
56. Abate, L.; Blanco, I.; Cicala, G.; La Spina, R.; Restuccia, C.L. Thermal and rheological behaviour of some random aromatic polyethersulfone/polyetherethersulfone copolymers. *Polym. Degrad. Stab.* **2006**, *91*, 924–930. [CrossRef]
57. Abate, L.; Blanco, I.; Cicala, G.; Recca, G.; Scamporrino, A. The influence of chain-ends on the thermal and rheological properties of some 40/60 PES/PEES copolymers. *Polym. Eng. Sci.* **2009**, *49*, 1477–1483. [CrossRef]
58. Abate, L.; Asarisi, V.; Blanco, I.; Cicala, G.; Recca, G. The influence of sulfonation degree on the thermal behaviour of sulfonated poly(arylene ethersulfone)s. *Polym. Degrad. Stab.* **2010**, *95*, 1568–1574. [CrossRef]
59. Blanco, I.; Abate, L.; Bottino, F.A.; Bottino, P. Synthesis, characterization and thermal stability of new dumbbell-shaped isobutyl-substituted POSSs linked by aromatic bridges. *J. Therm. Anal. Calorim.* **2014**, *117*, 243–250. [CrossRef]
60. Blanco, I.; Abate, L.; Bottino, F.A. Synthesis and thermal properties of new dumbbell-shaped isobutyl-substituted POSSs linked by aliphatic bridges. *J. Therm. Anal. Calorim.* **2014**, *116*, 5–13. [CrossRef]
61. Blanco, I.; Abate, L.; Bottino, F.A.; Bottino, P. Thermal behaviour of a series of novel aliphatic bridged polyhedral oligomeric silsesquioxanes (POSSs)/polystyrene (PS) nanocomposites: The influence of the bridge length on the resistance to thermal degradation. *Polym. Degrad. Stab.* **2014**, *102*, 132–137. [CrossRef]
62. Tanaka, K.; Adachi, S.; Chujo, Y. Structure–property relationship of octa-substituted POSS in thermal and mechanical reinforcements of conventional polymers. *J. Polym. Sci. Part A Polym. Chem.* **2009**, *47*, 5690–5697. [CrossRef]
63. Moore, B.M.; Ramirez, S.M.; Yandeka, G.R.; Haddad, T.S.; Mabry, J.M. Asymmetric aryl polyhedral oligomeric silsesquioxanes (ArPOSS) with enhanced solubility. *J. Organomet. Chem.* **2011**, *696*, 2676–2680. [CrossRef]
64. Huang, M.; Yue, K.; Huang, J.; Liu, C.; Zhou, Z.; Wang, J.; Wu, K.; Shan, W.; Shi, A.C.; Cheng, S.Z.D. Highly Asymmetric Phase Behaviors of Polyhedral Oligomeric Silsesquioxane-Based Multiheaded Giant Surfactants. *ACS Nano* **2018**, *12*, 1868–1877. [CrossRef] [PubMed]

polymers

MDPI

Article

Effects of an Organic-Inorganic Hybrid Containing Allyl Benzoxazine and POSS on Thermal Properties and Flame Retardancy of Epoxy Resin

Benben Liu, Huiling Wang, Xiaoyan Guo, Rongjie Yang and Xiangmei Li *

School of Materials Science, Beijing Institute of Technology, 5 Zhongguancun South Street, Haidian District, Beijing 100081, China; liuben0309@163.com (B.L.); 3120181135@bit.edu.cn (H.W.); gxy@bit.edu.cn (X.G.); yrj@bit.edu.cn (R.Y.)
* Correspondence: bjlglxm@bit.edu.cn; Tel.: +86-010-6894-3961

Received: 1 April 2019; Accepted: 25 April 2019; Published: 1 May 2019

Abstract: A novel organic-inorganic hybrid containing allyl benzoxazine and polyhedral oligomeric silsesquioxane (POSS) was synthesized by the thiol-ene (click) reaction. The benzoxazine (BOZ)-containing POSS (SPOSS-BOZ) copolymerized with benzoxazine/epoxy resin was used to prepare composites of SPOSS-PBZ-E nanocomposites(NPs). The polymerization behavior was monitored by FTIR and non-isothermal differential scanning calorimetry (DSC), which showed that the composites had completely cured with multiple polymerization mechanisms according to the oxazine ring-opening and epoxy resin (EP) polymerization. The thermal properties of the organic–inorganic polybenzoxazine (PBZ) nanocomposites were analyzed by DSC and thermogravimetric analysis (TGA). Furthermore, the X-ray diffraction analysis and the scanning electron microscopy (SEM) micrographs of the SPOSS-PBZ-E nanocomposites indicated that SPOSS was chemically incorporated into the hybrid nanocomposites in the size range of 80–200 nm. The flame retardancy of the benzoxazine epoxy resin composites was investigated by limiting oxygen index (LOI), UL 94 vertical burn test, and cone calorimeter tests. When the amount of SPOSS reached 10% or more, the vertical burning rating of the curing system arrived at V-1, and when the SPOSS-BOZ content reached 20 wt %, the thermal stability and flame retardancy of the material were both improved. Moreover, in the cone calorimeter testing, the addition of SPOSS-BOZ hindered the decomposition of the composites and led to a reduction in the peak heat release rate (pHRR), the average heat release rate (aHRR), and the total heat release (THR) values by about 20%, 25%, and 25%, respectively. The morphologies of the chars were also studied by SEM and energy dispersive X-ray spectroscopy (EDX), and the flame-retardant mechanism of POSS was mainly a condensed-phase flame retardant. The ceramic layer was formed by the enrichment of silicon on the char surface. When there are enough POSS nanoparticles, it can effectively protect the combustion of internal polymers.

Keywords: benzoxazine; POSS; organic-inorganic hybrids; epoxy resin; thermal properties; flame retardancy

1. Introduction

Benzoxazine has a stable aromatic ring structure and can be cross-linked under certain conditions to form a structure such as a phenolic resin. [1]. During curing, there is no need for strong acid catalysts. The monomers can be prepared from low-cost raw materials such as phenols, primary amines, and formaldehyde [2–5]. Compared with traditional phenolic resins, benzoxazines have many excellent properties as a class of high-performance thermal setting resins [6]. However, with the higher requirements imposed by the development of electronic technology on printed circuit board (PCB) material, some disadvantages of benzoxazines such as high ring-opening temperature, brittleness, insufficient heat resistance, and poor compatibility with composites also need to be further improved [7].

At the same time, epoxy resin is the most widely used PCB material. However, epoxy resin also decays the insulating property of the PCB. In light of the above disadvantages of benzoxazines and epoxy resin, an epoxy resin needs to be incorporated into the benzoxazines to obtain both good mechanical and thermal properties in the copolymers [8–10]. It has been shown that the modification of epoxy with benzoxazine not only improves flame retardancy, but also improves the mechanical properties [11]. The phenolic hydroxyl structure formed after the ring opening of the oxazine ring can react with the epoxy ring group to form benzoxazine/epoxy (BOZ/EP) cross-linking copolymerization, which is superior to the conventional single-system polymer. According to recent research, the introduction of a third component based on EP and BOZ composites, or the modification of either EP or BOZ, has received more attention [12].

Polyhedral oligomeric silsesquioxane (POSS) is a novel organic–inorganic hybrid filler with a cage structure, which is mainly composed of three-dimensional cage-like structures connected by Si–O–Si bonds and is very stable. The incorporation of POSS into benzoxazine monomers can improve the thermal, mechanical [13–16], and other unique properties [17,18] of polybenzoxazine (PBZ) resins. It consists of an inorganic siloxane group as the inner core and the organic groups as an outer layer; both organic components and inorganic POSS can display enhanced performance capability compared with that of their non-hybrid polymeric materials [19,20]. Moreover, they have well-defined structures, high-temperature stability, monodisperse molecular weights, and greater design flexibility relative to conventional fillers (e.g., clay, graphene, carbon nanotubes, and boron nitride). Generally, chemical copolymerization [21] and physical blending can incorporate the POSS into polymer materials. However, the linkage of covalent bonds [22] can avoid macro-phase separation between the polymer matrix and the POSS nanocomposites (NPs). Three approaches have been reported for the introduction of POSS into BOZ monomers, which contain mono-benzoxazine functionalized POSS and multi-benzoxazine functionalized POSS [23]. Lee Y J et al. [24] synthesized vinyl-terminated benzoxazine together with amine-containing POSS, formaldehyde, and phenol in THF solution at 90 °C to obtain POSS-BOZ. However, these synthetic steps are complicated and are performed under very harsh conditions, especially in the case of platinum as the catalyst, which requires a water-free environment. Therefore, considering the wide-ranging applications of organic–inorganic hybrids, it is necessary to develop a rapid, highly efficient method for the target-oriented synthesis of hybrid materials.

It has been increasingly recognized that "good" reactions for making POSS-related materials including the criteria of efficiency, versatility, and selectivity; therefore, "click" reactions are ideal tools [25]. Li et al. [26] reviewed the preparation of different functional group POSS nanocomposites using "click" reactions, where this concept has generated much interest in creatively preparing materials of choice. Additionally, the polymer chain composition and POSS surface chemistry can be tuned in a modular and efficient way by thiol–ene "click" chemistry [27,28]. Furthermore, sulfur-based POSS can be introduced into olefin material [29] or epoxy resin [30] by thiol-ene and thiol-epoxy click reactions, respectively, and they have excellent dispersibility. Wu and Kuo [31] synthesized POSS-BOZ nanocomposites with octa-azido functioned POSS(OVBN$_3$-POSS) and 3,4-dihydro-3-(prop-2-ynyl)-2H-benzoxazine (P-pa) by a click reaction. However, the preparation of octa-azido functionalized POSS is too complicated. In previous studies, no researchers used thiol-based POSS and allyl benzoxazine to obtain POSS-BOZ nanocomposites nor did they study their effects on the thermal stability and flame retardancy of BOZ/E composites.

Therefore, in this paper, an organic–inorganic hybrid benzoxazine was prepared by the thiol-ene (click) reaction between benzoxazine and sulfur-containing POSS. The nanocomposites were blended with benzoxazine containing allyl group (BOZ)/epoxy composites and cured. The effect of POSS on the thermal stability and flame retardancy of the material was studied.

2. Experimental Section

2.1. Materials and Methods

Allylamine was obtained from the Chengdu Huaxia Chemical Reagent Co. Ltd. (Chengdu, China), >95%. Paraformaldehyde was obtained from the Tianjin Fuchen Chemical Reagent Factory (AR, Tianjin, China). Phenol was obtained from the Beijing Chemical Plant (AR, Beijing, China). The SPOSS (TH-1550) was obtained from Hybrid Plastics Inc. (Hybrid Plastics®, Hattiesburg, MS, USA), >99%. DMPA was purchased from Aladdin Biochemical Technology Co. Ltd. (Shanghai, China), >99%. Tetrahydrofuran was obtained from Beijing Tongguang Fine Chemical Company (AR, Beijing, China). Epoxy resin was purchased from E-44, Beijing Tongguang Fine Chemical Company (Beijing, China).

2.2. Synthesis of 3-Allyl-3,4-Dihydro-2H-1,3-Benzoxazine Monomers (BOZ)

The synthesis route of BOZ is shown in Scheme 1 [1]. Paraformaldehyde (25.6 g, 0.8 mol) and allylamine (22.8 g, 0.4 mol) were stirred in toluene (60 mL) at room temperature for 30 min under a N_2 atmosphere. Then, phenol (37.6 g, 0.4 mol) was added into the reaction system. The mixture was stirred at 80 °C for 2 h. Subsequently, the solution was cooled to room temperature, and washed using 1 mol/L NaOH solution and distilled water several times. Then, the organic layer was dried over anhydrous Na_2SO_4. The solution was dried under vacuum to afford a light-yellow liquid. Yield: 56 g, 80%; ^{1}H NMR (600 MHz, CDCl$_3$, 298 K): δ 6.80–7.13 (s, 4H, Ar–H), 5.92 (tt, 1H, C–CH=C), 5.27–5.18 (m, 2H, =CH$_2$), 4.88 (s, 2H, O–CH$_2$–N), 4.00 (s, 2H, Ar–O–N), 3.40 (d, 2H, N–CH$_2$–C). FTIR (cm^{-1}): 3077 cm^{-1} (stretching of =C–H), 1843–1560 cm^{-1} (skeleton vibration of benzene ring), 1646 cm^{-1} (stretching of C=C), 914 cm^{-1} (symmetric stretching of C-O-C), 746 cm^{-1} (out-of-plane bending vibrations of olefinic =C–H). HRMS (MALDI-TOF): m/z 175 [M+, calcd 175].

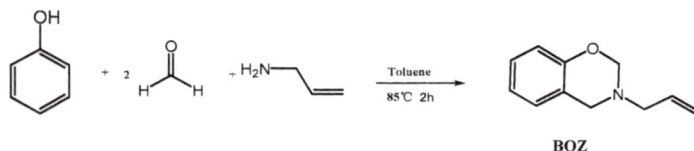

Scheme 1. Synthesis of benzoxazine (BOZ).

2.3. Preparation of the Nanocomposites Benzoxazine POSS (SPOSS-BOZ)

Benzoxazine POSS was synthesized using the procedure presented in Scheme 2. Both the benzoxazine containing allyl group (BOZ) and SPOSS were dissolved in tetrahydrofuran (THF) and this solution was stirred at room temperature for 10 min, then the initiator of DMPA (1% eq) was added and irradiated for 5 min under UV light. Then, the resultant product was collected once dried in vacuum to obtain SPOSS-BOZ as a light-yellow powder.

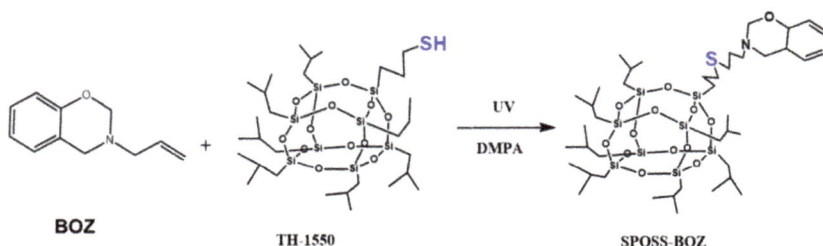

Scheme 2. Chemical structures and reaction schemes for the syntheses of benzoxazine polyhedral oligomeric silsesquioxane (POSS) (SPOSS-BOZ).

2.4. Preparation of the Composites SPOSS-PBZ-E

Composites (SPOSS-PBZ-E) containing polybenzoxazine, POSS, and epoxy resin were prepared by the procedure in Scheme 3 and the schematic diagram of the preparation method is shown in Figure 1. The desired amounts of SPOSS-BOZ, BOZ, and E-44 were mixed and stirred at 80 °C for 30 min. The residual solvent was removed in vacuum at 60 °C to afford homogenous mixtures. The mixture was poured into Teflon molds and cured in an air circulating oven (Yiheng Scientific Instrument Co., Ltd., Shanghai, China) by the following steps: 80 °C (1 h), 110 °C (1 h), 130 °C (1 h), 160 °C (1 h), 180 °C (1 h), and 210 °C (1 h). For these composites, the mass ratio of BOZ to E-44 was 1:1 and then we added a certain amount of SPOSS-BOZ. The cured product was transparent and had a red-wine color and the procedure was repeated to prepare other hybrid materials by varying the amount of SPOSS-BOZ (0, 5, 10, 15, and 20 wt %).

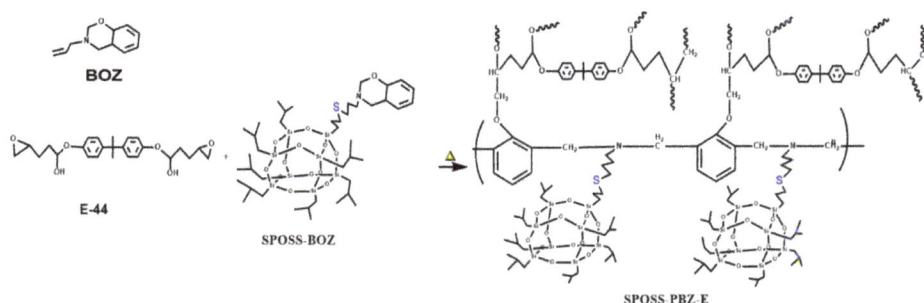

Scheme 3. Structure of the SPOSS-PBZ-E nanocomposites after thermal curing. PBZ, polybenzoxazine.

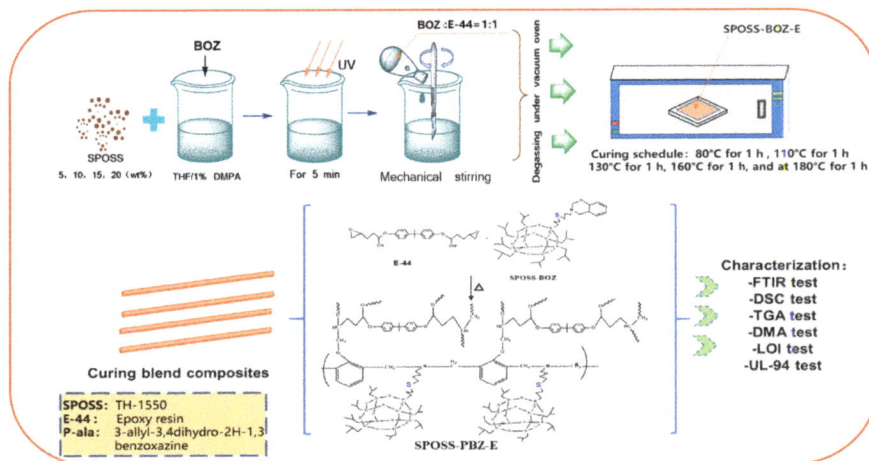

Figure 1. Schematic diagram of the preparation of the SPOSS-PBZ-E nanocomposites.

2.5. Measurements and Characterization

UV-curing: A HWUV0133X three-dimensional UV curing box (Zhonghe Machinery Equipment Manufacturing Co., Ltd, Baoding, China) was used for the experiment. The wavelength of the ultraviolet light was 300–400 nm, the lamp power was 400W, and the radiation intensity was 100 mW/cm^2.

Nuclear magnetic resonance (NMR) spectroscopy: ^{1}H-NMR spectra were recorded on a Bruker Avance 600 NMR spectrometer (BRUKER OPTICS, Beijing, China) operated in the Fourier transform mode. CDCl$_3$ was used as the solvent, and the solution was measured with tetramethyl silane (TMS) as an internal reference.

Fourier-transform infrared (FTIR) spectroscopy: FTIR spectra were recorded on a NICOLET 6700 IR spectrometer (BRUKER OPTICS, Beijing, China). The spectra were collected at 32 scans with a spectral resolution of 4 cm^{-1}.

Mass spectrometry: Matrix-assisted laser desorption/ionization time-of-flight mass spectrometry (MALDI-TOF MS) was performed using a Bruker BIFLEX III device (BRUKER OPTICS, Beijing, China) equipped with a pulsed nitrogen laser (k $\frac{1}{4}$ 337 nm, pulse width $\frac{1}{4}$ 3 ns, and average power $\frac{1}{4}$ 5 mW at 20 Hz). Samples were measured in positive ion modes with usually 50 spectra accumulated. For the MALDI mass spectrum, we used a-Cyano-4-hydroxycinnamic acid matrix and the salts were a mixture of NaCl and KCl.

X-ray diffraction (XRD). The samples were scanned at a speed of 8 °C/min at ambient temperature using an X-ray diffractometer (DX-2600, Rigaku corporation, Tokyo, Japan) at a generator voltage of 35 kV and a current of 25 mA. The data were collected from 2° to 40° intervals.

Thermogravimetric analysis (TGA) was performed with a NETZSCH 209 F1 thermal analyzer (NETZSCH, Bavarian, Germany) at a heating rate of 10 °C/min under nitrogen and air atmosphere, respectively. The temperature ranged from 40 to 800 °C.

Differential scanning calorimetry (DSC) curves of the EP composites were measured using a Netzsch 204 F1 differential scanning calorimeter (NETZSCH, Bavarian, Germany) with a pressure cell. Samples (5–10 mg) were tested at a heating rate of 10 °C/min and results from the second heating were in the range of 35–300 °C.

The dynamic mechanical(DMA) properties of the cured blends were performed using a METTLER TOLEDO SDTA861 instrument (Greifensee, Switzerland) with a sample dimension of 5 × 5 × 3 mm in a controlled strain tension mode and a temperature ramp rate of 4 °C/min from −70 to 200 °C at a frequency of 1.0 Hz.

The limiting oxygen index (LOI) was obtained using the standard ASTM D 2863 procedure, which involves measuring the minimum oxygen concentration required to support candle-like combustion of plastics. An oxygen index instrument (Rheometric Scientific Ltd., (Phoenix Instruments Co., Ltd, Suzhou, China) was used on barrel-shaped samples with the dimensions of 100 × 6.5 × 3 mm

Vertical burning tests were performed using the UL 94 standard on samples with the dimensions of 125 × 13 × 3.2 mm with the CZF-5A horizontal vertical burning tester by Jiangning Analytical Instrument Factory (Phoenix Instruments Co., Ltd, Suzhou, China). In this test, the burning grade of a material was classified as V-0, V-1, V-2, or NR (no rating), depending on its behavior (dripping and burning time).

Scanning electron microscopy (SEM) and energy dispersive X-ray spectroscopy (EDX) images were taken with a Hitachi SU8020 (Beijing, China) with a 20 kV accelerating voltage. The samples were coated with thin layers of gold to make the surface conductive.

Cone calorimeter measurements were performed at an incident radiant flux of 50 kW/m^2, according to the ISO 5660 protocol, using a Fire Testing Technology apparatus (Phoenix Instruments Co., Ltd, Suzhou, China) with a truncated cone-shaped radiator. The specimen (100 × 100 × 3 mm) was measured horizontally without any grids. Typical results from the cone calorimeter were reproducible within ±10%, and the reported parameters were the average of three measurements.

3. Results and Discussion

3.1. Characteristics of Benzoxazine Monomers (BOZ)

Different synthesis methods can be used to prepare benzoxazine monomers. Agag and Takeichi prepared monomers with a solventless method [1]; However, the solvent method minimizes the oligomer production. The chemical structure of BOZ was confirmed by ^{1}H NMR (Figure 2). The resonances of –CH*=C and =CH$_2^*$ protons appeared at 5.9 ppm and 5.2–5.25 ppm, respectively.

The resonances at 4.0 ppm and 4.87 ppm were assigned to the proton of ArCH$_2^*$–N and NCH$_2^*$–O, respectively, confirming the formation of the oxazine ring and indicating the high purity of BOZ. The FTIR spectrum of BOZ provided the information shown in Figure 3a. The bands at 3077 cm^{-1} and 1646 cm^{-1} represented the characteristic peaks of =C–H* and C=C, respectively. The characteristic absorptions of the benzoxazine ring were at 916 cm^{-1} (symmetric stretching of C–O–C). Moreover, the molecular quality of P-ala was obtained in Figure 3b. Based on these results, BOZ was effectively prepared.

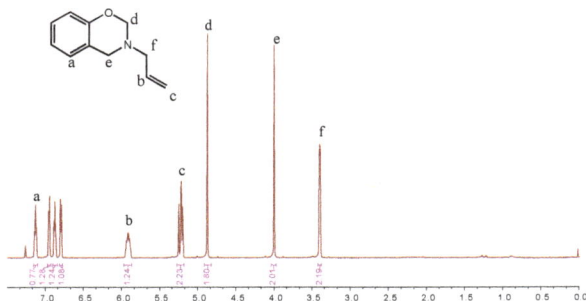

Figure 2. ^{1}H NMR spectra of BOZ.

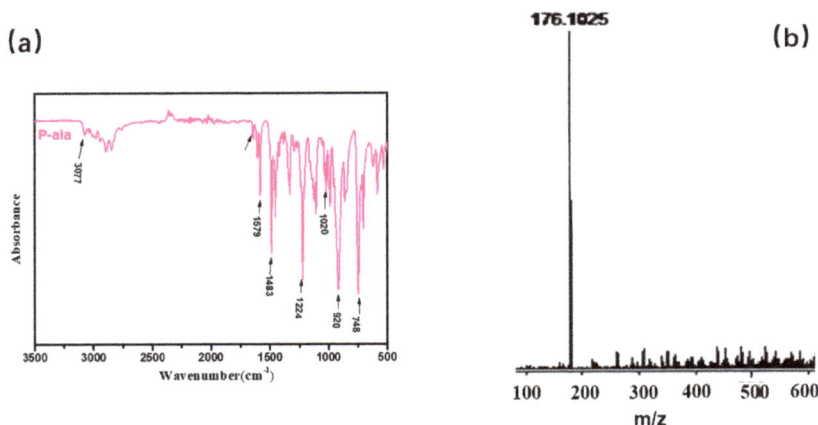

Figure 3. (**a**) Fourier-transform infrared (FTIR) and (**b**) matrix-assisted laser desorption/ionization time-of-flight mass spectrometry (MALDI-TOF MS) of BOZ.

3.2. Preparation of the Nanocomposites Benzoxazine POSS (SPOSS-BOZ)

The monofunctional POSS derivative is the most useful compound for copolymerization with other monomers through diverse functional groups, and a mono-functionalized BOZ ring containing POSS (BOZ-POSS) has been prepared using different methods [32,33]. However, we synthesized SPOSS-BOZ using the thiol-ene (click) reaction, which is an effective method with almost no side reactions [34]. SPOSS and P-ala were dissolved in tetrahydrofuran (THF) and this solution was irradiated only for 5 min under UV light to prepare the mono-functionalized BOZ ring containing POSS. It can be seen from the FTIR spectrum in Figure 4 that the S–H peak at 2700 cm^{-1} disappeared after SPOSS was modified by benzoxazine. The chemical structure of SPOSS-BOZ was confirmed by ^{1}H NMR (Figure 5). The resonance peaks at 6.8–7.2 ppm were assigned to the aromatic protons. The characteristic protons of the oxazine ring appeared at 4.8 ppm and 3.96 ppm (Peak c) and were assigned to –O–CH$_2$–N– and

–Ar– CH$_2$–N–, respectively. The resonance peaks at 0.68 ppm (Peak b), 0.93 ppm (Peak a), and 1.88 ppm were caused by the seven isobutyl hydrocarbon substituents of the POSS. Next, we magnified the spectra of the allyl moiety (Figure 5B). Furthermore, the resonances of –CH*=C and =CH$_2^*$ protons that appeared at 5.9 ppm and 5.2–5.25 ppm almost disappeared, respectively. Therefore, POSS was successfully incorporated into benzoxazine.

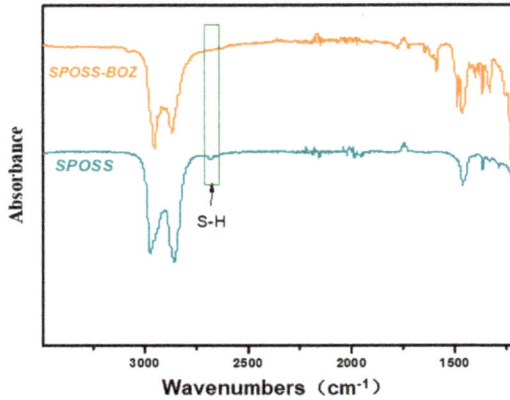

Figure 4. FTIR spectrum of SPOSS and SPOSS-BOZ.

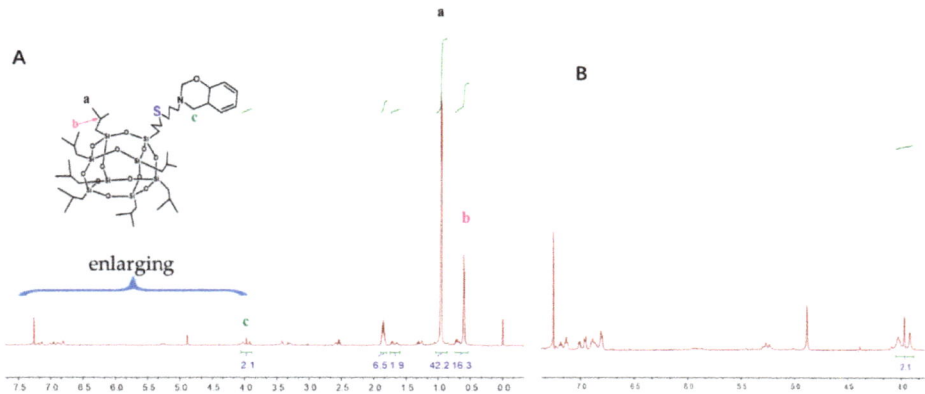

Figure 5. (**A**) ^{1}H NMR spectra of SPOSS-BOZ and (**B**) enlarged view between 4.0 and 7.5 of SPOSS-BOZ.

3.3. Curing Behavior of Polybenzoxazine/POSS/Epoxy (SPOSS-PBZ-E)

The thermally activated polymerization reaction of the copolymers was also studied by non-isothermal DSC, as shown in Figure S1, Supplementary Materials. The oxazine ring opening highly overlapped the allyl addition polymerization exotherm at the temperature range of 170–250 °C. To eliminate this effect, the extrapolation method was used to determine the peak temperature when the heating rate β was 0, and the curing temperature range was determined. The DSC data is shown in Table 1. It can be seen from Figure 6 that when β was 0, T_i, T_p, and T_f were 144 °C, 210 °C, and 240 °C, respectively, and the resin system started to react in the range of 144 to 210 °C. Combined with the relevant data [35], the curing process of the resin system was determined to be 80/1 h + 110/1 h + 130/1 h + 160/1 h + 180/1 h + 210/1 h. Furthermore, the step-wise curing process of the composites was studied by FTIR spectroscopy as shown in Figure 7. As the curing temperature increased, the bands at 1649 cm^{-1} and 916 cm^{-1} decreased due to the allyl and the characteristic peaks of the oxazine ring respectively, which indicated that the ring-opening reaction and polymerization of the oxazine ring had taken place. Meanwhile, the peak at 3203 cm^{-1} was due to the phenolic hydroxyl group that occurred at 160 °C/1 h. With curing proceeding, the peak at 1224 cm^{-1} vanished because the epoxy group decreased, whereas the peak at 1012 cm^{-1} increased due to the ether bond. The reaction between the phenolic hydroxyl group and epoxy group was found. Therefore, these results proved that the copolymer had been completely polymerized [9].

Table 1. Peak temperature at different heating rates.

β (°C·min^{-1})	T_i/°C	T_p/°C	T_f/°C
5	147	211	245
10	158	224	258
15	164	232	270
20	172	238	281
25	178	243	289

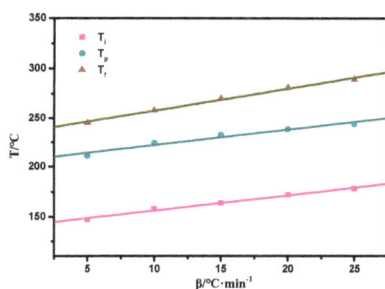

Figure 6. Curves of T versus β of the SPOSS-PBZ-E nanocomposites.

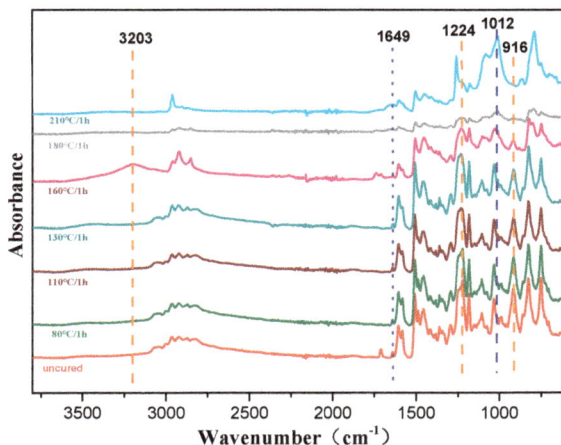

Figure 7. FTIR spectra of prepolymers (5 wt % POSS-PBZ -E) with different curing rates.

3.4. X-Ray Diffraction Analysis

Figure 8 displays the X-ray diffraction patterns for the SPOSS and SPOSS-PBZ-E nanocomposites. The sharp diffraction peaks seen for the SPOSS indicated a highly crystal structure consistent with the literature [36–40]. As can be seen in Figure 8, the XRD pattern for BOZ:E itself showed one broad peak at 2θ ≈ 20°, which confirmed its amorphous structure. After the incorporation of SPOSS nanoparticles into the BOZ:E matrix, the peaks observed in the pure POSS nanoparticle completely disappeared in the case of the 5 wt % SPOSS-PBZ-E, suggesting the molecular dispersion of the SPOSS nanoparticles in the SPOSS-PBZ-E nanocomposite, and also displayed a broad peak at $2\theta \approx 20°$. This indicated that SPOSS was chemically incorporated into the hybrid nanocomposite and formed a cross-linked network between the SPOSS and BOZ:E blend. However, a crystalline peak at $2\theta = 7.97°$, corresponding to the strong diffraction of the SPOSS monomer, was observed when the content of SPOSS was more than 10 wt %. Hence, this indicates that separate POSS domains were present in the SPOSS-PBZ-E nanocomposites.

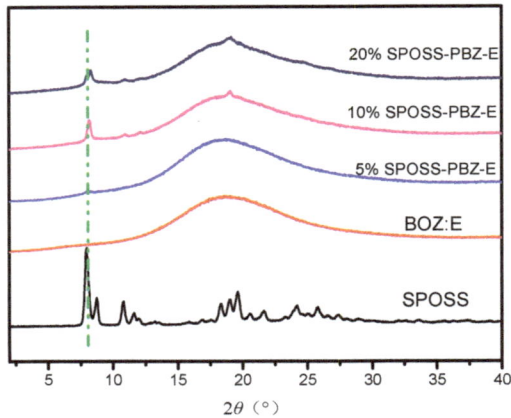

Figure 8. X-ray diffraction patterns of the SPOSS and SPOSS-PBZ-E nanocomposites.

3.5. Scanning Electron Microscopy Micrographs

Figure 9 displays the cross-sectional images of materials with various SPOSS-BOZ contents. The hybrid materials had a good distribution, and microphase separation was not observed in the matrix. As seen from the image of the benzoxazine/epoxy resin, the majority of the POSS particles are in the size range of 80–200 nm. However, when the amounts of SPOSS-BOZ reached 10 wt % and 20 wt %, the fracture surface was very rough. Furthermore, as shown in Figure S3, Supplementary Materials and according to the results in Table 2, an EDX analysis showed the presence of POSS particles. The different contents of silicon atoms indicate the different contents of POSS particles in the whole system.

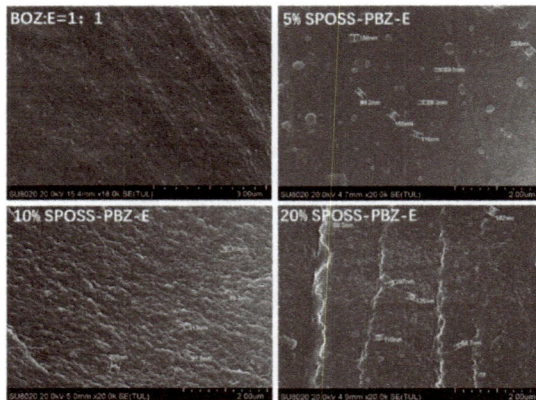

Figure 9. SEM cross-sectional analysis of the SPOSS-PBZ-E nanocomposites.

Table 2. EDX analysis results of the SPOSS-PBZ-E composites.

Sample	C (%)	O (%)	Si (%)
PBZ-E	54.7	45.3	-
5%POSS-PBZ -E	52.23	46.01	1.76
10%POSS-PBZ-E	58.81	39.07	2.13
20%POSS-PBZ-E	49.72	44.16	6.12

3.6. Thermal Properties of the SPOSS-PBZ-E Nanocomposites

TGA in the nitrogen and air atmospheres, derivative thermograms, and tan δ curves of the nanocomposites are shown in Figure 10, Figure 11 and Figure S2, Supplementary Materials, respectively. A summary of the thermal properties for each nanocomposite Is listed in Table 3.

From the DMA experimental data, the glass transition temperature (T_g) of the nanocomposites was slightly lower when compared with the BOZ:E matrix. As the POSS cage structure consisting of Si atoms and O atoms was bulky, this hindered the formation of the benzoxazine/epoxy resin cross-linked network to a certain extent and reduced the crosslink density [41]. The char yield of the SPOSS-PBZ-E nanocomposites increased gradually as the content of SPOSS-BOZ increased due to the increase in the silica and SiO_2 yield that was produced by the thermal degradation of POSS. The char yield in the air atmosphere was far less than in N_2. As can be seen from the thermograms, not only in the N_2 but also in the air atmosphere was the weight loss curve of the BOZ:E matrix not significantly altered by the presence of SPOSS. Similar weight loss traces were found from the TGA profiles. The TGA clearly showed that the 5% mass loss temperature ($T_{5\%}$) of SPOSS-PBZ-E nanocomposites was slightly lower than the BOZ:E matrix except for the modified composites with a 20 wt % addition of SPOSS-BOZ in the N_2 atmosphere Therefore, the decrease in $T_{5\%}$ and T_g was consistent. At a later period of the degradation process, the degradation rate slowed down with the increased POSS content and the T_{peak} of all the SPOSS-PBZ-E hybrid nanocomposites almost remained unchanged with respect to that of the BOZ:E matrix. The steric hindrance and interaction with the polymer chains caused by the incorporation of the POSS cages may result in a limit to the moving of polymer chains during the thermal degradation and lead to a small increase in thermal stability. Moreover, the thermal stability of the inorganic component (POSS) was very good and, during combustion, the POSS units could form a ceramic superficial layer because of the low surface energy of the siloxane structure of POSS [36]. Therefore, the char yield of the nanocomposites increased with the addition of the SPOSS-BOZ organic–inorganic hybrid. When the amount of SPOSS added was 20%, the mass residual ratio of the system increased by 31.8% and 27% compared with that before modification, which was 29.45% in N_2 and 2.8% in air, respectively.

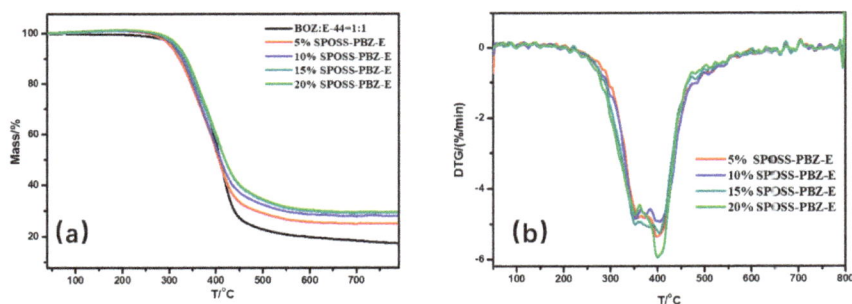

Figure 10. (a) TG and (b) DTG curves of the SPOSS-PBZ-E nanocomposites under N_2 atmosphere.

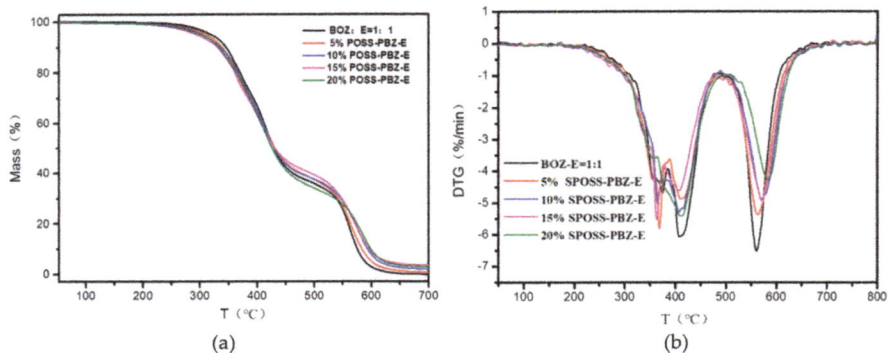

Figure 11. (**a**) TG and (**b**) DTG curves of the SPOSS-PBZ-E nanocomposites under air atmosphere.

Table 3. Thermal properties of the SPOSS-PBZ-E nanocomposites.

Sample	T_g	$T_{5\%}$ (°C)		T_{peak}/°C	DTG_{peak}/ (%/min)	Char Yield (wt %, 700 °C)	
	-	N_2	Air	N_2	N_2	N_2	Air
BOZ:E = 1:1	125	316	319	403	−6.97	22.4	0.1
5%POSS-PBZ-E	115	302	311	401	−6.12	24.9	0.7
10%POSS-PBZ-E	113	309	303	399	−5.24	27.9	2.0
15%POSS-PBZ-E	115	314	294	400	−4.95	29.1	3.6
20%POSS-PBZ-E	111	319	302	398	−5.41	29.5	2.8

3.7. Flame Retardancies of SPOSS-PBZ-E

The flame retardancies of the SPOSS-BOZ-E composites are presented in Table 4. Following the incorporation of SPOSS-BOZ, the LOI value of the blends increased slightly from 28.2% to approximately 28.9%. Thus, the silsesquioxane apparently played only a weak role in affecting the LOI value of the epoxy resin. However, the UL 94 vertical burn test includes the vertical burning level, the average afterflame time t_1 and t_2 after the first and second ignition, total afterflame time ($t_1 + t_2$), and cotton ignited by flaming particles or drops. The composites of PBZ-E without SPOSS-BOZ and 5% SPOSS-BOZ burned readily and had no rating; however, even with an incorporation of 10% of SPOSS, the burning behavior of blends altered and there was no dripping. The UL 94 V-1 rating was obtained and the total afterflame time ($t_1 + t_2$) was reduced in comparison to the 5% POSS-BOZ-E. During combustion, the SPOSS-BOZ nanocomposites formed a dense SiO_2 layer that could prevent the penetration of combustion. When the POSS content cannot reach this level, an effective SiO_2 layer will not be formed during combustion.

Table 4. The LOI and vertical combustion analysis of SPOSS-BOZ-E.

Sample	LOI	UL 94/3.2 mm	t_1/s	t_2/s	Dripping
PBZ-E	28.2	No Rating	14	-	No
5%POSS-PBZ-E	26.8	No Rating	16	15	No
10%POSS-PBZ-E	27.9	V-1	5	17	No
15%POSS-PBZ-E	28.1	V-1	9	11	No
20%POSS-PBZ-E	28.9	V-1	8	10	No

3.8. Cone Calorimeter Analysis

Cone calorimetry is used as a more promising technique to compare and evaluate the fire performances of polymeric materials. The fire performances of the composites were compared using time to ignition (TTI), peak heat release rate (pHRR), average HRR (aHRR), total heat release (THR), total smoke production (TSP), and total smoke release (TSR). These parameters are reported in Table 5 for the different nanocomposites.

Table 5. Cone calorimeter data for the SPOSS-BOZ-E nanocomposites. TTI, time to ignition; pHRR, peak heat release rate; aHRR, average HRR; THR, total heat release; TSP, total smoke production; TSR, total smoke release.

Sample	TTI (s)	pHRR (kW/m^2)	aHRR (kW/m^2)	THR (MJ/m^2)	TSP (m^2)	TSR (m^2/m^2)	Char Residue (%)
BOZ-E	28	1156	119	56	12.8	1450	2.2
5%SPOSS-PBZ-E	22	957	74	35	11.4	1289	2.0
10%SPOSS-PBZ-E	17	962	94	45	14.0	1481	2.3
15%SPOSS-PBZ-E	24	942	89	42	14.1	1487	4.3
20%SPOSS-PBZ-E	14	769	87	42	15.1	1712	4.5

TTI is used to determine the influence of flame retardant on ignitability. As shown in Table 5, the TTI of the flame-retarded epoxy resins with SPOSS-BOZ loading did not show any effective enhancement.

The HRR versus time curves for the SPOSS-BOZ-E nanocomposites are presented in Figure 12. It can be seen that all the composites burned rapidly, and the typical char-forming platform was not observed. For the neat benzoxazine/epoxy resin, the pHRR was 1156 kW/m^2, and the pHRR values of the 5% SPOSS-PBZ-E, 10% SPOSS-PBZ-E, 15% SPOSS-PBZ-E, and 20% SPOSS-PBZ-E were 957, 962, 942, and 769 kW/m^2, respectively. It is clear that SPOSS-BOZ apparently hinders the decomposition of the composites and improves the flame retardancy. The addition of 20 wt % SPOSS-BOZ caused reductions in the pHRR, aHRR, and THR values by about 33%, 27%, and 25%, respectively (Table 5). The reductions of these data resulted from the effect of the formation of a char barrier (Figure 13). During the cone calorimeter test, the organic group of POSS nanoparticles degrades and leaves the Si–O–Si-based ceramic layer [37,38]. As shown in Figure 13, when the content of SPOSS-BOZ increased, the residual carbon content of the benzoxazine/epoxy resin increased slightly, and when the content of SPOSS-BOZ reached 20 wt %, a thicker, continuous, and intumescent insulation layer with an integral structure was formed during combustion. However, the TSR of samples increased in the 15%SPOSS-PBZ-E and 20%SPOSS-PBZ-E samples with enlarging POSS content.

The char residues were further analyzed with SEM-EDX analysis. The external char residues of the BOZ-E, 5% SPOSS-PBZ-E, 10% SPOSS-PBZ-E, and 15% SPOSS-PBZ-E composites and the internal and external char of the 20% SPOSS-PBZ-E were analyzed by SEM-EDX (Figures 14 and 15). It can be seen that the char of BOZ-E was loose and porous, but for the nanocomposites containing SPOSS-BOZ, a dense and compact external layer can be seen in Figure 14. From Figure 15, the presence of the Si atom can be observed on the char surface with the POSS nanoparticles, and the content of Si reaches its highest value with the 20 wt % SPOSS-BOZ loaded from the EDX, but there is less silicon in the internal char. It is thought that the increase in fire protection from using SPOSS-BOZ arises from the formation of a ceramic layer due to the enrichment of silicon on the char surface. When there are enough POSS nanoparticles, it can effectively protect the combustion of internal polymers. Therefore, the flame-retardant mechanism of POSS is mainly a condensed-phase flame retardant.

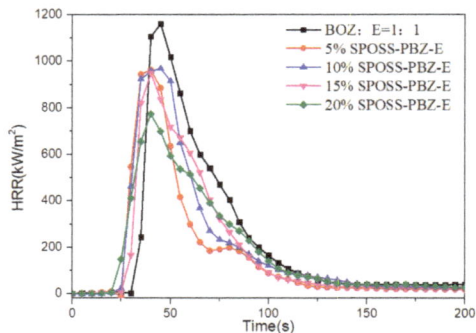

Figure 12. Heat release rate curves of the SPOSS-BOZ nanocomposites.

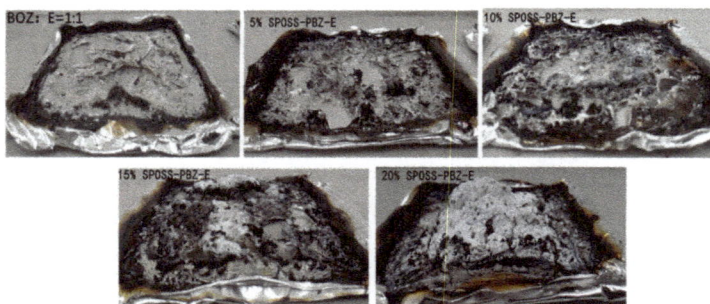

Figure 13. The char photos of nanocomposites with 0, 5, 10, 15, and 20 wt % SPOSS-BOZ after the cone calorimeter test.

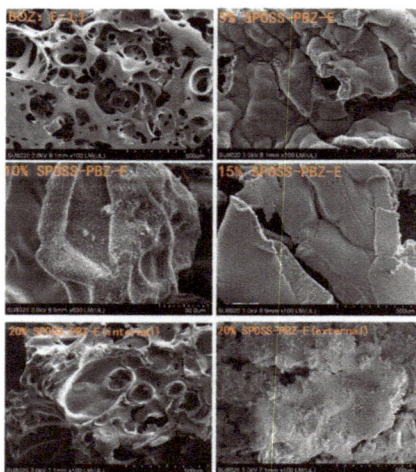

Figure 14. SEM micrographs of the char residues remaining after the cone calorimeter test of the SPOSS-BOZ-E nanocomposites.

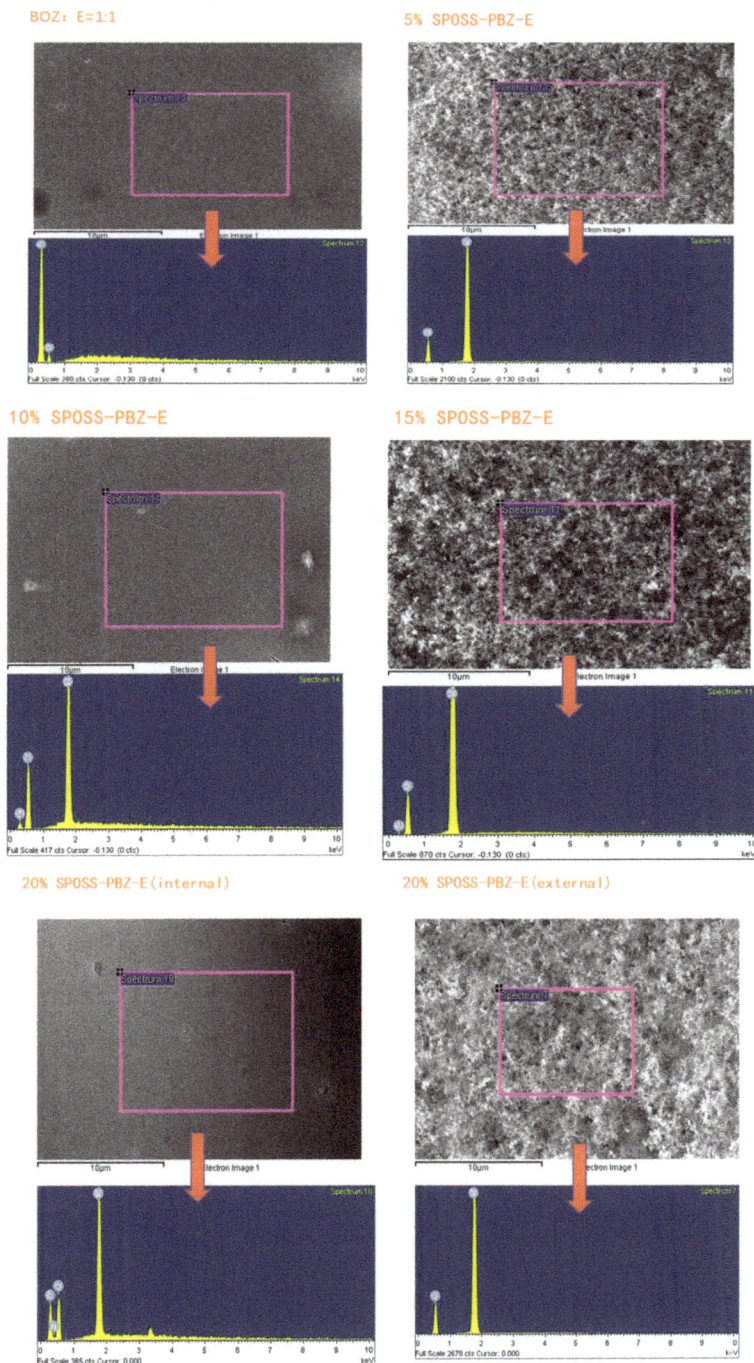

Figure 15. EDX analysis of the char residues remaining after the cone calorimeter test of the SPOSS-BOZ-E nanocomposites.

4. Conclusions

A novel organic–inorganic hybrid containing allyl benzoxazine and polyhedral oligomeric silsesquioxane (SPOSS) was synthesized by the thiol-ene (click) reaction. The curing process of SPOSS/polybenzoxazine/epoxy has multiple polymerization mechanisms according to the oxazine ring-opening and epoxy resin (EP) polymerization, which was characterized by FTIR, and showed that the phenolic hydroxyl group was formed after the ring opening of the oxazine ring, which promotes the ring-opening polymerization of epoxy. The X-ray diffraction analysis and the SEM micrographs of the SPOSS-PBZ-E nanocomposites indicated that SPOSS was chemically incorporated into the hybrid nanocomposites in the size range of 80–200 nm. The thermal properties and the char yield of the nanocomposites increased, and the highest maximum rate of weight loss decreased with respect to the addition of the SPOSS-BOZ organic–inorganic hybrid. At the same time, the flame retardancy of nanocomposites was improved when compared with the pure benzoxazine/epoxy blend. When the amount of SPOSS reached 10 wt % or more, the UL 94 vertical burning rating reached V-1, and when the SPOSS content reached 20 wt %, both the thermal stability and the flame retardancy of the curing system were improved. In the cone calorimeter testing, the addition of SPOSS-BOZ hindered the decomposition of the composites and the addition of 20 wt % SPOSS-BOZ caused reductions in the pHRR, aHRR, and THR values of approximately 33%, 27%, and 25%, respectively. Then, the morphologies of the chars were also studied by SEM and EDX, which indicated that during the combustion process of the nanocomposites, the POSS nanoparticles accumulated on the surface and formed a ceramic layer in the exterior char residue.

Supplementary Materials: The following are available online at http://www.mdpi.com/2073-4360/11/5/770/s1, Figure S1: The DSC curves of benzoxazine at different heating rates. Figure S2: Tan δ curves of nanocomposites containing various SPOSS-BOZ contents. Figure S3: EDX analysis of the fracture surface of SPOSS-PBZ-E.

Author Contributions: X.L. conceived and designed the experiments; B.L. and H.W. performed the experiments; B.L. and H.W. synthesized monomer and composites, B.L., H.W., R.Y., and X.G. analyzed the data; B.L. wrote the paper.

Acknowledgments: The authors are grateful for the financial support of the National Key R&D Program of China (Grant Nos. 2016YFB0302100 and 2016YFB0302101).

Conflicts of Interest: The authors declare no conflict of interest.

References

1. Agag, T.; Takeichi, T. Synthesis and characterization of novel benzoxazine monomers containing allyl groups and their high performance thermosets. *Macromolecules* **2003**, *36*, 6010–6017. [CrossRef]
2. Burke, W.J. 3,4-Dihydro-1,3,2H-Benzoxazines. Reaction of p-substituted phenols with N. N-dimethylolamines. *J. Am. Chem. Soc.* **1949**, *71*, 609–612. [CrossRef]
3. Burke, W.J.; Weatherbee, C. 3,4-Dihydro-1,3,2H-Benzoxazines. Reaction of Polyhydroxybenzenes with N-Methylolamines1. *J. Am. Chem. Soc.* **1950**, *72*, 4691–4694. [CrossRef]
4. Kumar, K.S.; Nair, C.R.; Radhakrishnan, T.S.; Ninan, K. Bis allyl benzoxazine: Synthesis, polymerisation and polymer properties. *Eur. Polym. J.* **2007**, *43*, 2504–2514. [CrossRef]
5. Nebhani, L.; Barner-Kowollik, C. Erratum: (Journal of Polymer Science, Part A: Polymer Chemistry). *J. Polym. Sci. Part A Polym. Chem.* **2010**, *48*, 6053–6071.
6. Ning, X.; Ishida, H. Phenolic materials via ring-opening polymerization: Synthesis and characterization of bisphenol-A based benzoxazines and their polymers. *J. Polym. Sci. Part A Polym. Chem.* **1994**, *32*, 1121–1129. [CrossRef]
7. Rimdusit, S.; Ishida, H. Development of new class of electronic packaging materials based on ternary systems of benzoxazine, epoxy, and phenolic resins. *Polymer* **2000**, *41*, 7941–7949. [CrossRef]
8. Patil, D.M.; Phalak, G.A.; Mhaske, S.T. Enhancement of anti-corrosive performances of cardanol based amine functional benzoxazine resin by copolymerizing with epoxy resins. *Prog. Org. Coat.* **2017**, *105*, 18–28. [CrossRef]

9. Kimura, H.; Matsumoto, A.; Hasegawa, K.; Ohtsuka, K.; Fukuda, A. Epoxy resin cured by bisphenol A based benzoxazine. *J. Appl. Polym. Sci.* **1998**, *68*, 1903–1910. [CrossRef]
10. Sahila, S.; Jayakumari, L.S. Development of in situ generated graphene/benzoxazine-epoxy nanocomposite for capacitor applications. *Polym. Compos.* **2015**, *36*, 1–7. [CrossRef]
11. Lu, S.Y.; Hamerton, I. Recent developments in the chemistry of halogen-free flame retardant polymers. *Prog. Polym. Sci.* **2002**, *27*, 1661–1712. [CrossRef]
12. Spontón, M.; Ronda, J.C.; Galià, M.; Cádiz, V. Cone calorimetry studies of benzoxazine–epoxy systems flame retarded by chemically bonded phosphorus or silicon. *Polym. Degrad. Stab.* **2009**, *94*, 102–106. [CrossRef]
13. Ohashi, S.; Pandey, V.; Arza, C.R.; Froimowicz, P.; Ishida, H. Simple and low energy consuming synthesis of cyanate ester functional naphthoxazines and their properties. *Polym. Chem.* **2016**, *7*, 2245–2252. [CrossRef]
14. Zhang, K.; Zhuang, Q.; Zhou, Y.; Liu, X.; Yang, G.; Han, Z. Preparation and properties of novel low dielectric constant benzoxazole-based polybenzoxazine. *J. Polym. Sci. Part A Polym. Chem.* **2012**, *50*, 5115–5123. [CrossRef]
15. Cui, H.W.; Kuo, S.W. Nanocomposites of polybenzoxazine and exfoliated montmorillonite using a polyhedral oligomeric silsesquioxane surfactant and click chemistry. *J. Polym. Res.* **2013**, *20*, 114. [CrossRef]
16. Huang, J.M.; Kuo, S.W.; Huang, H.J.; Wang, Y.X.; Chen, Y.T. Preparation of VB-a/POSS hybrid monomer and its polymerization of polybenzoxazine/POSS hybrid nanocomposites. *J. Appl. Polym. Sci.* **2009**, *111*, 628–634. [CrossRef]
17. Leu, C.M.; Chang, Y.T.; Wei, K.H. Synthesis and dielectric properties of polyimide-tethered polyhedral oligomeric silsesquioxane (POSS) nanocomposites via POSS-diamine. *Macromolecules* **2003**, *36*, 9122–9127. [CrossRef]
18. Phillips, S.H.; Haddad, T.S.; Tomczak, S.J. Developments in nanoscience: Polyhedral oligomeric silsesquioxane (POSS)-polymers. *Curr. Opin. Solid State Mater. Sci.* **2004**, *8*, 21–29. [CrossRef]
19. Lee, K.M.; Knight, P.T.; Chung, T.; Mather, P.T. Polycaprolactone–POSS Chemical/Physical Double Networks. *Macromolecules* **2008**, *41*, 4730–4738. [CrossRef]
20. Kuo, S.W. Building Blocks Precisely from Polyhedral Oligomeric Silsesquioxane Nanoparticles. *ACS Cent. Sci.* **2016**, *2*, 62. [CrossRef] [PubMed]
21. Choi, J.; Harcup, J.; Yee, A.F.; Zhu, Q.; Laine, R.M. Organic/inorganic hybrid composites from cubic silsesquioxanes. *J. Am. Chem. Soc.* **2001**, *123*, 11420–11430. [CrossRef]
22. Striolo, A.; McCabe, C.; Cummings, P.T. Organic-inorganic telechelic molecules: Solution properties from simulations. *J. Chem. Phys.* **2006**, *125*, 104904. [CrossRef]
23. Mohamed, M.G.; Kuo, S.W. Polybenzoxazine/polyhedral oligomeric silsesquioxane (POSS) nanocomposites. *Polymer* **2016**, *8*, 225. [CrossRef] [PubMed]
24. Lee, Y.J.; Kuo, S.W.; Su, Y.C.; Chen, J.K.; Tu, C.W.; Chang, F.C. Syntheses, thermal properties, and phase morphologies of novel benzoxazines functionalized with polyhedral oligomeric silsesquioxane (POSS) nanocomposites. *Polymer* **2004**, *45*, 6321–6331. [CrossRef]
25. Kolb, H.C.; Finn, M.G.; Sharpless, K.B. Click chemistry: Diverse chemical function from a few good reactions. *Angew. Chem. Int. Ed.* **2001**, *40*, 2004–2021. [CrossRef]
26. Li, Y.; Dong, X.H.; Zou, Y.; Wang, Z.; Yue, K.; Huang, M.; Zhang, W.B. Polyhedral oligomeric silsesquioxane meets "click" chemistry: Rational design and facile preparation of functional hybrid materials. *Polymer* **2017**, *125*, 303–329. [CrossRef]
27. Li, Y.; Dong, X.H.; Guo, K.; Wang, Z.; Chen, Z.; Wesdemiotis, C.; Cheng, S.Z. Synthesis of shape amphiphiles based on POSS tethered with two symmetric/asymmetric polymer tails via sequential "grafting-from" and thiol–ene "click" chemistry. *ACS Macro Lett.* **2012**, *1*, 834–839. [CrossRef]
28. Hou, K.; Zeng, Y.; Zhou, C.; Chen, J.; Wen, X.; Xu, S.; Pi, P. Facile generation of robust POSS-based superhydrophobic fabrics via thiol-ene click chemistry. *Chem. Eng. J.* **2018**, *332*, 150–159. [CrossRef]
29. Luo, A.; Jiang, X.; Lin, H.; Yin, J. "Thiol-ene" photo-cured hybrid materials based on POSS and renewable vegetable oil. *J. Mater. Chem.* **2011**, *21*, 12753–12760. [CrossRef]
30. Lungu, A.; Ghitman, J.; Cernencu, A.I.; Serafim, A.; Florea, N.M.; Vasile, E.; Iovu, H. POSS-containing hybrid nanomaterials based on thiol-epoxy click reaction. *Polymer* **2018**, *145*, 324–333. [CrossRef]
31. Wu, Y.C.; Kuo, S.W. Synthesis and characterization of polyhedral oligomeric silsesquioxane (POSS) with multifunctional benzoxazine groups through click chemistry. *Polymer* **2010**, *51*, 3948–3955. [CrossRef]

32. Lee, Y.J.; Kuo, S.W.; Huang, C.F.; Chang, F.C. Synthesis and characterization of polybenzoxazine networks nanocomposites containing multifunctional polyhedral oligomeric silsesquioxane (POSS). *Polymer* **2006**, *47*, 4378–4386. [CrossRef]
33. Meng, X.; Edgar, K.J. "Click" reactions in polysaccharide modification. *Prog. Polym. Sci.* **2016**, *53*, 52–85. [CrossRef]
34. Fan, Y.; Quan, X.; Zhao, H.; Chen, S.; Yu, H.; Zhang, Y.; Zhang, Q. Poly(vinylidene fluoride). Poly (vinylidene fluoride) hollow-fiber membranes containing silver/graphene oxide dope with excellent filtration performance. *J. Appl. Polym. Sci.* **2017**, *134*, 44713. [CrossRef]
35. Ishida, H.; Allen, D.J. Mechanical characterization of copolymers based on benzoxazine and epoxy. *Polymer* **1996**, *37*, 4487–4495. [CrossRef]
36. Xue, Y.; Liu, Y.; Lu, F.; Qu, J.; Chen, H.; Dai, L. Functionalization of graphene oxide with polyhedral oligomeric silsesquioxane (POSS) for multifunctional applications. *J. Phys. Chem. Lett.* **2012**, *3*, 1607–1612. [CrossRef]
37. Kodal, M. Polypropylene/polyamide 6/POSS ternary nanocomposites: Effects of POSS nanoparticles on the compatibility. *Polymer* **2016**, *105*, 43–50. [CrossRef]
38. Wang, Y.; Liu, F.; Xue, X. Synthesis and characterization of UV-cured epoxy acrylate/POSS nanocomposites. *Prog. Org. Coat.* **2013**, *76*, 863–869. [CrossRef]
39. Lin, H.C.; Kuo, S.W.; Huang, C.F.; Chang, F.C. Thermal and surface properties of phenolic nanocomposites containing octaphenol polyhedral oligomeric silsesquioxane. *Macromol. Rapid Commun.* **2006**, *27*, 537–541. [CrossRef]
40. Du, B.; Ma, H.; Fang, Z. How nano-fillers affect thermal stability and flame retardancy of intumescent flame retarded polypropylene. *Polym. Adv. Technol.* **2011**, *22*, 1139–1146. [CrossRef]
41. Xuan, S.; Hu, Y.; Song, L.; Wang, X.; Yang, H.; Lu, H. Synergistic effect of polyhedral oligomeric silsesquioxane on the flame retardancy and thermal degradation of intumescent flame retardant polylactide. *Combust. Sci. Technol.* **2012**, *184*, 456–468. [CrossRef]

![polymers logo] *polymers*

MDPI

Article

Improving the Damping Properties of Nanocomposites by Monodispersed Hybrid POSS Nanoparticles: Preparation and Mechanisms

Wei Wei, Yingjun Zhang, Meihua Liu *, Yifan Zhang, Yuan Yin, Wojciech Stanislaw Gutowski, Pengyang Deng and Chunbai Zheng *

CAS Key Laboratory of High-Performace Synthetic Rubber and its Composite Materials, Changchun Institute of Applied Chemistry, Renmin Street 5625, Changchun 130022, China; weiwei@ciac.ac.cn (W.W.); zyj13125823573@163.com (Y.Z.); yfzhang@ciac.ac.cn (Y.Z.); yuany@ciac.ac.cn (Y.Y.); voytek.gutowski@csiro.au (W.S.G.); pydeng@ciac.ac.cn (P.D.)
* Correspondence: liumh@ciac.ac.cn (M.L.); zhengcb@ciac.ac.cn (C.Z.)

Received: 3 April 2019; Accepted: 6 April 2019; Published: 9 April 2019

Abstract: In this work, a series of heptaphenyl siloxane trisilanol/polyhedral oligomeric silsesquioxane (T_7-POSS) modified by polyols with different molecular weights were synthesized into liquid-like nanoparticle–organic hybrid materials using the grafted-from method. All grafted POSS nanoparticles changed from solid powders to liquid at room temperature. Polyurethane (PU) nanocomposites with POSS contents ranging from 1.75 to 9.72 wt % were prepared from these liquefied polyols-terminated POSS with polyepichlorohydrin (POSS–PECH). Transmission electron microscopy (TEM), scanning electron microscopy (SEM) and energy dispersive spectroscopy (EDS) were used to characterize the morphology of the POSS–PECH/PU nanocomposites. The results showed that the polyol-terminated POSS particles overcame the nanoagglomeration effect and evenly disperse in the polymeric matrix. The damping factor (tan δ) of resultant nanocomposites increased from 0.90 to 1.16, while the glass transition temperature decreased from 15.8 to 9.4 °C when POSS contents increased from 0 to 9.75 wt %. The gel content, tensile strength and Fourier transform infrared (FTIR) analyses demonstrated that the molecular thermal movement ability of the polyurethane (PU) matrix increased with increasing POSS hybrid content. Therefore, the improvement of the damping properties of the composites was mainly due to the friction-related losses occurring in the interface region between the nanoparticles and the matrix.

Keywords: monodisperse; nanocomposites; damping; POSS; liquefied

---

1. Introduction

In recent decades, nanocomposites have been intensely investigated by academy and industry researchers in the field of new materials development. A large number of nanocomposites have been successfully developed and applied, becoming the key driving force for the rapid development of advanced materials for applications in aviation, aerospace, transportation and energy technologies [1–8].

Nanocomposites often exhibit excellent strength, modulus of elasticity and toughness due to the reinforcement effects of nanofillers. In these commodity engineering applications, traditional theories of polymer mechanics are usually effective in predicting and guiding the static elastic properties of composites.

It is well known that the viscous properties of polymer composites can be expressed by the loss function known as 'tangent delta' (tan δ), which is attributed to 1) intra-molecular friction and molecular relaxation, 2) friction between the polymer chain and filler, 3) friction between adjacent filler particles, and 4) flexibility of the polymer chain. The value of tan δ is mainly determined by its viscoelastic behavior in the glass temperature (T_g) transitional region. In this region, macromolecule chain segments tend to vibrate in phase with external vibrations. If numerous types of interactions exist between the polymer and filler, a broader tan δ transition region would be desirable. The higher the internal friction, the higher the tan δ value and the broader the transition region will be; thus, excellent energy dissipation (damping) performance of such composites will be observed [9–14]. However, increasing the strength and modulus of elasticity of nanocomposites usually leads to a decrease in damping properties. Researchers active in this field believe that this is because high modulus nanofillers (especially inorganic nanofillers) restrict the thermal movement of the chains in the polymer matrix.

Since 2005, many theoretical studies have demonstrated that fracturing (detaching) molecular chains from the surface of nanofillers can cause changes in conformational entropy. The associated mechanical energy loss is transformed into heat energy [15–20]. This implies that the larger the effective interface region between polymer chains and nanofillers, the greater the damping capacity of composites to attenuate vibrations or noise. As far as we know, no studies to date have shown that the damping properties of nanocomposites would continuously improve with the increase of nanofillers in addition to the matrix, although numerous studies have demonstrated that the uniform distribution of nanosized filler particles in the polymer matrix leads to reasonably good interfacial bonding strength [21–25]. This may be due to the serious nanoagglomeration effect in traditional nanocomposites, which leads to a decrease in the area of the interface region with the increase of nanofillers.

In this study, solid T_7-POSS particles were liquefied by polyepichlorohydrin (PECH) oligomer grafted by a graft-from method. POSS–PECH/polyurethane (PU) nanocomposites, all exhibiting homogeneous morphology of monodispersed POSS nanocores in the PU matrix were then fabricated and characterized. Their damping performance is starkly contrasting that observed in traditional silicon dioxide (SiO_2), T_7-POSS or T_8-POSS composites. The mobility of PU chain segments increases with the content of POSS–PECH. The observed high tan δ and broad transition regions were attributed to increased internal friction and other interactions between the polymer chain and surface-modified POSS filler.

2. Materials and Methods

2.1. Materials

PECH was synthesized from ethylene glycol with epichlorohydrin (ECH) used as the monomer and boron trifluoride etherate (BF_3-etherate) as the catalyst (laboratory grade). Castor oil was purchased from Tongliao Chemical Reagent Factory (Tongliao, Neimenggu, China). Polyaryl polymethylene isocyanate (PAPI) was provided by Bayer (NCO 30.5-32.0 wt %, Leverkusen, Germany). T_7-POSS was received from Hybrid Plastics, CA (Hattiesburg, MS, USA).

2.2. Synthesis of POSS–PECH

The synthesis route of POSS–PECH utilizing T_7-POSS as the starting agent is shown in Figure 1. The ring-opening polymerization of epichlorohydrin was carried out in a 500 mL triple-neck flask equipped with a mechanical paddle, a temperature gauge and a dropping funnel. To the stirred solution of T_7-POSS in dichloromethane, a precisely measured volume of boron trifluoride ether was then added as a catalyst at room temperature, and stirring was continued for 30 min. The reaction flask temperature was brought down to 18 °C using ice. Epichlorohydrin was then slowly added to the reaction mixture over a period of 4 h using a dropping funnel. The reaction mixture was then stirred for 4 h at 40 °C. The synthesized polymer solution was subsequently treated with saturated $NaHCO_3$ solution and distilled water to remove unreacted diol and the initiator when the reaction was completed. The reaction mixture was then filtrated, and the solvent and small-molecule compounds were removed by vacuum distillation at 110 °C for 2 h, yielding a viscous fluid (see Figure 2b). Thus, four synthesized POSS–PECH compounds with different molecular weights were produced and are listed in Table 1.

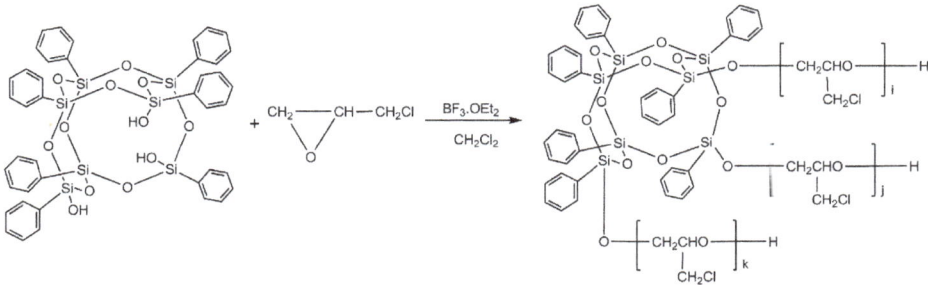

Figure 1. Synthetic route of POSS–PECH.

Table 1. Feed molar ratio of POSS–PECH with different molecular weights.

PECH Samples	T$_7$-POSS (mol)	ECH (mol)	BF$_3$·OEt$_2$ (mol)	Mw of POSS–PECH
POSS-1	1	10	0.6	1573
POSS-2	1	25	0.6	2762
POSS-3	1	40	0.6	3533
POSS-4	1	55	0.6	4136

2.3. Sample Preparation

POSS-based polyurethane formulations (see Table 2 for details) were prepared by a one-step method. Pre-weighted quantities of POSS–PECH, PECH and castor oil were added to a Teflon cup and then placed in the oven preheated to 85 °C for 30 min. The melt curing agent for the PUs, PAPI, was then added and mechanically stirred to achieve homogeneity. The resultant mixture was degassed under vacuum, poured into preheated glass molds coated with a mold release agent and then cured at 100 °C for 24 h. In this study, a series of polyurethane samples was prepared by changing the isocyanate index (R) and the addition of POSS–PECH.

Table 2. Formulations of PU compositions

PU Samples	Castor Oil (g)	PECH (g)	POSS1 (g)	POSS2 (g)	POSS3 (g)	POSS4 (g)	PAPI (g)
Pure-PECH/PU	10	10					6.09
1.75%-POSS1/PU	10	10	2				6.28
4.54%-POSS1/PU	10	10	6				6.66
7.57%-POSS1/PU	10	10	12				7.22
9.72%-POSS1/PU	10	10	18				7.79
5.55%-POSS2/PU	10	10		18			7.68
4.35%-POSS3/PU	10	10			18		7.60
3.74%-POSS4/PU	10	10				18	7.48
POSS1/PU-0.90	10	10	18				6.67
POSS1/PU-0.95	10	10	18				7.05
POSS1/PU-1.00	10	10	18				7.42
POSS1/PU-1.05	10	10	18				7.79

2.4. Gel Extraction Experiments

Samples weighing approximately 0.2 g, were cut into pieces and weighed with an accuracy of ±0.5 mg, and designated W$_0$. They were covered with filter paper, placed on a nickel mesh, and their total weight was determined as W_1. These were placed in a triple-neck flask, refluxed with dichloromethane for 48 h with the solvent changed every 24 h, then washed with ethanol and placed in an oven until constant weight, W_2, was reached. The gel content, G [%], was determined using the following formula:

$$G = [1- (W_1 - W_2)/W_0] \times 100\% \qquad (1)$$

2.5. Characterization

Dynamic mechanical analysis (DMA) measurements were performed on a DMA +450 analyzer (01db-Metravib, Paris, France). The PU film was cut into $4 \times 1 \times 0.2$ cm rectangular strips and tested in tensile mode over the temperature range from -40 to 60 °C. The heating rate was fixed at 3 °C/min. The frequencies were and 0.5, 1, 3 and 5 Hz respectively.

FTIR spectra were recorded with a VERTEX 70 spectrometer (Bruker Daltonics Inc., Billerica, MA, USA) in the range of 4000–500 cm^{-1} at a resolution of 4 cm^{-1}; eight scans were collected per sample.

TEM images were recorded with a Tecnai G2 transmission electron microscope with 200 kV accelerating voltage (FEI Co., Hillsboro, OR, USA). Specimens with a thickness of ca. 50 nm were prepared by ultra-cryomicrotomy at -35 °C using a Leica UCT microtome.

SEM and EDS images were taken using a Philips XL-30 FEG (Amsterdam, The Netherlands). The samples were broken in liquid nitrogen, and all samples were coated with an ultrathin Au film by high-vacuum evaporation before observing the cross sections. At least three sections were observed to show a representative image.

Tensile strength was determined by universal testing machine INSTRON 1121 (INSTREAM Corporation, Boston, MA, USA), according to GB/T1040.1-2006. The samples (dumbbell shape, 50×2 mm, gauge length 4 mm) were tested at a strain rate of 100 mm/min.

Gel permeation chromatography (GPC) analyses were carried out using PL-GPC-120 from Polymer Laboratories (Lanarkshire, UK). The solvent used was DMF at 80 °C with a flow rate of 1.0 mL/min.

^{1}H-NMR and ^{13}C-NMR analyses were conducted on a BRUKER 400 MHz NMR spectrometer (Bruker Daltonics Inc.) in CDCl$_3$ solvent with tetramethylsilane as the internal standard.

3. Results

3.1. Chemical Structural Analysis of POSS–PECH

At room temperature, T_7-POSS transformed from the solid to liquid state after polymerization (see Figure 2a,b). The FTIR spectrum of POSS–PECH presented in Figure 2 shows a peak at approximately 3442 cm^{-1} that was assigned to the O–H stretching band. The peaks at 2958 and 2876 cm^{-1} represent a C–H stretching band. The characteristic peaks at approximately 1430, 1130 and 747 cm^{-1} correspond to deforming of the C–H, C–O–C stretching band and C–Cl stretching band, respectively. The peak at 1596 cm^{-1} is characteristic of phenyl on POSS substituents.

NMR spectra were useful for determining the material composition. Figure 2 presents ^{1}H-NMR (2d) and ^{13}C-NMR(2e) spectra of POSS–PECH. The peaks at 3.62, 3.72 and 3.99 ppm in Figure 2d correspond to the main chain hydrogen atom of POSS–PECH. The areas of signals at 7.10–7.95 ppm can be related to phenyl on POSS substituents. The ^{13}C-NMR spectrum of POSS–PECH is shown in Figure 2e. The assignment of individual peaks is as follows: δ(CH$_2$–Cl) is at 43.13–45.07 ppm, δ(O–CH$_2$) is at 68.98–71.13 ppm, δ(O–CH) is at 78.60–78.98 ppm, and δ(phenyl) is at 127.53–133.92 ppm.

The above analyses of FTIR and NMR spectra prove that the synthesized product was an oligomer of hydroxyl-terminated polyepichlorohydrin grafted to the POSS structure. The molecular weights of alternative POSS–PECH samples that were synthesized are shown in Table 3. The hydroxyl value of POSS–PECH, as seen in Table 4 and measured by the acetic anhydride–pyridine method, decreases with increasing molecular weight.

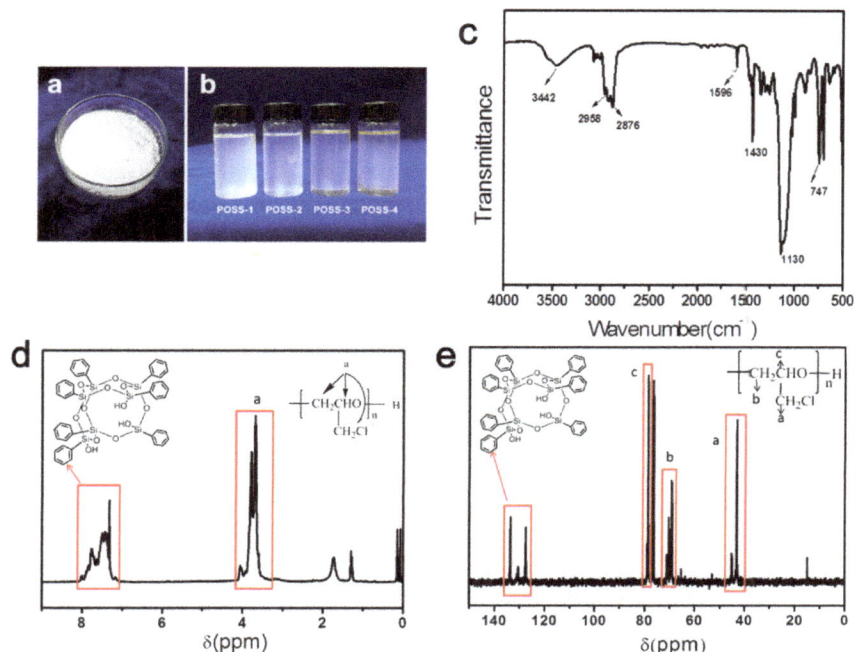

Figure 2. (**a**) Photograph of T$_7$-POSS before liquefaction. (**b**) Photograph of T$_7$-POSS after liquefaction. (**c**) FTIR spectra of POSS–PECH. (**d**) ^{1}H NMR spectra of POSS–PECH. (**e**) ^{13}C NMR spectra of POSS–PECH.

Table 3. Summary of gel permeation chromatography (GPC) data of POSS–PECH.

PECH Samples	Mp	Mn	Mv	Mw	Mz	Mz + 1	PD
POSS-1	1607	1348	1538	1573	1812	2057	1.1669
POSS-2	2120	1960	2628	2762	3737	4698	1.4092
POSS-3	3841	2439	3365	3533	4689	5730	1.4485
POSS-4	4871	2614	3894	4136	5849	7420	1.5822

Table 4. Hydroxyl value of POSS–PECH with different molecular weights.

PECH Samples	Hydroxyl Value (mol/g)
POSS-1	6.64×10^{-4}
POSS-2	6.20×10^{-4}
POSS-3	5.90×10^{-4}
POSS-4	5.42×10^{-4}

3.2. Structure and Phase Morphology Analysis of POSS-Based Polyurethane Nanocomposites

To investigate the chemical structure of the PUs with respect to various contents of POSS–PECH and pure-PECH, their FTIR spectra were obtained and are shown in Figure 3a. The absorption peak of the –NH bond appears at 3329 cm^{-1}, and the stretching vibration peak of the C=O bond in the urethane bond appears at 1727 cm^{-1}. The characteristic absorption peak of the ether bond C–O–C appears at approximately 1112 cm^{-1}. These three features indicate that the characteristic functional group of polyurethane carbamate, NHCOO, was successfully formed. The absence of an absorption peak from isocyanate groups at 2270 cm^{-1} reflects the complete curing reaction.

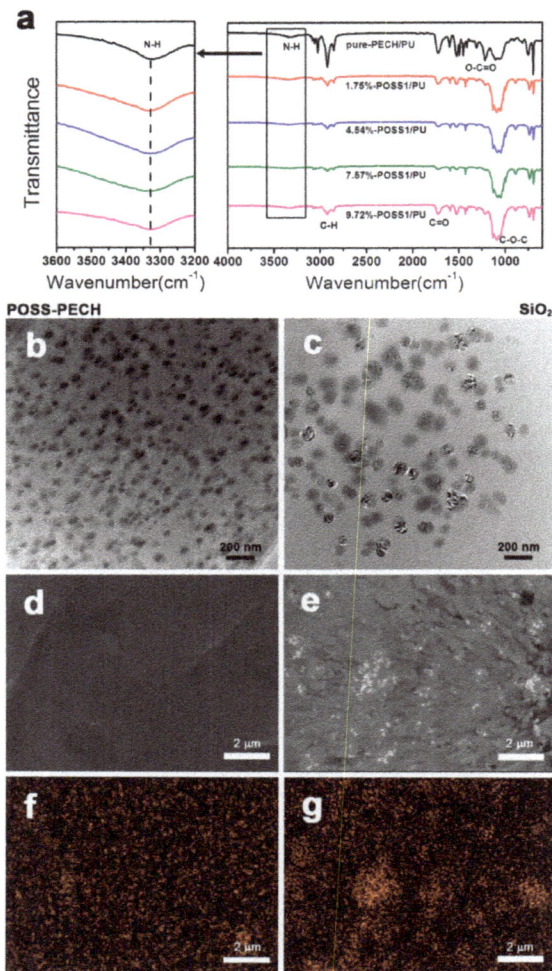

Figure 3. (**a**) FTIR spectra of PUs with different POSS contents. (**b**) TEM images of 9.72% POSS-modified PU. (**c**) TEM images of 9.74% SiO$_2$-modified PU. (**d**) SEM images of 9.72% POSS-modified PU. (**e**) SEM images of 1.96% SiO$_2$-modified PU. (**f**) Element mapping images of Si 9.72% POSS-modified PU. (**g**) Element mapping images of Si 1.96% SiO$_2$-modified PU.

The dispersion of POSS nanoparticles in the PU composites was observed by TEM. The results show that even when the POSS content reached 9.72 wt % of the total weight of the composites, no POSS particles were agglomerated, and all were uniformly distributed in the composite resin with a nearly identical size. The fracture surface of the composite presented in Figure 3d demonstrates a homogeneous morphology with no phase separation observed between the POSS particles and matrix resin. The results of EDS (see Figure 3f) are in good agreement with those of TEM; that is, the silicon elements in the POSS cage are uniformly distributed throughout the cross section of the sample, with no obvious agglomeration observed.

The above results are substantially different from the traditional morphology and structure of nano-SiO$_2$–PU composites. Figure 3c shows a typical TEM micrograph of nanocomposites comprising nanosilica in the PU matrix. The results demonstrate that even if the content of nano-SiO$_2$ is only

1.96 wt % of the total composite weight, agglomeration always occurs, and the cluster size after agglomeration is between 40 and 150 nm. It is also observed that the particle dispersion within clusters is not uniform. Consequently, the morphology of fracture surfaces in silica–PU composites (Figure 3e) is notably different from those containing modified POSS. The sharp interface between agglomerated SiO_2 particles and matrix resin is clearly visible, with phase separation consistent with the agglomeration and non-uniform dispersion of elemental silicon clearly shown in EDS in Figure 3g.

The above results show that the PECH oligomeric chain grafted onto the POSS cage effectively changed the particle surface structure, overcoming the agglomeration problem of POSS particles and facilitating uniform dispersion of POSS inorganic cages within the polyurethane polymer in a monodispersed state [26,27].

3.3. Damping Properties

The dynamic properties of materials, including their damping performance, are commonly studied by DMA, in which the storage modulus (E') and the loss modulus (E'') of the sample under an oscillating load are monitored against time, temperature and frequency of the oscillation. These moduli change with frequency and temperature as the molecular motions within the polymer change. The ratio of $E'/E'' = \tan \delta$, which defines the inherent energy dissipation ability of the material, is commonly used to characterize its damping ability. A high $\tan \delta$ value and large temperature range indicates excellent damping performance of the system.

In general, the energy dissipation reaches a maximum near the glass transition temperature of the polymer (T_g) because in this regime, without considering the chemical structure of the molecular chain, the following molecular chain motions simultaneously transform mechanical energy into thermal energy. When movable molecular chains are relatively short, under the action of external strain, there will be a large conformational change leading to the conformational entropy of materials. In the glassy state, the distance between chains is short and the interaction between chains is strong. Therefore, when the free chains move, the interaction force between the molecules that needs to be overcome is high. When the free chains move relative to each other, the internal friction between the molecular chains that needs to be overcome is large, and the internal friction between the molecular chain and the filler is also large.

3.3.1. Effect of POSS1 Content on the Damping Properties of PU Composites

The effect of POSS1 content on $\tan \delta$ of PU nanocomposites is shown in Figure 4a and Table 5. The transition peaks in the DMA spectra of PU nanocomposites with different POSS1 contents indicate that POSS–PECH has good compatibility with the matrix material. Compared with pure polyurethane polymer, the $\tan \delta$ of POSS-based PU nanocomposites increased from 0.91 to 1.16 with the increasing POSS content (an increase of 27.5%), whereas the damping temperature range ($\tan \delta > 0.3$) increased from $\Delta T = 34.4$ to 44.0 °C, thus increasing by 28.1%. Meanwhile, the T_g and the initial temperature of the damping temperature range (T_1) of the PU–POSS nanocomposites gradually decreased respectively: T_g to 9.4 °C from 15.8 °C and T_1 to –6.9 °C from –1.1 °C. These results demonstrate that monodispersed nano-POSS particles effectively improved the damping properties of the composites, importantly demonstrating that the increased POSS content did not restrict the thermal mobility of molecular chains.

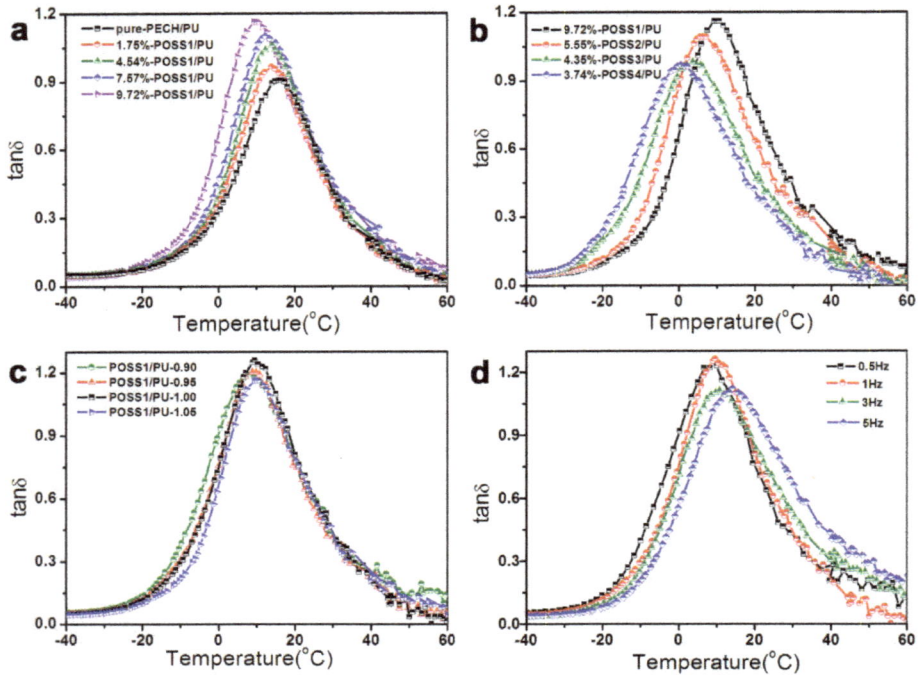

Figure 4. This loss factor (tan δ) curves vs temperature for: (**a**) The neat PU matrix and PUs with different POSS contents (POSS–PECH with same molecular weight), (**b**) PUs with different POSS contents (POSS–PECH with different molecular weights), (**c**) PUs with different isocyanate indexes (R) and (**d**) PUs response to different strain frequencies.

Table 5. Summary of DMA data of PUs with different POSS contents (POSS–PECH with same molecular weight).

PU Samples	Inorganic Core Content (%)	tan δ	T_g/°C	Damping Temperature Field (tan δ > 0.3)		
				T_1/°C	T_2/°C	ΔT°C
Pure-PECH/PU	0	0.9064	15.8	-1.1	33.3	34.4
1.75%-POSS1/PU	1.75	0.9659	14.0	-2.0	32.6	34.6
4.54%-POSS1/PU	4.54	1.058	14.1	-2.7	34.2	36.9
7.57%-POSS1/PU	7.57	1.099	12.3	-4.0	37.4	41.4
9.72%-POSS1/PU	9.72	1.164	9.4	-6.9	37.1	44.0

M_w = 1573, R = 1.05, the percentage content is the content of inorganic POSS nanoparticles. T_1 is the starting temperature of tan δ > 0.3, and T_2 is the terminating temperature of tan δ > 0.3.

3.3.2. Effect of Molecular Weight Change of POSS–PECH on Damping Property of Composites

The effects of POSS–PECH polyols with different molecular weights on the damping properties of nanocomposites were investigated, as shown in Figure 4b and Table 6. The results demonstrate that the tan δ and ΔT of all POSS–PECH composites were higher than those of the unmodified bulk material. The tan δ decreased gradually with increasing molecular weight of POSS–PECH; although the ΔT was almost unchanged, the T_1 of the damping temperature range decreased gradually with increasing molecular weight.

At the same time, the T_g of composites gradually decreased with increasing molecular weight of POSS–PECH. The initial T_g was 15.8 °C for pure PU–PECH and decreased to 0.6 °C for composites with POSS4 modified with PECH and a M_w = 4136; this is the highest range for the materials investigated in this work. This trend highlights the fact that the size of the interfacial region surrounding the POSS–PECH nanoparticles appears to be a key factor controlling the viscous properties of the system and thus the composite damping properties.

Table 6. Summary of DMA data of PUs with different molecular weights of POSS–PECH.

PU Samples	Inorganic Core Content (%)	$\tan \delta\ T_g/^\circ$C		Damping Temperature Field (tan $\delta > 0.3$)		
				$T_1/^\circ$C	$T_2/^\circ$C	ΔT°C
Pure PECH/PU	0	0.964	15.8	−1.1	33.3	34.4
9.72%-POSS1/PU	9.72	1.164	9.4	−6.9	37.1	44.0
5.55%-POSS2/PU	5.55	1.091	6.5	−10.3	34.0	44.3
4.35%-POSS3/PU	4.35	0.9813	3.7	−15.1	28.2	43.3
3.74%-POSS4/PU	3.74	0.9681	0.6	−18.1	25.9	44.0

T_1 is the starting temperature of tan $\delta > 0.3$, and T_2 is the terminating temperature of tan $\delta > 0.3$.

3.3.3. Effect of R Value on Damping Property of Polyurethane Polymer

The R value is the molar ratio of isocyanate groups (–NCO) to hydroxyl groups (–OH). By adjusting the R value, the crosslinking density and structure of composites can be custom-tailored, and hence, it was anticipated that the damping properties could also be adjusted. The numerical data in Table 7 and those graphically presented in Figure 4c provide insight into the damping properties of POSS-based polyurethane composites with different R-values and almost constant contents of POSS nanoparticles. The results demonstrate that the damping properties of POSS-based PU nanocomposites, that is, the tan δ and ΔT were almost independent of the R value. Only at R = 1.00 was the loss factor tan $\delta = 1.26$, thus showing some improvement of damping properties. However, with the increase in the R value, the ΔT was shifted towards higher temperatures, that is, from (−12.3 ↔ +35.8 °C) to (−6.9 °C ↔ +37.1 °C). This is because the content of the hard segment increased slightly with the increase in R value, resulting in the ΔT moving towards a high temperature. These results also indicate that the improvement of the damping properties of POSS–PECH based nanocomposites is mainly due to the change in PECH-grafted POSS particle content.

Table 7. Summary of DMA data of PUs with different molecular weights of POSS–PECH.

PU Samples [POSS/PU–R]	Inorganic Core Content (%)	$\tan \delta\ T_g/^\circ$C		Damping Temperature Field (tan $\delta > 0.3$)		
				$T_1/^\circ$C	$T_2/^\circ$C	ΔT°C
POSS1/PU-0.90	9.96	1.186	7.5	−12.3	35.8	48.1
POSS1/PU-0.95	9.88	1.201	9.6	−10.4	33.6	44.0
POSS1/PU-1.00	9.80	1.262	9.6	−9.7	35.1	44.8
POSS1/PU-1.05	9.72	1.164	9.4	−6.9	37.1	44.0

T_1 is the starting temperature of tan $\delta > 0.3$ and T_2 is the terminating temperature of tan $\delta > 0.3$.

3.3.4. Effect of Different Testing Frequencies on Damping Property of Polyurethane Polymer

The effects of different test frequencies on the damping properties of polyurethane polymers were analyzed using data sets presented in Figure 4d and Table 8. The results show that with increasing test frequency, the tan δ of the polymer first increased and then decreased. The T_g gradually increased, the ΔT moved towards high temperature, and the range of tan $\delta > 0.3$ increased slightly.

Table 8. Summary of DMA data of PUs with different test frequencies.

Test Frequency (Hz)	Tan δ T_g/°C	Damping Temperature Field (tan δ > 0.3)		
		T_1/°C	T_2/°C	ΔT°C
0.5	1.234	9.9	−12.3	34.5	46.8
1	1.262	9.6	−9.7	35.1	44.8
3	1.111	10.7	−8.7	40.1	48.8
5	1.119	14.1	−6.8	48.5	55.3

M_w = 1573, R = 1.05, PU with 9.72% inorganic nanomaterials content was tested by DMA at different frequencies.

4. Discussion

The analysis of the results presented above yields the following information:

When the POSS nanoparticles surface-grafted with pendent chains of PECH are distributed in the polymer matrix in the form of a monodispersed phase, the damping properties of the nanocomposites increase with the increase the nano-POSS content. This effect is completely opposite to the performance of traditional nanocomposites, such as nano-SiO$_2$–PU and unmodified POSS–PU composites (for details see Supplementary Materials), whose damping properties are presented in Table 9 and depicted in Figure 5. These results clearly demonstrate that increasing the nano-SiO$_2$ content results in the deterioration of the damping properties of SiO$_2$–PU nanocomposites. The tan δ is shown to be gradually reduced by 25.3% from the initial value of tan δ equal to 0.91 for the unmodified (pure PECH–PU) material to tan δ equal to 0.68 exhibited by the nano-SiO$_2$–PU composite containing 9.74% of SiO$_2$ nanoparticles. This loss function value equals only 50% of the damping capability of our composites, with POSS1–PU containing a similar concentration of surface-grafted POSS nanoparticles. The ΔT is almost unchanged compared with that related to the original material, but the T_1 and the T_g of the SiO$_2$-composite system both increase with the increasing amount of silica.

A number of earlier studies [28,29] have shown that the decrease in damping performance of traditional SiO$_2$–PU composite systems is due to nanoparticles being dispersed in the soft matrix and acting as 'physical crosslinks', which limit the thermal movements of molecular chains in the soft matrix. These restrictions become stronger as the number of nanoparticles increases, and hence, the number of 'physical crosslinks' is increased. Consequently, any restrictions on the molecular thermal motion ability will lead to a decrease in the damping property of the composite material and an increase in T_1 and T_g.

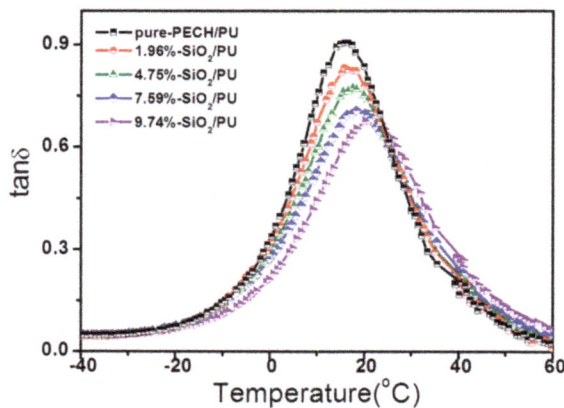

Figure 5. Loss factor (tan δ) curves vs temperature for PU composites with the addition of traditional nano-SiO$_2$ (0 to 9.74%)

Table 9. Summary of DMA data of PU Composites with traditional nano- SiO_2.

| PU Samples | tan δ $T_g/°C$ | | Damping Temperature Field (tan δ > 0.3) | | |
			$T_1/°C$	$T_2/°C$	$\Delta T°C$
pure-PECH/PU	0.9064	15.75	−1.1	33.3	34.4
1.96%-SiO$_2$/PU	0.8319	15.70	−0.8	34.8	35.6
4.75%-SiO$_2$/PU	0.7729	17.50	0.2	34.8	34.6
7.59%-SiO$_2$/PU	0.7065	18.25	1.2	36.8	35.6
9.74%-SiO$_2$/PU	0.6761	21.00	4.4	38.7	34.3

4.1. Analysis of the Mechanism(s) of T_g and T_1 Changes in POSS-Based Polyurethane Nanocomposites

For the amorphous POSS–PU nanocomposite systems, the change of T_1 and T_g is closely related to the crosslinking density, molecular weight between crosslinking points and the interaction between molecular chains. For this reason, we determined the following important properties of our materials: (i) The crosslinking density of the systems, which was tested by gel extraction (see Section 2.4 for details), and (ii) the change of cohesive energy density (CED) of POSS–PU and SiO$_2$–PU composite systems, which is used to estimate the intermolecular interaction in polymers and can be estimated by formula (2) [30,31]:

$$E = 13.3 \, \delta^2, \tag{2}$$

where E is the modulus of elasticity (as determined through mechanical testing of our composites); see data in Tables 10 and 11.

Table 10. Mechanical properties and gel content of polyurethane polymers with different POSS–PECH contents.

PU Samples	Tensile Strength (MPa)	Modulus of Elasticity (MPa)	Elongation at Break (%)	Critical Fracture Stress (MPa)	Cohesive Energy Density (MPa)	Gel Content (%)
Pure-PECH/PU	8.32	10.90	142.5	8.32	0.82	96.28
1.75%-POSS1/PU	8.18	9.97	160	8.18	0.75	90.20
4.54%-POSS1/PU	5.70	7.54	140	7.86	0.57	83.18
7.57%-POSS1/PU	4.84	6.26	150	5.21	0.47	78.47
9.72%-POSS1/PU	3.89	5.29	145	3.89	0.40	70.10

Table 11. Mechanical properties and gel content of PU composites with different nano-SiO$_2$ contents.

PU Samples	Tensile Strength (MPa)	Modulus of Elasticity (MPa)	Elongation at Break (%)	Critical Fracture Stress (MPa)	Cohesive Energy Density (MPa)	Gel Content (%)
Pure-PECH/PU	8.32	10.90	142.5	8.32	0.82	96.28
1.96%-SiO$_2$/PU	5.76	12.18	94.0	5.76	0.92	98.24
4.75%-SiO$_2$/PU	6.87	17.00	86.7	6.87	1.28	96.67
7.59%-SiO$_2$/PU	10.21	23.93	84.7	10.21	1.80	97.39
9.74%-SiO$_2$/PU	15.83	31.93	102	15.83	2.40	96.96

The analysis of data in Tables 10 and 11 highlights the following:

The gel content of the PU composite decreased with the increase of POSS1 content. This indicated that the chemical crosslinking degree of the system was also decreasing. However, SiO$_2$ with similar content had little influence on the gel content of the system. Therefore, compared with SiO$_2$–PU composites, the molecular chain thermal motions of the POSS–PU system increased with the increase of increasing POSS content due to the limitation of the decreased limitations of chemical and physical crosslinking decrease.

The CED of the PU composites decreased with increasing POSS1 content. The CED of the POSS–PU system decreased from 820 KPa (for an unmodified PU) to 400 KPa (9.72 wt % POSS1). However, the CED (a measure of the strength of molecular chain interactions) of the SiO$_2$ composite system with similar content increased to 2400 KPa. The decreasing CED indicated that in the monodispersed state, even if the POSS particles content reached 9.72 wt %, it will not hinder the thermal motion of PU molecular chains.

The influence of gel content and CED on the thermal motion ability of the molecular chains of the POSS–PU system was consistent. That is, with the increase of POSS content, chemical and physical restrictions in the system were constantly reduced. Hence, the thermal motion ability of the PU molecular chain segment was enhanced. As a result, T_1 and T_g decreased with the increasing POSS content.

4.2. Analysis of the Damping Mechanism of the POSS-Based Polyurethane Composite System

The damping properties of polymeric materials arise from the internal friction caused by molecular chain motions and the formation and dissociation of intermolecular hydrogen bonds. The structural chemical characteristics of our nanocomposite systems were determined through FTIR, which identifies the hydrogen bonds in the analyzed systems. For the PUs, hydrogen bonding appeared in the N–H stretching region (3500–3100 cm^{-1}) and the C=O stretching region (1740–1710 cm^{-1}), in which two absorption peaks are observed. The spectra of the stretching vibrations of N–H and –C=O in polymers with different POSS–PECH contents at the same molecular weight of POSS–PECH are shown in Figure 6. As shown in Figure 6, the stretching vibration peaks of weakly hydrogen bonded N–H and non-hydrogen bonded –C=O appeared at 3329 cm^{-1} and 1727 cm^{-1}, respectively, and the peak position did not shift with the change of POSS–PECH content.

In other words, the increase in the POSS core contents did not cause a change in the hydrogen bonding between the molecular chains. Therefore, the damping factor of the POSS–PU composite material was mainly caused by the friction between the PU molecular chain and the inorganic nano-POSS cage surface.

The fraction of interfacial area between our PECH-functionalized hybrid nanoparticles and the polymer matrix increased with increasing such POSS content. Under the action of periodic external forces, the total amount of PU molecular chains in the interfacial region that were repeatedly stripped and re-adsorbed also kept increasing, which results in the damping factor increasing with the increase of our PECH–POSS content.

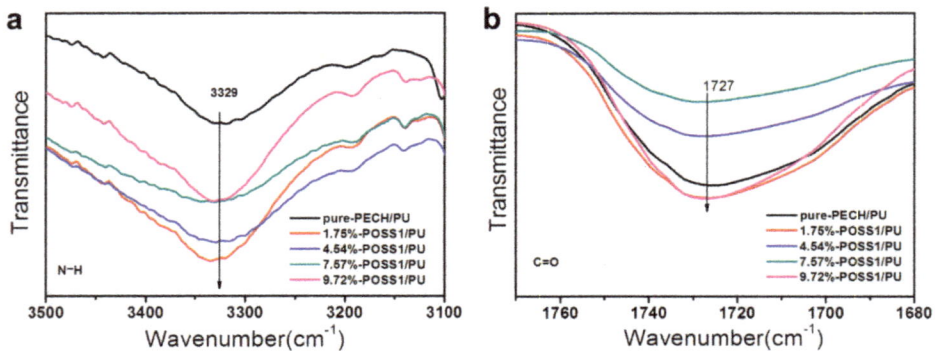

Figure 6. FTIR spectra of the N–H stretching regions and the C=O stretching regions of PUs with different contents of POSS–PECH.

5. Conclusions

Although polymers often exhibit good damping ability near the T_g, they generally exhibit a relatively low (tan δ) and narrow ΔT because they lack extreme and continuous composition heterogeneity and thus cannot serve as damping materials in a broad temperature range. Considering the above issues, in this work we accomplished alleviating the above problems through: (i) Preparation

Polymers **2019**, *11*, 647

of homogeneous monodispersed polymer nanocomposites utilizing bio-based PU resin as the matrix and (ii) significant broadening of the damping temperature range.

We demonstrated here that POSS surface-grafting with oligomeric chains of variable molecular weight yields liquefied hydroxyl-terminated polyepichlorohydrin with a POSS structure. The new organic–inorganic hybrid material enables facile preparation of mono-dispersed nanocomposite systems achieving (i) total dispersion of nanoparticles providing excellent polymer nanocomposite homogeneity for POSS contents in the range 1.75 to 9.75%, (ii) excellent damping properties achieved through improvement of viscoelastic properties of interfacial zone achieved through shielding of inorganic (POSS) core by surface-grafted oligomeric molecular chains, and (iii) significant reduction of T_g, combined with broadening of damping temperature field. The proposed approach lays the foundation for a novel means of engineering the glass transition breadth and damping properties of elastomeric nanocomposites over a broadened range of application temperatures.

Supplementary Materials: The following are available online at http://www.mdpi.com/2073-4360/11/4/647/s1, Figure S1 ^{29}Si NMR spectra of POSS1–PECH Figure S2: FTIR spectra of POSS2–PECH, Figure S3: FTIR spectra of POSS3–PECH, Figure S4: FTIR spectra of POSS4–PECH, Figure S5: ^{1}H NMR spectra of POSS2–PECH, Figure S6: ^{1}H NMR spectra of POSS3–PECH, Figure S7: ^{1}H NMR spectra of POSS4–PECH, Figure S8: ^{13}C NMR spectra of POSS2–PECH, Figure S9: ^{13}C NMR spectra of POSS3–PECH, Figure S10: ^{13}C NMR spectra of POSS4–PECH, Figure S11: ATR-IR spectra of PUs with different POSS–PECH, Figure S12: ATR-IR spectra of PUs with different isocyanate index, Figure S13: ATR-IR spectra of PUs with different SiO_2 content S14: (**a**) SEM images of 1.75% POSS modified PU. (**b**) Element mapping images of Si (1.75% POSS modified PU). (**c**)SEM images of 4.54% POSS modified PU. (**d**)Element mapping images of Si (4.54%POSS modified PU). (**e**)SEM images of 7.57% POSS modified PU. (**f**)Element mapping images of Si (7.57% POSS modified PU), Figure S15: (**a**) SEM images of 4.75% SiO_2 modified PU. (**b**) Element mapping images of Si (4.75% SiO_2 modified PU). (**c**) SEM images of 7.59% SiO_2 modified PU. (**d**) Element mapping images of Si (7.59% SiO_2 modified PU). (**e**) SEM images of 9.74%SiO_2 modified PU. (**f**) Element mapping images of Si (9.74% SiO_2 modified PU), Figure S16: Loss factor(tan δ) curves vs temperature for PUs with different T_7-POSS contents, Figure S17: (**a**) SEM images of 1.77% T_7-POSS modified PU. (**b**) SEM images of 4.52%T_7-POSS modified PU. (**c**) SEM images of 7.55% T_7-POSS modified PU. (**d**) SEM images of 9.73% T_7-POSS modified PU, Figure S18: TEM images of 1.77% T_7-POSS modified PU, Figure S19: Loss factor(tan δ) curves vs temperature for PUs with different T_8-POSS contents, Figure S20: (**a**) SEM images of 1.76% T_8-POSS modified PU. (**b**) SEM images of 4.55% T_8-POSS modified PU. (**c**) SEM images of 7.57% T_8-POSS modified PU. (**d**) SEM images of 9.75% T_8-POSS modified PU, Figure S21: TEM images of 1.76% T_8-POSS modified PU, Table S1: Formulations of PU compositions, Table S2: Summary of DMA data of PUs with different T_7-POSS contents, Table S3: Mechanical properties and gel content of PU Composites with different T_7-POSS content, Table S4: Formulations of PU compositions, Table S5: Summary of DMA data of PUs with different T_8-POSS contents, Table S6: Mechanical properties and gel content of PU Composites with differentT_7-POSS content.

Author Contributions: Investigation, W.G. and M.L.; writing—original draft preparation, W.W.; writing—review and editing, P.D. and C.Z.; formal analysis, Y.Z. (Yingjun Zhang) and Y.Z. (Yifan Zhang); software, Y.Y.

Funding: This research received no external funding.

Acknowledgments: The authors are grateful for the financial support from the Chinese National Natural Science Foundation (Project Nos. 51603201, 51603202, 51803208).

Conflicts of Interest: The authors declare no conflict of interest.

References

1. Tjong, S.C. Structural and mechanical properties of polymernanocomposites. *Mater. Sci. Eng. R. Rep.* **2006**, *53*, 72–197. [CrossRef]

2. Krishnamoorti, R.; Vaia, R.A. Polymer nanocomposites. *J. Polym. Sci. Part B Polym. Phys.* **2007**, *45*, 3252–3256. [CrossRef]

3. Meszaros, L. Polymer matrix hybrid composites: The efficient way of improved performance. *Express Polym. Lett.* **2014**, *8*, 790. [CrossRef]

4. Lee, J.K.Y.; Chen, N.; Peng, S.J.; Li, L.; Tian, L.; Thakor, N.; Ramakrishna, S. Polymer-based composites by electrospinning: Preparation &functionalization with nanocarbons. *Prog. Polym. Sc.* **2018**, *86*, 40–84.

5. Machrafi, H.; Lebon, G.; Iorio, C.S. Effect of volume-fraction dependent agglomeration of nanoparticles on the thermal conductivity of nanocomposites: Applications to epoxy resins, filled by SiO_2, AlN and MgO nanoparticles. *Compos. Sci. Technol.* **2016**, *130*, 78–87. [CrossRef]

6. Lee, D.W.; Yoo, B.R. Advanced silica/polymer composites: Materials and applications. *J. Ind. Eng. Chem.* **2016**, *38*, 1–12. [CrossRef]

7. Chruściel, J.J.; Leśniak, E. Modification of epoxy resins with functional silanes, polysiloxanes, silsesquioxanes, silica and silicates. *Prog. Polym. Sci.* **2015**, *41*, 67–121. [CrossRef]

8. Mittal, G.; Dhand, V.; Rhee, K.Y.; Park, S.J.; Lee, W.R. A review on carbon nanotubes and graphene as fillers in reinforced polymer nanocomposites. *J. Ind. Eng. Chem.* **2015**, *21*, 11–25. [CrossRef]

9. Zhou, X.Q.; Yu, D.Y.; Shao, X.Y.; Zhang, S.Q.; Wang, S. Research and applications of viscoelastic vibration damping materials: A review. *Compos. Struct.* **2016**, *136*, 460–480. [CrossRef]

10. Sun, L.Y.; Gibson, R.F.; Gordaninejad, F.; Suhr, J. Energy absorption capability of nanocomposites: A review. *Compos. Sci. Technol.* **2009**, *69*, 2392–2409. [CrossRef]

11. Zhu, G.L.; Han, D.; Yuan, Y.; Chen, F.; Fu, Q. Improving Damping Properties and Thermal Stability of Epoxy/Polyurethane Grafted Copolymer by Adding Glycidyl POSS. *Chin. J. Polym. Sci.* **2018**, *36*, 1297–1302. [CrossRef]

12. Ristić, I.S.; Simendić, J.B.; Krakovsky, I.; Valentova, H.; Radičević, R.; Cakić, S.; Nikolić, N. The properties of polyurethane hybrid materials based on castor oil. *Chin. Mater. Chem. Phys.* **2012**, *132*, 74–81. [CrossRef]

13. Sahoo, S.; Kalita, H.; Mohanty, S.; Nayak, S.K. Shear Strength and Morphological Study of Polyurethane-OMMT Clay Nanocomposite Adhesive Derived from Vegetable Oil-Based Constituents. *J. Renew. Mater.* **2018**, *6*, 117–125. [CrossRef]

14. Li, Y.; Jiao, H.Y.; Pan, G.Q. Mechanical and damping properties of carbon nanotube-modified polyisobutylene-based polyurethane composites. *J. Compos. Mater.* **2016**, *50*, 929–936. [CrossRef]

15. Michael, E.M.; Anish, T.; Phillip, M.D. General strategies for nanoparticle dispersion. *Science* **2006**, *311*, 1740–1743.

16. Gupta, S.; Zhang, Q.L.; Emrick, T. Entropy-driven segregation of nanoparticles to cracks in multilayered composite polymer structures. *Nat. Mater.* **2006**, *5*, 229–233. [CrossRef]

17. Anna, C.B.; Emrick, T.; Russell, T.P. Nanoparticle polymer compositesn: Where two small worlds meet. *Science* **2006**, *314*, 1107–1110.

18. Guo, Z.R.; Chang, T.C.; Guo, X.M. Thermal-Induced Edge Barriers and Forces in Interlayer Interaction of Concentric Carbon Nanotubes. *Phys. Rev. Lett.* **2011**, *107*, 105502. [CrossRef]

19. Chang, T.C.; Zhang, H.W.; Guo, Z.R. Nanoscale directional motion towards regions of higher stiffness. *Phys. Rev. Lett.* **2015**, *114*, 015504. [CrossRef] [PubMed]

20. Li, J.X.; Zhang, H.W.; Guo, Z.R.; Chang, T.C. Edge Forces in Contacting Graphene Layers. *J. Appl. Mech.* **2015**, *82*, 101011. [CrossRef]

21. Suhr, J.; Koratkar, N.; Keblinski, P.; Ajayan, P. Viscoelasticity in carbon nanotube composites. *Compos. Sci. Technol.* **2005**, *4*, 134–137. [CrossRef]

22. Chen, S.; Wang, Q.; Wang, T. Damping, thermal, and mechanical properties of carbon nanotubes modified castor oilbased polyurethane/epoxy interpenetrating polymer network composites. *Mater. Design.* **2012**, *38*, 47–52. [CrossRef]

23. Chen, S.; Wang, Q.; Wang, T. Damping, thermal, and mechanical properties of montmorillonite modified castor oilbased polyurethane/epoxy graft IPN composites. *Mater. Chem. Phys.* **2011**, *130*, 680–684. [CrossRef]

24. Chen, S.; Wang, Q.; Wang, T.; Pei, X. Preparation, damping and thermal properties of potassium titanate whiskers filled castor oil-based polyurethane/epoxy interpenetrating polymer network composites. *Compos. Sci. Technol.* **2011**, *32*, 803–807. [CrossRef]

25. Blanco, I.; Bottino, F.A.; Cicala, G.; Cozzo, G.; Latteri, A.; Recca, A. Synthesis and thermal characterization of new dumbbell shaped POSS/PS nanocomposites: Influence of the symmetrical structure of the nanoparticles on the dispersion/aggregation in the polymer matrix. *Polym. Compos.* **2014**, *36*, 1394–1400. [CrossRef]

26. Blanco, I. The Rediscovery of POSS: A Molecule Rather than a Filler. *Polymers* **2018**, *10*, 904. [CrossRef]

27. Tanaka, K.; Adachi, S.; Chujo, Y. Structure-property relationship of octa-substituted POSS in thermal and mechanical reinforcements of conventional polymers. *J. Polym. Sci.* **2009**, *47*, 5690–5697. [CrossRef]

28. Trakulsujaritchok, T.; Hourston, D.J. Damping characteristics and mechanical properties of silica Wlled PUR/PEMA simultaneous interpenetrating polymer networks. *Eur. Pol. J.* **2006**, *42*, 2968–2976. [CrossRef]

29. Zhang, H.W.; Wang, B.; Li, H.T.; Yan, J.; Wang, J.Y. Synthesis and characterization of nanocomposites of silicon dioxide and polyurethane and epoxy resin interpenetrating network. *Polym. Int.* **2003**, *52*, 1493–1497. [CrossRef]

Polymers **2019**, *11*, 647

30. Robert, M.E.; Tyler, R.L.; Erich, D.B.; Joseph, L.L.; Timothy, W.S. Mechanics and nanovoid nucleation dynamics: Effects of polar functi- onality in glassy polymer networks. *Soft Matter* **2018**, *14*, 8895–8911.

31. Fu, Z. Strength theory of polymer materials. In *Strength and Failure Behavior of Polymer Materials*, 1st ed.; Chemical Industry Press: Beijing China, 2005; pp. 41–42.

polymers MDPI

Article

POSS-Derived Synthesis and Full Life Structural Analysis of Si@C as Anode Material in Lithium Ion Battery

Ziyu Bai [1], Wenmao Tu [1,*], Junke Zhu [1], Junsheng Li [2,*], Zhao Deng [1], Danpeng Li [1] and Haolin Tang [1,*]

[1] State Key Laboratory of Advanced Technology for Materials Synthesis and Processing, Wuhan University of Technology, Wuhan 430070, China; 18971142560@163.com (Z.B.); bebetterzjk@163.com (J.Z.); dengzhao@whut.edu.cn (Z.D.); 18435138790@163.com (D.L.)
[2] School of Chemistry, Chemical Engineering and Life Sciences, Wuhan University of Technology, Wuhan 430070, China
* Correspondence: tuwm@whut.edu.cn (W.T.); li_j@whut.edu.cn (J.L.); thln@whut.edu.cn (H.T.)

Received: 15 March 2019; Accepted: 26 March 2019; Published: 29 March 2019

Abstract: Polyhedral oligomeric silsesquioxane (POSS)-derived Si@C anode material is prepared by the copolymerization of octavinyl-polyhedral oligomeric silsesquioxane (octavinyl-POSS) and styrene. Octavinyl-polyhedral oligomeric silsesquioxane has an inorganic core ($-Si_8O_{12}$) and an organic vinyl shell. Carbonization of the core-shell structured organic-inorganic hybrid precursor results in the formation of carbon protected Si-based anode material applicable for lithium ion battery. The initial discharge capacity of the battery based on the as-obtained Si@C material Si reaches 1500 mAh g^{-1}. After 550 charge-discharge cycles, a high capacity of 1430 mAh g^{-1} was maintained. A combined XRD, XPS and TEM analysis was performed to investigate the variation of the discharge performance during the cycling experiments. The results show that the decrease in discharge capacity in the first few cycles is related to the formation of solid electrolyte interphase (SEI). The subsequent rise in the capacity can be ascribed to the gradual morphology evolution of the anode material and the loss of capacity after long-term cycles is due to the structural pulverization of silicon within the electrode. Our results not only show the high potential of the novel electrode material but also provide insight into the dynamic features of the material during battery cycling, which is useful for the future design of high-performance electrode material.

Keywords: Octavinyl-POSS; organic-inorganic crosslinking; Si@C anode; lithium ion battery; mechanism analysis

---

1. Introduction

Owing to the aggravation of energy crisis, the demand for new energy conversion and storage devices is growing continuously in recent years. As a promising chemical power supply for electronic device and electric vehicles [1,2], the lithium ion battery has found enormous applications in a variety of applications due to its advantages of high energy storage density and high open circuit voltage [3,4]. At present, graphite carbon is used as negative electrode in the commercialized lithium ion battery system. The theoretical capacity of graphite is only 372 mAh g^{-1} and replacing graphite in the anode with a robust material of high capacity may lead to a battery with higher energy density [5–9]. The theoretical specific capacity of silicon is up to 4200 mAh g^{-1}, one order of magnitude higher than that of graphite anode material. Furthermore, the insertion/deinsertion potential of lithium ion for silicon is moderate [10,11]. Due to these advantages, a silicon-based material is an ideal choice for the next generation of lithium ion battery anode. However, during the alloying reaction of silicon with lithium, silicon material will suffer severe volume expansion, which can easily lead to rapid pulverization and detachment of the active material from current collector, which causes the rapid decline of the battery performance [12]. Meanwhile, the silicon material cannot form stable solid electrolyte interface (SEI). The newly exposed silicon surface due to the pulverization will continuously form a new SEI film, which leads to the decrease of charge and discharge efficiency and the acceleration of capacity attenuation [13].

In order to tackle the above-mentioned problems, the following strategies have been made in the past few years: (1) preparation of nano-sized silicon materials, such as the silicon nanoparticle [14,15], silicon nanowire [16,17], silicon nanotube [18,19] and silicon film [20]. The silicon nanomaterial provides large surface area and short ion diffusion path. The characteristics of high peristalsis and high plasticity of silicon nanomaterials can alleviate the volume effect at a certain extent and improve the cycling stability of the materials. Xinliang Feng et al. used block copolymers to construct independent two-dimensional structures model [21,22]. However, the existence and conservation of two-dimensional structure is very difficult; it is easy to agglomerate and form three-dimensional (3D) large particles. Therefore, the maintenance of two-dimensional structure is of great significance for the nanocrystallization of materials. (2) Preparation of silicon composites by introduction of the second phase, such as silicon/carbon composite material [23–25] and silicon-based metal complex [26,27]. The volume change of silicon during battery cycling is suppressed by the excellent mechanical properties of the second phase, and the conductivity of the composite can also be increased because of the introduction of the second phase. Thus, the rate performance of electrode will be improved. A combination of these two strategies is expected to be a promising approach toward high-performance silicon-based anode material. However, it is difficult to realize the controllable and large-scale synthesis of nanostructured silicon/carbon composite because low cost and scalable fabrication method free of using highly reactive silane species are rare [28,29]. Dong Jin Yoo et al. prepared a doped three-dimensional/small particle composite structure by reuniting with graphene oxide and platinum particles [30,31]. The composite has good catalytic activity and high stability in the field of fuel cell catalysts. Therefore, the three-dimensional structure and nano-size of materials are of great significance for improving the properties of materials.

In this work, octavinyl-polyhedral oligomeric silsesquioxane is used as a silicon source and inorganic-organic crosslink agent for the synthesis of nano Si@C material. Because the polyhedral oligomeric silsesquioxane has a three-dimensional cage-like framework [32,33] and its size is at the nanometer level, the thus-developed Si@C negative electrode material prepared here preserves the three-dimensional pore structure of the silicon polyoxane. Such a unique morphology can alleviate the volume expansion of silicon material well. After magnesium heat reduction treatment, a conformable carbon coating is formed on the surface of silicon nanostructures. The coated carbon can also buffer the volume changes of silicon species during the cycling. At the same time, due to the existence of the pore structure in the developed Si@C material, the carbon phase in the composite would re-organize during the charge-discharge process, thus increasing the cycle life. Finally, the Si@C negative electrode material structure change along with the charge/discharge cycles of this synthesized material is investigated in order to get the mechanism of the capacity variation upon this kind of material during the whole life of charge-discharge cycles.

2. Materials and Methods

Octavinyl-POSS(polyhedral oligomeric silsesquioxane), styrene and initiator AIBN (azobisisobutyronitrile) were purchased from Sigma-Aldrich Co. Ltd (Shanghai, China). Octavinyl-POSS, styrere and AIBN were added in a two-necked flask. After stirring for 10 min, the flask was placed in liquid nitrogen. During the freeze-pump-thaw cycles, oxygen was driven away. Then, the flask was transferred into the oil bath with mild magnetic stirring at 65 °C under nitrogen atmosphere. When the solution became transparent gel, copolymerization reaction finished. The production was vacuum dried at 80 °C for 9 h. After that, the poly(POSS-styrene) was carbonized at 900 °C for 3 h under the protection of Ar with a heating rate of 3 °C min^{-1}. The as-obtained carbonized product was treated with magnesiothermic reduction at 650 °C for 2 h under the protection of Ar. The end-product was washed with 1 mol/L HCl and distilled water for several times and dried at 80 °C for 12 h. The Si@C anode material was then obtained. For comparison, the carbon material without Si was also prepared by etching the Si@C with analytic grade HF for 3 h, followed by repeated washing with ethanol six times.

The Si@C material was mixed with acetylene black and poly(vinylidene fluoride) (PVDF) to form a slurry at a mass ratio of 7:2:1 in N-Methyl pyrrolidone (NMP). After stirring for 12 h, the slurry was coated on copper foil by using the 90um spreader. Then, the slurry was dried at 70 °C for 12 h. The Si@C anode was then incorporated into coin cell in an argon filled glove box with a lithium foil as counter electrode, a Celgard separator and 1 M LiPF$_6$ in a 1:1 ethyl carbonate (EC): dimethyl carbonate (DMC) solvent as electrolyte.

All the electrochemical tests were accomplished at room temperature. The galvanostatic charge-discharge tests were carried out using a Land CT2001A (whshland Co. Ltd., Wuhan, China) between 0.001–3 V. The cyclic voltammetry (CV) tests were carried out between 0.001–3 V by using a CHI660D electrochemical workstation (CH Instruments Co. Ltd., Shanghai, China) with a scan rate of 0.1 mVs^{-1}. And the electrochemical impedance spectroscopy (EIS) tests were carried out between 0.001 Hz–10 KHz by using CorrTest CS310 electrochemical workstation (Wuhan Corrtest Instrument Co. Ltd., Wuhan, China) and the AC signal amplitude was 5 mV.

The material morphology and EDS mapping were observed by the scanning electron microscope (SEM, JSM-IT300) and the transmission electron microscope (TEM, Talos F200S, FEI, Waltham, MA, USA). The material structure was performed by the X-ray diffractometer (XRD, D8 Advance), the RENISHAW Raman microscope (Raman, RENISHAW InVia, Wotton-under-Edge, UK) and the X-ray photoelectron spectroscopy (XPS, ESCALAB 250Xi, Thermo Fisher Scientific, Waltham, MA, USA).

3. Results

Figure 1a shows the synthesis process for the POSS-derived Si@C anode material. In this synthesis, octavinyl-POSS, as an organic-inorganic silicon source, introduces polyhedral silicon framework and nanoscale pores to the material system. Styrene is selected as a carbon source and a cross-linker to form 3D copolymer network as shown in Figure 1b. The formation of copolymer is based on copolymerization of vinyl monomers with a thermal initiator. After high temperature carbonization and magnesium reduction of the copolymer, carbon coated silicon species (Si@C) will be formed. The macroscopic morphology of the polymerized gel product and the final carbonized product are shown in Figure 1c.

The poly (POSS-styrene) retains the 3D skeleton of octavinyl-POSS well, as shown in Figure 2a. The magnified SEM image of the poly(POSS-styrene) is shown in Figure 2b. After carbonization and reduction, the product exhibits a typical loose porous feature (Figure 2c). Element mapping of the final products reveals that carbon and silicon are evenly distributed in the sample (Figure 2d). Figure S1a,b (supplementary material) show XRD and Raman of POSS monomer and the polymer. The peak position of the polymer is roughly the same as the peak position of the monomer, indicating that the polymerization does not affect the crystal structure of POSS.

Several broadening peaks are detected by the XRD measurement in Figure 2e, confirming the low degree of graphitization of carbon skeleton in Si@C. The XRD pattern shows that the magnesiothermic reduction process can promote the formation of Si@C composite structure. The broad peak around 25° can be attributed to MHSiO$_2$ in amorphous, and another wide peak at 40° can be attributed to MHSiO$_2$ complexed with amorphous carbon [34,35]. Raman in Figure 2f shows the broadening silicon peak at 400–500 cm^{-1} and 800–900 cm^{-1} [36]. In these Raman spectra, the D-band is located at 1340 cm^{-1} to characterize the defects of carbon atomic crystals and the G-band is located at 1556 cm^{-1} to characterize the plane expansion vibration of carbon atom SP2 hybrid [37,38]. The intensity ratio I_D/I_G close to 1 further shows the low degree of graphitization of Si@C anode material.

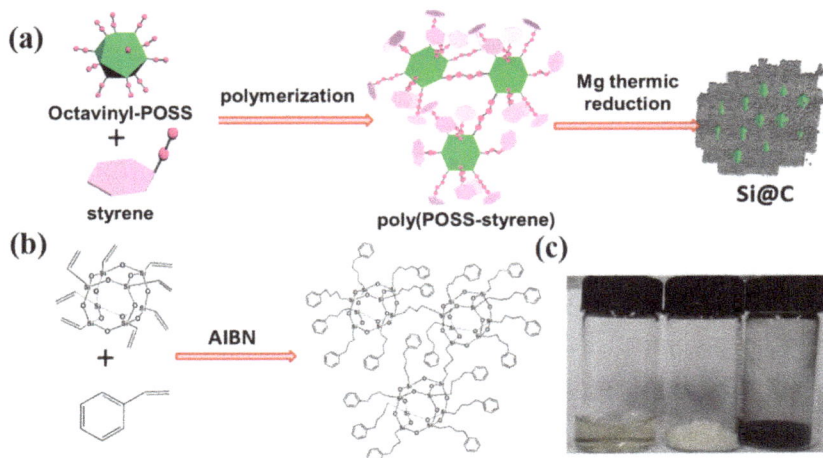

Figure 1. (**a**) Schematic illustration of synthesis of polyhedral oligomeric silsesquioxane (POSS)-derived Si@C anode material. (**b**) Formation of poly(POSS-styrene) between Octavinyl-POSS and styrene cross-linker via Vinyl copolymerization. (**c**) The photos of poly(POSS-styrene), dry-poly(POSS-styrene) and Si@C.

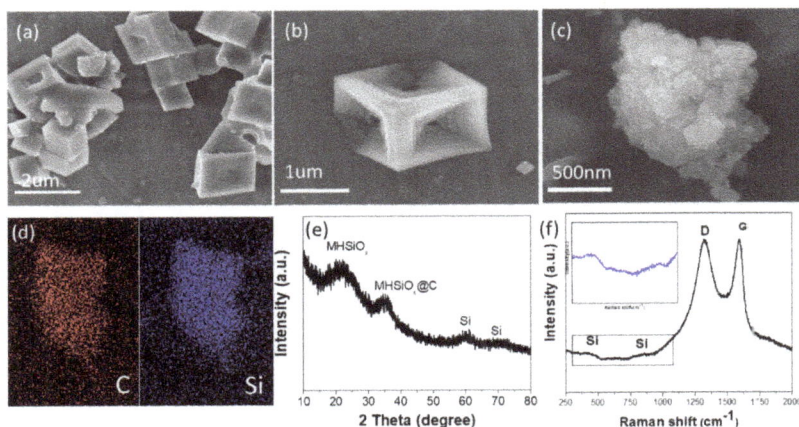

Figure 2. (**a,b**) Scanning electron microscope (SEM) images of poly(POSS-styrene) at different magnification. (**c,d**) SEM image and elemental mapping of Si@C. (**e**) X-ray diffractometer (XRD) patterns of Si@C. (**f**) Raman spectra of Si@C.

The TEM image and elemental mapping of POSS monomer in Figure 3a show the octahedron morphology, confirm that the distribution of element in monomer is uniform. According to the area of elemental mapping of poly (POSS-styrene) in Figure 3b, carbon coats on the silicon surface after cross-linking and the integrity of the silicon skeleton well preserves. Figure 3c shows the morphology and elemental distribution of Si@C. It illustrates that even if the silicon skeleton collapses, silicon is still coated with carbon. This characteristic can help to buffer the expansion of silicon during charging and discharging.

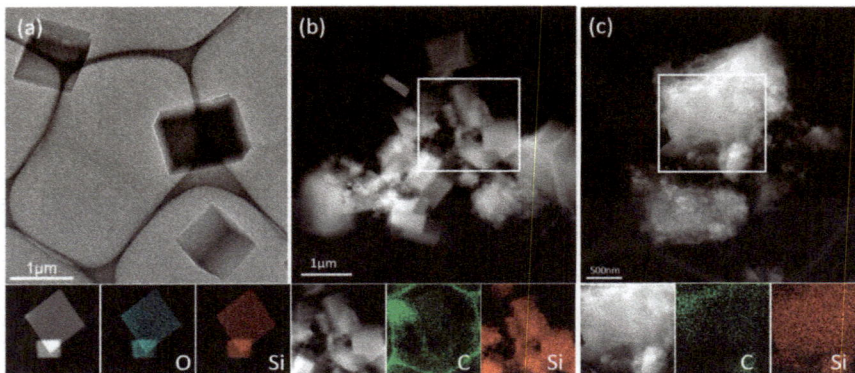

Figure 3. Transmission electron microscope (TEM) image and an area elemental mapping of poss (**a**), polymer (**b**) and Si@C (**c**).

The chemical states of POSS, Polymer and Si@C materials are analyzed by XPS. The accurate silicon content within the Si@C, determined by atomic absorption spectroscope, is 39.27 wt.%. As shown in Figure 4a, the full-range XPS surveys present the existence of silicon, carbon and oxygen. The high-resolution spectra of Si 2p and C 1s are shown in Figure 4b,c. After copolymerization, the peak position of SiOx located at 102.6 eV shifts between POSS and polymer [39]. The Si 2p spectra of Si@C exhibits four peaks located at 100.4, 102.6, 102.8 and 103.6 eV, which correspond to Si-C, Si^{2+}, Si^{3+} and Si^{4+}, respectively. The presence of these Si species demonstrates the multiple oxidation states of Si in the composite and carbon phase is in close vicinity of Si species after carbonization and magnesiothermic reduction. In Figure 4b, to compare the spectra of C 1s between POSS and Polymer, the peak located at 284.5 and 284.6 eV corresponding to C=C and C-C indicates the copolymerization of vinyl radical. The spectra of C 1s of Si@C shows the peak of C-Si located at 283.5 eV, which is consistent with the peak of Si-C located at 100.4 eV.

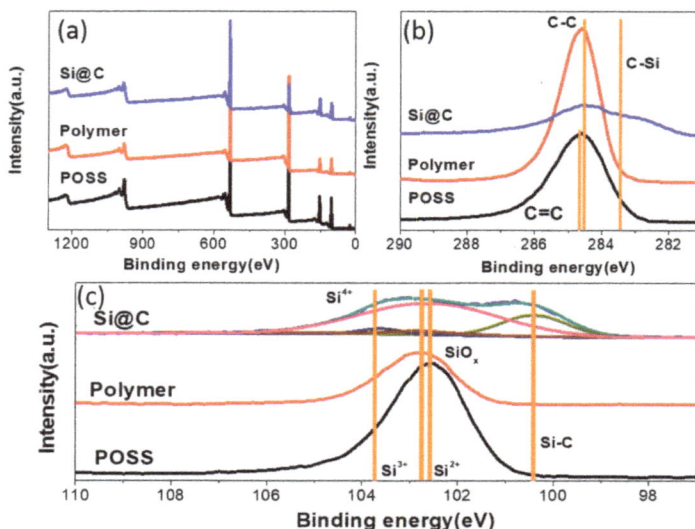

Figure 4. (**a**) X-ray photoelectron spectroscopy (XPS) survey spectra, and XPS spectra for the (**b**) C 1s and Si 2p (**c**) of POSS, polymer and Si@C.

Electrochemical characteristics of synthesized carbon material without Si are shown in Figure 5. In the first three CV curves, as shown in Figure 5a, the redox peaks and potential plateaus are not obvious. As shown in Figure 5b, the capacity of the battery below 0.5 V was mainly contributed by lithium insertion into the graphitic layers. The appearance of a potential slope indicates the disordered stacking of the graphitic layers, which leads to electrochemically and geometrically nonequivalent active lithium sites [40,41]. The first cycle discharge capacity of synthesized carbon material without Si is about 1250 mAh g^{-1}. After a couple of cycles, the discharge capacity attenuates drastically to 380mAh g^{-1} (in Figure 5b,c).

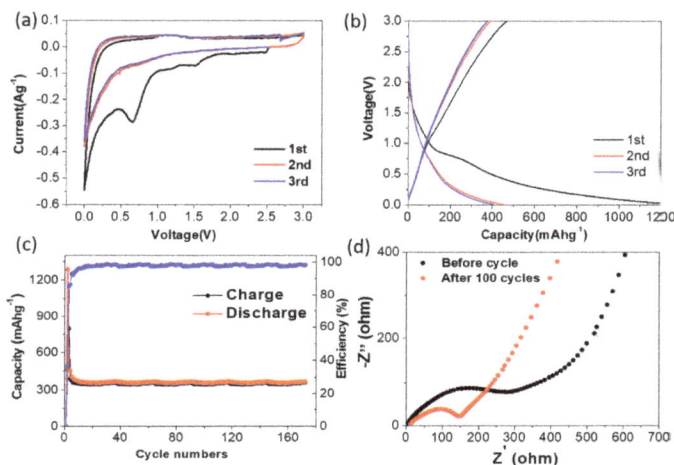

Figure 5. (**a**) First three CV curves of synthesized carbon material without Si at a scan rate of 0.1 mVs^{-1}. (**b**) First three galvanostatic charge-discharge curves of synthesized carbon material without Si. (**c**) Cycling performance of synthesized carbon material without Si at a rate of 200 mAh g^{-1}. (**d**) Electrochemical impedance spectra of synthesized carbon material without Si before and after cycling.

The electrochemical characteristics of Si@C are shown in Figure 6. During the CV and charge-discharge curves (Figure 6a,b), the broad reduction peak (about 1.5 V) confirms the preliminary decomposition of the electrolyte and the formation of solid electrolyte interface (SEI) films. At about 1.3 V, the broad oxidation peaks are observed in the first three anodic scans, illustrating the reversible oxidation of some SEI components. At a charge-discharge rate of 200 mA g^{-1}, the first cycle discharge and charge capacity of the Si@C electrode is 580 mAh g^{-1} and 1500Ahg^{-1}, respectively as shown in Figure 6c. The large irreversible capacity exists here because of the formation of the solid electrolyte interphase (SEI) and some irreversible reactions upon the electrode and/or electrolyte [42]. There is an interesting phenomenon during charge-discharge process that the Si-C anode undergoes through a quick capacity drop-off at initial several cycles, and then the capacity creep up is observed along subsequent cycles; Finally, as the cycle continues, the capacity reaches the initial capacity (Figure 6c). Furthermore, the Si@C anode shows the superior rate performance as shown in Figure 6c. For instance, the Si@C anode presents relatively high reversible capacities of 570 mAh g^{-1}, 500 mAh g^{-1} and 450 mAh g^{-1} at the rate of 1 A g^{-1}, 2 A g^{-1} and 5 A g^{-1}, respectively. When the current density goes back to 200 mA g^{-1}, the reversible capacity can recover back to the initial value of 780 mAh g^{-1}. The possible reason to the excellent rate performance of Si@C anode could be that its hollow cage structure offers a better electrode-electrolyte contact for the fast transmit of Li^{+} into Si@C material. In order to explore the reasons for the capacity change during the cycling, four representative analysis points were selected (number Si@C-@1, Si@C-@2, Si@C-@3 and Si@C-@4 shown in Figure 6c) and investigated for the mechanistic discussion. The Si@C-@1 point represents the fresh battery without cycling. The Si@C-@2 represents the battery after 20 cycles when the capacity reaches the lowest in the full cycle life. The Si@C-@3 represents the battery after 550 cycles when the capacity is the highest in the full cycle life. The Si@C-@4 represents the end point of the full cycle life.

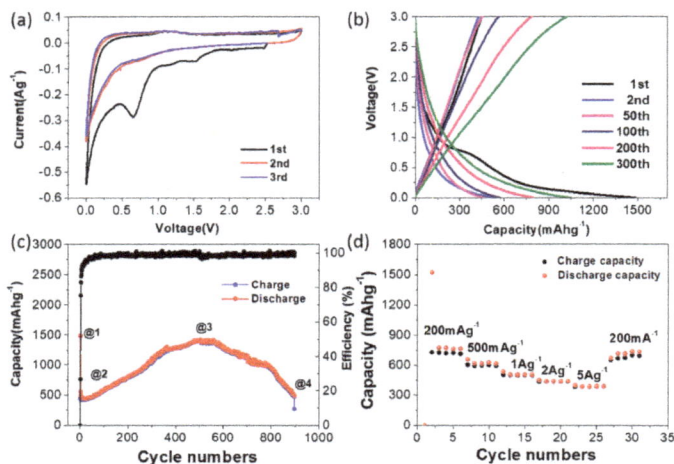

Figure 6. (a) First three CV curves of Si@C at a scan rate of $0.1 mVs^{-1}$. (b) Galvanostatic charge-discharge curves of Si@C at a rate of $200 mAg^{-1}$. (c) Full life cycling performance of Si@C at a rate of $200 mAg^{-1}$. (d) Rate performance of Si@C at different rates.

Electrochemical impedance measurements are carried out to discuss the electrode kinetic information of Si@C anode at these four analysis points (Figure 7). The Nyquist plot includes three parts: (1) a semicircle at high frequency region which characterizes partial de-solvation and adsorption of lithium ions onto the surface of the electrode; (2) a semicircle at intermediate frequency region representing the desolvated lithium ions into the lattice of active materials; (3) an oblique line at low frequency region relating to the solid-state diffusion of lithium ions [43–45]. In general, the electrode process is mainly controlled by the charge transfer and diffusion processes and the Nyquist diagram is composed of the semicircle in the high frequency region and the sloped line of the low frequency region. The resistance R_{ct} represents the charge transfer resistance of the electrochemical reaction and the charge transfer resistance has a great relationship with the electrochemical activity of the electrode material. The resistance R_{ct} of Si@C-@3 (~40 Ω) is minimum that may indicate the maximum activation of silicon and mutual embedment between silicon and carbon. The R_{ct} of Si@C-@2 (~300 Ω) is significantly larger than in the Si@C-@1 (~100 Ω) that illustrates the formation of solid electrolyte interphase (SEI) and the initial irreversible reaction during initial charge-discharge cycles. The resistance R_{ct} of Si@C-@4 (~500 Ω) is maximum that may indicate the structural collapse and the pulverization of grain after long-term charge-discharge. The obvious reduced resistance between Si@C-@2 and Si@C-@3 is beneficial for promoting the transport and storage of lithium ions, which explains the increased lithium storage capacity for as-obtained Si@C anode from 20 cycles to 550 cycles. The phenomenon of the increase on capacity after cycling had been reported for carbonaceous material. However, few reports explain the cause and mechanism of the rise of the capacity [46–48].

Furthermore, four representative cells cycled to the desired analysis points (in the complete delithiation condition) were disassembled in the glove box and Si@C anode material in these cells were characterized to further explain the mechanism for the capacity variation. Figure 8 shows the morphology and element distribution of the Si@C anode material in these four analysis points. Compared with Si@C-@1 (Figure 8), fluorine can be observed in Si@C-@2 (Figure 8). This result suggests that during the first charge and discharge process, the electrolyte diffuses into the sample to form a solid electrolyte membrane, causing an irreversible capacity change, which agrees with the rapid decline of the capacity at this time. In Si@C-@3 (Figure 8), the silicon surface is coated with a dense layer of carbon compared to Si@C-@1 and Si@C-@2. During the following cycles, carbon gradually penetrates into the silicon frame structure, relieves the volume expansion caused by the silicon itself in the process of charging and discharging and enhances the cyclic stability of the battery. At the same time, the continuous activation of silicon increases the lithium intercalation site and increases the lithium storage capacity, resulting in a capacity up to 1430 mAh g^{-1}. In Figure 8, after long-term charge-discharge cycling, most of the carbon has been separated from the silicon surface and the structure collapsed, leading to a decline of capacity in later period.

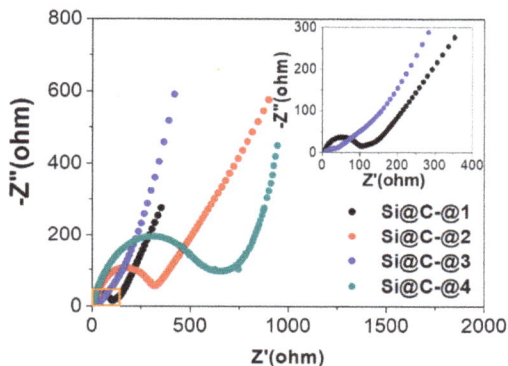

Figure 7. Electrochemical impedance spectra obtained at Si@C-@1, Si@C-@2, Si@C-@3 and Si@C-@4 points.

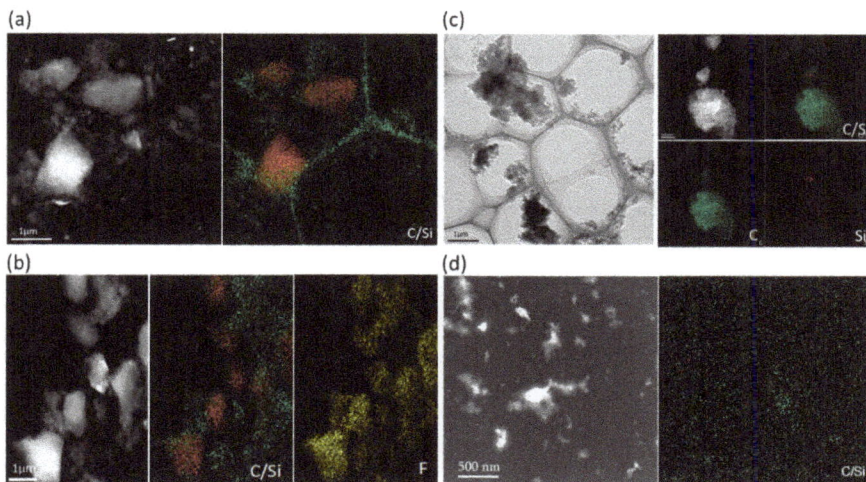

Figure 8. TEM image and elemental mapping of (**a**) Si@C-@1, (**b**) Si@C-@2, (**c**) Si@C-@3 and (**d**) Si@C-@4.

In Figure 9, by analyzing the composition of Si@C anode material in the four analysis points, the mechanism of capacity variation can be further unveiled. According to the small angle X-ray diffraction in Figure 9a, the peak of silicon (~46°) in Si@C-@4 is clearly observed due to the pulverization and abscission of silicon particles after a long-term charge and discharge. In the XPS survey spectra

(Figure 9b), it is observed that the F 1s peak intensity in Si@C-@2 is the highest within four analysis points, indicating that the effect of the solid electrolyte membrane is dominant in Si@C-@2, which can explain the sharp capacity decline in the first few charge and discharge cycles. In Si 2p spectra (Figure 9c), compared to Si@C-@2, the chemical state of silicon in Si@C-@3 is mainly the chemical bond with carbon, indicates a conformable carbon coating on silicon, agreeing well the distribution of elements shown in Figure 9c. In Si@C-@4, the Si 2p spectra are dominated by silicon-silicon bonds. The similar trend can be also observed in XPS spectra for the C 1s (Figure 9d). Compared with Si@C-@3 and Si@C-@4, the intensity of Carbon-Fluorine peak is found to be the strongest in Si@C-@2. In Si@C-@4, the disappearance of the carbon-silicon peak matches well with the XPS spectra of the Si2p in the Si@C-@4 (Figure 9c). And long charge-discharge cycles, the electrolyte will decompose and volatilize to produce free HF, leading to insufficient electrolyte for complete charge-discharge reaction, thus resulting in the decline of the actual capacity. The appearance of free HF can damage the anode material and metal ion (Al, Mg, Ba, Li, Ca) in electrolyte can eliminate the free HF from the decomposition of $LiPF_6$. The peak of the C-M appeared in Si@C-@4, indicating the destroying of the charge-discharge system.

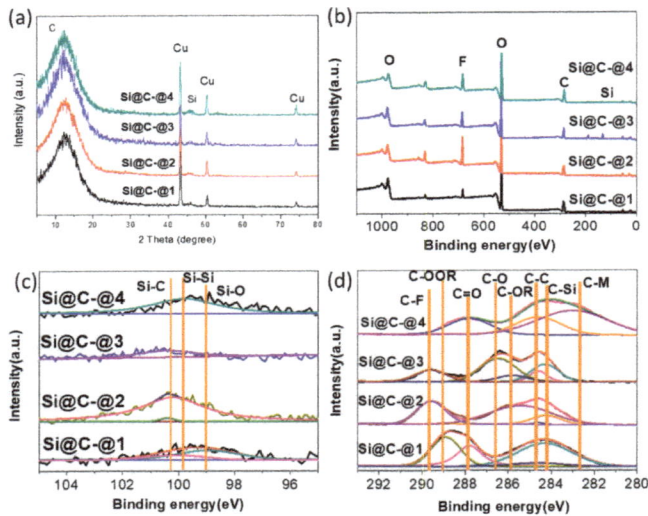

Figure 9. (a) XRD patterns of Si@C-@1, Si@C-@2, Si@C-@3 and Si@C-@4. (b) XPS survey spectra for Si@C-@1, Si@C-@2, Si@C-@3 and Si@C-@4. XPS spectra for the (c) Si 2p and (d) C 1s of Si@C-@1, Si@C-@2, Si@C-@3 and Si@C-@4. (M represents alkali metal or alkaline earth metal existed in the lithium salt and electrolyte additives.).

4. Conclusions

In summary, POSS-derived Si@C was developed by carbonization of the precursor, followed by magnesium thermal reduction. As an LIBs anode material, it shows a good performance of lithium storage performance due to its unique nanostructure. The half battery with Si@C anode delivered an initial discharge capacity up to 1500 mAh g^{-1} and a high discharge capacity of 1430 mAh g^{-1} was retained after 550 cycles. In addition, we observed a rapid fade in discharge capacity of the electrode after 550 cycles. To understand such a discharge behavior, a combined analysis of the electrode with XRD, XPS, TEM and ex-situ impedance measurements were performed with Si@C anode at different cycling states. Our results show that the change in the discharge capacity is closely related to the changes of electrode composition induced by the electrolyte decomposition, agreeing with previous finds that the electrochemical properties of an electrode is closely related to the nanostructure of the material [49]. Our results demonstrate that the novel Si@C material is a promising alternative to existing graphite anode. In addition, the mechanism for the capacity change revealed in this study may provide useful guidance for future development of electrode materials.

Supplementary Materials: The following are available online at http://www.mdpi.com/2073-4360/11/4/576/s1, Figure S1: XRD and Raman spectra of POSS, Figure S2: electrochemical performance of the electrode.

Author Contributions: Conceptualization, M.T. and H.T.; methodology, Z.B., J.Z., Z.D. and D.L.; analysis, Z.B., J.Z., J.L, Z.D. and D.L.; investigation, D.L.; analysis, Z.B., J.Z., J.L., Z.D. and D.L.; writing—original draft preparation, Z.B.; writing—review and editing, J.L.

Funding: This research was funded by the National Natural Science Foundation of China (51472187), Fundamental Research Funds for the Central Universities (WUT: 2018-IB-026) and the State Key Laboratory of Advanced Technology for Material Synthesis and Processing (Wuhan University of Technology, 2019-KF-10).

Conflicts of Interest: The authors declare no conflict of interest.

References

1. Vinothkannan, M.; Kim, A.R.; Kumar, G.G.; Yoo, D.J. Sulfonated graphene oxide/Nafion composite membranes for high temperature and low humidity proton exchange membrane fuel cells. *RSC Adv.* **2018**, *8*, 7494–7508. [CrossRef]
2. Vinothkannan, M.; Kim, A.R.; Kumar, G.G.; Yoon, J.-M.; Yoo, D.J. Toward improved mechanical strength, oxidative stability and proton conductivity of an aligned quadratic hybrid (SPEEK/FPAPB/Fe 3 O 4 -FGO) membrane for application in high temperature and low humidity fuel cells. *RSC Adv.* **2017**, *7*, 39034–39048. [CrossRef]
3. Choi, J.W.; Aurbach, D. Promise and reality of post-lithium-ion batteries with high energy densities. *Nat. Rev. Mater.* **2016**, *1*, 16013. [CrossRef]
4. Lewis, N.S. Research opportunities to advance solar energy utilization. *Science* **2016**, *351*. [CrossRef] [PubMed]
5. Cui, Q.; Zhong, Y.; Pan, L.; Zhang, H.; Yang, Y.; Liu, D.; Teng, F.; Bando, Y.; Yao, J.; Wang, X. Recent Advances in Designing High-Capacity Anode Nanomaterials for Li-Ion Batteries and Their Atomic-Scale Storage Mechanism Studies. *Adv. Sci.* **2018**, *5*, 1700902. [CrossRef] [PubMed]
6. Chu, Q.; Yang, B.; Tong, W.; Wang, X.; Liu, X.; Chen, J.; Wang, W. Fabrication of a Stainless-Steel-Mesh-Supported Hierarchical Fe 2 O 3 @NiCo 2 O 4 Core-Shell Tubular Array Anode for Lithium-Ion Battery. *ChemistrySelect* **2016**, *1*, 5569–5573. [CrossRef]
7. Ma, N.; Jiang, X.-Y.; Zhang, L.; Wang, X.-S.; Cao, Y.-L.; Zhang, X.-Z. Novel 2D Layered Molybdenum Ditelluride Encapsulated in Few-Layer Graphene as High-Performance Anode for Lithium-Ion Batteries. *Small* **2018**, *14*, 1703680. [CrossRef]
8. Tang, Q.; Su, H.; Cui, Y.; Baker, A.P.; Liu, Y.; Lu, J.; Song, X.; Zhang, H.; Wu, J.; Yu, H.; et al. Ternary tin-based chalcogenide nanoplates as a promising anode material for lithium-ion batteries. *J. Sources* **2018**, *379*, 182–190. [CrossRef]
9. Wu, Y.; Huang, L.; Huang, X.; Guo, X.; Liu, D.; Zheng, D.; Zhang, X.; Ren, R.; Qu, D.; Chen, J. A room-temperature liquid metal-based self-healing anode for lithium-ion batteries with an ultra-long cycle life. *Environ. Sci.* **2017**, *10*, 1854–1861. [CrossRef]
10. Zuo, X.; Zhu, J.; Müller-Buschbaum, P.; Cheng, Y.-J. Silicon based lithium-ion battery anodes: A chronicle perspective review. *Nano Energy* **2017**, *31*, 113–143. [CrossRef]
11. Shen, X.; Tian, Z.; Fan, R.; Shao, L.; Zhang, D.; Cao, G.; Kou, L.; Bai, Y. Research progress on silicon/carbon composite anode materials for lithium-ion battery. *J. Chem.* **2018**, *27*, 1067–1090. [CrossRef]
12. Hertzberg, B.; Alexeev, A.; Yushin, G. Deformations in Si−Li Anodes Upon Electrochemical Alloying in Nano-Confined Space. *J. Am. Chem. Soc.* **2010**, *132*, 8548–8549. [CrossRef]
13. An, S.J.; Li, J.; Daniel, C.; Mohanty, D.; Nagpure, S.; Wood, D.L.; Iii, D.L.W. The state of understanding of the lithium-ion-battery graphite solid electrolyte interphase (SEI) and its relationship to formation cycling. *Carbon* **2016**, *105*, 52–76. [CrossRef]
14. Kovalenko, I.; Zdyrko, B.; Magasinski, A.; Hertzberg, B.; Milicev, Z.; Burtovyy, R.; Luzinov, I.; Yushin, G. A Major Constituent of Brown Algae for Use in High-Capacity Li-Ion Batteries. *Science* **2011**, *334*, 75–79. [CrossRef]

15. Haro, M.; Singh, V.; Steinhauer, S.; Toulkeridou, E.; Grammatikopoulos, P.; Sowwan, M. Nanoscale Heterogeneity of Multilayered Si Anodes with Embedded Nanoparticle Scaffolds for Li-Ion Batteries. *Adv. Sci.* **2017**, *4*, 1700180. [CrossRef]

16. Salvatierra, R.V.; Raji, A.-R.O.; Lee, S.-K.; Ji, Y.; Li, L.; Tour, J.M. Silicon Nanowires and Lithium Cobalt Oxide Nanowires in Graphene Nanoribbon Papers for Full Lithium Ion Battery. *Adv. Mater.* **2016**, *6*, 1600918. [CrossRef]

17. Song, H.; Wang, S.; Song, X.; Yang, H.; Du, G.; Yu, L.; Xu, J.; He, P.; Zhou, H.; Chen, K.J. A bottom-up synthetic hierarchical buffer structure of copper silicon nanowire hybrids as ultra-stable and high-rate lithium-ion battery anodes. *J. Mater. Chem. A* **2018**, *6*, 7877–7886. [CrossRef]

18. Kim, Y.-Y.; Kim, H.-J.; Jeong, J.-H.; Lee, J.; Choi, J.-H.; Jung, J.-Y.; Lee, J.-H.; Cheng, H.; Lee, K.-W.; Choi, D.-G.; et al. Facile Fabrication of Silicon Nanotube Arrays and Their Application in Lithium-Ion Batteries. *Adv. Eng. Mater.* **2016**, *18*, 1349–1353. [CrossRef]

19. Wang, W.; Gu, L.; Qian, H.; Zhao, M.; Ding, X.; Peng, X.; Sha, J.; Wang, Y. Carbon-coated silicon nanotube arrays on carbon cloth as a hybrid anode for lithium-ion batteries. *J. Sources* **2016**, *307*, 410–415. [CrossRef]

20. Suresh, S.; Wu, Z.P.; Bartolucci, S.F.; Basu, S.; Mukherjee, R.; Gupta, T.; Hundekar, P.; Shi, Y.; Lu, T.-M.; Koratkar, N. Protecting Silicon Film Anodes in Lithium-Ion Batteries Using an Atomically Thin Graphene Drape. *ACS Nano* **2017**, *11*, 5051–5061. [CrossRef] [PubMed]

21. Liu, S.; Zhang, J.; Dong, R.; Gordiichuk, P.; Zhang, T.; Zhuang, X.-D.; Mai, Y.; Liu, F.; Herrmann, A.; Feng, X. Two-Dimensional Mesoscale-Ordered Conducting Polymers. *Angew. Chem. Int. Ed.* **2016**, *55*, 12516–12521. [CrossRef] [PubMed]

22. Liu, S.; Gordiichuk, P.; Wu, Z.-S.; Liu, Z.; Wei, W.; Wagner, M.; Mohamed-Noriega, N.; Wu, D.; Mai, Y.; Herrmann, A.; et al. Patterning two-dimensional free-standing surfaces with mesoporous conducting polymers. *Nat. Commun.* **2015**, *6*, 8817. [CrossRef] [PubMed]

23. Feng, K.; Li, M.; Liu, W.; Kashkooli, A.G.; Xiao, X.; Cai, M.; Chen, Z. Silicon-Based Anodes for Lithium-Ion Batteries: From Fundamentals to Practical Applications. *Small* **2018**, *14*, 1702737. [CrossRef]

24. Long, W.; Fang, B.; Ignaszak, A.; Wu, Z.; Wang, Y.-J.; Wilkinson, D. Biomass-derived nanostructured carbons and their composites as anode materials for lithium ion batteries. *Chem. Soc. Rev.* **2017**, *46*, 7176–7190. [CrossRef] [PubMed]

25. Jiang, Y.; Zhang, Y.; Yan, X.; Tian, M.; Xiao, W.; Tang, H. A sustainable route from fly ash to silicon nanorods for high performance lithium ion batteries. *Chem. Eng. J.* **2017**, *330*, 1052–1059. [CrossRef]

26. Fukata, N.; Mitome, M.; Bando, Y.; Wu, W.; Wang, Z.L. Lithium ion battery anodes using Si-Fe based nanocomposite structures. *Nano Energy* **2016**, *26*, 37–42. [CrossRef]

27. Park, H.; Choi, S.; Lee, S.-J.; Cho, Y.-G.; Hwang, G.; Song, H.-K.; Choi, N.-S.; Park, S. Design of an ultra-durable silicon-based battery anode material with exceptional high-temperature cycling stability. *Nano Energy* **2016**, *26*, 192–199. [CrossRef]

28. Chen, X.; Li, X.; Ding, F.; Xu, W.; Xiao, J.; Cao, Y.; Meduri, P.; Liu, J.; Graff, G.L.; Zhang, J.-G. Conductive Rigid Skeleton Supported Silicon as High-Performance Li-Ion Battery Anodes. *Nano Lett.* **2012**, *12*, 4124–4130. [CrossRef]

29. Li, X.; Tian, X.; Yang, T.; Wang, W.; Song, Y.; Guo, Q.; Liu, Z. Silylated functionalized silicon-based composite as anode with excellent cyclic performance for lithium-ion battery. *J. Sources* **2018**, *385*, 84–90. [CrossRef]

30. Ramakrishnan, S.; Karuppannan, M.; Vinothkannan, M.; Ramachandran, K.; Kwon, O.J.; Yoo, D.J. Ultrafine Pt Nanoparticles Stabilized by MoS2/N-Doped Reduced Graphene Oxide as a Durable Electrocatalyst for Alcohol Oxidation and Oxygen Reduction Reactions. *ACS Appl. Mater. Interfaces* **2019**. [CrossRef]

31. Ravi, A.; Mohanraj, V.; Rhan, K.A.; Jin, Y.D. Cumulative effect of bimetallic alloy, conductive polymer and graphene toward electrooxidation of methanol: An efficient anode catalyst for direct methanol fuel cells. *J. Alloys Compd.* **2019**, *771*, 477–488.

32. Liu, D.; Cheng, G.; Zhao, H.; Zeng, C.; Qu, D.; Xiao, L.; Tang, H.; Deng, Z.; Li, Y.; Su, B.-L. Self-assembly of polyhedral oligosilsesquioxane (POSS) into hierarchically ordered mesoporous carbons with uniform microporosity and nitrogen-doping for high performance supercapacitors. *Nano Energy* **2016**, *22*, 255–268. [CrossRef]

33. Tang, H.; Zeng, Y.; Gao, X.; Yao, B.; Liu, D.; Wu, J.; Qu, D.; Liu, K.; Xie, Z.; Zhang, H.; et al. Octa(aminophenyl)silsesquioxane derived nitrogen-doped well-defined nanoporous carbon materials: Synthesis and application for supercapacitors. *Electrochimica Acta* **2016**, *194*, 143–150. [CrossRef]

34. Zhuang, X.; Zhang, Y.; He, L.; Zhu, Y.; Shi, Q.; Wang, Q.; Song, G.; Yan, X.; Li, L. Strategy to form homogeneously macroporous Si as enhanced anode material of Li-ion batteries. *J. Alloys Compd.* **2018**, *731*, 1–9. [CrossRef]

35. An, W.; Fu, J.; Su, J.; Wang, L.; Peng, X.; Wu, K.; Chen, Q.; Bi, Y.; Gao, B.; Zhang, X. Mesoporous hollow nanospheres consisting of carbon coated silica nanoparticles for robust lithium-ion battery anodes. *J. Sources* **2017**, *345*, 227–236. [CrossRef]

36. Guo, L.; He, H.; Ren, Y.; Wang, C.; Li, M. Core-shell SiO@F-doped C composites with interspaces and voids as anodes for high-performance lithium-ion batteries. *Chem. Eng. J.* **2018**, *335*, 32–40. [CrossRef]

37. Han, F.-D.; Bai, Y.-J.; Liu, R.; Yao, B.; Qi, Y.-X.; Lun, N.; Zhang, J.-X. Template-Free Synthesis of Interconnected Hollow Carbon Nanospheres for High-Performance Anode Material in Lithium-Ion Batteries. *Adv. Mater.* **2011**, *1*, 798–801. [CrossRef]

38. Yi, Y.; Lee, G.-H.; Kim, J.-C.; Shim, H.-W.; Kim, D.-W. Tailored silicon hollow spheres with Micrococcus for Li ion battery electrodes. *Chem. Eng. J.* **2017**, *327*, 297–306. [CrossRef]

39. Li, Z.; Li, Z.; Zhong, W.; Li, C.; Li, L.; Zhang, H. Facile synthesis of ultrasmall Si particles embedded in carbon framework using Si-carbon integration strategy with superior lithium ion storage performance. *Chem. Eng. J.* **2017**, *319*, 1–8. [CrossRef]

40. Wang, Z.-L.; Xu, D.; Wang, H.-G.; Wu, Z.; Zhang, X.-B. In Situ Fabrication of Porous Graphene Electrodes for High-Performance Energy Storage. *ACS Nano* **2013**, *7*, 2422–2430. [CrossRef]

41. Hu, C.; Wang, L.; Zhao, Y.; Ye, M.; Chen, Q.; Feng, Z.; Qu, L. Designing nitrogen-enriched echinus-like carbon capsules for highly efficient oxygen reduction reaction and lithium ion storage. *Nanoscale* **2014**, *6*, 8002. [CrossRef] [PubMed]

42. Bruce, P.; Saidi, M. The mechanism of electrointercalation. *J. Electroanal. Chem.* **1992**, *322*, 93–105. [CrossRef]

43. Kobayashi, S.; Uchimoto, Y. Lithium Ion Phase-Transfer Reaction at the Interface between the Lithium Manganese Oxide Electrode and the Nonaqueous Electrolyte. *J. Phys. Chem. B* **2005**, *109*, 13322–13326. [CrossRef] [PubMed]

44. Nakayama, M.; Ikuta, H.; Uchimoto, Y.; Wakihara, M. Study on the AC Impedance Spectroscopy for the Li Insertion Reaction of LixLa1/3NbO3at the Electrode−Electrolyte Interface. *J. Phys. Chem. B* **2003**, *107*, 10603–10607. [CrossRef]

45. Srivastav, S.; Xu, C.; Edstrom, K.; Gustafsson, T.; Brandell, D. Modelling the morphological background to capacity fade in Si-based lithium-ion batteries. *Electrochimica Acta* **2017**, *258*, 755–763. [CrossRef]

46. Li, X.; Zhu, X.; Zhu, Y.; Yuan, Z.; Si, L.; Qian, Y. Porous nitrogen-doped carbon vegetable-sponges with enhanced lithium storage performance. *Carbon* **2014**, *69*, 515–524. [CrossRef]

47. Qie, L.; Chen, W.-M.; Wang, Z.-H.; Shao, Q.-G.; Yuan, L.-X.; Hu, X.-L.; Zhang, W.-X.; Huang, Y.-H.; Chen, W.; Wang, Z.; et al. Nitrogen-Doped Porous Carbon Nanofiber Webs as Anodes for Lithium Ion Batteries with a Superhigh Capacity and Rate Capability. *Adv. Mater.* **2012**, *24*, 2047–2050. [CrossRef] [PubMed]

48. Song, H.; Li, N.; Cui, H.; Wang, C. Enhanced storage capability and kinetic processes by pores- and hetero-atoms- riched carbon nanobubbles for lithium-ion and sodium-ion batteries anodes. *Nano Energy* **2014**, *4*, 81–87. [CrossRef]

49. Liu, Y.; Yan, X.; Yu, Y.; Yang, X. Self-improving anode for lithium-ion batteries: continuous interlamellar spacing expansion induced capacity increase in polydopamine-derived nitrogen-doped carbon tubes during cycling. *J. Mater. Chem. A* **2015**, *3*, 20880–20885. [CrossRef]

![polymers logo] *polymers*

MDPI

Article

POSS Hybrid Robust Biomass IPN Hydrogels with Temperature Responsiveness

Yi Chen [1,2,*], Yueyun Zhou [1], Wenyong Liu [2,3], Hejie Pi [1] and Guangsheng Zeng [2,3,*]

[1] Hunan Provincial Key Laboratory of Comprehensive Utilization of Agricultural and Animal Husbandry Waste Resources, College of Urban and Environmental Sciences, Hunan University of Technology, Zhuzhou 412007, China; zyenn@21cn.com (Y.Z.); pihejie@163.com (H.P.)
[2] Hunan Provincial Engineering Laboratory of Key Technique of Non-metallic Packaging Waste Resources Utilization, Hunan University of Technology, Zhuzhou 412007, China; lwy@iccas.ac.cn
[3] Hunan Provincial Key Laboratory of Biomass Fiber Functional Materials, Hunan University of Technology, Zhuzhou 412007, China
* Correspondence: yiyue514@aliyun.com (Y.C.); guangsheng_zeng@126.com (G.Z.); Tel.: +86-183-7332-8823 (Y.C.); +86-139-7337-6415 (G.Z.)

Received: 15 February 2019; Accepted: 15 March 2019; Published: 20 March 2019

Abstract: In order to improve the performance of traditional sodium alginate (SA) hydrogels cross-linked by Ca^{2+} ions to meet greater application demand, a strategy was designed to structure novel SA-based gels (named OP-PN gels) to achieve both stimulus responsiveness and improved mechanical strength. In this strategy, the SA chains are co-cross-linked by $CaCl_2$ and cationic octa-ammonium polyhedral oligomeric silsesquioxane (Oa-POSS) particles as the first network, and an organically cross-linked poly(N-isopropyl acrylamide) (PNIPA) network is introduced into the gels as the second network. Several main results are obtained from the synthesis and characterization of the gels. For OP-PN gels, their properties depend on the content of both uniformly dispersed Oa-POSS and PNIPA network directly. The increased Oa-POSS and PNIPA network content significantly improves both the strength and resilience of gels. Relatively, the increased Oa-POSS is greatly beneficial to the modulus of gels, and the increased PNIPA network is more favorable to advancing the tensile deformation of gels. The gels with hydrophilic PNIPA network exhibit better swelling ability and remarkable temperature responsiveness, and their volume phase transition temperature can be adjusted by altering the content of Oa-POSS. The deswelling rate of gels increases gradually with the increase of POSS content due to the hydrophobic Si–O skeleton of POSS. Moreover, the enhanced drug loading and sustained release ability of the target drug bovine serum albumin displays great potential for this hybrid gel in the biomedical field.

Keywords: octa-ammonium POSS; sodium alginate; hydrogels; temperature responsiveness

---

1. Introduction

Hydrogels are physically or chemically cross-linked three-dimensional hydrophilic polymeric networks capable of absorbing large amounts of water (or biological fluids) and swelling. These soft materials have similar properties to human tissue and show responsiveness to special stimuli, attracting growing interest in recent years due to their potential application in tissue engineering [1,2], soft robotics [3,4], biosensing [5,6], flexible displays [7,8], and drug delivery [9,10]. In the synthesis of hydrogels, both natural and synthetic polymers are used based on their respective advantages. Natural polymers, including cellulose, chitin, chitosan, sodium alginate, gelatin, and pectin, have received more attention for their low cytotoxicity as well as their biocompatibility and biodegradability, which are advantageous in the biomedical field [11–13]. Among natural polymers, sodium alginate (SA)—one of the most abundant polysaccharides, which is extracted from brown seaweed—is very promising and has been widely exploited in designing oral delivery of protein or peptide drugs [14–16]. The mucoadhesive property of alginate has also made it a favored formulation excipient in the pharmaceutical industry [17]. More interestingly, it can undergo sol–gel change under special conditions by physical cross-linking. SA undergoes supramolecular assembly in acid at pH below 3, forming acid gel, or can form ionotropic physical gel by cooperative binding with multivalent cations, typically Ca^{2+} [18,19]. SA hydrogel has intrinsic biological advantages. For example, SA hydrogels contain carboxylate anions on their hydrophilic surface similar to the characteristics of cell surfaces in vertebrates, which could minimize the effect on immunity, thus advancing biocompatibility [20,21]. Moreover, because of their swelling ability, robust network, fast forming ability, and structural similarity to macromolecular-based components in natural tissue, SA hydrogels have been used as sustained drug delivery systems for active biomacromolecules, as biomimetic extracellular matrices for cell attachment in tissue engineering applications, as biomaterials for protein and cell encapsulation, and as raw material for 3D printing [22–24].

Though SA-based hydrogels have achieved great advancements in fundamental and laboratory research, they are still far from being used in industrial or clinical applications, since many challenges still exist. First, their mechanical properties are still poorer than those of synthetic polymer gels, and this deficiency limits their application in some areas such as cell scaffold, which needs enough load capacity to allow cell proliferation. Second, the absence of stimulus responsiveness makes it impossible for gels to achieve controllable drug release under special conditions, which is important in in vivo drug delivery. Third, the new functionality of alginate gels still needs to be exploited and developed in order to meet the demands of complex therapy.

To further improve the mechanical properties and introduce stimulus responsiveness into SA gels, in this work, new nanocomposite SA hydrogels with an interpenetrating polymer network (IPN) structure were designed. Considering that the Ca^{2+} ions cross-link SA through ion exchange, another water-soluble cationic nanoparticle, octa-ammonium polyhedral oligomeric silsesquioxane (Oa-POSS), was chosen to use as a co-cross-linker together with $CaCl_2$. POSS has a compact hybrid structure with an inorganic core made up of silicon and oxygen and eight corner arm groups: $g[RSiO3/2]n$ (with n = 8, 10, and 12); R is H, alkyl, alkylene, aryl, aromatic alkylene, or their derivative. The corner arms are reactive or inertial functional groups, which endow POSS with diversified functionality and reactivity [25–27]. According to previous research, introducing POSS particles into polymer-based materials and gels could improve the mechanical properties or endow the matrix with special functions. For example, polyethylene glycol (PEG)-POSS multiblock polyurethane hydrogels have higher stiffness than normal polyurethane hydrogels, in which the POSS nanocrystals serve as physical cross-linking points [28]. Poly(2-hydroxyethyl methacrylate) (PHEMA) and poly(ethylene glycol) dimethacrylate/poly(ethylene glycol) monomethacrylate (PEGDM/PEGMM) hydrogels cross-linked by methacryloxy-multifunctionalized POSS showed improved surface properties, swelling ability, and mechanical properties in the swollen state [29]. POSS particles were also used to prepare poly(N-isopropyl acrylamide) (PNIPA) gels as cross-linker or additive. Zeng prepared a series of rapid thermoresponsive POSS-containing PNIPA hydrogels by using POSS with various long flexible chains, and obvious hydrophobic nanodomains in gels were observed by atomic force micrography [30]. Octa (propylglycidyl ether)-POSS, mercapto-POSS, and octavinyl-POSS were also used in PNIPA gels to improve the deswelling rate and mechanical properties [31–33]. Moreover, POSS was proven to have good biocompatibility, which is more promising for use as a biomedical material [34,35]. The IPN structure was applied here to introduce responsiveness and further improve the properties. Previously, many IPN gels were designed and achieved excellent mechanical properties due to their special matched network structure and strong network entanglement for stress dispersion [36]. For example, Gong and co-workers developed IPN hydrogels consisting of poly(2-acrylamido-2-methylpropane sulfonic acid) (PAMPS) for the first network and polyacrylamide (PAAm) for the second network, and the gels showed extremely high mechanical strength greatly beyond the original strength of the respective networks [37]. Besides, many functions of the respective network are brought into the new IPN gels or combined to achieve complementarity [38,39].

Based on the above strategy, novel hydrogels were fabricated by combining two different networks: the first one is the SA co-cross-linked by $CaCl_2$ and Oa-POSS collectively, and the second network is the organically cross-linked PNIPA. The structure of this gel was comprehensively investigated by Fourier transform infrared spectroscopy (FTIR), X-ray diffraction (XRD), differential scanning calorimetry (DSC), thermal gravimetric analysis (TGA), scanning electron microscopy (SEM), and optical microscopy (OM), and the properties of gels—including mechanical properties, swelling, and deswelling behavior—were measured. Moreover, to determine its potential in drug release, bovine serum albumin (BSA) was used as the target drug to analyze the loading and releasing ability of the gels.

2. Materials and Methods

2.1. Materials

NIPA monomer, potassium peroxodisulfate (KPS), N,N′-methylenebisacrylamide (BIS), and N, N, N, N-tetramethylethylenediamine (TEMED) were provided by TCI Co., Tokyo, Japan. NIPA was purified by recrystallization from a toluene/n-hexane mixture (2/1 *w/w*) and dried under vacuum at 40 °C. Sodium alginate (SA) and calcium chloride and were purchased from Aladdin Co., Shanghai, China, and used without further purification. The viscosity of SA (1 wt %, 20 °C) is 400 ± 50 mPa.s. Oa-POSS was provided by Hybrid Plastics Co., Irvine, California, USA. BSA was purchased from J&K Chemical Co., Shanghai, China.

2.2. Methods

2.2.1. Synthesis of OPn-PNm Gels

The POSS hybrid SA/PNIPA IPN gels (OPn-PNm gels) were prepared by a two-step reaction, and the process of synthesis was as follows: A given amount of sodium alginate, KPS, and TEMED was mixed with deionized water in a flask and stirred by a magnetic stirrer for 1 h at room temperature to get a homogeneous solution, then a mixed aqueous solution with monomer NIPA and cross-linker BIS was added to the sodium alginate solution. The solution was stirred quickly for 2 min and then transferred to a mold with certain shape and sealed. Free-radical polymerization was carried out at 20 °C for 24 h to form the first network gels. In the gels, the concentration of sodium alginate was fixed at 2%; the mass ratio of water, NIPA, initiator KPS, and catalyzer TEMED was 1000:100:1.02:0.755; and the molar ratio of BIS and NIPA was fixed at 4%. Then, the formed gels were immersed into supersonic 3 wt % $CaCl_2$ and Oa-POSS aqueous solution at 20 °C for 48 h to form the second network inside the gels. Finally, the prepared gels were swollen and deswollen repeatedly in deionized water to remove the unreacted monomer as purification.

In this research, the resulting prepared gels are expressed as OPn-PNm gels, using a simplified numerical value of Oa-POSS content and the ratio of both networks. For OPn-PNm gels, n is defined as $1/4 \times$ mass percentage of Oa-POSS in total $CaCl_2$ and Oa-POSS for a simple expression, and m corresponds to the mole ratio of NIPA to a single sodium alginate molecule. For example, OP3-PN3 gel means that the mass of Oa-POSS in aqueous solution is 12% of all $CaCl_2$ and Oa-POSS mass, the mole ratio of NIPA is 3 times that of SA. In this research, n and m are equal to or less than 5 to guarantee the homogeneity of gels.

2.2.2. Characterization of OPn-PNm Gels

FTIR spectra of the gels were taken under ambient conditions by a Nicolet 380 FT-IR spectrometer (Thermo Fisher Scientific Inc., Waltham, MA, USA) with milled, dried, unpurified hydrogels by the conventional KBr disk tablet method.

XRD patterns were conducted on a D8 advance X-ray diffractometer (Bruker AXS Inc., Karlsruhe, Germany) using Cu Kα radiation at 40 kV. The diffraction data were collected with a 2θ angle in the range of 2° to 60° at a scanning rate of 5/min.

TGA was analyzed by a Q50 thermogravimetric analyzer (TA Instruments Inc., Newcastle, DE, USA) at a heating rate of 20 °C/min in the range of 20 °C to 600 °C under a nitrogen atmosphere, with a flow rate of nitrogen of 30 mL/min.

DSC was used to determine the volume phase transition temperature (VPTT) of the samples by using a Q20-Series differential scanning calorimeter (TA Instruments Inc., Newcastle, DE, USA). All samples used in DSC measurement were immersed in deionized water at room temperature and allowed to swell for at least 72 h to reach the equilibrium state. The calorimetric analysis was performed from 20 °C to 60 °C at a heating rate of 1 °C/min under a dry nitrogen atmosphere, and the flow rate of nitrogen was 20 mL/min.

The morphology of the gels' network structure was observed by a S3000-N low-vacuum scanning electron microscope (Hitachi High Technologies America Inc., Chandler, AZ, USA). Each sample for observation was rapidly frozen and subsequently freeze-dried for 48 h under vacuum at −60 °C. The dispersion of Oa-POSS in the gels was also observed by SEM, and the samples for observation were etched by 25 wt % hydrofluoric acid (HF) solution to dissolve and eliminate the inorganic particles and then dried under vacuum at 30 °C. The surface morphology of gels was observed by a LV100POL/50iPOL polarizing microscope (Nikon Inc., Tokyo, Japan).

2.2.3. Mechanical Measurements

Tensile mechanical measurements were performed on gels of the same size, $10 \times 3 \times 80$ mm (width × thickness × length), using a UTM6000 universal mechanical tester (Suns Inc., Shenzhen, China). The sample length between the jaws was 35 mm and the crosshead speed was 10 mm/min. An initial cross-section of 20 mm^2 was used to calculate the tensile strength and modulus, and the tensile strain was taken as the length change relative to the initial length of the sample. The tensile modulus was calculated from the increase in load detected between elongation of 10% and 45%. Compression tests were carried out using samples of the same size, $10 \times 10 \times 10$ mm (width × thickness × length), on the UTM6000 universal mechanical tester (Suns Inc., Shenzhen, China). The compression property of all gels was obtained under the following conditions: compression speed 5 mm/min and compression distance 8 mm (80% strain). All gels for testing maintained the same water/polymer ratio (8/1 (*w/w*)).

2.2.4. Dynamic Rheological Experiments

The rheological tests were carried out on a stress-controlled AR2000 rheometer (TA company, Newcastle, DE, USA) in dynamic mode at 20 °C. The linear viscoelasticity region was confirmed by dynamic strain sweeps at a constant frequency of 1 rad/s. The storage modulus G′ and loss modulus G″ were characterized by dynamic frequency sweeps from 0.1 to 200 rad/s, conducted in fixed strain γ = 1%, which is within the linear regime confirmed by dynamic strain sweeps.

2.2.5. Swelling Experiment

The swelling experiment was performed by immersing the dried gels in water at 20 °C for at least 48 h to reach equilibrium. The weight of the gel was measured after wiping off the excess water on the surface with moistened filter paper. The swelling ratio was calculated from the equation

$$\text{Swelling ratio (SR)} = (Ws - Wd) / Wd \qquad (1)$$

where Ws is the weight of swollen hydrogel and Wd is the dry weight of the hydrogel.

2.2.6. Deswelling Kinetics Experiment

The deswelling kinetics of the hydrogels after a temperature jump from the equilibrated swollen state at 20 °C to hot water at 50 °C were measured after wiping off the excess water on the surface with moistened filter paper. The water retention corresponding to the deswelling ratio was calculated from the equation

$$\text{Water retention (WR)} = (Wt - Wd) \,/\, Ws \times 100\% \tag{2}$$

where Wt is the weight of swollen hydrogel at a specific time.

2.2.7. Drug Load and Release Experiment

Synthesis of hydrogel–drug conjugate was carried out. In the first step of synthesis, BSA was added to sodium alginate aqueous solution to form a homogeneous solution to achieve imbedding. Aside from this step, the process was consistent with the synthesis of blank gels. The loading percentage of BSA was tested by a difference value between added mass and dissolving mass of BSA in reaction condition (CaCl$_2$ and Oa-POSS aqueous solution), BSA concentration was measured by UV spectroscopy (Lambda 950 UV/vis spectrophotometer, Perkin-Elmer Inc., Waltham, MA, USA), and absorbance was monitored at 280 nm [40].

The protein-loaded gels were placed into a 200 mL aqueous solution containing 0.02 w/v % sodium azide as a bacteriostatic agent at 25 °C. At a certain time interval, the gels were removed from the solution and the solution was mixed to homogeneity by a rotary shaker at a speed of 100 rpm. The release content was calculated by the concentration of BSA in the solution, which was measured by UV spectroscopy.

3. Results and Discussion

3.1. Synthesis and Structure of OPn-PNm Gels

When it contacted the multivalent cations, the sodium alginate (SA) could form the ionotropic physical gel rapidly by ion exchange. To structure IPN hydrogels based on SA, NIPA was chosen to be cross-linked first to form a precursor gel with SA, and then the SA network was formed inside the precursor gel by immersing the gel into the solution with abundant exchangeable ions. Polymerization yields, evaluated from the weights of dried gels, were nearly 100% in all cases, revealing a complete reaction. Our concern here is the successful introduction of POSS particles and their link type inside the gels, and several characterizations were used to analyze this problem. First, from the appearance of gels (Figure 1a), it could be observed that gels within the range of the synthetic formula (n ≤ 5) stayed transparent without obvious phase separation or precipitation. Relatively, when the content of Oa-POSS becomes higher (n > 6), some degree of heterogeneity appears inside the gels.

Moreover, when adding Oa-POSS aqueous solution to the SA aqueous solution as the single cross-linker, the viscosity of the solution increases quickly, accompanied by a great amount of white flocculates, and the integral forming of gels is unsuccessful, meaning that the Oa-POSS particles could not be used as the single cross-linker to achieve the preparation of SA gels. However, this phenomenon indicates that the ion exchange between Oa-POSS and SA happens in the system. According to the molecular structural formula of Oa-POSS, there are eight arms with monovalent ions in the corner of the Si–O cage; a high concentration of Oa-POSS should bring about an increased charge density, which makes it easier to cause aggregation of SA around the POSS particles instead of uniform cross-linking.

Figure 1b shows the FTIR spectra of the Oa-POSS, OP0-PN3, and OP3-PN3 gels. Comparing these absorption curves, the curve of OP3-PN3 gel contains the same characteristic absorption peaks as the curve of Oa-POSS, in which a broad peak between 2800 and 3000 cm^{-1} corresponding to the stretching vibration of (NH3$^+$), 1624; 1501 cm^{-1} corresponding to the bending vibration of (N–H); and 1139 cm^{-1} corresponding to the stretching vibration of (Si–O–Si) could be observed. This feature proves the existence of Oa-POSS inside the gels [41].

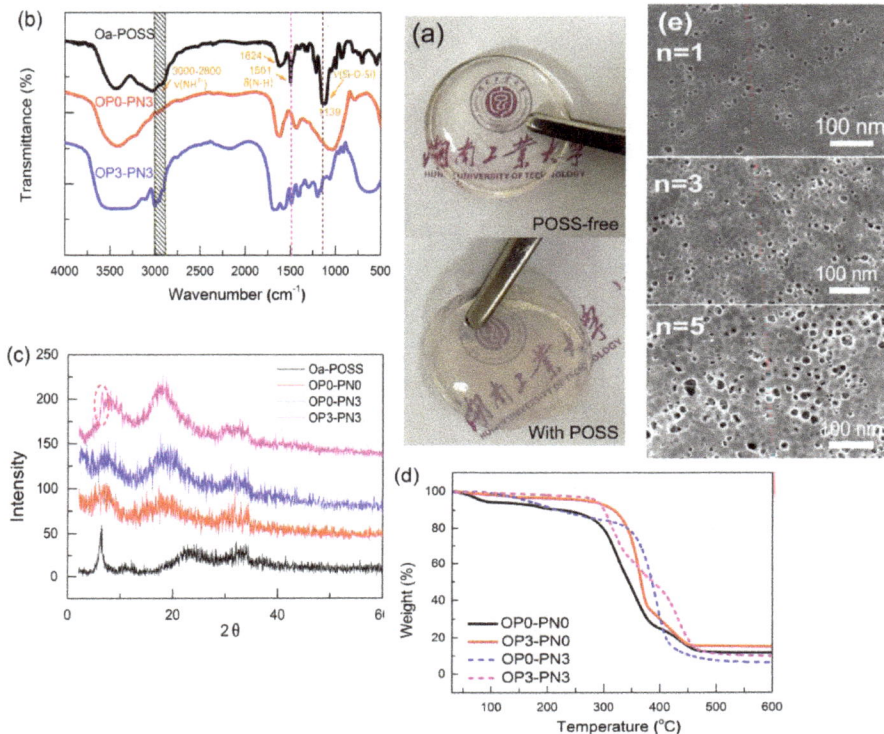

Figure 1. Appearance and structure analysis of OPn-PNm gels: (**a**) photos of interpenetrating polymer network (IPN) gels with and without octa-ammonium polyhedral oligomeric silsesquioxane (Oa-POSS); (**b**) FTIR spectra of Oa-POSS particles and OPn-PNm gels; (**c**) X-ray profiles of Oa-POSS particles and OPn-PNm gels; (**c**) thermogravimetric curves of OPn-PNm gels; (**d**) SEM images of etched OPn-PN3 gels (n = 1, 3, 5).

Moreover, as shown in the X-ray profiles of Oa-POSS, pure SA, OP0-PN3, and OP3-PN3 gels in Figure 1c, the peak of 100 crystal plane belonging to Oa-POSS particles appears at about $2\theta = 5°$. From the profile of OP3-PN3 gel, this characteristic peak can still be found but it becomes weaker, indicating the existence of slight crystals of Oa-POSS inside the gel. It was also observed that the diffraction peak at $2\theta = 8°$ belonging to SA was strengthened after adding Oa-POSS, which may be attributed to the tighter combination of SA around the Oa-POSS particles promoting the crystallization of SA.

Thermogravimetric analysis was also used for the gels. Two groups of TG curves belonging to SA and IPN gels with or without Oa-POSS were compared. As shown in Figure 1d, for SA gels (OP0-PN0, OP3-PN0), the TG curve of POSS hybrid gel shifts to the higher temperature more than that of neat gels, meaning there was better thermostability caused by the addition of Oa-POSS. Furthermore, the residual mass of OP3-PN0 gel was higher than that of OP0-PN0 gel. Obviously, the co-cross-linking by Oa-POSS led to a tighter network structure, which is conducive to thermostability, and the Oa-POSS particles had a higher thermolysis residue ratio than other ingredients in the gels, resulting in a high residual mass. A similar phenomenon was found in IPN gels (OP0-PN3, OP3-PN3), with a slight difference. The POSS hybrid gels had better thermostability in the low temperature region, and the temperature corresponding to their maximum decomposition rate was higher than that of neat gels. This may be attributed to a structural feature of POSS hybrid IPN gels, in which a part of PNIPA polymer chains are wrapped by the SA network co-cross-linked by Oa-POSS, and the tighter network causes harder heat transfer, resulting in a higher weight loss temperature. Moreover, higher residue mass was also found in POSS hybrid gels. Compared to the SA gels cross-linked only by $CaCl_2$, the gels cross-linked by POSS particles had a higher ratio of solid residue (e.g., if 1 mol of $CaCl_2$ or Oa-POSS corresponds to 2 mol SA equally, the percentage of residue mass is 14.7% and 33.1%, respectively; if 1 mol Oa-POSS corresponds to 4 mol SA, the residue mass percentage is 26.7%; the reality may be between these situations). Though the actual situation is complex, the combination type (as cross-linker or pure additive) and the cross-linking degree (1 mol Oa-POSS cross-links 1–8 mol SA) are hard to define, and the increased residue mass could still prove the successful introduction of Oa-POSS particles.

Figure 1e shows the surface morphology of OPn-PN3 gels (n = 1, 3, 5) etched by HF solution to dissolve the POSS particles. As shown in the figure, the holes, corresponding to the dissolved Oa-POSS particles, are dispersed uniformly in the gel, essentially revealing uniform dispersion of Oa-POSS particles inside the gel. The quantity of pores increases with the increased Oa-POSS concentration used in the reaction, indicating that the Oa-POSS content inside the gels directly depends on the Oa-POSS concentration in the reaction process. However, the size of the holes (10–30 nm) is bigger than the original size of Oa-POSS (2–4 nm), especially when the content of Oa-POSS is high, indicating a certain degree of aggregation inside the gels. Based on the above analysis, it can be concluded that the Oa-POSS content is proportional to Oa-POSS concentration in the reaction solution.

3.2. Mechanical Properties of OPn-PNm Gels

Although pure SA hydrogel is not easy to break under compression and small deformation, its strength and resilience still have room to advance in order to meet application demands. The tensile curves of different OPn-PNm gels are shown in Figure 2a. As shown in the figure, the interpenetrating network structure significantly improves the strength and elongation at breakage of gels. The introduction of Oa-POSS particles could further increase the strength, while it should cause earlier breakage of gels in tensile tests. However, the POSS hybrid IPN gels still could be stretched uniformly, showing ductile deformation (Figure 2b).

From the specific data shown in Figure 2c–e, it can clearly be seen that the IPN gels show higher strength with increased PNIPA network content; the tighter network structure is more beneficial to stress loading, while the modulus increases slightly or even decreases when the PNIPA content is high. Comparing the nature of the two networks inside the gels, the SA network has higher strength than the PNIPA network, revealing rigid characteristics. Therefore, the initial stress should mainly be loaded by the SA network; the PNIPA network only works as a collaborator to share part of the stress. When the content of PNIPA increases and takes on the main role for stress loading, the modulus is inclined to close to the feature of the PNIPA network. Moreover, elongation at breakage of gels increases monotonously with the increased PNIPA network content; obviously, the entanglement of cross-linked polymer is helpful to the deformation of the polymer network.

On the other hand, introducing Oa-POSS truly improves the strength and modulus of both pure SA gels and IPN gels, especially IPN gels. The modulus of OP5-PN3 gels is advanced almost three times that of OP0-PN3 gels. This is ascribed to the overlapped effect of tighter network structure and better stress dispersion by nanoparticles. However, a mass of Oa-POSS should decrease elongation at breakage of gels; the dispersed POSS particles acting as stress concentration points in the gel network lead to increased heterogeneity of the gel, which brings about earlier heterogeneous fracture.

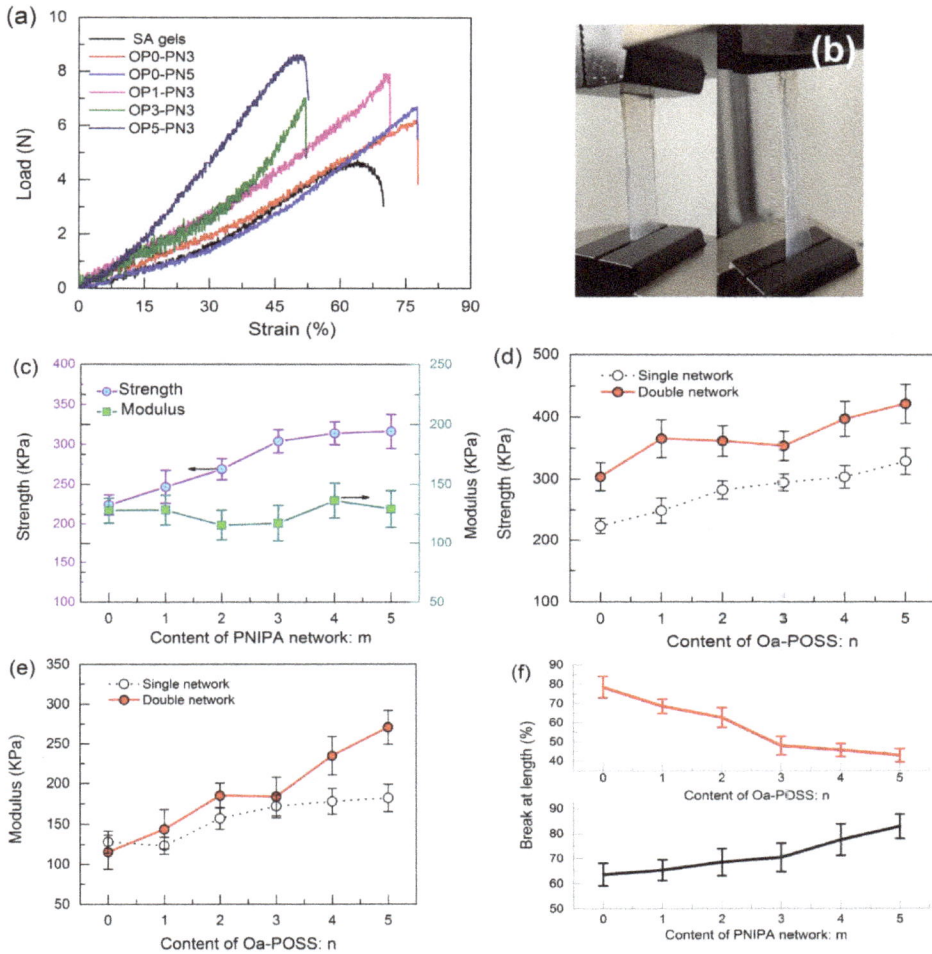

Figure 2. Tensile properties of OPn-PNm gels: (**a**) load–strain curves of gels in tensile tests; (**b**) images of OP3-PN3 gel in tensile tests; (**c**) calculated strength and modulus of OP0-PNm gel; (**d**) calculated strength of OPn-PN0 and OPn-PN3 gels; (**e**) calculated modulus of OPn-PN0 and OPn-PN3 gels; (**f**) elongation at breakage of OP1-PNm and OPn-PN3 gels.

The compression properties of pure SA gels and OPn-PNm gels were measured by using a uniaxial compression test. As shown in Figure 3, all gels could bear great compression deformation and stay unbroken even under greater than 90% strain (Figure 3b). The corresponding stress–strain curves of OPn-PNm gels with different n and m under compression (up to 80% compression) are shown in Figure 3a. As shown in the figure, the introduction of IPN structure and POSS particles increases the rigidity of the gel network, resulting in higher compression strength. Notably, the gels with Oa-POSS show better resilience under compression (Figure 3c) than SA gels, revealing that the interpenetrating network structure and Oa-POSS endow the polymer network with better elasticity.

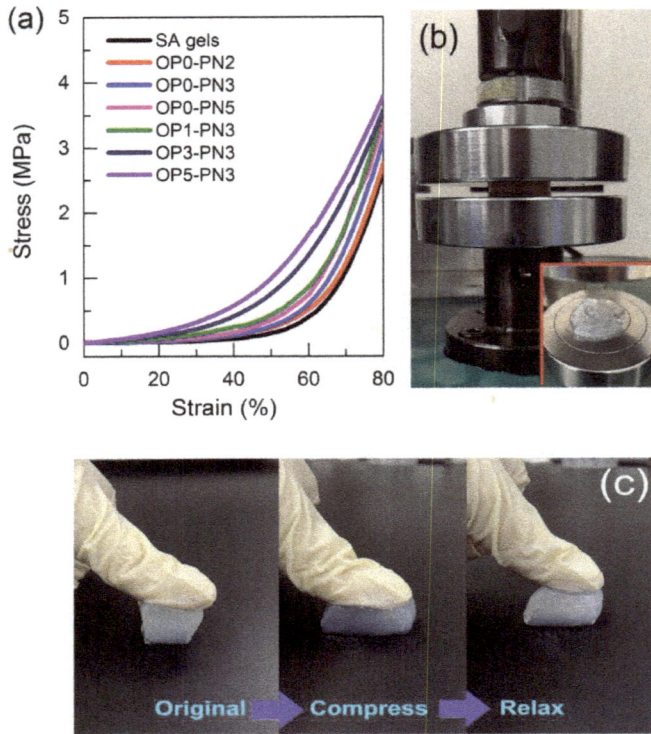

Figure 3. Compression properties of OPn-PNm gels: (**a**) stress–strain curves of sodium alginate (SA) and OPn–PNm gels under compression tests; (**b**) OP3-PN3 gels under 95% strain compression; (**c**) compression and resilience behavior of OP3-PN3 gels.

3.3. Viscoelastic Behavior of OPn-PNm Gels

The structural characteristics of different gels were further confirmed by rheology tests, and several rheological features are shown in Figure 4. First, for all tested hydrogels, the storage modulus (G′) was higher than the loss modulus (G″) over the entire frequency range (0.1 to 200 rad/s), with G′ in the order of 10^3–10^4 Pa, indicating elastic solid behavior. Second, G′ increased gradually with the increase of frequency for both SA and OPn-PNm gels, indicating that elasticity increases with increased frequency, while G″, corresponding to the viscosity of the network, was stable under low frequency and increased at high frequency. Third, the IPN structure and increased Oa-POSS content led to higher G′, revealing higher strength and better elasticity, which is consistent with the results of tensile tests. The same was found for the G″ evolution of gels, meaning greater network viscosity caused by IPN structure and POSS co-cross-linking. Finally, the evolution of the damping factor (tan(δ)) for all gels with shear frequency is shown in Figure 4c. Totally, the tan(δ) of OPn-PNm gels was higher than that of pure SA gels, revealing a better damping effect due to the IPN structure. Relatively, the effect of Oa-POSS content change on tan(δ) was slight. The introduction of the PNIPA network tightens the network structure of gels, and the interface between the two kinds of polymer chains causes increased viscosity, resulting in greater energy consumption, which could also be proved by the increased tan(δ) of IPN gels at high frequency.

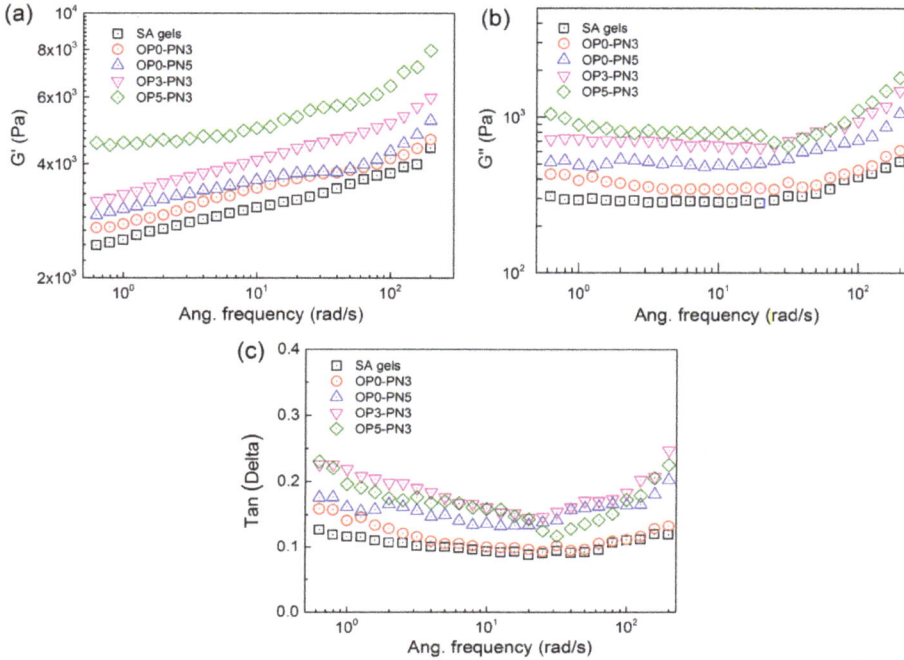

Figure 4. Rheological behavior of SA and OPn-PNm gels measured under a dynamic frequency sweep at fixed strain, $\gamma = 1\%$, which is in linear viscoelastic region LVE. (**a**,**b**) G′ and G″ of gels; (**c**) tan(Delta) of gels.

3.4. Swelling and Responsiveness Properties of OPn-PNm Gels

After introducing the hydrophilic PNIPA network, the swelling ability of gels underwent a remarkable change, and at the same time temperature responsiveness was also brought into the gels. Figure 5a shows the equilibrium swelling ratio of OPn-PNm gels as a function of temperature; it was found that the swelling ratio under low temperature was markedly differentiated. When the content of the PNIPA network increased in the gels, the swelling ability improved gradually; the swelling ratio of OP0-PN5 gels could reach about 18, nearly two times that of pure SA gels. On the contrary, adding Oa-POSS led to a great decrease in swelling ability; the swelling ratio of OP5-PN5 gels decreased to about 10, only half that of OP0-PN5 gels. Although the Oa-POSS is water-soluble due to the existence of ionic groups in the corner of the macromolecule, its Si–O–Si skeleton still exhibits hydrophobic features, which makes it oppose the combination with water. This can be supported by the test results, including water contact angle (CA) on both the surface and break section of gels. As shown in Figure 5b, with increased Oa-POSS in the gels, the CA on the gel surface increased slightly, while the CA on the break section increased significantly. As described in the experimental section, the gels were synthesized and formed in an aqueous environment, and the hydrophilic segment should tend toward the water and build up a more hydrophilic surface. Therefore, the break section is more representative of the true status inside the gels. Moreover, introducing the second network and adding Oa-POSS as the co-cross-linker should change the cross-linking density and style of the gel network. As shown in Figure 6a–c, SEM images of the typical network morphology of pure SA gels, IPN gels, and POSS hybrid IPN gels show that the pore size of IPN gels is smaller than that of single network gels, revealing a tighter network structure. The addition of Oa-POSS intensifies this phenomenon, and a more compact network could be considered as further evidence to prove the role of Oa-POSS as co-cross-linker.

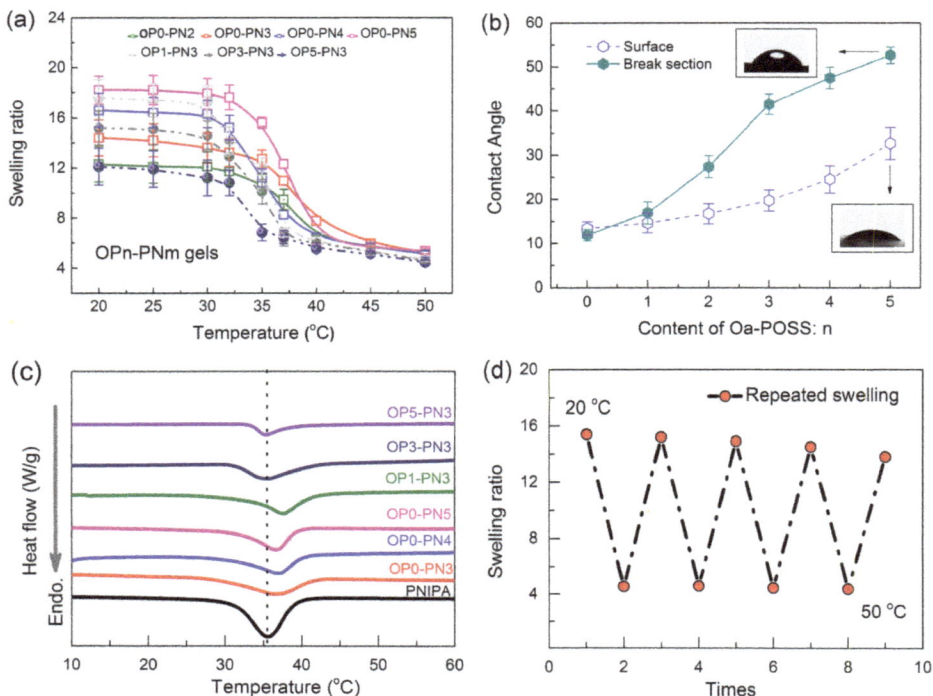

Figure 5. Swelling properties of OPn-PNm gels: (**a**) equilibrium swelling ratio of OPn-PNm in pure water at different temperatures; (**b**) water contact angle of OPn-PN3 gels; (**c**) differential scanning calorimetry (DSC) curves of OP0-PNm and OPn-PN3 gels; (**d**) repeated swelling ability of OP3-PN3 gels. Gels were tested until reaching equilibrium in 20 and 50 °C water alternatively.

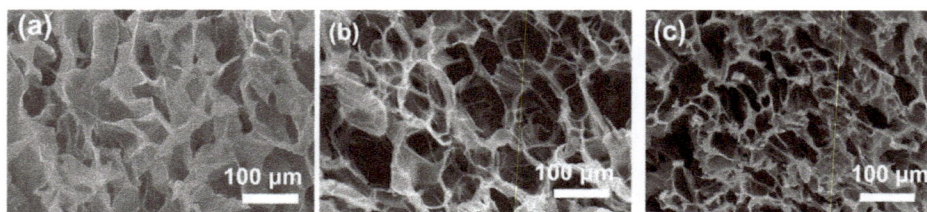

Figure 6. SEM images of SA, OP0-PN3, and OP3-PN3 gels. Network morphology of gels for observation was prepared by freeze-drying. (**a**) SA gels; (**b**) OP0-PN3 gels; (**c**) OP3-PN3 gels.

Moreover, according to the swelling characteristics, all gels had a volume phase transition temperature (VPTT) corresponding to the drastic decrease of swelling ratio with increased temperature. Although the lower critical solution temperature (LCST) of PNIPA macromolecule chains is unique, the OPn-PNm gels showed differential VPTT due to the interaction between both networks. Thermoanalysis was used to display the VPTT of gels more directly. Figure 5c shows DSC thermograms of OPn-PNm gels; here, temperature at the maximum point of the endotherm is referred to the VPTT as illustrated in the figure for a better comparison. As we can see, all OPn-PNm gels exhibit a VPTT near 35 °C, which is close to that of pure PNIPA hydrogels. IPN gels without Oa-POSS show higher VPTT, about 37–38 °C, which may be attributed to the rigid SA network impeding the shrinking PNIPA network. With increased Oa-POSS content, the VPTT gradually returns to lower temperature. The reason may be that the hydrophobic cage skeleton of POSS nanoparticles works as a core to promote the shrinking of the network when the phase transformation happens. Another phenomenon should be noted, that the shrinking of gels is incomplete with reserved swelling ratio of about 4–5. Obviously, the rigid and hydrophilic SA network prevents the PNIPA network from completely shrinking. Finally, the repeated swelling ability of OPn-PNm gels was tested by swelling and deswelling the gels in 20 and 50 °C water to reach equilibrium alternatively. The result in Figure 5d shows that the gels have good swelling–deswelling repeatability.

3.5. Deswelling Properties of OPn-PNm Gels

The addition of the PNIPA network endows the gels with responsiveness. The response rate of gels, as one of the most important factors that must be considered for the application of thermosensitive hydrogels in many fields, was analyzed by testing the deswelling kinetics.

Figure 7a shows the water retention change of OPn-PNm hydrogels after a temperature jump from equilibrated swollen state at 20 °C to ultrapure water at 50 °C. It can be clearly seen that the deswelling rate of OPn-PNm gels increases significantly with the increased PINPA network. The deswelling rate of OP0-PN5 gels is close to that of pure PNIPA gels, revealing only slight obstruction from the SA network.

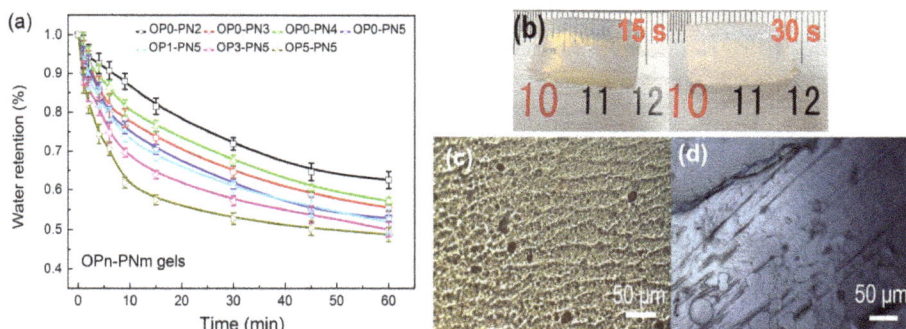

Figure 7. Deswelling properties of OPn-PNm gels: (**a**) water retention of OPn-PNm gels as a function of time in 50 °C water; (**b**) photo of OP5-PN5 gels after being immersed in 50 °C water for 15 s and 30 s; (**c**) optical microscopy (OM) image of OP5-PN5 gel surface in deswelling process; (**d**) OM image of OP0-PN5 gel surface in deswelling process.

Relatively, the addition of Oa-POSS increases the deswelling rate of gels significantly. OP5-PN5 gels exhibit fast transparency change as a sign of phase transformation (Figure 7b); they can shrink and lose water to over 40 wt % within 10 min. The increased response rate may be attributed to several reasons. First, the hydrophobic Oa-POSS particle and its aggregates could form a microhydrophobic area inside the gel, which acts as a core to promote shrinking of the whole network. Second, small pores appear in the gel network due to the increased heterogeneity caused by Oa-POSS, as shown in Figure 6c, which is beneficial to the exclusion of water from pores. Moreover, as shown in OM images (Figure 7c,d), the surfaces of gels in the deswelling process show some small pores and wrinkles, while a similar phenomenon is not found in the surfaces of gels without POSS. This characteristic caused by the inhomogeneous shrinkage around the POSS particles also favors dehydration.

3.6. Structure Simulation of OPn-PNm Gels

Based on the results obtained from comprehensive structure and performance characterization, the structure of the POSS hybrid IPN hydrogels was simulated, as shown in Figure 8. Considering the introduction of Oa-POSS particles in polymerization, combined with the improved strength, elasticity, and tighter network structure observed in SEM images, we can expect that at least part of the Oa-POSS particles work as a cross-linker to link SA macromolecular chains, and another part is only additive in the matrix of IPN gels. Moreover, Oa-POSS has eight ionic groups, and if all of them take part in cross-linking through ion exchange, the increased network density and heterogeneous structure should lead to great nonuniformity inside the gels, and the gels should be opaque, which is inconsistent with the facts. Therefore, it could be reasonably speculated that only one to four ionic groups in each Oa-POSS macromolecule participate in ion exchange with SA, and other ionic groups are free or combine with the NH groups of the matrix by hydrogen bonding. Based on the X-ray analysis, considering that there are many small pores in the observed network morphology and surface images of POSS hybrid hydrogels (SEM and OM images), which corresponds to the inhomogeneity feature caused by POSS particles, it can be assumed that although the dispersion of Oa-POSS inside the gels is uniform on the whole, there is still slight aggregation.

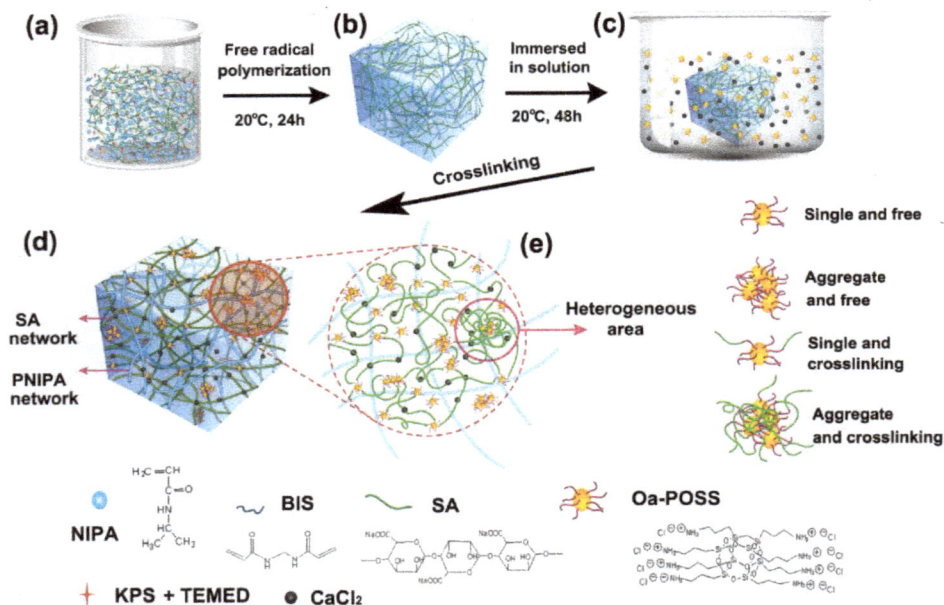

Figure 8. Schematic representation of synthesis process and structure model of OPn-PNm gels: (a–d) synthesis process of gels; (e) speculated microstructure of network, in which Oa-POSS particles have four states of existence: single and free (existing as single particle without cross-linking), aggregate and free (aggregate without cross-linking), single with cross-linking (single particle with cross-linking point, n = 1–4), aggregate and cross-linking (aggregate with cross-linking). The aggregate and free, and aggregate and cross-linking states cause heterogeneous areas inside the gels.

3.7. Drug Load and Release of OPn-PDm Gels

In order to determine the application potential of these kinds of gels, a common drug-delivery macromolecule, bovine serum albumin (BSA), was used as a target to test drug loading and sustained release ability. Figure 9 exhibits the content of loaded drug in OPn-PNm gels as a function of POSS content. As shown in the figure, the interpenetrating network is quite beneficial to the drug loading, and the Oa-POSS also has a positive effect on drug loading. The micrograph of drug-loaded gels (Figure 9b) shows that the network of gels is filled with the drug. BSA has some amino acid groups that are easy to combine with the amino groups in NIPA and Oa-POSS. Moreover, the hydrophobic chains in drug macromolecules are inclined to combine with the hydrophobic POSS cage. Therefore, for drug loading, establishing the IPN structure and introducing POSS particles are effective in improving the loading ability.

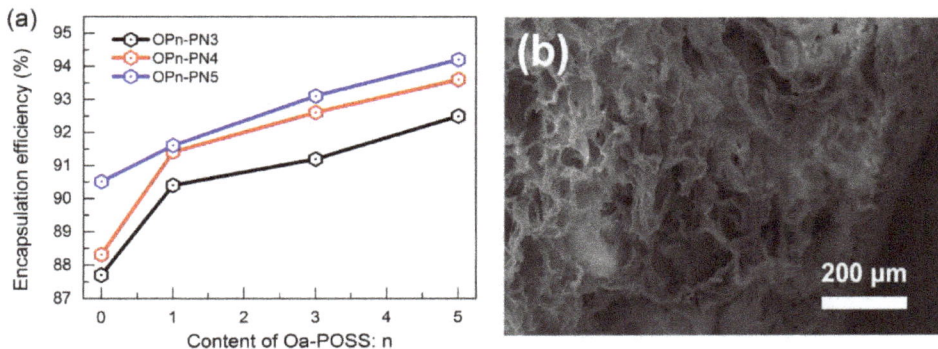

Figure 9. Drug-loading character of OPn-PNm gels: (**a**) drug loading content as a function of n and m for OPn-PNm gels; (**b**) SEM image of drug-loaded OP5-PN5 gels treated by freeze-drying.

The profiles of drug release for SA gels and OPn-PNm gels are shown in Figure 10. As shown in Figure 10a, in pure water, all gels show a burst release of BSA initially for about 5 d, then the drug releases sequentially from the matrix within the studied period of 20 d, indicating the effective sustained release of drug. Moreover, a better release effect was found with increased PNIPA network and Oa-POSS content. The reason for this is the same as that for the better drug loading. When the release condition was changed to Tris-HCl buffer solution (25 mM, pH = 7.3), as shown in Figure 10b, although the total release character and rule are similar to those in pure water, the release rate of all gels becomes faster and the burst release phenomenon is weakened in POSS hybrid gels. Tris buffer solution is good for solving the protein, which accelerates the dissolution of BSA out of the gels. The POSS particles have better combination ability with BSA, resulting in a weaker burst release. Although the loading mass of OPn-PNm gels with high n and m is greater than that of pure SA gels, the release rate is still slower, showing a better sustained release effect.

It needs to be pointed out that the drug loading and release exploration here is only one specific example to display the potential of this gel as a drug-release system. As we know, different drugs with special molecular structure would have different interactions with the gel matrix, which would to be analyzed individually. For the OPn-PNm gels, the introduction of IPN structure and Oa-POSS could adjust the hydrophilic–hydrophobic character of the gels and improve their ability to be combined with certain drugs. On the other hand, the temperature responsiveness of this gel could also be applied for drug delivery in vitro. Correlative experiments were carried out in a simulated human body environment by using the micromolecule drug flutamide as the target drug. It was found that, compared to pure PNIPA gels, OPn-PNm gels had better drug loading and sustained release ability.

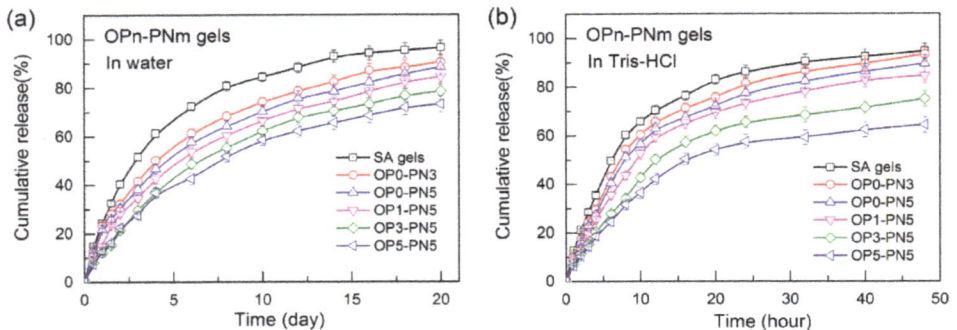

Figure 10. Release profiles of bovine serum albumin (BSA) in OPn-PNm gels in different solutions: (a) deionized water; (b) Tris-HCl buffer solution, 25 mM, pH = 7.3.

4. Conclusions

In this work, to develop novel biomass-based stimulus-responsive hydrogels with excellent properties, double network POSS hybrid hydrogels (named OPn-PNm gels) were synthesized by using Oa-POSS cationic particles to partly substitute CaCl$_2$ as the co-cross-linking agent of SA and introducing a temperature-responsive PNIPA network into the SA hydrogels simultaneously, to obtain a series of transparent and homogeneous gels without obvious phase separation or aggregates. Through comprehensive analysis of the gels by FTIR, X-ray, TGA, DSC, SEM, and property characterization, it could be confirmed that the Oa-POSS was introduced into the gels successfully, and its content in the gels was determined by the concentration of POSS in a reaction in aqueous solution. Moreover, the addition of a second PNIPA network and Oa-POSS had a significant effect on the properties of gels, and several main conclusions are obtained.

First, the mechanical properties of OPn-PNm gels showed great change, which depended on the content of Oa-POSS and PNIPA network inside the gels. With the increase of PNIPA content, the strength and elongation at breakage of gels increased considerably, while the modulus was affected slightly. This is attributed to the stress loading mode of both networks: the rigid SA network as the main stress loader determines the modulus in initial deformation, the second PNIPA network helps to share more stress and keep the gels unbroken in bigger deformation. The increased Oa-POSS improved the strength and modulus of gels significantly, while the elongation at breakage of gels exhibited a considerable decrease. This is ascribed to the stress concentration caused by the increased heterogeneity of gels due to microaggregation of Oa-POSS particles. In addition, the compression resistance and resilience of gels were improved significantly by adding the second network and Oa-POSS, revealing that the network became stiffer. This network characteristic could also be proved by the rheological measurement results. Under the same frequency region, with the increase of the second PNIPA network and Oa-POSS content, the storage modulus and loss modulus of gels increased at different levels, indicating a more elastic network and stronger viscous behavior. Also, the damping effect of OPn-PNm gels was enhanced with the increased PNIPA network content. There was a stronger interface between the two polymer networks, resulting in greater energy consumption.

Second, through the research related to the swelling and deswelling properties of gels, it was found that the PNIPA network improved the swelling ability of gels significantly due to its hydrophilic nature, while Oa-POSS had an inverse effect due to the increased hydrophobicity, which could be supported by the results of water contact angle tests. The temperature responsiveness became obvious with the increase of PNIPA content. The VPTT of OPn-PNm gels was close to that of PNIPA gels and should shift to lower temperature with the increase of Oa-POSS content. However, the complete shrinking of OPn-PNm gels at high temperature was inhibited by the robust and hydrophilic SA

Polymers **2019**, *11*, 524

network. Moreover, the OPn-PNm gels exhibited good swelling–deswelling repeatability under low–high temperature circulation due to the resilience of the gel network.

Third, the deswelling rate of OPn-PNm gels increased with the increased PNIPA and Oa-POSS content, the PNIPA network was the main driving force causing the shrinking of the network, and the Si–O skeleton of Oa-POSS acted as a hydrophobic core to accelerate the shrinking of polymer chains. Moreover, the introduction of Oa-POSS particles created some small pores inside the gel network, and the inhomogeneous shrinkage around POSS also formed small pores in the gel surface, all of which improved the deswelling rate.

Finally, to investigate the potential of these hybrid gels as a drug-release system, BSA was used as a target drug to analyze the drug-loading and -release properties of the OPn-PNm gels. As the results showed, the addition of PNIPA network and Oa-POSS particles could improve the drug-loading ability effectively due to their strong interaction with drug macromolecules. Although the burst release of drug still existed for all gels, the increased Oa-POSS content significantly improved the sustained release effect of gels in both water and Tris-HCl buffer solution.

Author Contributions: Data curation, W.L. and H.P.; Formal analysis, G.Z.; Investigation, Y.Z.; Methodology, Y.C.; Project administration, Y.C.; Resources, G.Z.; Writing—review and editing, Y.C.

Funding: Hunan Province Funds for Distinguished Young Scientists (2019JJ20007), Key Research and Development Plan of Hunan Province (2016SK2077), Key Project of Education Department of Hunan Province (18A262).

Acknowledgments: We acknowledge financial support from the Hunan Province Funds for Distinguished Young Scientists (2019JJ20007), the Key Research and Development Plan of Hunan Province (2016SK2077) and the Key Project of Education Department of Hunan Province (18A262).

Conflicts of Interest: The authors declare no conflict of interest.

References

1. Dunlop, J.W.C.; Weinkamer, R.; Fratzl, P. Artful interfaces within biological materials. *Mater. Today* **2011**, *14*, 70–78. [CrossRef]
2. Sun, J.Y.; Zhao, X.; Illeperuma, W.R.K.; Chaudhuri, O.; Oh, K.H.; Mooney, D.J.; Vlassak, J.J.; Suo, Z. Highly stretchable and tough hydrogels. *Nature* **2012**, *489*, 133–136. [CrossRef]
3. Wehner, M.; Truby, R.L.; Fitzgerald, D.J.; Mosadegh, B.; Whitesides, G.M.; Lewis, J.A.; Wood, R.J. An integrated design and fabrication strategy for entirely soft, autonomous robots. *Nature* **2016**, *536*, 451–455. [CrossRef]
4. Li, T.F.; Li, G.R.; Liang, Y.M.; Cheng, T.Y.; Dai, J.; Yang, X.X.; Liu, B.Y.; Zeng, Z.D.; Huang, Z.L.; Luo, Y.W.; et al. Fast-moving soft electronic fish. *Sci. Adv.* **2017**, *3*, e1602045. [CrossRef] [PubMed]
5. Cao, J.; Lu, C.; Zhuang, J.; Liu, M.; Zhang, X.; Yu, Y.; Tao, Q. Multiple hydrogen bonding enables the self-healing of sensors for human-machine interaction. *Angew. Chem. Int. Ed.* **2017**, *56*, 8795–8800. [CrossRef]
6. Boland, C.S.; Khan, U.; Ryan, G.; Barwich, S.; Charifou, R.; Harvey, A.; Backes, C.; Li, Z.; Ferreira, M.S.; Möbius, M.E.; et al. Sensitive electromechanical sensors using viscoelastic graphene-polymer nanocomposites. *Science* **2016**, *354*, 1257–1260. [CrossRef]
7. Yang, C.H.; Chen, B.; Zhou, J.; Chen, Y.M.; Suo, Z. Electroluminescence of giant stretchability. *Adv. Mater.* **2016**, *28*, 4480–4484. [CrossRef] [PubMed]
8. Larson, C.; Peele, B.; Li, S.; Robinson, S.; Totaro, M.; Beccai, L.; Mazzolai, B.; Shepherd, R. Highly stretchable electroluminescent skin for optical signaling and tactile sensing. *Science* **2016**, *351*, 1071–1074. [CrossRef] [PubMed]
9. Das, S.; Subuddhi, U. Cyclodextrin mediated controlled release of naproxen from pH-sensitive chitosan/poly(vinyl alcohol) hydrogels for colon targeted delivery. *Ind. Eng. Chem. Res.* **2013**, *52*, 14192–14200. [CrossRef]
10. Vashist, A.; Shahabuddin, S.; Gupta, Y.K.; Ahmad, S. Polyol induced interpenetrating networks: Chitosan-methyl methacrylate based biocompatible and pH responsive hydrogels for drug delivery system. *J. Mater. Chem. B* **2013**, *1*, 168–178. [CrossRef]

11. Li, P.; Poon, Y.F.; Li, W.; Zhu, H.-Y.; Yeap, S.H.; Cao, Y.; Qi, X.; Zhou, C.; Lamrani, M.; Beuerman, R.W.; et al. A polycationic antimicrobial and biocompatible gydrogel with microbe membrane suctioning ability. *Nat. Mater.* **2011**, *10*, 149–156. [CrossRef]

12. Cavallaro, G.; Lazzara, G.; Milioto, S.; Parisi, F.; Evtugyn, V.; Rozhina, E.; Fakhrullin, R. Nanohydrogel formation within the halloysite lumen for triggered and sustained release. *ACS Appl. Mater. Interfaces* **2018**, *10*, 8265–8273. [CrossRef]

13. Cavallaro, G.; Lazzara, G.; Konnova, S.; Fakhrullin, R.; Lvov, Y. Composite films of natural clay nanotubes with cellulose and chitosan. *Green Mater.* **2014**, *2*, 232–242. [CrossRef]

14. Draget, K.I.; SkjakBraek, G.; Smidsrod, O. Alginate based new materials. *Int. J. Biol. Macromol.* **1997**, *21*, 47–55. [CrossRef]

15. Orive, G.; Carcaboso, A.M.; Hernandez, R.M.; Gascon, A.R.; Pedraz, J.L. Biocompati- bility evaluation of different alginates and alginatebased microcapsules. *Biomacromolecules* **2005**, *6*, 927–931. [CrossRef]

16. Chan, G.; Mooney, D.J. New materials for tissue engineering: Towards greater control over the biological response. *Trends Biotechnol.* **2008**, *26*, 382–392. [CrossRef]

17. Bernkop-Schnurch, A.; Kast, C.E.; Richter, M.F. Improvement in the mucoadhesive properties of alginate by the covalent attachment of cysteine. *J. Control. Release* **2001**, *71*, 277–285. [CrossRef]

18. Chan, A.W.; Whitney, R.A.; Neufeld, R.J. Kinetic controlled synthesis of pH-responsive network alginate. *Biomacromolecules* **2008**, *9*, 2536–2545. [CrossRef]

19. Draget, K.I.; Stokke, B.T.; Yuguchi, Y.; Urakawa, H.; Kajiwara, K. Small-angle X-ray scattering and rheological characterization of alginate gels. 3. Alginic acid gels. *Biomacromolecules* **2003**, *4*, 1661–1668. [CrossRef]

20. Chan, A.W.; Whitney, R.A.; Neufeld, R.J. Semisynthesis of a controlled stimuli- responsive alginate hydrogel. *Biomacromolecules* **2009**, *10*, 609–616. [CrossRef]

21. Lee, K.Y.; Mooney, D.J. Alginate: Properties and biomedical applications. *Prog. Polym. Sci.* **2012**, *37*, 106–126. [CrossRef] [PubMed]

22. Bidarra, S.J.; Barrias, C.C.; Granja, P.L. Injectable alginate hydrogels for cell delivery in tissue engineering. *Acta Biomater.* **2014**, *10*, 1646–1662. [CrossRef]

23. Vlierberghe, S.V.; Dubruel, P.; Schacht, E. Biopolymer-based hydrogels as scaffolds for tissue engineering applications: A review. *Biomacromolecules* **2011**, *12*, 1387–1408. [CrossRef]

24. Markstedt, K.; Mantas, A.; Tournier, I.; Ávila, H.M.; Hägg, D.; Gatenholm, P. 3D bioprinting human chondrocytes with nanocellulose-alginate bioink for cartilage tissue engineering applications. *Biomacromolecules* **2015**, *16*, 1489–1496. [CrossRef]

25. Pawlak, T.; Kowalewska, A.; Zgardzińska, B.; Potrzebowski, M.J. Structure, dynamics, and host-guest interactions in POSS functionalized cross-linked nanoporous hybrid organic-inorganic polymers. *J. Phys. Chem. C* **2015**, *119*, 26575–26587. [CrossRef]

26. Cordes, D.B.; Lickiss, P.D.; Rataboul, F. Recent Developments in the Chemistry of Cubic Polyhedral Oligosilsesquioxanes. *Chem. Rev.* **2010**, *110*, 2081–2173. [CrossRef] [PubMed]

27. Blanco, I. The rediscovery of POSS: A molecule rather than a filler. *Polymers* **2018**, *10*, 904. [CrossRef]

28. Wu, J.; Ge, Q.; Mather, P.T. PEG−POSS multiblock polyurethanes: Synthesis, characterization, and hydrogel formation. *Macromolecules* **2010**, *43*, 7637–7649. [CrossRef]

29. Prządka, D.; Andrzejewska, E.; Marcinkowska, A. Multimethacryloxy-POSS as a crosslinker for hydrogel materials. *Eur. Polym. J.* **2015**, *72*, 34–49. [CrossRef]

30. Zeng, K.; Fang, Y.; Zheng, S.X. Origanic-inorganic hybrid hydrogels involving poly(N-isopropylacrylamide) and polyhedral oligomeric silsequioxane: Preparation and rapid thermoresponsive properties. *J. Polym. Sci. Part B Polym. Phys.* **2011**, *47*, 504–516. [CrossRef]

31. Chen, Y.; Xiong, Y.Q.; Peng, C.; Liu, W.Y.; Peng, Y.; Xu, W.J. Synthesis and characterization of POSS hybrid co-cross-linked p(NIPA-co-DMAEMA) hydrogels. *J. Polym. Sci. Part B Polym. Phys.* **2013**, *51*, 1494–1504. [CrossRef]

32. Xu, Z.Q.; Ni, C.H.; Yao, B.L.; Tao, L.; Zhu, C.Q.; Han, Q.B.; Mi, J.Q. The preparation and properties of hybridized hybridized hydrogels based on cubic thiol-functionalized silsequioxane covalently linked with poly(N-isopropylacrylamide). *Colloid Polym. Sci.* **2011**, *289*, 1777–1782. [CrossRef]

33. Wang, J.; Sntti, A.; Wang, X.G.; Lin, T. Fast responsive and morphologically robust thermo-responsive hydrogel nanofibers from poly(N-isopropylacrylamide) and POSS crosslinker. *Soft Matter* **2011**, *7*, 4364–4369. [CrossRef]

34. Pramudya, I.; Rico, C.G.; Lee, C.; Chung, H. POSS-containing bioinspired adhesives with enhanced mechanical and optical properties for biomedical applications. *Biomacromolecules* **2016**, *17*, 3853–3861. [CrossRef] [PubMed]

35. Shen, C.H.; Yuemei Han, Y.M.; Wang, B.L.; Tang, J.M.; Chen, H.; Lin, Q.K. Ocular biocompatibility evaluation of POSS nanomaterials for biomedical material applications. *RSC Adv.* **2015**, *5*, 53782–55378. [CrossRef]

36. Dragan, E.S. Design and applications of interpenetrating polymer network hydrogels. A review. *Chem. Eng. J.* **2014**, *243*, 572–590. [CrossRef]

37. Gong, J.P.; Katsuyama, Y.; Kurokawa, T.; Osada, T. Double-network hydrogels with extremely high mechanical strength. *Adv. Mater.* **2003**, *15*, 1155–1158. [CrossRef]

38. Yan, Y.; Li, M.N.; Yang, D.; Wang, Q.; Liang, F.X.; Qu, X.Z.; Qiu, D.; Yang, Z.Z. Construction of injectable double-network hydrogels for cell delivery. *Biomacromolecules* **2017**, *18*, 2128–2138. [CrossRef]

39. Karami, P.; Wyss, C.S.; Khoushabi, A.; Schmocker, A.; Broome, M.; Moser, C.; Bourban, P.E.; Pioletti, D.P. Composite double-network hydrogels to improve adhesion on biological surfaces. *ACS Appl. Mater. Interfaces* **2018**, *10*, 38692–38699. [CrossRef]

40. Dai, C.Y.; Wang, B.C.; Zhao, H.W.; Li, B. Factors affecting protein release from microcapsule prepared by liposome in alginate. *Colloid Surf. B* **2005**, *42*, 253–258. [CrossRef]

41. Blanco, I.; Abate, L.; Bottino, F.A.J. Variously substituted phenyl hepta cyclopentyl- polyhedral oligomeric silsesquioxane (ph,hcp-POSS)/polystyrene (PS) nanocomposites. *Therm. Anal. Calorim.* **2013**, *112*, 421–428. [CrossRef]

polymers

MDPI

Article

Anchor Effect in Polymerization Kinetics: Case of Monofunctionalized POSS

Agnieszka Marcinkowska [1,*], Dawid Przadka [1], Beata Dudziec [2], Katarzyna Szczesniak [3] and Ewa Andrzejewska [1,*]

[1]   Faculty of Chemical Technology, Poznan University of Technology, Poznan, Berdychowo 4, 60-965 Poznan, Poland; daw-prza@wp.pl
[2]   Faculty of Chemistry and Centre for Advanced Technologies, Adam Mickiewicz University in Poznan, Umultowska 89 B and C, 61-614 Poznan, Poland; beata.dudziec@gmail.com
[3]   NanoBioMedical Center, Adam Mickiewicz University, Umultowska 85, 61-614 Poznan, Poland; k.szczesniak@amu.edu.pl
*   Correspondence: agnieszka.marcinkowska@put.poznan.pl (A.M.); ewa.andrzejewska@put.poznan.pl (E.A.); Tel.: +48-61-665-36-05 (A.M.); +48-61-665-36-37 (E.A.)

Received: 4 March 2019; Accepted: 13 March 2019; Published: 19 March 2019

Abstract: The effect of the anchoring group on the detailed polymerization kinetics was investigated using monomethacryloxy-heptaisobutyl POSS (1M-POSS). This compound was copolymerized with lauryl methacrylate (LM) as the base monomer, at various molar ratios. The process was initiated photochemically. The polymerization kinetics were followed by photo-DSC and photorheology while the polymers were characterized by nuclear magnetic resonance (NMR), gel permeation chromatography (GPC), transmission electron microscopy (TEM), and differential scanning calorimetry (DSC). For comparison, a methacrylate containing the branched siloxy-silane group (TSM) was also studied. It was found that the modifiers with a bulky substituent have a dual effect on the termination process: (i) At low concentrations, they increase the molecular mobility by increasing the free volume fraction, which leads to an acceleration of the termination and slows the polymerization; while (ii) at higher concentrations, they retard molecular motions due to the "anchor effect" that suppresses the termination, leading to acceleration of the polymerization. The anchor effect can also be considered from a different point of view: The possibility of anchoring a monomer with a long substituent (LM) around the POSS cage, which can further enhance propagation. These conclusions were derived based on kinetic results, determination of polymerization rate coefficients, and copolymer analysis.

Keywords: Monomethacryloxy POSS; lauryl methacrylate; polymerization kinetics; anchor effect

1. Introduction

In recent years, there has been growing interest in high-performance inorganic–organic hybrid polymer materials. Both components (the inorganic modifier and the polymer matrix) can be linked together physically or by chemical links. Good compatibility between the components can be provided by functionalization of the inorganic filler with organic substituents. A widely studied novel class of nanofillers, in which the inorganic core can be functionalized by a wide range of organic substituents, are polyhedral oligomeric silsesquioxanes (POSSs). POSSs are organic–inorganic hybrid compounds constituted by an inorganic silica cage described by the general formula $Rn(SiO_{1.5})_n$ or T_n, where n is the number of repeat units (n = 8, 10, or 12, T is $RSiO_{1.5}$) and R is H or an organic substituent. This results in the POSS molecule having a unique shape: the inorganic core is surrounded by an organic outer layer. The size of the POSS cage is ~1.5 nm [1].

The substituents, R, can be reactive (e.g., containing double bonds, epoxy groups) or nonreactive, of any structure. These attached functional groups improve the solubility/compatibility of POSS with polymers and lead to the improvement of many properties, such as the thermal stability, thermomechanical and electrical properties, flame retardancy as well as mechanical strength [2–5]. POSS cages are considered as the smallest possible form of silica [5]. POSS compounds can be soluble in the polymer matrix; therefore, they can be dispersed on the molecular level, which enables them to control the motions of the chains and to control the physical properties of the base polymer. In such a case, true hybrid materials are formed. However, POSS molecules or POSS moieties have a tendency to agglomerate (or even form crystallites), which in turn leads to the formation of (nano)composites. POSSs can be introduced into the polymer matrix by physical blending or via copolymerization (covalent bonding into a polymer backbone) [6].

Copolymerization is the most common approach used to obtain polymer/POSSs materials. When POSS contains only one reactive substituent, it can be incorporated into the polymer as a pendant unit; if it bears more reactive groups, it will form cross-linking points in the arising network. Widely investigated systems are methacryloxy-based POSS-containing materials. Copolymerization of various methacrylates with methacryloxy-substituted POSS has been described in a number of papers. Both thermal and photochemical initiation were applied [7].

UV-curable polymer systems are of particular interest in the development of new materials. This technology offers a number of advantages, namely, ultrafast curing, ambient temperature operation, and spatial and temporal control of the process. Additionally, it has an environment-friendly aspect (no emission of volatile organic compounds, low energy consumption). UV-curing has many important modern applications, e.g., in the coatings industry, dentistry, lithography and microlithography, microelectronics, optoelectronics, etc. [7,8].

Existing articles have been devoted to various aspects of the POSS effect, mainly to the improvement of mechanical and thermal properties as well as to morphology and, in general, to structure–property relationships. However, it is very important to know the polymerization kinetics, because they determine, to a large extent, the properties of the material, provide information on the optimal monomer/modifier ratios, and enable the selection of the technological parameters of curing.

Despite many publications being devoted to methacryloxy-based POSS-containing materials, the number discussing the influence of POSS compounds on the polymerization kinetics is very limited. There is a lack of deeper insight into the kinetics reaction; only a general discussion was presented in a few reports, e.g., [9–16] both using monomethacryloxy (1M, linear monomer) as well as octamethacryloxy (8M, a crosslinking monomer) POSS. Mainly the effect of the addition of functionalized POSS on the polymerization rate and double bond conversion of methacrylate-based formulations was studied, but the results and possible explanations differed; e.g., a decrease in the final conversion (both with 8M- and 1M-POSS, explained by the steric hindrance and the reduction of the polymer network mobility associated with the inorganic part of these POSSs) [9], reduction of both the double bond conversion and the polymerization rate (with 8M-POSS) [11], acceleration and enhanced conversion or slowdown and reduced conversion, depending on the reaction stage (due to both an enhancement of the gel effect and blocking light absorption by the photoinitiator, octavinyl-POSS) [12], and an increase in the double bond conversion and polymerization rate due to an increase in the concentration of double bonds (8M-POSS [15] and 1M-POSS [16]). The more detailed study describing the effect of 8M-POSS on the formation of a rigid polymer (poly(2-hydroxyethyl methacrylate)) or an elastomer (poly (oxyethylene glycol (di)methacrylate)) in a wide range of its concentration was presented in [7]. The final double bond conversion decreased with the increase of the POSS content, but the polymerization rate first increased and then decreased, reaching its highest value in the presence of about 15–25 wt % of the modifier (depending on the monomer). These phenomena were related to the influence of the POSS cage and the polymer cross-linking on the diffusional ability of the reacting species and resulting changes in the termination mechanisms.

However, it is obvious that when 8M-POSS is used, its crosslinking effect will dominate the polymerization kinetics and the effect of the POSS cage will be much less pronounced. Therefore, the best model POSS monomer would be a monofunctionalized compound, which, after incorporation into the polymer, will form a linear copolymer.

It was indicated that POSS moiety, incorporated into the polymer chain as a massive and bulky pendant group, serves as an anchor point bound to the chain. The presence of POSS moieties (nondiffusive "anchors") along the backbone of polymer chains may dramatically alter the diffusion of polymer chains and slow down molecular motions [17–19]. This, in turn, can also have a significant effect on the polymerization kinetics.

The aim of the work was to investigate the possible anchoring effect exerted by the POSS cage on the polymerization kinetics of the UV-initiated process. The model system consisted of a monomethacryloxy substituted POSS (methacryloxypropyl-heptaisobutyl POSS, denoted as 1M-POSS) and lauryl methacrylate (LM, the only popular and non-volatile methacrylate monomer miscible with 1M-POSS) at various comonomers ratios. The investigated system is linear; therefore, its polymerization kinetics will not be complicated by crosslinking effects. For comparison, also another methacrylate having a large volume substituent (corresponding to an open part of the POSS cage, which may also serve as an anchor) was investigated, namely 3-[tris(trimethylsiloxy)silyl]propyl methacrylate (TSM) (Figure 1). This provides better insight into the possible "anchor effect". The polymerization was initiated photochemically, which enables precise control of the reaction (the initiator derived radicals begin to form when the light is switched on and their formation is stopped at the break of irradiation; the polymerization occurs only in the irradiated area [20]).

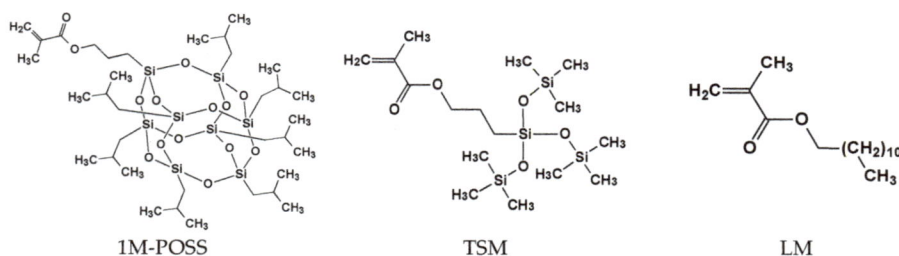

Figure 1. Structures of the monomers used.

2. Materials and Methods

2.1. Materials

TSM (purity 98%) and LM (purity 96%) were delivered by Sigma-Aldrich; they were purified by column chromatography (aluminiumoxid 90 Activ basisch, Merck, Darmstadt, Germany) before use and were stored over molecular sieves. The photoinitiator, 2,2-dimethoxy-2-phenylacetophenone, was also supplied by Sigma-Aldrich and was used in conc. 1 wt %.

1M-POSS Preparation

1M-POSS (full name: 1-methacryloxypropyl-3,5,7,9,11,13,15-hepta(isobutyl)pentacyclo-[9.5.1.13, 9.15,15.17,13] octasiloxane) was prepared in a two-step procedure, i.e., condensation of 1,3,5,7,9,11,14-heptaisobutyltricyclo [7.3.3.15,11] heptasiloxane-endo-3,7,14-triol with trichlorosilane, resulting in 1-hydro-3,5,7,9,11,13,15-hepta(isobutyl)pentacyclo [9.5.1.13,9.15,15.17,13] octasiloxane and its subsequent hydrosilylation with allyl methylacrylate using a Karstedt catalyst [21]. All solvents and liquid reagents were purchased from Merck and ABCR (Karlsruhe, Germany), and were dried and distilled under argon prior to use. The structure of the obtained compound was confirmed by spectroscopic analysis [22]. ^{1}H, ^{13}C, ^{29}Si NMR spectra were recorded on Bruker Avance 300 MHz and 400 MHz (Bruker Scientific LLC, Billerica, MA, USA) in CDCl$_3$.

^{1}H NMR (CDCl$_3$, 400 MHz, δ, ppm): 0.59–62 (m, 16H, SiCH$_2$), 0.95 (dd, 42H, CH$_3$), 1.73–1.79 (m, 2H, SiCH$_2$CH$_2$), 1.81–1.89 (m, 7H, CH$_2$CHCH$_3$), 1.95 (s, 3H, CH$_3$), 4.11 (t, JH-H = 6.6Hz, 2H, CH$_2$CH$_2$CO), 5.55, 6.11 (s, 2H, CH$_2$=).

^{13}C NMR (CDCl$_3$, 100 MHz, δ, ppm): 8.88 (SiCH$_2$), 18.79 (CH$_2$CH$_2$), 22.98 (CH$_2$CH(CH$_3$)$_2$), 24.33 (CH$_2$CH(CH$_3$)$_2$), 26.16 (CH$_2$CH(CH$_3$)$_2$), 67.07 (CH$_2$O), 125.63 (=CH$_2$), 137.01 (=CCH$_3$), 167.94 (C=O).
^{29}Si NMR (CDCl$_3$, 79 MHz, δ, ppm): −67.70, −68.56, −68.85 (SiO$_3$).

2.2. Methods

2.2.1. Viscosity

The viscosity of the photocurable systems was measured with a DV-II+ PRO Brookfield Digital Viscometer (Brookfield, Toronto, Ontario, Canada) at the polymerization temperature.

2.2.2. Glass Transition

Glass transition temperature, T_g, was measured with a Mettler Toledo DSC1 instrument (Mettler Toledo GmbH, Schwerzenbach, Switzerland) under a nitrogen atmosphere at a heating rate of 20 °C/min in the temperature range from −80 °C to 180 °C. The T_g value was determined from the second run of the DSC measurement and was taken as an average value from three measurements. The reproducibility of the results was about ±4%.

2.2.3. NMR Study

^{1}H and ^{13}C NMR spectra were performed on Bruker Ultra Shield 400 and 300 spectrometers (Bruker Scientific LLC, Billerica, MA, USA) using toluene-d$_8$ and CDCl$_3$ as solvents. Chemical shifts are reported in ppm with reference to the residual solvents' (C$_7$H$_8$, CHCl$_3$) peaks for ^{1}H and ^{13}C NMR.

2.2.4. Photopolymerization Kinetics

Reaction rate profiles were monitored by DSC using a Pyris 6 instrument (Perkin–Elmer, Waltham, MA, USA) equipped with a lid especially designed for photochemical measurements. The 15 mg samples were polymerized in open aluminum pans (diameter 6.6 mm) under isothermal conditions (40 °C) in high-purity argon atmosphere (<0.0005% of O$_2$). The polymerization was initiated by the UV light (365 nm, light intensity 2 mW·cm^{-2}) from the LED lamp (LC-L1, Hamamatsu Photonics, Hamamatsu, Japan). All photopolymerization experiments were conducted at least in triplicate. The reproducibility of the kinetic results was about ±3%. For computations, the heat of the polymerization of the methacrylate group, 56 kJ·mol^{-1}, was taken [23].

The experimental data for the calculations of the polymerization rate coefficients: Propagation rate coefficient (k_p) and bimolecular termination rate coefficient ($k_t{}^b$) in the expression for the polymerization rate (Equation (1)) [20]:

$$R_p = \frac{k_p}{\left(k_t^b\right)^{0.5}} \cdot [M] \cdot (\phi \cdot I_a)^{0.5} \tag{1}$$

where $[M]$ is the double bond concentration, ϕ denotes the quantum yield of initiation, and I_a is intensity of the light absorbed, and were obtained from postpolymerization processes (non-steady-state measurements), which were registered after stopping the irradiation at various degrees of the double bond conversion. The calculations were performed over the first 10 s of the dark reaction. The rate coefficients were determined according to the bimolecular termination model [20]:

$$\frac{[M]_t}{(R_p)_t} = \frac{2 \cdot k_t^b}{k_p} \cdot t + \frac{[M]_0}{(R_p)_0} \tag{2}$$

where $(R_p)_t$, $(R_p)_0$ are the polymerization rates at time, t, of the dark reaction and at the moment of breaking the irradiation, resp., and $[M]_t$ and $[M]_0$ are the concentrations of double bonds after the time, t, of the dark reaction and at the moment of breaking the irradiation. Equation (2) allows the determination of the k_t^b/k_p ratio. Using Equation (2) (non-stationary conditions) and Equation (3) (from steady-state measurements):

$$(R_p)_0 = \frac{k_p}{\left(k_t^b\right)^{0,5}} \cdot [M]_0 \cdot (F)^{0,5} \tag{3}$$

where $F = \phi \cdot I_a$, we can determine the polymerization rate coefficients in the form of $k_p \times F$ and $k_t^b \times F$. It was assumed that F is constant in the range of the conversions studied. A detailed procedure of the calculations was described in [20,24].

It should be emphasized that the measured parameters concern the copolymerization process as a whole; the double bond conversion involves the conversion of both comonomers while the polymerization rate is the overall rate of the C=C conversion of the two comonomers.

2.2.5. Photorheology

Real-time photorheology experiments were performed on an Anton Paar MCR 301 rheometer (Anton Paar, Graz, Austria) using parallel plate geometry of a quartz plate and metallic plate (disposable D-PP25-SN0 measuring system, Anton Paar, Graz, Austria) with a diameter of 25 mm and a gap thickness of 0.2 mm. The measurements were carried out at 40 °C (in the heating chamber); the rheometer operated at a frequency of 10 Hz (oscillatory mode). Polymerization was initiated 30 s after starting the measurement by switching on the UV light (320–500 nm, 1 mW·cm^{-2} from an OmniCure S 1000 high-pressure mercury lamp (OmniCure, London, England) projected via a waveguide into the sample through the bottom quartz window. During the measurements, the change in the normal force along with the storage modulus, G′, and loss modulus, G″, were recorded. The measurement for each sample was run for 2400 s.

2.2.6. GPC

GPC analyses were performed using Agilent 1260 Infinity system (Agilent Technologies, Santa Clara, Utah, USA) equipped with RI detector and Phenogel 10 μm Linear(2) 300 × 7.8 mm column. THF was used as a mobile phase in a flow rate of 1.0 mL·min^{-1}. Temperatures of RI detector and column were set at 35 °C. Time of analysis was 15 min. Molecular weights (number average, M_n; weight average, M_w) and polydispersity index (PDI) values were calculated based on the calibration curve using polystyrene standards (Shodex) in a range of 1000–3,500,000, using Agilent Software GPC/SEC – 1260 GPC set.

2.2.7. TEM

TEM images were recorded on a JEM-1400 microscope made by JEOL (Tokyo, Japan). The accelerating voltage was 120 kV. A small amount of the samples was applied on a copper grid (Formvar/Carbon 200 mesh made by TedPella Inc., Redding, CA, USA).

3. Results and Discussion

The photopolymerization kinetics of the LM/1M-POSS and LM/TSM systems were followed at 40 °C in a wide range of the comonomer ratios. The kinetics were investigated both in general terms (kinetic curves) as well as by determination of the polymerization rate coefficients. Due to the fact that the effects of two different monomers with bulky substituents were compared, the concentrations of the formulation components were set in mole percent. 1M-POSS showed a limited solubility with LM (up to 12 mol %), whereas TSM was soluble in every ratio; therefore, the range of the investigated (base monomer)/modifier compositions was determined by the limits of miscibility. The composition of the investigated formulations (given in mol % and wt %) is given Table 1. 1M-POSS is a solid (m.p. 160 °C from DSC), therefore, its homopolymerization could not be followed.

Table 1. Investigated compositions.

LM/1M-POSS		LM/TSM	
1M-POSS Concentration		**TSM Concentration**	
mol %	wt %	mol %	wt %
0.0	0.0	0.0	0.0
0.5	1.8	2.0	3.3
1.0	3.6	4.0	6.5
2.0	7.0	8.0	12.6
4.0	13.4	12.0	18.5
6.0	19.1	20.0	29.4
8.0	24.4	50.0	62.4
10.0	29.2	80.0	86.9
12.0	33.6	100.0	100.0

3.1. Viscosity of Formulations

The initial viscosity of a polymerizable composition (along with that growing during the polymerization) is the important factor influencing the termination rate coefficient (an inverse proportionality: $k_t^b \sim \eta^{-1}$). Thus, it is necessary to consider its changes when the ratio of the comonomers is changed.

The dependence of the viscosity on the modifier content for the two systems (LM/1M-POSS and LM/TSM) is given in Figure 2. The viscosity of the LM/1M-POSS system rapidly increases with the 1M-POSS content, indicating the existence of strong interactions between the POSS-cages (not yet tethered to the polymer chain) and the LM monomer.

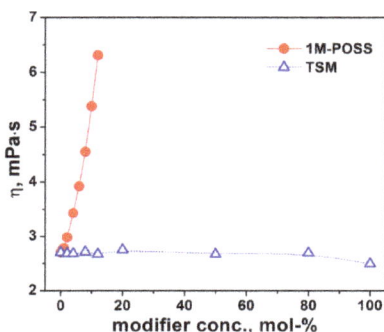

Figure 2. Viscosity of LM/1M-POSS and LM/TSM systems at 40 °C as a function of the Si-containing modifier concentration. The lines are eye guides.

On the other hand, the viscosity of the LM/TSM system is low and changes very slightly (LM: η^{40} = 2.7 mPa·s, TSM: η^{40} = 2.49 mPa·s). This proves changes in viscosity due to changes in the monomer ratio will practically not affect the polymerization.

3.2. NMR Analysis

^{1}H and ^{13}C NMR spectra of the monomers and (co)polymers containing 8 mol % of 1M-POSS or 50 mol % of TSM are shown in Figures 3 and 4, respectively.

In Figure 3, the ^{1}H NMR stacked spectra disclose the correlations between the ^{1}H NMR spectra of LM, 1M-POSS, and their copolymer (Figure 3a) and LM, TSM, and their copolymer (Figure 3b).

The resonance lines originating to specific types of groups of both comonomers are visible. There are new signals arising from the -CH$_2$- (polymer chain) at 1.63 ppm (wide resonance line) and 1.35 ppm, along with the changes in the region between 4.10–4.00 ppm that is assigned to the -OCH$_2$- groups deriving from the copolymer and monomers. The surrounding -OCH$_2$- chemical is slightly changed in the fragments derived from LM and 1M-POSS. This region is the most susceptible to changes and they are obviously more noticeable with the rising 1M-POSS content in the copolymer. In the case of the TSM-containing copolymer, there is a resemblance in the matter of the -OCH$_2$- moiety and new signals arising from the -CH$_2$- fragments.

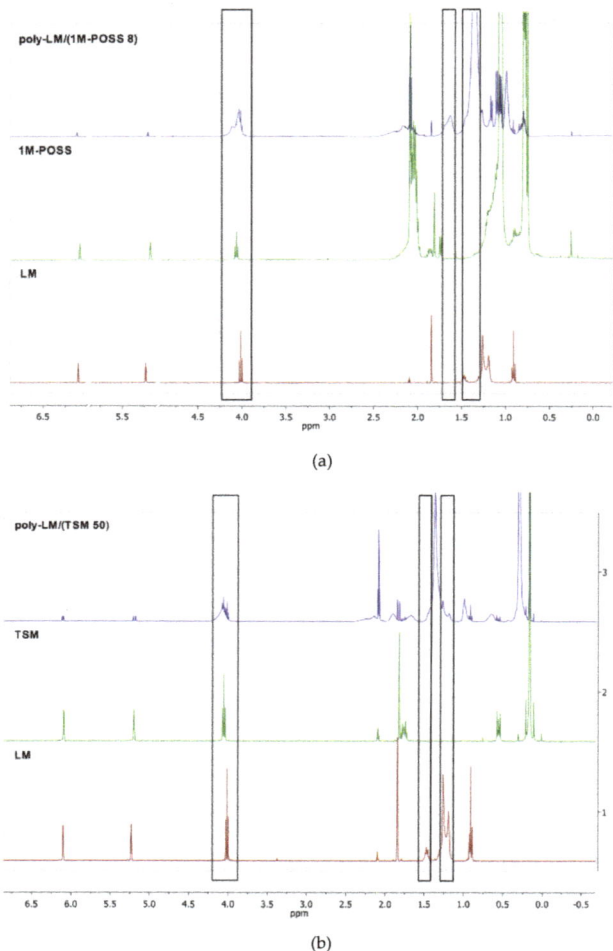

Figure 3. ^{1}H NMR stacked spectra of the LM monomer, the modifiers, and the copolymers containing (a) 8 mol % of 1-M POSS, (b) 50 mol % of TSM.

The ^{13}C NMR spectra confirm the appearance of new resonance lines at ca. 177.3 ppm that can be attributed to the carbonyl moiety (C=O) in the newly formed copolymer and this region is shifted towards higher values of ppm when compared to the LM monomer and 1M-POSS (166.7 ppm) (Figure 4a). In addition, there are also slight changes of signals in the region of 64-66 ppm originating from -OCH$_2$- groups and a new signal appears at 65.16 ppm, which can be attributed to the copolymer, while the LM and 1M-POSS monomers show 64.6 and 66.4 ppm resonance lines, respectively. An analogous comparison was made for LM, TSM, and poly-LM/(TSM 50) (Figure 4b).

There is a similarity with regard to the new resonance lines at about 177.49 ppm derived from carbonyl groups (C=O), as well as the region corresponding to the -OCH$_2$- moiety, i.e., the presence of new peaks at 67.75 and 65.53 ppm from the poly-LM/(TSM 50) copolymer together with residual signals at 67.15 and 65.11 ppm assigned to the LM and TSM monomers.

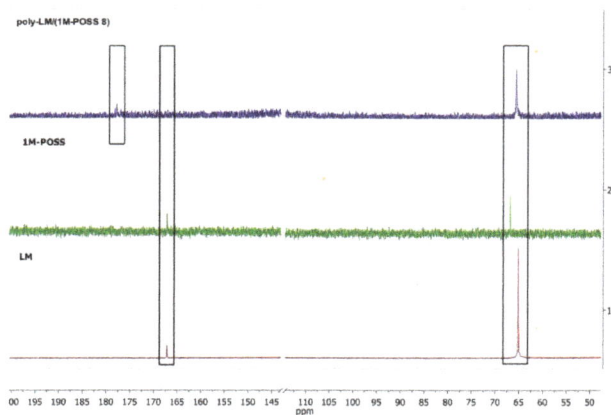

(a)

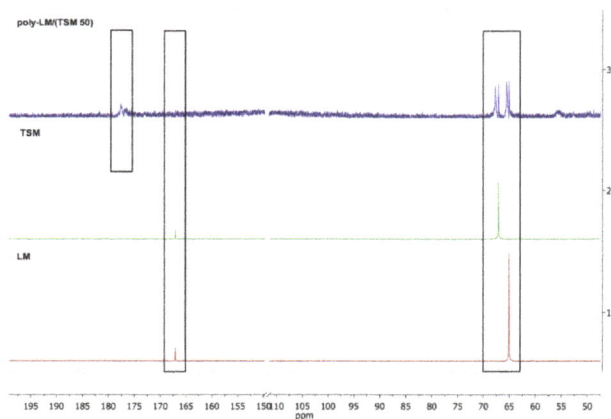

(b)

Figure 4. ^{13}C NMR stacked spectra of the LM monomer, the modifiers, and the copolymers containing (a) 8 mol % of 1-M POSS, (b) 50 mol % of TSM.

The ^{1}H NMR spectra of the copolymers reveal the presence of chemical shifts (as residual resonance lines) of the methylene protons at 6.11 and 5.21 ppm derived from the double bonds of the monomers.

From the spectra of the (co)polymers with different modifier contents, the conversions of the monomers were calculated by an evaluation of the integration ratio of methylene protons with the rest of the protons' contribution before and the after (co)polymerization. The results are given in Table S1. They are discussed in Section 3.4.1.

3.3. Photopolymerization Kinetics

3.3.1. DSC Study

Polymerization traces of neat LM and LM/1M-POSS mixtures, expressed as the dependence of the polymerization rate, R_p, on the irradiation time, t, and on the degree of the double bond conversion, p, are shown in Figure 5. The corresponding curves for the LM/TSM formulations are presented in Figure 6.

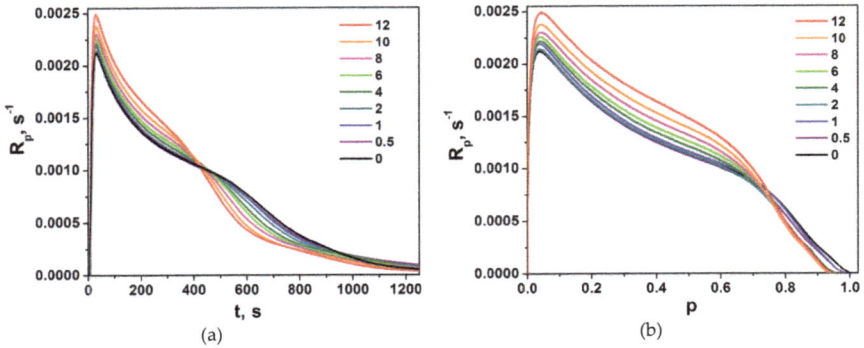

Figure 5. Polymerization rate, R_p, of LM/1M-POSS compositions as a function of (**a**) the irradiation time, t, and (**b**) double bond conversion, p, at 40 °C. The numbers indicate the 1M-POSS content (mol %) in the mixture.

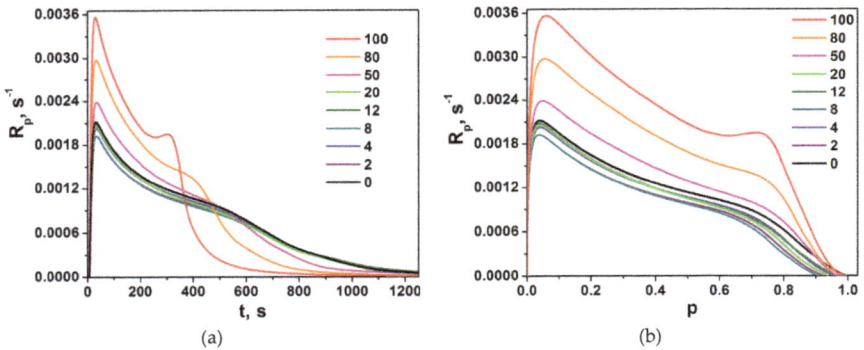

Figure 6. Polymerization rate, R_p, of LM/TSM compositions as a function of (**a**) the irradiation time, t, and (**b**) double bond conversion, p, at 40 °C. The numbers indicate the TSM content (mol %) in the mixture.

The homopolymerization of neat LM (as other methacrylates with a longer ester group) occurs, with a poorly marked gel effect appearing at higher conversions [25]). Consequently, the polymerization rate at the maximum of this effect is much lower than at the beginning of the polymerization, in contrast to short-chain methacrylates, such as methyl methacrylate MMA (e.g., [26]), butyl methacrylate, or 2-hydroxyethyl methacrylate [7]. Therefore, in our case, the maximum polymerization rate, R_p^{max}, is defined as the highest reaction rate, not the rate at the maximum of the gel effect.

TSM homopolymerizes faster than LM, and the gel effect is more visible (its maximum occurs at about $p \sim 0.7 \div 0.8$ (Figure 6). As mentioned above, homopolymerization of 1M-POSS in bulk could not be followed because this compound is solid at 40 °C.

Figures 5 and 6 suggest that the introduction of both modifiers into the base monomer increases the polymerization rate. Usually, the first explanation that comes to mind (and is most often right) is the reduction of the termination rate. This can result both from the increase in the initial viscosity of the formulation, as well as from the expected effect of the bulky substituent affecting the diffusion. These both factors can influence the polymerization of 1M-POSS-containing compositions, but the effect of TSM must result only from the bulkiness of the substituent (no influence of the formulation viscosity). It was speculated, for example, that much faster polymerization of 1-adamantyl methacrylate (AdMA) than other methacrylates (MMA, tert-butyl methacrylate, cyclohexyl methacrylate) is caused by the reduction of k_t (not determined in the paper cited) due to the steric effect of the substituents, i.e., the mobility of the segment around the radical center [27]. In this case, the adamantyl group acts as an anchor and sterically blocks termination, which results in the increased concentration of propagating radicals.

The second possibility of the acceleration by the modifiers is an increase in the propagation rate coefficient, k_p, and this will be discussed in Section 3.3.3. Such a possibility was rather excluded in the case of AdMA [27].

A more detailed analysis of Figures 5 and 6 indicates, however, that the increase in the polymerization rate in the presence of the modifiers does not occur in the whole range of their concentrations. Figure 7a presents R_p^{max} values as functions of the content of the Si-containing monomers. It is clearly visible that the addition of their small amounts slows down the reaction. This effect is well observable in the case of LM/TSM compositions; the lowest R_p^{max} value appears at a TSM concentration about 8 mol %; at higher concentrations, the polymerization rate begins to increase. For the LM/1M-POSS system, the clear minimum is not visible due to the scattering of the data and only a kind of a plateau up to 2 mol % of 1M-POSS content can be observed (the minimum appears at 1 mol % of the 1M-POSS content as can be found from rheological studies, see Section 3.3.2). This tendency is reflected also by the behavior of the final conversion of double bonds, p^f, Figure 7b (the determination of p^f is always subject to a greater error, hence the considerable scatter of points). Further discussion of these phenomena first requires consideration of the behavior of the gel effect.

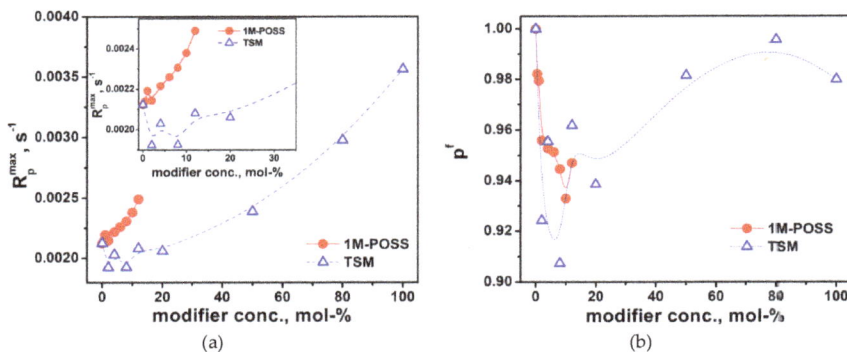

Figure 7. The dependence of (**a**) the maximum polymerization rate, R_p^{max}, and (**b**) the final conversion of double bonds, p^f, at 40 °C on the modifier content. The lines are eye guides.

As mentioned above, the gel effect in LM-based formulations is poorly marked. We can observe it better on the plots showing the dependence of the concentration of radicals on the conversion degree. The polymerization rate is expressed by Equation (4):

$$R_p = k_p[M][M^\cdot] = k_p[M^\cdot](1-p) \qquad (4)$$

where $[M^\cdot]$ is the concentration of radicals, and k_p is a constant up to high conversions. Therefore, the corresponding changes in the radical concentrations during the polymerization can be expressed readily by the function $R_p/(1-p) = f(p)$ (Figure 8). As could be expected, for the LM/1M-POSS system, the beginning of the gel effect (the beginning of the increase in the $R_p/(1-p)$ value) as well as the maximum of the effect shift to lower conversions with an increasing modifier concentration, which can result to a high degree from the increase in the viscosity of the formulations.

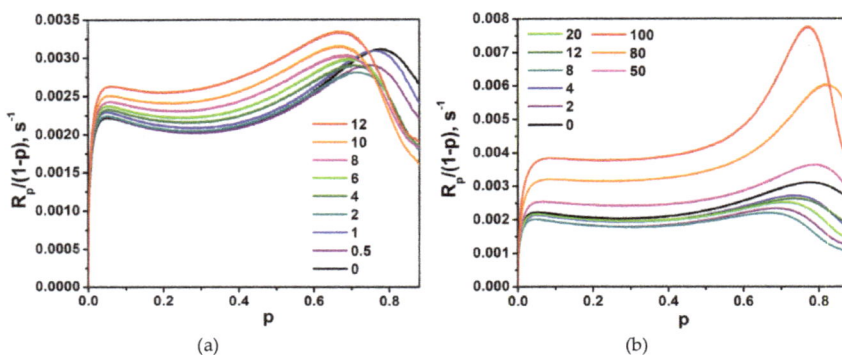

Figure 8. $R_p/(1-p)$ as a function of the double bond conversion, p, at 40 °C for (**a**) LM/1M-POSS and (**b**) LM/TSM systems. The numbers indicate the modifier content (mol %) in the composition.

However, the addition of small amounts of 1M-POSS (<~4 mol %) at first weakens the gel effect (a decrease in the value of the $R_p/(1-p)$ function); the enhancement is observed above this concentration. An analogous phenomenon, even more pronounced, occurs in the case of the LM/TSM system. The weakening of the gel effect, which in this case is also associated with the decrease in the polymerization rate and decrease in the radical concentration in the steady-state region, occurs for compositions containing up to at least 12 mol %; above this concentration, an enhancement of the gel effect is observed.

The results presented show that the effect of adding a comonomer containing an anchor group to the base monomer is twofold: At low comonomer concentrations, it leads to a decrease in the reaction rate, but after exceeding a certain concentration threshold, the trend is reversed and the polymerization rate begins to increase, reaching, at higher modifier contents, significantly higher values compared to the polymerization of the base monomer.

The weakening of the gel effect indicates an enhanced mobility of the radical centers (both by segmental and translational diffusion of macroradicals), which results in the acceleration of termination and leads to the drop of R_p. In the case of TSM-containing formulations, this slowing down effect is stronger than in the case of 1M-POSS probably due to practically no change in the initial viscosity of the formulations. For the LM/1M-POSS system, the enhancement of macroradical mobility by the introduction of 1M-POSS moieties overcomes the effect of the increased viscosity, which slows down the translational diffusion and accelerates the reaction.

It has been suggested that introduction of methacryloxypropyl-heptaisobutyl POSS into the polymer chain can increase molecular mobility by increasing the free volume fraction through the bulky substituent and the plasticizing effect exerted by isobutyl groups [19,28]. We can assume that analogous effects are responsible for the kinetic behavior of the formulations containing small amounts of both modifiers. On the other hand, it is often indicated that interchain interactions between the massive inorganic POSS cages ("anchor objects") [19] incorporated as pendant groups are responsible for the retardation and slow down of molecular motions [18,29]. This effect, which suppresses the termination, may in turn be one of the reasons for the acceleration of the polymerization in the presence of a larger amount of the modifiers. After exceeding a threshold concentration (about 2–4 mol %), the "anchor effect" begins to overcome the previously dominant effect of the increasing chain mobility. Therefore, minima appear on the dependencies, $R_p^{max} = f(modifier\ conc.)$ and $p^f = f(modifier\ conc.)$. The threshold concentration is higher for TSM due to the smaller "anchor group" and lower viscosity. However, it should be remembered that the anchor effect can only be one of the reasons for the increase in the polymerization rate.

An additional complicating, but important parameter is the reactivity ratio of the two comonomers. From the data provided by hybrid plastics [30,31], in a mixture of methyl methacrylate (MMA) and monomethacryl-based POSS, the POSS reactivity ratio, r_{POSS}, is below 1 while the reactivity ratio of MMA, r_{MMA}, is above 1. In the case of methacryloxypropyl-heptaisobutyl POSS, the reactivity ratios for POSS and MMA are 0.584 and 1.607, respectively. This means that MMA is preferentially incorporated into the copolymer and the amount of POSS introduced increases as the reaction progresses. The difference in the reactivity of the two comonomers can be explained by the significant difference in their molecular size and in the steric hindrance they exert [32]. Very recently, a higher reactivity of MMA and butyl methacrylate in copolymerization with methacryloxypropyl-heptaisobutyl POSS was also reported [33]. Interestingly, the opposite situation was indicated for the monomer pair, AdMA-MMA, with reactivity ratios of $r_{AdMA} = 1.25$ and $r_{MMA} = 0.79$ [27].

However, in the case of LM/1M-POSS and LM/TSM pairs, the situation may be different. Re-inspection of Figures 6 and 8 indicates that the course of the kinetic curves is very similar in the steady state region for all formulations (the kinetic curves are almost parallel) and the possible compositional drift does not seem important. Therefore, based on the behavior of the kinetic curves, we can conclude that the dominant effect of the monomers with bulky substituents on the termination kinetics is rather the anchor effect.

3.3.2. Photorheological Study

The second technique used to follow the polymerization kinetics was photorheology. The curing of the investigated formulations was monitored, with the rheometer registering the storage and loss modulus, G' and G'', respectively, as a function of the irradiation time at 40 °C (Figures S1 and S2). Because the polymers and copolymers are linear and their physical state can be described as viscous liquids [34], the moduli do not intersect (intersection corresponds to the gel point). Discussion of the photopolymerization kinetics is usually based on the evolution of the storage modulus [35–37] and the final values, G'_f, of the storage modulus, G' (at the plateau region), are a measure of the final conversion [37]. For our systems, changes in G'_f occurring as the modifiers' content increases reflect both the effect of changes in the double bond conversions (chemical effect) and to a much greater extent the effect of the increasing amounts of pendant anchor groups (physical effect).

The dependence of G'_f on the modifier concentration is shown in Figure 9. The final values of the modulus are very low, which results from the physical state of the copolymers. These values are reduced at low concentrations of the modifiers (compared to poly-LM), which confirms their plasticizing effect, but increase drastically at higher concentrations mainly due to the stiffering effect of the anchor substituents. An analogous effect is shown by the final values of the complex viscosity, η^*_f (Figure S3a). The changes in complex viscosity during the polymerization are shown in Figure S3b–d; the general picture of the obtained dependences is analogous to the changes in moduli. The modulus of neat poly-TSM is about 500 times higher than that of poly-LM.

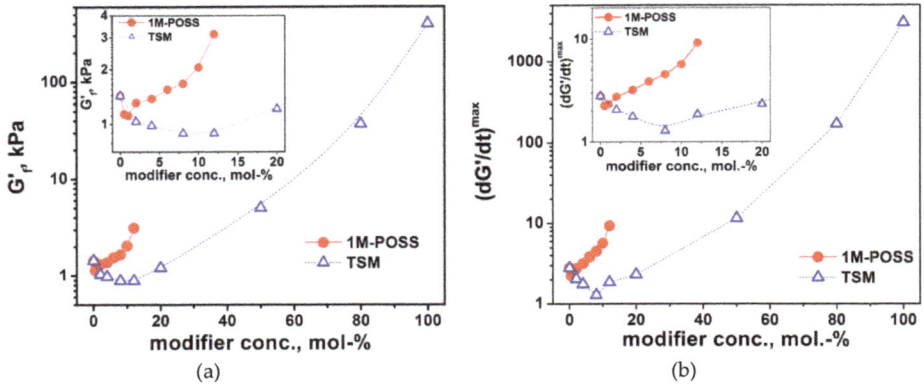

Figure 9. (**a**) Final storage modulus, G_f', and (**b**) the maximum value of G' derivative $(dG'/dt)^{max}$ as a function of the modifier content. Reaction temperature is 40 °C; the lines are eye guides.

However, the derivative of the modulus with respect to time, dG'/dt, corresponds to the polymerization rate, and its changes over time reflect changes in the reaction rate. Thus, the maximum value of the derivative, $(dG'/dt)^{max}$, corresponds to the maximum polymerization rate (at the inflection point of the curves in Figure S2). Figure 9b presents the $(dG'/dt)^{max}$ dependence on the 1M-POSS and TSM contents. The plot of the function, $(dG'/dt)^{max} = f$ *(modifier conc.)*, exactly corresponds to the behavior of the R_p^{max} values determined by the DSC method (compared with Figure 7a). Due to the greater accuracy of the photorheological measurement, the minima on the plots are clearly visible at 1 and 8 wt % of 1M-POSS and TSM, respectively.

3.3.3. Polymerization Rate Coefficients

Any change in the polymerization rate reflects changes in the polymerization rate coefficients. As mentioned earlier, the accelerating effect of the additives may be caused by a suppression of termination (reduction of k_t^b) and/or acceleration of the propagation (increase in k_p), compared to Equation (1). The discussion presented above suggested the conclusion that the accelerating effect of 1M-POSS and TSM is associated with the anchor effect that inhibits termination. To find out whether our modifiers affect only the termination or also the propagation, the rate coefficients (in the form of $k_p \times F$ and $k_t^b \times F$) were calculated for the neat monomers (LM and TSM) and for compositions containing 10 mol % of 1M-POSS and 80 mol % of TSM. The compositions were selected to obtain large differences in the polymerization rates with respect to the neat LM. It is noteworthy that a similar increase in R_p^{max}, as with the addition of 10 mol % 1M-POSS, requires the addition of up to 50 mol % TSM.

Parameters related to the rate coefficients of propagation, $(k_p \times F)$, and termination, $(k_t^b \times F)$, as functions of the double bond conversion, p, are shown in Figure 10. These parameters were only to describe the tendencies of changes in the actual k_p and k_t^b coefficients after adding modifiers with bulky substituents.

For the mixtures of the monomers, the calculated values were the resultants of the homopolymerization and copolymerization rate coefficients of the individual comonomers.

The $k_p \times F$ and $k_t^b \times F$ parameters show a classical dependence on double bond conversion [20]. The $k_p \times F$ values practically do not change to high conversions (propagation is not diffusion controlled); their slight increase at medium conversions (in the region of the gel effect) results from the deficiency of the model used for calculations. Regarding the $k_t^b \times F$ parameter, at the steady state region (termination controlled by segmental diffusion), its values show a plateau up to conversions of about $0.3 \div 0.4$. When the gel effect begins (termination becomes controlled by translational diffusion), $k_t^b \times F$ starts to decrease with the increase in conversion. The k_t^b behavior in different reaction stages is better visible on the $k_t^b/k_p = f(p)$ dependence (Figure S4). Because the gel effect occurs at relatively high conversions, the reaction stage at which the termination is reaction-diffusion controlled is practically not observed; a second plateau corresponding to this stage is visible only at the highest conversions on the two curves in Figure S4 (from $p \sim 0.7 \div 0.8$). The practical lack of the reaction diffusion controlled termination region is probably associated with the fact that poly-LM is a highly viscous liquid even at room temperature.

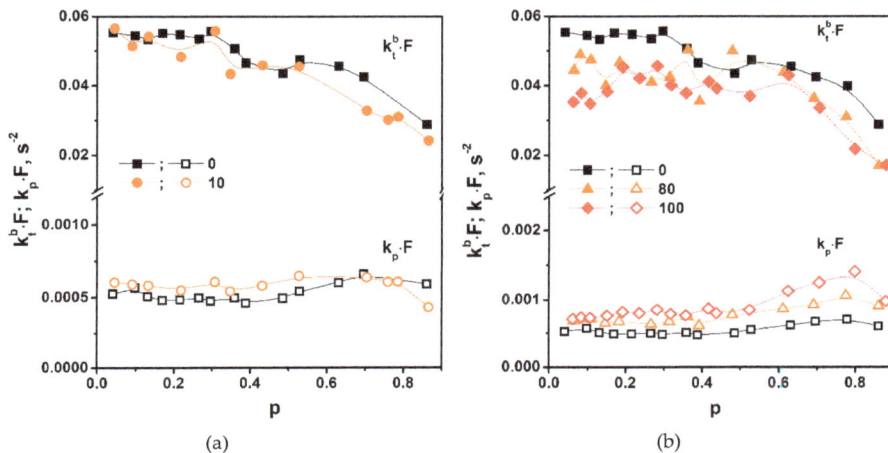

Figure 10. Parameters related to propagation ($k_p \cdot F$) and termination ($k_t^b \cdot F$) rate coefficients as a function of the double bond conversion (**a**) LM/1M-POSS system and (**b**) the LM/TSM system. The numbers indicate the modifier content (mol %) in the mixture. Polymerization temperature: 40 °C. The lines are eye guides.

Concerning the effect of the modifiers, it is clear that—as expected—they reduce the k_t^b coefficient. The reduction is moderate—for the LM/(1M-POSS 12 mol %) formulation by about 3.6% and in the case of LM/(TSM 80 mol %) by about 14%. The unexpected result is that k_p of neat TSM is about 1.7 times higher than k_p of LM. This clearly indicates that TSM reactivity is higher than that of LM. We can expect a similar result for the LM/1M-POSS pair. The $k_p \times F$ parameters in the two copolymerization processes are higher by 43% for TSM and by 26% for 1M-POSS compared to LM. Because the polymerization rate is inversely proportional to the square root of k_t^b and directly proportional to k_p, the obtained results lead to the important conclusion that the accelerating effect of the modifiers (at concentrations higher than a few mol %) on LM polymerization results, to a greater extent, from the increase in the propagation rate rather than from suppression of termination.

An explanation of this result is difficult. Both anchor groups cause high steric hindrance, which should impede the access of the reactive group to the growing macroradical. In the published literature, the k_t^b coefficient of branched tert-butyl methacrylate (tert-BMA) has been reported to be lower than that of linear n-butyl methacrylate (n-BMA), as attributed to the effects of chain mobility hindered to a greater extent in tert-BMA than in n-BMA polymerization [38]. On the other hand, the k_p values of methacrylates with cyclic ester groups (e.g., benzyl, cyclohexyl or isobornyl methacrylate) are clearly higher than k_p of LM [39]. This is consistent with the results of the homopolymerization of the three monomers. It seems also that in our case the cross-propagation is not slower, because the modifiers enhance the polymerization rate. We can speculate that the apparently higher reactivity of the monomers with "anchor" groups results from the tendency of these groups to associate [18], which increases the monomer concentration in the close vicinity of the radical center. It is possible that in the case of the LM/modifier mixture, the physical interactions between the monomers lead to a similar effect, which may be another manifestation of the anchor effect (and its other interpretation)—anchoring the monomer with a long substituent around the POSS cage.

3.4. Copolymer Characterization

3.4.1. GPC Analysis

The results of GPC analysis (number average, M_n, and weight average, M_w, molecular weights, polydyspersity index, PDI, and fraction contents) of selected copolymerization products containing 4 and 8 mol % of 1M-POSS or 4, 8, and 50 mol % of TSM are given in Table S2. The dependence of the copolymer molecular weight (MW) and polydispersity index, PDI, on the modifier content is shown in Figure 11a. Interestingly, the plots show minima, analogously to the results obtained from the kinetic measurements. This seems to confirm the previous assumption that at low concentrations of the modifiers, the termination increases, while at higher contents, the anchor effect promotes chain elongation.

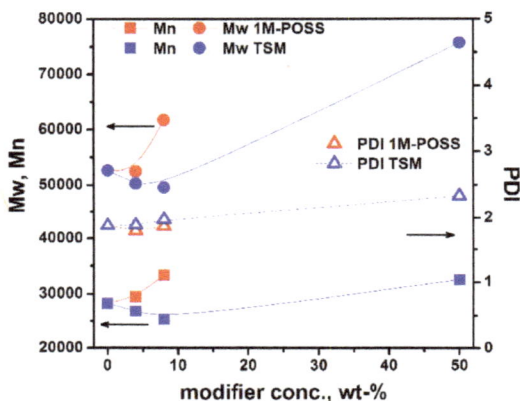

(a)

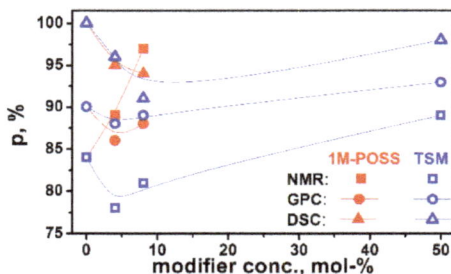

(b)

Figure 11. The dependence of (**a**) the copolymer molecular weight (determined by GPC) and (**b**) the copolymer yield (determined by various methods) on the modifier content. The lines are eye guides.

GPC analysis also enables the determination of the copolymer and monomer/oligomer residues' contents. The results of the copolymer yield as a dependence of the modifier content are given in Figure 11b. The results are compared with the results obtained by other methods (DSC, NMR).

Again, the plots show the same feature: The existence of a minimum at low amounts of the introduced modifier. The difference in the p^f values obtained from various methods results from the differences between the methods and parameters they measure directly. The decrease in the conversion with increasing modifier content is associated with the enhanced termination, while the increase in the conversion is with suppression of termination (and/or enhancement of propagation), which allows more monomer molecules to react.

3.4.2. TEM Analysis

Figure 12 presents the TEM micrographs of copolymers containing 12 and 0.5 mol % of 1M-POSS.

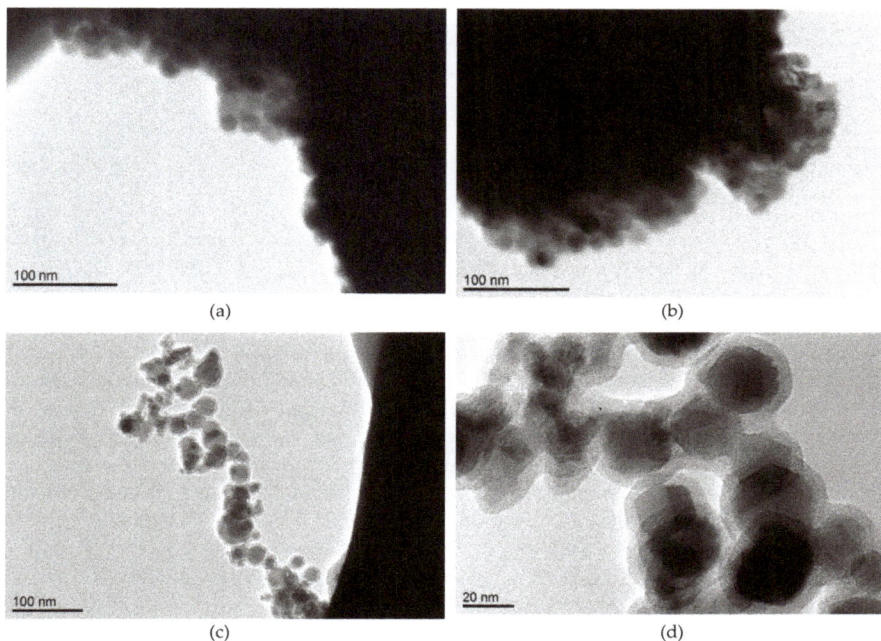

Figure 12. TEM images of (**a**) and (**b**) poly-LM/(1M-POSS 12); (**c**) and (**d**) poly-LM/(1M-POSS 0.5).

TEM images presented in Figure 12a–b show the existence of spherical copolymer structures, which were probably formed by wrapping polymer chains around the aggregates of POSS cages. This can be better observed in Figure 12c–d. Inside spherical copolymer structures, the aggregates of POSS cages are clearly visible, forming, in many cases, hexagonal structures (octaalkyl-substituted cubic silsesquioxane nanoparticles form hexagonal, or equivalently rhombohedral, crystal structures [40]). This means that during the polymerization part of the pendant, POSS cages tend to form nanocrystals (on the order of 20 nm), and the polymer chains wrap around them. Due to the limited solubility of 1M-POSS in LM, such aggregation and wrapping (also of the monomer) probably begins at the early polymerization stages, which can positively affect the propagation rate.

3.5. Glass Transition Temperature, T_g

The above-described effect of bulky substituents on the mobility of polymer chains should also be reflected in the values of glass temperature. Figure 13 shows T_g plots against the modifier content. The minimum on the curve for the LM/TSM system appears again, whereas for the LM/1M-POSS system, only a drop in T_g values is observed; however, we cannot rule out that with higher POSS contents, there would be an increase in T_g.

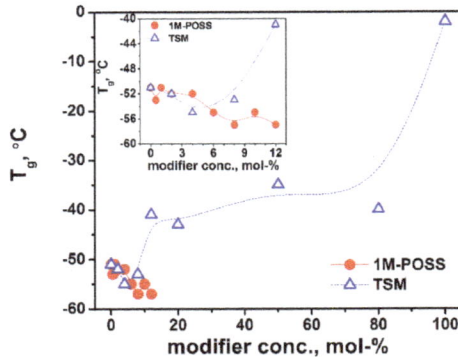

Figure 13. Glass transition temperature, T_g, of LM/1M-POSS and LM/TSM copolymers as a function of the modifier content. The lines are eye guides.

It was indicated that POSS copolymerized with a base monomer can induce modifications to cooperative molecular motions typical of the glass transition, leading to retardation or acceleration of the dynamics. Therefore, the change in the glass transition temperature relative to the base polymer can be negative or positive and it is not correlated with the molecular weight [41]. In the case of MMA/1M-POSS copolymers (POSS concentrations up to 45 wt %), the significant negative deviations from T_g of neat poly-MMA were observed in the entire range of comonomers ratios (as in our case). A similar effect was reported also for the copolymer of styrene and methacryloxystyrene-POSS functionalized with isobutyl groups [42]. Because the variation of the glass transition temperature in the random copolymers is the net result of several effects, free volume fraction, steric barrier, and POSS-polymeric segment interactions, the authors concluded that in their case, POSS-segment interactions are dominated by the internal plasticization (by isobutyl groups) and local free volume addition. Similarly, we can expect that in our case, the steric hindrance of the POSS cages that causes difficulties in the physical movements of the polymer chain is overcome by the flexibility of isobutyl groups. A monotonic decrease of the glass-transition temperature occurred also in the case of MMA copolymers with increasing amounts of methacrylcyclohexyl POSS and this was attributed to an increased free volume [43].

For LM/TSM copolymers, we observed both a negative and a positive change in T_g, which confirms the effect of the branched siloxy-silane group on the mobility of the polymer chains, as discussed above (Section 3.3.1). This influence is analogous to that observed for the storage modulus and polymerization rate. A similar effect (decrease and increase in T_g) for copolymers of styrene and methacryloxystyrene-POSS containing cyclohexyl vertex groups was described in [42]. In this case, intermolecular POSS-PS segment interactions were important for the T_g values of the copolymers, with competition between the free volume and intermolecular interactions. We can assume a similar explanation for the behavior of our LM/TSM copolymers.

4. Conclusions

The effect of the anchoring group on the detailed polymerization kinetics was investigated by DSC and photorheology using photochemical initiation. The monomer with the bulky substituent was monomethacryloxy POSS (1M-POSS); for comparison, a methacrylate with a branched siloxy-silane group (TSM) was also applied. Both monomers were copolymerized with LM at various molar ratios.

The addition of the comonomers containing "anchor objects" had a dual effect on termination in LM/modifier copolymerization:

An increase of the molecular mobility by increasing the free volume fraction through the bulky substituent (and the plasticizing effect exerted by isobutyl groups in 1M-POSS), which results in an enhanced mobility of radical centers, leads to an acceleration of termination, a weakening of the gel effect, and a slowing down of the polymerization; these effects dominated at low concentrations of the modifiers.

Retardation and slowdown of molecular motions due to the "anchor effect" suppress the termination, leading to the acceleration of the polymerization; these effects dominated at higher modifier contents.

As an effect, the dependencies of R_p^{max}, p^f (determined by NMR, GPC, and DSC), MW, and PDI on the modifier content showed the minima. The threshold concentration was higher for TSM due to the smaller "anchor group" and the lower viscosity of the TSM-containing formulations.

Determination of the propagation, k_p, and termination, k_t^b, rate coefficients led to the conclusion that the accelerating effect of the modifiers on LM polymerization results both from the suppression of termination as well as from the increase in propagation The increase in k_p led to an additional interpretation of the anchor effect: Anchoring the monomer with a long substituent (LM) around the POSS cage; this additionally enhanced the propagation by a local increase in the concentration of double bonds.

TEM images suggest that during polymerization, some of the pendant POSS cages formed nanocrystals (on the order of 20 nm), and polymer chains wrap around them. This could partially confirm the above supposition.

The T_g behavior of TSM-containing copolymers was similar to the behavior of the polymerization rate (a minimum in the plot of T_g vs. the modifier content), resulting from the influence of TSM on polymer chain mobility. In the case of 1M-POSS, only a plasticizing effect was observed.

In conclusion, we can say that the anchoring effect in the polymerization kinetics (for the monomers used in this work) affects the termination and propagation steps. The direction of this influence on termination depends on the concentration of the monomer containing the anchor object.

Supplementary Materials: The following are available online at http://www.mdpi.com/2073-4360/11/3/515/s1, Figure S1: (a) Storage modulus G' and (b) loss modulus G'' as functions of irradiation time t at 40 °C for LM/1M-POSS system; Figure S2 (a, b) Storage modulus G' and (c, d) loss modulus G'' as functions of irradiation time t at 40 °C for LM/TSM system; Figure S3: Final complex viscosity η^*f as a function of the modifier content (a) and complex viscosity as a function of the reaction time for (b) LM/1M-POSSsystem, (c) and (d) LM/TSM system; Figure S4: Dependence of the k_t^b/k_p ratio on double bond conversion: (a) LM/1M-POSS system and (b) LM/TSM system; Table S1: Total monomer conversion determined from 1H NMR spectra and Table S2: Average molecular weights, PDI and fractions contents of the base polymer and copolymers.

Author Contributions: Conception of the work: A.M., D.P. and E.A., POSS synthesis: B.D., investigation: A.M. and D.P., NMR and GPC measurements: B.D., TEM measurements: K.S., data analysis and interpretation: A.M., D.P. and E.A., Writing—Original Draft Preparation: A.M. and E.A.; Writing—Review & Editing: A.M. and E.A.

Acknowledgments: This work was supported by the Polish Ministry of Science and Higher Education and the European Regional Development Fund under the Innovative Economy Operational Programme for 2007-2013, Priority 1, Action 1.3. (Project No.UDA-POIG.01.03.01-30-173/09 Nanosil).

Conflicts of Interest: The authors declare no conflict of interest.

References

1. Wang, H.; Zhong, W. PMMA-, PAN-, and Acrylic-Based Polymer Nanocomposites. In *Recent Advances in Polymer Nano-composites*; Thomas, S., Zaikov, G.E., Valsaraj, S.V., Eds.; CRC Press: Leiden, The Netherlands; Boston, MA, USA, 2009; pp. 155–181.
2. Zhou, H.; Ye, Q.; Xu, J. Polyhedral oligomeric silsesquioxane-based hybrid materials and their applications. *Mater. Chem. Front.* **2017**, *1*, 212–230. [CrossRef]
3. Michałowski, S.; Hebda, E.; Pielichowski, K. Thermal stability and flammability of polyurethane foams chemically reinforced with POSS. *J. Therm. Anal. Calorim.* **2017**, *130*, 155–163. [CrossRef]
4. Blanco, I. The rediscovery of POSS: A molecule rather than a filler. *Polymers* **2018**, *10*, 904. [CrossRef]
5. Hartmann-Thompson, C. (Ed.) *Applications of Polyhedral Oligomeric Silsesquioxanes*; Springer New York, NY, USA, 2011.

6. Pielichowski, K.; Njuguna, J.; Janowski, B.; Pielichowski, J. Polyhedral Oligomeric Silsesquioxanes (POSS)-Containing Nanohybrid Polymers. *Adv. Polym. Sci.* **2006**, *201*, 225–296. [CrossRef]

7. Prządka, D.; Andrzejewska, E.; Marcinkowska, A. Multimethacryloxy-POSS as a crosslinker for hydrogel materials. *Eur. Polym. J.* **2015**, *72*, 34–49. [CrossRef]

8. Andrzejewska, E. Photoinitiated polymerization in ionic liquids and its application. *Polym. Int.* **2017**, *66*, 366–381. [CrossRef]

9. Wang, W.; Sun, X.; Huang, L.; Gao, Y.; Ban, J.; Shen, L.; Chen, J. Structure-property relationships in hybrid dental nanocomposite resins containing monofunctional and multifunctional polyhedral oligomeric silsesquioxanes. *Int. J. Nanomed.* **2014**, *9*, 841–852. [CrossRef]

10. Amerio, E.; Sangermano, M.; Colucci, G.; Malucelli, G.; Messori, M.; Taurino, R.; Fabbri, P. UV curing of organic-inorganic hybrid coatings containing polyhedral oligomeric silsesquioxane blocks. *Macromol. Mater. Eng.* **2008**, *293*, 700–707. [CrossRef]

11. Fong, H.; Dickens, S.H.; Flaim, G.M. Evaluation of dental restorative composites containing polyhedral oligomeric silsesquioxane methacrylate. *Dent. Mater.* **2005**, *21*, 520–529. [CrossRef]

12. Wang, Y.; Liu, F.; Xue, X. Synthesis and characterization of UV-cured epoxy acrylate/POSS nanocomposites. *Prog. Org. Coat.* **2013**, *76*, 863–869. [CrossRef]

13. Florea, N.M.; Damian, C.M.; Ionescu, C.; Lungu, A.; Vasile, E.; Iovu, H. Designing of polyhedral oligomeric silsesquioxane (POSS)-based dithiol/dimethacrylate nano-hybrids. *Polym. Bull.* **2018**, *75*, 3897–3916. [CrossRef]

14. Lungu, A.; Florea, N.M.; Iovu, H. Dimethacrylic/epoxy interpenetrating polymer networks including octafunctional POSS. *Polymer* **2012**, *53*, 300–307. [CrossRef]

15. Wang, Y.; Liu, F.; Xue, X. Morphology and properties of UV-curing epoxy acrylate coatings modified with methacryl-POSS. *Prog. Org. Coat.* **2015**, *78*, 404–410. [CrossRef]

16. Norouzi, S.; Mohseni, M.; Yahyaei, H. Preparation and characterization of an acrylic acid modified polyhedral oligomeric silsesquioxane and investigating its effect in a UV curable coating. *Prog. Org. Coat.* **2016**, *99*, 1–10. [CrossRef]

17. Matejka, L. Epoxy-silica/silsesquioxane nanocomposites. In *Hybrid Nanocomposites for Nanotechnology. Electronic, Optical, Magnetic and Biomedical Applications*; Merhari, L., Ed.; Springer Science & Business Media LLC.: New Delhi, India, 2009.

18. Romo–Uribe, A.; Mather, P.T.; Haddad, T.S.; Lichtenhan, J.D. Viscoelastic and morphological behavior of hybridstyryl-based polyhedral oligomeric silsesquioxane (POSS) copolymers. *J. Polym. Sci. Part Polym. Phys.* **1998**, *36*, 1857–1872. [CrossRef]

19. Bizet, S.; Galy, J.; Gerard, J.F. Structure-Property Relationships in Organic-Inorganic Nanomaterials Based on Methacryl-POSS and Dimethacrylate Networks. *Macromolecules* **2006**, *39*, 2574–2583. [CrossRef]

20. Andrzejewska, E. Photopolymerization kinetics of multifunctional monomers. *Prog. Polym. Sci.* **2001**, *26*, 605–665. [CrossRef]

21. Walczak, M.; Januszewski, R.; Franczyk, A.; Marciniec, B. Synthesis of monofunctionalized POSS through hydrosilylation. *J. Organomet. Chem.* **2018**, *872*, 73–78. [CrossRef]

22. Saito, H. Process for Production of Powder of Cage Silsesquioxane Compound. US Patent 2010/0081837 A1, 2010.

23. Odian, G. *Principles of Polymerization*, 4th ed.; Wiley-Interscience, John Wiley & Sons, Inc.: Hoboken, NY, USA, 2004.

24. Andrzejewska, E.; Podgórska-Golubska, M.; Stępniak, I.; Andrzejewski, M. Photoinitiated polymerization in ionic liquids: Kinetics and viscosity effects. *Polymer* **2009**, *50*, 2040–2047. [CrossRef]

25. Lekishvili, N.G.; Nadareishvili, N.I.; Khananashvili, Z.M.; Zaikov, G.E. Polymers for fiber optics. In *Chemical and Physical Properties of Polymers*; Zaikov, G.E., Kozlowski, R., Eds.; Nova Science Publishers, Inc.: New York, NY, USA, 2005.

26. Jaso, V.; Stoiljkovic, D.; Radicevic, R.; Bera, O. Kinetic modeling of bulk free-radical polymerization of methyl methacrylate. *Polym. J.* **2013**, *45*, 631–636. [CrossRef]

27. Matsumoto, A.; Tanaka, S.; Otsu, T. Synthesis and Characterization of Poly (l-adamantyl methacrylate): Effects of the Adamantyl Group on Radical Polymerization Kinetics and Thermal Properties of the Polymer. *Macromolecules* **1991**, *24*, 4017–4024. [CrossRef]

28. Wu, J.; Haddad, T.S.; Kim, G.-M.; Mather, P.T. Rheological Behavior of Entangled Polystyrene-Polyhedral Oligosilsesquioxane (POSS) Copolymers. *Macromolecules* **2007**, *40*, 544–554. [CrossRef]

29. Kuo, S.-W.; Chang, F.C. POSS related polymer nanocomposites. *Prog. Polym. Sci.* **2011**, *36*, 1649–1696. [CrossRef]
30. Bizet, S.; Galy, J.; Gerard, J.-F. Molecular dynamics simulation of organic-inorganic copolymers based on methacryl-POSS and methyl methacrylate. *Polymer* **2006**, *47*, 8219–8227. [CrossRef]
31. Zhang, H.-X.; Lee, H.-Y.; Shin, Y.-J.; Lee, D.-H. Preparation and properties of mma/1-propylmethacrylate-POSS copolymer with atom transfer radical polymerization. *Chin. J. Polym. Sci.* **2008**, *26*, 533–537. [CrossRef]
32. Molina, D.; Levi, M.; Turri, S.; Penso, M. Self-Assembly of Methacrylic Nanostructured Copolymers Containing Polyhedral Oligomeric Silsesquioxanes. *e-Polymers* **2007**, *7*, 1–13. [CrossRef]
33. Taherzadeh, H.; Ishida, Y.; Kameyama, A. Phase-separated structures of random methacrylate copolymers with pendant POSS moieties. *J. Appl. Polym. Sci.* **2018**. [CrossRef]
34. Raja, S.N.; Bekenstein, Y.; Koc, M.A.; Fischer, S.; Zhang, D.; Lin, L.; Ritchie, R.O.; Yang, P.; Alivisatos, A.P. Encapsulation of Perovskite Nanocrystals into Macroscale Polymer Matrices: Enhanced Stability and Polarization. *ACS Appl. Mater. Interfaces* **2016**, *8*, 35523–35533. [CrossRef]
35. Cook, W.D.; Chausson, S.; Chen, F.; Le Pluart, L.; Bowman, C.N.; Scott, T.F. Photopolymerization kinetics, photorheology and photoplasticity of thiol–ene–allylic sulfide networks. *Polym. Int.* **2008**, *57*, 469–478. [CrossRef]
36. Ligon, S.C.; Seidler, K.; Gorsche, C.; Griesser, M.; Moszner, N.; Liska, R. Allyl Sulfides and a-Substituted Acrylates as Addition–Fragmentation Chain Transfer Agents for Methacrylate Polymer Networks. *J. Polym. Sci. Part Polym. Chem.* **2016**, *54*, 394–406. [CrossRef]
37. Gorsche, C.; Harikrishna, R.; Baudis, S.; Knaack, P.; Husar, B.; Laeuger, J.; Hoffmann, H.; Liska, R. Real Time-NIR/MIR-Photorheology: A Versatile Tool for the in Situ Characterization of Photopolymerization Reactions. *Anal. Chem.* **2017**, *89*, 4958–4968. [CrossRef]
38. Buback, M.; Junkers, T. Termination Kinetics of tert-Butyl Methacrylate and of n-Butyl Methacrylate Free-Radical Bulk Homopolymerizations. *Macromol. Chem. Phys.* **2006**, *207*, 1640–1650. [CrossRef]
39. Beuermann, S.; Buback, M.; Davis, T.P.; Garcıa, N.; Gilbert, R.G.; Hutchinson, R.A.; Kajiwara, A.; Kamachi, M.; Lacık, I.; Russell, G.T. Critically Evaluated Rate Coefficients for Free-Radical Polymerization, 4, Propagation Rate Coefficients for Methacrylates with Cyclic Ester Groups. *Macromol. Chem. Phys.* **2003**, *204*, 1338–1350. [CrossRef]
40. Zheng, L.; Hong, S.; Cardoen, G.; Burgaz, E.; Gido, S.P.; Coughlin, E.B. Polymer Nanocomposites through Controlled Self-Assembly of Cubic Silsesquioxane Scaffolds. *Macromolecules* **2004**, *37*, 8606–8611. [CrossRef]
41. Romo-Uribe, A. Viscoelasticity and microstructure of POSS-methyl methacrylate nanocomposites. Dynamics and entanglement dilution. *Polymer* **2018**, *148*, 27–38. [CrossRef]
42. Wu, J.; Haddad, T.S.; Mather, P.T. Vertex Group Effects in Entangled Polystyrene-Polyhedral Oligosilsesquioxane (POSS) Copolymers. *Macromolecules* **2009**, *42*, 1142–1152. [CrossRef]
43. Amir, N.; Levina, A.; Silverstein, M.S. Nanocomposites Through Copolymerization of a Polyhedral Oligomeric Silsesquioxane and Methyl Methacrylate. *J. Polym. Sci. Part Polym. Chem.* **2007**, *45*, 4264–4275. [CrossRef]

Article

polymers

MDPI

Mechanically Robust Hybrid POSS Thermoplastic Polyurethanes with Enhanced Surface Hydrophobicity

Xiuhuan Song [1], Xiaoxiao Zhang [1], Tianduo Li [1], Zibiao Li [2,*] and Hong Chi [1,*]

[1]   Shandong Provincial Key Laboratory of Molecular Engineering, School of Chemistry and Pharmaceutical Engineering, Qilu University of Technology (Shandong Academy of Sciences), Jinan 250353, China; 18366102716@163.com (X.S.); zhang11071111@163.com (X.Z.); ylpt6296@vip.163.com (T.L.)
[2]   Institute of Materials Research and Engineering, A*STAR (Agency for Science, Technology and Research), 2 Fusionopolis Way, Innovis, #08-03, Singapore 138634, Singapore
*   Correspondence: lizb@imre.a-star.edu.sg (Z.L.); ch9161@gmail.com (H.C.); Tel.: +86-135-5319-7297 (H.C.)

Received: 24 January 2019; Accepted: 18 February 2019; Published: 20 February 2019

Abstract: A series of hybrid thermoplastic polyurethanes (PUs) were synthesized from bi-functional polyhedral oligomeric silsesquioxane (B-POSS) and polycaprolactone (PCL) using 1,6-hexamethylene diisocyanate (HDI) as a coupling agent for the first time. The newly synthesized hybrid materials were fully characterized in terms of structure, morphology, thermal and mechanical performance, as well as their toughening effect toward polyesters. Thermal gravimeter analysis (TGA) and differential scanning calorimetry (DSC) showed enhanced thermal stability by 76 °C higher in decomposition temperature (T_d) of the POSS PUs, and 22 °C higher glass transition temperature (T_g) when compared with control PU without POSS. Static contact angle results showed a significant increment of 49.8° and 53.4° for the respective surface hydrophobicity and lipophilicity measurements. More importantly, both storage modulus (G′) and loss modulus (G″) are improved in the hybrid POSS PUs and these parameters can be further adjusted by varying POSS content in the copolymer. As a biodegradable hybrid filler, the as-synthesized POSS PUs also demonstrated a remarkable effect in toughening commercial polyesters, indicating a simple yet useful strategy in developing high-performance polyester for advanced biomedical applications.

Keywords: bi-functional POSS; hydrophobic modification; thermoplastic polyurethane; mechanical performance

1. Introduction

Polyurethane (PU) is a kind of thermoplastic elastomeric polymer, consisting of both hard and soft segments. The soft segments are usually composed of flexible polyester or polyether and the hard segments are usually composed of diisocyanates with benzyl structure [1]. Such alternating structure endows PU with good shape recovery property as well as high resistance to abrasion, stretchability [2], high adhesiveness and eases of processability because the hard phase can form a force center which holds the soft phase and retains the original shape [3]. Therefore, PU is widely applied in many fields including coating, foaming, adhesive, tissue engineering and holds a unique importance in daily life [4]. Despite the advantages as mentioned above, PU does show several drawbacks including low mechanical strength and short shelf-life. PU exhibits hydroscopic tendencies due to the hydrophilic urethane group. The poor hydrophobicity of polyurethane products could shorten their shelf-life as they absorb water and gradually decompose and lose their performance. Efforts have been devoted to improve the mechanical strength and hydrophobicity of PU by introducing various components to obtain organic-inorganic hybrid materials. The fillers can be graphene [5,6] small molecules [7], inorganic nanoparticles [8,9] and carbon nanotubes etc. [10]. For example, H. Jerry Qi and co-workers prepared photo-curable PUs made of aliphatic urethane diacrylate. Nanoparticles, such as fumed silica (200–300 nm), were added as a rheology modifier that imparts a shear thinning effect to uncured ink. Such a composition of the ink allows good printability as well thermal stability after printing [11]. Rigoberto C Advincula and coworkers used graphene oxide (GO) to further tailor their properties of polyurethanes and found that the addition of GO could enhance the mechanical property and thermal stability of PUs [12].

In the past few decades, polyhedral oligomeric silsesquioxanes (POSS) were reported to be a rising star in hybrid materials [13,14]. POSS represents the smallest molecular silica with the dimension in the range of 1–3 nm [15], and the general chemical formula of $(RSiO_{1.5})_n$ (n= 6, 8, 12, etc.), where R could be a hydrogen atom/organic group, such as alkyl, aryl, vinyl, acrylate, epoxide group, to name a few [16]. The first merit to incorporate POSS into polymers chains is the super-hydrophobic tendency of POSS which mainly results from the surrounding R group [17–20]. The second merit is the biocompatibility and non-toxicity of POSSs [21–24]. Moreover, according to our previous results, not only could POSS enhance thermal stability [25,26], but it could also strengthen the mechanical properties [27] and retain the optical stability of the materials [28,29] due to its special composition and cage-liked nanostructure. In addition, POSS can be introduced into a polymer matrix by using covalently bonding as a pendent side group [30–33] or as end groups [34,35]. For example, Mya KY et al. [36] and Kun Wei et al. [37] reported star-shape POSS grafted PUs, which exhibited enhanced thermal stability and mechanical properties. Turri et al. [38,39] and Choudhury et al. [40] studied the surface properties of linear PUs which were modified with mono-functional POSS. It was found that the PU/POSS nanocomposites could significantly reduce the surface free energy and improve the surface hydrophobicity of the hybrids. In these ways, the properties of the polymer were less affected because the main chains of the polymers remain unchanged.

Herein, a series of hybrid PUs with POSS in the main chains were designed and synthesized. The obtained PUs showed enhanced thermal stability and mechanical property due to the existence of bi-functional POSS. Moreover, PCL was used as chain extender due to its excellent flexibility of the polymer chains [41]. Nuclear magnetic resonance (^{1}H-NMR), fourier transform infrared spectroscopy (FTIR) and gel permeation chromatography (GPC) verified the successful synthesis of PUs. Static contact angle tests showed that the respective hydrophobicity and lipophilicity were remarkably increased by 49.8° and 53.4° compared with the results obtained on neat PU. Scanning electron microscopy (SEM) showed that there are microspores on the surface of PU films. Thermogravimetric analysis (TGA) and differential scanning calorimetry (DSC) indicated 76 °C higher decomposition temperature (T_d) and 22 °C higher glass transition temperature (T_g) compared to neat PU. Rheology tests suggested that both storage modulus (G′) and loss modulus (G″) are improved in the hybrid PUs with the increasing ratio of POSS.

2. Materials and Methods

2.1. Materials

Phenyltrimethoxysilane (98%), methyldichlorosilane (97%), allyloxytrimethylsilane (98%), ε-caprolactone (99%) (ε-CL), Karstedt catalyst tin (II) 2-ethythexanoate (Sn(Oct)$_2$) (96%) were all obtained from Alfa Aesar (Shanghai, China). Isopropyl alcohol, sodium hydroxide, triethylamine, dichloromethane, hexamethylene diisocyanate (HDI), polyethylene glycol (PEG: M_w = 1500) were purchased from Aladdin Reagent (Shanghai, China). PLGA (M_w = 100,000) was purchased from Sigma-Aldrich (Shanghai, China). All other solvents were purchased from Sino-pharm Chemical Reagent Co. Ltd. (Shanghai, China). Toluene, tetrahydrofuran and methanol were distilled before use, while other chemicals were used without further purification.

2.2. Synthesis of 3,13-dihydrooctaphenyl B-POSS

3,13-dihydrooctaphenyl B-POSS was synthesized as per our previous report [29]. Phenyltrimethoxysilane (15 mL, 0.08 mol), isopropyl alcohol (80 mL), deionized water (1.7 g, 0.09 mol) and sodium hydroxide (2.13 g, 0.05 mol) were charged to a three-necked flask under N_2 atmosphere to synthesize octaphenyldicycloocatasiloxane tetrasodium silanolate $Na_4O_{14}Si_8(C_6H_5)_8$. The reaction was allowed to reflux at 96 °C for 4 h. After cooling down to room temperature, the mixture was stirred for additional 15 h. Solvent was removed via rotary evaporation and white powder was obtained after drying at 60 °C for 12 h in a vacuum oven. 3,13-dihydrooctaphenyl B-POSS was then obtained through the reaction between $Na_4O_{14}Si_8(C_6H_5)_8$ and methyldichlorosilane. Typically, a mixture of $Na_4O_{14}Si_8(C_6H_5)_8$ (11.24 g, 9.7 mmol), triethylamine (2.92 mL, 28.8 mmol) and anhydrous tetrahydrofuran (100 mL) in a flask was cooled down in an ice-water bath before the addition of methyldichlorosilane (3.385 g, 28.8 mmol) in tetrahydrofuran (10 mL). The reaction was conducted at 0 °C for one hour and 70 °C for another 3 h under dry nitrogen. White powder was collected via rotary evaporation and washed with 100 mL methanol three times. Finally, 4.57 g product with yield of 41 wt % was obtained. ^{1}H-NMR (δ ppm, CDCl$_3$): 0.38 (d, 6H, CH$_3$–Si), 4.98 (d, 2H, Si–H), 7.14–7.50 (m, 40H, protons of aromatic rings).

2.3. Synthesis of 3,13-di(trimethylsilyl)oxypropyloctaphenyl B-POSS

3,13-di(trimethylsilyl)oxypropyloctaphenyl B-POSS was synthesized as per Wei et al.'s study [37]. 3,13-di(trimethylsilyl)oxypropyloctaphenyl B-POSS was synthesized via the hydrosilylation reaction between 3,13-dihydrooctaphenyl B-POSS and allyloxytrimethylsilane. Typically, 3,13-dihydrooctaphenyl B-POSS (5.44 g, 4.7 mmol), anhydrous toluene (50 mL), allyloxytrimethylsilane (3.67 g, 28.2 mmol) and the Karstedt catalyst were charged to a flask and the reaction was carried out at 95 °C for 36 h with vigorous stirring. Then, 6.76 g product with a yield of 82 wt % was collected after removing the solvent via rotary evaporation. ^{1}H-NMR (δ ppm, CDCl$_3$): 0.00 [s, 9H, –CH$_2$CH$_2$CH$_2$OSi(CH$_3$)$_3$], 0.30 [s, 3H, –OSiCH$_3$CH$_2$CH$_2$CH$_2$OSi(CH$_3$)$_3$], 0.71 [t, 2H, –OSiCH$_3$CH$_2$CH$_2$CH$_2$OSi(CH$_3$)$_3$], 1.62 [m, 2H, –OSiCH$_3$CH$_2$CH$_2$– CH$_2$OSi(CH$_3$)$_3$] and 3.44 [t, 2H, –OSiCH$_3$CH$_2$CH$_2$CH$_2$OSi(CH$_3$)$_3$].

2.4. Synthesis of 3,13-dihydroxypropyloctaphenyl B-POSS

3,13-dihydroxypropyloctaphenyl B-POSS (B-POSS) was synthesized as per Wei et al.'s study [37]. 3,13-dihydroxypropyloctaphenyl B-POSS (B-POSS) was obtained through the deprotection reaction of 3,13-di(trimethylsilyl)oxypropyloctaphenyl B-POSS. 3,13-di(trimethlylsilyl)oxypropyloctaphenyl B-POSS (3.0 g, 2.12 mmol), anhydrous methanol (90 mL) and anhydrous CH$_2$Cl$_2$ (90 mL) in a flask were stirred and purged with nitrogen. Methyltrichlorosilane (0.68 g, 6.26 mmol) in anhydrous methanol (10 mL) was added dropwise to the solution within 30 min. The reaction was carried out at room temperature for 5 h, and then solvents were removed by rotary evaporation. Finally, 2.25 g powder was collected with yield of 93 wt % through washing the product with a mixture of THF and hexane (50/50 vol/vol) and dried in a vacuum oven. ^{1}H-NMR (δ ppm, CDCl$_3$): 0.31 (s, 3H, –OSiCH$_3$CH$_2$CH$_2$CH$_2$OH), 0.74 [t, 2H, –OSiCH$_3$CH$_2$CH$_2$CH$_2$OH], 1.63 [m, 2H,–OSiCH$_3$CH$_2$CH$_2$–CH$_2$OH], 3.47 [t, 2H, OSiCH$_3$CH$_2$CH$_2$CH$_2$OH]; MALDI-TOFMASS (product + Na+): 1292.1 Da (calculated: 1292.36 Da).

2.5. Synthesis of O-4000, O-8000, O-10000, O-12000 and O-14000

The B-POSS incorporated oligomers of polycarprolactone (PCL) were synthesized with B-POSS and ε-CL by varying ε-CL content as shown in Figure 1. In a typical reaction, a three-necked round bottom flask charged with B-POSS (1.5 g, 1.18 mmol), ε-CL (8.07 g, 70.79 mmol), Sn(Oct)$_2$ (375 μL) and anhydrous toluene (150 mL), vigorously stirred at room temperature for 5 min and then reacted at 110 °C for another 20 h under dry nitrogen. After that, the solvent was eliminated via rotary evaporation and the mixture was precipitated by adding an excessive amount of methanol.

Then, the precipitate was obtained through centrifugation and washed with methanol three times. The product was dried in a vacuum oven at 40 °C for 24 h and then 8.13 g product of O-8000 was obtained with a yield of 85%. ^{1}H-NMR (δ ppm, CDCl$_3$): 1.4–1.6 (m, 6H –CH$_2$CH$_2$CH$_2$–), 2.3 (t, 2H –COOCH$_2$–), 4.1 (t, 2H –CH$_2$OH–), 7.14–7.50 (m, 5H aromatic protons). FTIR (KBr, cm^{-1}): 1536 (–NH–), 1730 (–COO–). GPC (PS standard, THF): M_w = 10,100, M_w/M_n = 1.35. O-4000, O-10000, O-12000 and O-14000 were prepared using the same manner, simply varying the ε-CL feed content.

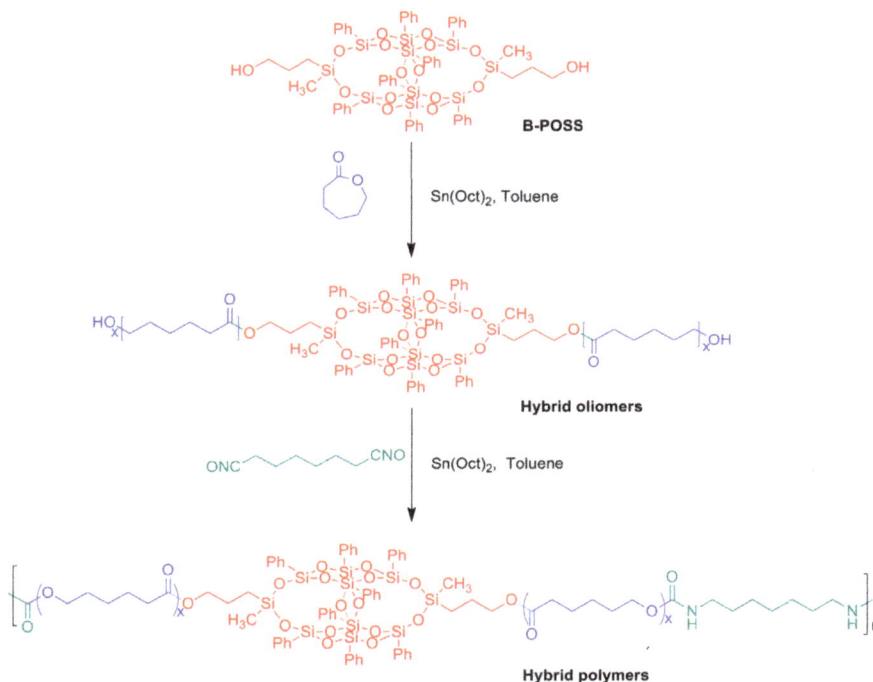

Figure 1. Synthesis of organic–inorganic hybrid PUs with B-POSS in the main chains.

2.6. Synthesis of PUs (PU, PU-4000, PU-8000, PU-10000, PU-12000 and PU-14000)

To synthesize the designed linear PUs, the POSS-PCL oligomers are further reacted with HDI through stoichiometric and temperature control to produce five types of PUs with various physical properties as shown in Figure 1. In a typical procedure, O-8000 (6.5 g, 0.81 mmol) was dissolved in toluene and refluxed at 110 °C for 4 h. After cooling to 70 °C, Sn(Oct)$_2$ (0.5 wt %, relative to the reactant) was added into solution and HDI (0.27 g, 1.62 mmol) in anhydrous toluene (10 mL) was added dropwise to the solution. The reaction was allowed to proceed at 70 °C for another 3 h under nitrogen. After that, the solvent was removed by rotary evaporation and PU-8000 was collected by precipitation with a large amount of ethyl ether. Finally, the product was obtained after drying at 40 °C for 24 h in a vacuum oven, and 5.65 g product was obtained with a yield of 83%. PU-4000, PU-10000, PU-12000 and PU-14000 were prepared through the same synthetic route with yields of 84%, 81%, 85% and 86%, respectively. Neat PU was obtained using PCL and HDI in the same way.

All the polyurethanes with B-POSS in the main chains were subjected to gel permeation chromatography (GPC) to measure their molecular weights. In all cases, high-molecular-weight products were obtained. It is noted that the polydispersity index (M_w/M_n) for these hybrid polyurethanes was slightly higher than that of the control polyurethane. The increased values of the polydispersity index could be attributed to the high steric hindrance of B-POSS. It is possible that the nanoscale size of B-POSS restricted the motion of the POSS macromer during the polymerization and that this effect was increasingly pronounced while its concentration was sufficiently high.

The synthetic route of B-POSS-PUs is shown in Figure 2. To adjust the content ratio of B-POSS incorporated into the main chain of PU, different amounts of ε-CL were used as chain extender and grafted onto B-POSS, and the products were denoted as O-xxx to differentiate the molecular weight of polycaprolatone (PCL), i.e., O-4000 represents the PCL grafts with M_w of 4000 g/mol.

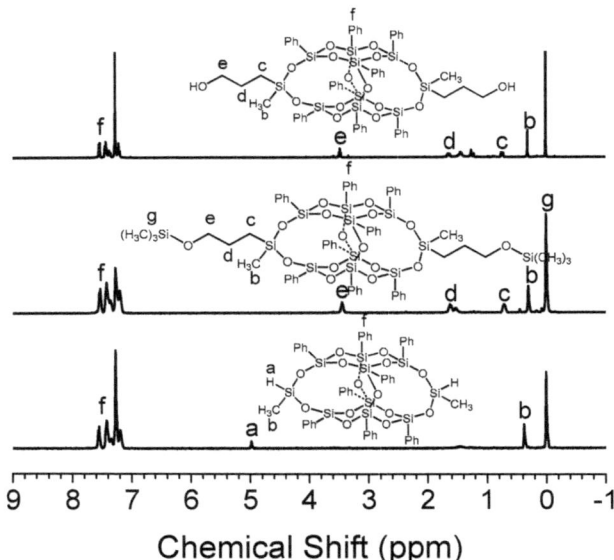

Figure 2. ^{1}H-NMR spectra of 3,13-dihydroxypropyloctaphenyl B-POSS (B-POSS), 3,13-di(trimethylsilyl) oxypropyloctaphenyl B-POSS and 3,13-dihydrooctaphenyl B-POSS from top to bottom.

2.7. Fabrication of Poly(lactic-co-glycolic acid) (PLGA) Composite Films

The as synthesized B-POSS, oligomers and PUs can be used as filler and doped into PLGA matrix by film casting with chloroform as a solvent. PLGA and B-POSS were dissolved in 100 mL of chloroform at room temperature. The mixture of the polymer and nanoparticles was placed in a Teflon dish, and the solution was evaporated at room temperature for 24 h. The formed film was vacuum-dried at room temperature for 24 h. The PLGA/B-POSS film was removed from the Teflon dish using liquid nitrogen. The film thickness was measured using a micrometer. PLGA/O-14000 and PLGA/PU-14000 were prepared through the same synthetic route.

3. Characterization

^{1}H-NMR spectra were carried out on a Bruker AVANCE II 400 spectrometer (Bruker, Karlsruhe, Germany) using tetramethylsilane (TMS; d = 0 ppm) as the internal standard. Chemical shifts were reported in parts per million (ppm) and CDCl$_3$ was used as solvent for all the samples.

Matrix-assisted ultraviolet laser desorption/ionization time-of-flight mass spectroscopy (MALDI-TOF-MS) (ultrafleXtreme MALDI-TOF/TOF, Bruker, Karlsruhe, Germany) with gentisic acid (2,5-dihydroxybenzoic acid, DHB) as the matrix and THF as the solvent was used to measure the molecular weight of B-POSS.

FTIR spectra were tested at room temperature via an IR Prestige-21 spectrometer (Bruker, Karlsruhe, Germany) operated at a resolution of 4 cm^{-1} with scan number of 32.

Thermogravimetric analysis was carried out using a thermogravimetric analyzer (TGA, SDT Q 600, TA Instruments, New Castle, DE, USA) at room temperature to 800 °C at 10 °C/min under an air atmosphere.

The thermal behavior test of the polymer was measured using a DSC (Q10) differential scanning calorimeter (TA Instruments, New Castle, DE, USA). The polymer was heated from room temperature to 150 °C at a heating rate of 15 °C/min under air atmosphere.

The molecular weight and molecular weight distribution of the polymer were determined using THF as the eluent and polystyrene as the standard gel permeation chromatography (GPC, Shimadzu, Kyoto, Japan).

Scanning electron microscopy was used to characterize the surface morphology using SEM (QUANTA 200, FEI, Eindhoven, the Netherlands).

The rheological measurements of the polymer were carried out using a TA DHR-2 rheometer (TA Instruments, New Castle, DE, USA) on parallel plates of 20 mm diameter.

The tensile test was carried out using a microcomputer-controlled electronic universal testing machine (Hensgrand, WDW-02, Jinan, China) at a crosshead speed of 10 mm/min^{-1} and 25 °C.

4. Results and Discussion

4.1. Characterization of B-POSS

In this work, B-POSS was synthesized (Figure S1) and introduced into the main chains of PUs. ^{1}H-NMR spectra of 3,13-dihydroxypropyloctaphenyl B-POSS (B-POSS), 3,13-di(trimethylsilyl) oxypropyloctaphenyl B-POSS and 3,13-dihydrooctaphenyl B-POSS is shown in Figure 2. The peaks of 3,13-dihydrooctaphenyl B-POSS appeared at 0.38 ppm (b), 3.41 ppm (e) and 7–8 ppm (f) were assigned to the protons of –Si–CH$_3$, –CH$_2$–OH and phenyl groups, respectively. The integration area ratio between b, e and f is 3.01: 2.04: 20.03, which is close to the theoretical ratios calculated from the structural formula. In Figure 2, the signals of 3,13-di(trimethylsilyl)oxypropyloctaphenyl B-POSS appeared at 0.38 ppm (b), 0.71 ppm (c), 1.62 ppm (d), 3.44 ppm (e) and 7–8 ppm (f), assigned to the protons of –Si–CH$_3$, –CH$_2$, –CH$_2$O and benzene rings, respectively. In addition, the peaks of Si–H protons at 4.98 ppm disappeared completely, indicating the successful hydrosilylation reaction. For B-POSS, the methyl protons peak of trimethylsilyloxypropyl at 0.02 ppm also completely disappeared after the deprotection reaction of 3,13-di(trimethylsilyl)oxypropyloctaphenyl B-POSS. Moreover, the successful synthesis of B-POSS was further confirmed by MALDI-TOF mass spectroscopy, where the molecular weight of B-POSS was determined to be 1292.1 (B-POSS + Na$^+$), which is in good agreement with the calculated value (1269.1) from the molecular structure (Figure S2).

4.2. Characterization of the B-POSS-PUs

The typical ^{1}H-NMR spectra of O-8000 and PU-8000 were presented as an example for the structure analysis of the as-synthesized PUs. As shown in Figure 3, the chemical shifts at 4.1 ppm (l), 2.3 ppm (h), 1.6 ppm (I, k), 1.4 ppm (j) were assigned to the proton of CH$_2$ on PCL in O-8000, indicating that PCL was grafted onto B-POSS successfully. As for PU-8000, the chemical shifts at 3.5 (m), 1.2 ppm (o, p, q, r) was assigned to the protons of CH$_2$ in the coupling regent HDI. The phenyl groups of B-POSS moieties were detected at 7.14–7.50 ppm in the spectrum, suggesting that B-POSS have been successfully incorporated into the main chain of PUs.

Through FTIR, the characteristic peak of –NCO in HDI at 2268 cm^{-1} disappeared after chain extension through urethane reaction (Figure S3, ESI). In addition, the appearance of strong bands at 1536 cm^{-1} and 1730 cm^{-1} assigned to –NH and –COO vibrations in the newly formed urethane linkage indicated the successful of reaction of POSS-PCL with HDI. ^{1}H-NMR, FTIR together with GPC indicated the successful synthesis of organic–inorganic hybrid PU copolymers.

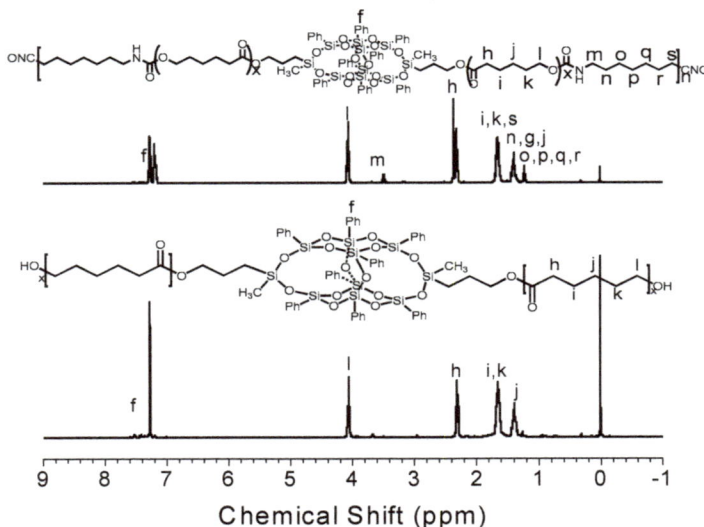

Figure 3. ^{1}H-NMR spectra of O-8000 and PU-8000 in CDCl$_3$.

4.3. Thermal Stability

Thermal stability and transitions of polymers were characterized by TGA and DSC analyses. TGA curves of PU, PU-4000, PU-8000 and PU-12000 are shown in Figure 4. One can see that the degradation temperatures (T_d) are determined as 207, 238, 246, 266, 280 and 283 °C for neat PU, PU-4000, PU-8000, PU-10000, PU-12000 and PU-14000, respectively (Table 1). According to the molecular structure shown in Figure 1, different amount of ε-CL were reacted with B-POSS to adjust the chain length of PCL as well as the ratio of B-POSS in hybrid PUs. Therefore, the number of B-POSS on each polymer chain could be estimated based on the feeding ratio of ε-CL and the M_w of hybrid PUs, which was in the order of PU-14000 > PU-12000 > PU-10000 > PU-8000 >PU-4000 > PU, indicating that the thermal stability could be improved by B-POSS. The addition of POSS was reported to have significant effect in retarding the random breakage of the polymer backbone as well [38]. Comparison of the length of PCL in the hybrids with similar ratio of B-POSS indicated that PCL may also help to enhance thermal stability of the PUs because T_d also depends on the molecular weight (M_n) of the polymer drastically [36], and the long-chain oligomers with POSS could enhances the thermal stability of the polyurethane by better dispersion [25–27,29]. The copolymer and neat PU exhibited similar TGA profiles, indicating that the incorporation of B-POSS had no significant effect on the degradation mechanism of the PUs backbone. The oxidation and decomposition of organic segments in pure PU and hybrid PUs are two stages of degradation. The residues in hybrid PUs were higher than those of neat PU which is probably due to the inorganic components in POSS. The formation of silica from thermal degradation of POSS is responsible for the residues after TGA run as reported from the FTIR analysis by Blanco and co-workers [42]. The silica coated on the surface of the material would protect the inner material from further decomposing due to its low thermal conductivity. Moreover, the weight percentage of B-POSS was in the order of PU-4000 > PU-8000 > PU-12000 > PU according to the molecular structure. Therefore, the decomposition yield of the residues could further confirm the better stability of the PUs due to the role of the B-POSS.

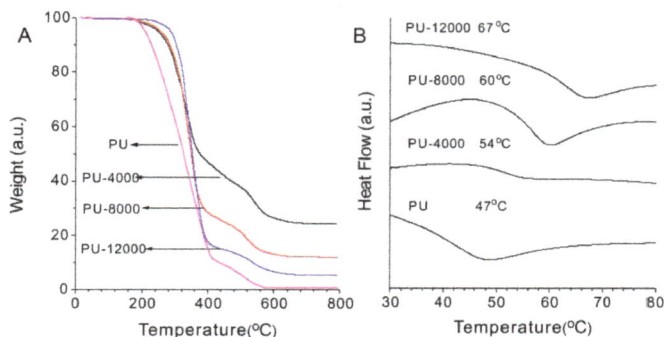

Figure 4. TGA curves of PU, PU-4000, PU-8000 and PU-12000 (**A**); DSC thermograms of PU, PU-4000, PU-8000 and PU-12000 (**B**).

Table 1. Characteristics of PU, PU-2000, PU-4000 and PU-10000.

Polyurethane	M_w (g/mol)	M_n (g/mol)	M_w/M_n	T_d (°C)	T_g (°C)
PU	22,600	22,100	1.02	207	47
PU-4000	11,100	7930	1.39	238	54
PU-8000	23,100	15,100	1.53	246	60
PU-10000	29,900	19,200	1.56	266	66
PU-12000	36,400	22,700	1.60	280	67
PU-14000	45,200	27,100	1.67	283	69

DSC curves of PU, PU-4000, PU-8000 and PU-12000 displayed in Figure 4B indicated that the glass transition temperatures (T_g) are 47, 54, 60, 64, 67 and 69 °C for PU, PU-4000, PU-8000, PU-10000, PU-12000 and PU-14000 (Table 1), respectively. Compared with neat PU, hybrid PUs showed enhanced T_g. This is probably due to rigid structure of POSS cages which could restrict the movement of the chain, thereby reducing the free volume; hence, a higher temperatures is required to provide the thermal energy to induce a glass transition in the polymers [43].

4.4. Surface Properties

Surface hydrophobicity of the hybrid PUs was investigated by static contact angle test. It has been reported that the incorporation of POSS cages can be used to enhance the hydrophobicity of polymers [27]. The hybrid PU polymers we synthesized in this study also exhibited hydrophobic feature as well as low surface free energy (Table 2). Static contact angles were tested with both water and ethylene glycol as probe liquids. Compared with neat PU, the water contact angle and ethylene glycol contact angles of the hybrid PUs were obviously improved. For example, the water contact angle of PU-14000 was as high as 108.9° (Figure 5), suggesting that the hydrophobicity and lipophilicity of PUs could be adjusted effectively by varying the B-POSS content in the copolymers.

Table 2. Static contact angles and surface free energy of PU and B-POSS.

Sample	Static Contact Angle		Surface Free Energy (mN/m)		
	Θ_{H2O}	$\Theta_{ethylene\ glycol}$	γ^p_s	γ^d_s	γ_s
PU	59.1 ± 0.6	40.2 ± 0.6	29.48	11.90	41.38
PU-4000	80.2 ± 0.8	63.2 ± 0.8	12.96	14.06	27.02
PU-8000	96.8 ± 0.9	77.1 ± 0.6	3.31	17.47	20.78
PU-10000	103.9 ± 0.6	86.8 ± 0.8	2.43	13.10	15.53
PU-12000	105.7 ± 0.8	88.9 ± 0.7	2.13	12.45	14.58
PU-14000	108.9 ± 0.9	93.6 ± 0.6	1.96	10.17	12.13

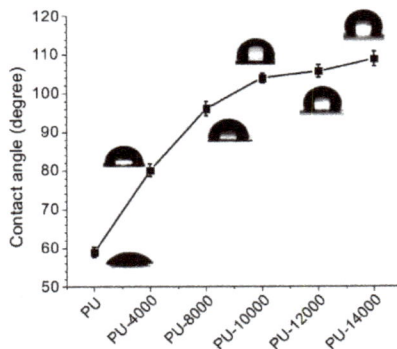

Figure 5. Plot of surface water contact angles of the organic–inorganic hybrid PUs.

The surface free energies of the organic-inorganic hybrid PUs were calculated according to the geometric mean model [44]:

$$\cos\theta = 2/\gamma_L [(\gamma^d{}_L \times \gamma^d{}_s)^{1/2} + (\gamma^P{}_L \times \gamma^P{}_s)^{1/2}] \tag{1}$$

$$\gamma_s = \gamma^d{}_s + \gamma^P{}_s \tag{2}$$

where θ is the contact angle and γ_L is the liquid surface tension; $\gamma^d{}_L$ and $\gamma^P{}_L$ are the dispersive and polar components of γ_L; γ_s is the solid surface tension, $\gamma^d{}_s$ and $\gamma^P{}_s$ are the dispersed and polar components of γ_s respectively. From this, the total surface free energies range of the organic-inorganic hybrid Pus are calculated to be in the range of 12.13 to 27.02 mN/m. The low surface energy endows the linear hybrids with potential application of highly hydrophobic coating materials. The presence of B-POSS cages could accumulate at the surface of the copolymers and thus minimize the polymer-air surface tension.

Surface morphology of the as-synthesized PUs were investigated by SEM. Figure S4 shows the SEM images of neat PU (A), PU-10000 (B), PU-12000 (C) and PU-14000 (D). Neat PU exhibited smooth surface morphology, whereas uniform pore distributions with the diameter of 1.08 μm were observed in PU-12000. Such an ordered self-assembly structure could be attributed to the phase separation between B-POSS and alkyl chains in PUs. This unique structural formation has been reported to be useful in lowering the dielectric constant value [45]. However, needle-like surface morphology was observed in PU-14000, which could be explained by the more effective separation of B-POSS from longer PCL chains. However, spindle-like surface morphology was observed in PU-14000, which could be explained by the longer chain of PCLs resulted packing.

4.5. Mechanical Properties

The effect of different ratio of B-POSS on the mechanical property of PUs was investigated by measurement. A viscoelastic region was obtained through dynamic strain sweep test. As shown in Figure 6, the investigated region of storage modulus was viscoelastic and independent of strain. As the B-POSS content increased, the viscoelastic linear region in each sample becomes shorter because of the higher rigidity and brittleness of the sample as evidenced by the Payne effect [46]. Due to the disentanglement of polymer macromolecules, storage modulus (G′) decreases as strain increases in the slope region [47,48]. The presence of B-POSS may restrict the chain mobility of the PLGA. Dynamic frequency sweep was further studied, and the viscoelastic linear region of the materials was analyzed.

Dynamic frequency sweep was conducted at a strain of 0.5% to ensure the sufficient liner viscoelasticity region and sensitivity (Figure S5). G′ is higher than G″ for all samples, suggesting that the samples are solid-like. Compared with neat PU, the hybrid PUs exhibited higher G′ and G″ values and the increment is proportional to the POSS content in the copolymer. Rheology results indicated that the incorporation of B-POSS could improve the mechanical property and the mechanical strength of PU.

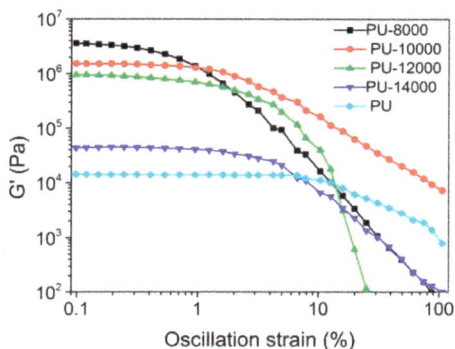

Figure 6. Dynamic strain sweep of: G' for neat PU, PU-8000, PU-10000, PU-12000 and PU-14000.

4.6. Influence on Mechanical Property of Poly(lactic-co-glycolic acid) (PLGA)

Due to the hybrid nature, the synthesized B-POSS, oligomers and hybrid PUs can be used as filler and doped into other polymer matrix to modulate their mechanical performances. To demonstrate the toughening effect of hybrid PUs, poly(lactic-co-glycolic acid) (PLGA) was selected as target polymer. As it is well-known, PLGA is a biodegradable functional polymer with good biocompatibility, non-toxicity and film-forming properties. It has been widely used in pharmaceutical, medical engineering materials and modern industrial fields [48]. However, the poor mechanical performance of PLGA has limited its application in more advanced areas. Interestingly, we found that the mechanical performance of PLGA could be greatly improved by blending with hybrid PUs in this study. The mechanical properties were investigated by tensile tests. The typical stress-strain curves in Figure S6 showed that the strength of the PLAG/POSS blends increased significantly whereas its high extensibility of PLGA was retained. After doping with 6.5 wt% of PU-14000, the mechanical strength was increased by 41.6% with a marginal sacrifice of the elongation-at-break (Table 3). This may result from the combined effects of high molecular weight of hybrid PUs and the good distribution of B-POSS in the blends. In addition, the B-POSS exhibited lowest toughness towards PLGA, probably because the rigid POSS materials could restrict the motion of PLGA chains and thus cause an increase in tensile strength and low strain. Among the tested samples, PLGA/O-14000 shows the highest toughness towards PLGA, as well as an improved elongation at breakage. Considering the use of organotins and diisocyanates precursors on the safety for pharmaceutical applications, Caracciolo, P.C. et al. [49] prepared a polyurethane elastomer network for drug transportation with DBTDL (dibutyltin dilaurate) as catalyst and diisocyanates as monomers. Gorna, K. et al. [50] reported using dibutyltin dilaurate as a catalyst and aliphatic hexamethylene diisocyanate, PCL and isosorbide glycol as precursors in preparing a biodegradable polyurethane scaffold for tissue repair and regeneration. Bonzani, I. C. et al. [51] also prepared a two-component injectable polyurethane for bone tissue engineering using pentaerythritol and glycolic acid 2,6-diisocyanate (ELDI) as a starting material under the catalysis of stannous octoate. Thus, such a simple yet useful strategy in modulating the mechanical performance of PLGA by hybrid PUs is promising in developing PLGA-based materials for more advanced biomedical applications.

Polymers **2019**, *11*, 373

Table 3. Tensile strength and elongation at break of PLGA, PLGA/B-POSS, PLGA/O-14000 and PLGA/PU-14000.

Sample	Tensile Strength (MPa)	Elongation at Break (%)	Toughness (MJ/m^3)
PLGA	0.36 ± 0.03	953.16 ± 22.3	241.85 ± 0.67
PLGA/B-POSS	0.44 ± 0.02	707.03 ± 14.2	229.10 ± 0.28
PLGA/O-14000	0.48 ± 0.01	833.62 ± 16.8	291.98 ± 0.17
PLGA/PU-14000	0.51 ± 0.03	725.52 ± 24.7	270.64 ± 0.74

5. Conclusions

In summary, a series of hybrid thermoplastic polyurethanes (PUs) incorporated with bi-functional polyhedral oligomeric silsesquioxane in the polymer backbone were successfully synthesized and fully characterized. With the introduction of B-POSS into the main chain, the surface hydrophobicity and thermal and mechanical stability of the hybrid PUs were significantly increased. The structure-property correlations were instigated by comparing with neat PU. As a new kind of thermoplastic filler, the hybrid PUs can also be used to improve the mechanical strength of PLGA, exhibiting promising application in biomedical devices.

Supplementary Materials: The following are available online at http://www.mdpi.com/2073-4360/11/2/373/s1, Figures S1–S6: characterization of B-POSS, hybrid PUs.

Author Contributions: Funding acquisition, H.C.; Investigation, X.S.; Resources: T.L.; Supervision, H.C., Z.L. and T.L.; Writing—original draft, X.S. and X.Z.; Writing—review & editing, H.C. and Z.L.

Funding: This work was financially supported by the National Natural Science Foundation of China (Grant No. 51702178), International Cooperation Research Special Fund Project (QLUTGJHZ2018001), the Taishan Scholar Program of Shandong (Grant No. 0308150303) and the Program for Scientific Research Innovation Team in Colleges and Universities of Shandong Province. Z.L would like to thank A*STAR for financial support.

Conflicts of Interest: The authors declare no conflict of interest.

References

1. Delebecq, E.; Pascault, J.-P.; Boutevin, B.; Ganachaud, F. On the versatility of urethane/urea bonds: Reversibility, blocked isocyanate, and non-isocyanate polyurethane. *Chem. Rev.* **2012**, *113*, 80–118. [CrossRef]
2. Fan, X.; Tan, B.H.; Li, Z.; Loh, X.J. Control of PLA Stereoisomers-Based Polyurethane Elastomers as Highly Efficient Shape Memory Materials. *ACS Sustain. Chem. Eng.* **2017**, *5*, 1217–1227. [CrossRef]
3. Yao, H.B.; Ge, J.; Wang, C.F.; Wang, X.; Hu, W.; Zheng, Z.J.; Ni, Y.; Yu, S.H. A flexible and highly pressure-sensitive graphene–polyurethane sponge based on fractured microstructure design. *Adv. Mater.* **2013**, *25*, 6692–6698. [CrossRef] [PubMed]
4. Janik, H.; Marzec, M. A review: Fabrication of porous polyurethane scaffolds. *Mater. Sci. Eng. C* **2015**, *48*, 586–591. [CrossRef] [PubMed]
5. Wang, X.; Hu, Y.; Song, L.; Yang, H.; Xing, W.; Lu, H. In situ polymerization of graphene nanosheets and polyurethane with enhanced mechanical and thermal properties. *J. Mater. Chem.* **2011**, *21*, 4222–4227. [CrossRef]
6. Liu, H.; Dong, M.; Huang, W.; Gao, J.; Dai, K.; Guo, J.; Zheng, G.; Liu, C.; Shen, C.; Guo, Z. Lightweight conductive graphene/thermoplastic polyurethane foams with ultrahigh compressibility for piezoresistive sensing. *J. Mater. Chem. C* **2017**, *5*, 73–83. [CrossRef]
7. Zhang, X.; Gong, T.; Chi, H.; Li, T. Nanostructured polyurethane perylene bisimide ester assemblies with tuneable morphology and enhanced stability. *Roy. Soc. Open Sci.* **2018**, *5*, 171686. [CrossRef]
8. Kong, L.; Li, Y.; Qiu, F.; Zhang, T.; Guo, Q.; Zhang, X.; Yang, D.; Xu, J.; Xue, M. Fabrication of hydrophobic and oleophilic polyurethane foam sponge modified with hydrophobic Al$_2$O$_3$ for oil/water separation. *J. Ind. Eng. Chem.* **2018**, *58*, 369–375. [CrossRef]
9. Yang, G.; Song, J.; Hou, X. Fabrication of highly hydrophobic two-component thermosetting polyurethane surfaces with silica nanoparticles. *Appl. Surf. Sci.* **2018**, *439*, 772–779. [CrossRef]

10. Wang, H.; Wang, E.; Liu, Z.; Gao, D.; Yuan, R.; Sun, L.; Zhu, Y. A novel carbon nanotubes reinforced superhydrophobic and superoleophilic polyurethane sponge for selective oil–water separation through a chemical fabrication. *J. Mater. Chem. A* **2015**, *3*, 266–273. [CrossRef]
11. Kuang, X.; Chen, K.; Dunn, C.; Wu, J.; Li, V.; Qi, H.J. 3D Printing of Highly Stretchable, Shape-Memory, and Self-Healing Elastomer toward Novel 4D Printing. *ACS Appl. Mater. Interfaces* **2018**, *10*, 7381–7388. [CrossRef] [PubMed]
12. Chen, Q.; Mangadlao, J.D.; Wallat, J.; De, L.A.; Pokorski, J.K.; Advincula, R.C. 3D printing biocompatible polyurethane/poly (lactic acid)/graphene oxide nanocomposites: Anisotropic properties. *ACS Appl. Mater. Interfaces* **2017**, *9*, 4015–4023. [CrossRef] [PubMed]
13. Li, Z.; Tan, B.H.; Jin, G.; Li, K.; He, C. Design of polyhedral oligomeric silsesquioxane (POSS) based thermo-responsive amphiphilic hybrid copolymers for thermally denatured protein protection applications. *Polym. Chem.* **2014**, *5*, 6740–6753. [CrossRef]
14. Li, Z.; Kong, J.; Wang, F.; He, C. Polyhedral oligomeric silsesquioxanes (POSSs): An important building block for organic optoelectronic materials. *J. Mater. Chem. C* **2017**, *5*, 5283–5298. [CrossRef]
15. Fan, X.; Cao, M.; Zhang, X.; Li, Z. Synthesis of star-like hybrid POSS-(PDMAEMA-b-PDLA)$_8$ copolymer and its stereocomplex properties with PLLA. *Mater. Sci. Eng. C* **2017**, *76*, 211–216. [CrossRef]
16. Uner, A.; Doganci, E.; Tasdelen, M.A.; Yilmaz, F.; Gürek, A.G. Synthesis, characterization and surface properties of star-shaped polymeric surfactants with polyhedral oligomeric silsesquioxane core. *Polym. Int.* **2017**, *66*, 1610–1616. [CrossRef]
17. Zhang, W.; Müller, A.H.E. Architecture, self-assembly and properties of well-defined hybrid polymers based on polyhedral oligomeric silsequioxane (POSS). *Prog. Polym. Sci.* **2013**, *38*, 1121–1162. [CrossRef]
18. Hirai, T.; Leolukman, M.; Liu, C.C.; Han, E.; Kim, Y.J.; Ishida, Y.; Hayakawa, T.; Kakimoto, M.; Nealey, N.F.; Gopalan, P. One-Step Direct-Patterning Template Utilizing Self-Assembly of POSS-Containing Block Copolymers. *Adv. Mater.* **2009**, *21*, 4334–4338. [CrossRef]
19. Zheng, L.; Hong, S.; Cardoen, G.; Burgaz, E.; Gido, S.P.; Coughlin, E.B. Polymer nanocomposites through controlled self-assembly of cubic silsesquioxane scaffolds. *Macromolecules* **2004**, *37*, 8606–8611. [CrossRef]
20. Liu, Y.; Yu, C.; Jin, H.; Jiang, B.; Zhu, X.; Zhou, Y.; Lu, Z.; Yan, D. A supramolecular Janus hyperbranched polymer and its photoresponsive self-assembly of vesicles with narrow size distribution. *J. Am. Chem. Soc.* **2013**, *135*, 4765–4770. [CrossRef]
21. Raghunath, J.; Zhang, H.; Edirisinghe, M.J.; Darbyshire, A.; Butler, P.E.; Seifalian, A.M. A new biodegradable nanocomposite based on polyhedral oligomeric silsesquioxane nanocages: Cytocompatibility and investigation into electrohydrodynamic jet fabrication techniques for tissue-engineered scaffolds. *Biotechnol. Appl. Bioc.* **2009**, *52*, 1–8. [CrossRef] [PubMed]
22. Wang, B.; Lin, Q.; Shen, C.; Han, Y.; Tang, J.; Chen, H. Synthesis of MA POSS–PMMA as an intraocular lens material with high light transmittance and good cytocompatibility. *RSC Adv.* **2014**, *4*, 52959–52966. [CrossRef]
23. Kannan, R.Y.; Salacinski, H.J.; De Groot, J.; Clatworthy, I.; Bozec, L.; Horton, M.; Butler, P.E.; Seifalian, A.M. The antithrombogenic potential of a polyhedral oligomeric silsesquioxane (POSS) nanocomposite. *Biomacromolecules* **2006**, *7*, 215–223. [CrossRef] [PubMed]
24. Ata, S.; Banerjee, S.L.; Singha, N.K. Polymer nano-hybrid material based on graphene oxide/POSS via surface initiated atom transfer radical polymerization (SI-ATRP): Its application in specialty hydrogel system. *Polymer* **2016**, *103*, 46–56. [CrossRef]
25. Zhang, X.; Chi, H.; Li, T.; Wang, F.; Chin, W.S.; Xu, J. Energy transfer along a sequence controlled hybrid polymer. *J. Polym. Sci. A* **2018**, *56*, 1225–1233. [CrossRef]
26. Chi, H.; Mya, K.Y.; Lin, T.; He, C.; Wang, F.; Chin, W.S. Thermally stable azobenzene dyes through hybridization with POSS. *New J. Chem.* **2013**, *37*, 735–742. [CrossRef]
27. Chi, H.; Wang, M.; Xiao, Y.; Wang, F.; Joshy, K.S. Self-Assembly and Applications of Amphiphilic Hybrid POSS Copolymers. *Molecules* **2018**, *23*, 2481. [CrossRef]
28. Han, Y.; Wang, F.; Lim, C.Y.; Chi, H.; Chen, D.; Wang, F.; Jiao, X. High-Performance Nano-Photoinitiators with Improved Safety for 3D Printing. *ACS Appl. Mater. Interfaces* **2017**, *9*, 32418–32423. [CrossRef]

29. Chi, H.; Lim, S.L.; Wang, F.; Wang, X.; He, C.; Chin, W.S. Pure Blue-Light Emissive Poly(oligofluorenes) with Bifunctional POSS in the Main Chain. *Macromol. Rapid Commun.* **2014**, *35*, 801–806. [CrossRef]

30. Liu, X.; Fan, X.; Jiang, L.; Loh, X.J.; Wu, Y.-L.; Li, Z. Biodegradable Polyester Unimolecular System as Emerging Materials for Therapeutic Applications. *J. Mater. Chem. B* **2018**, *6*, 5488–5498. [CrossRef]

31. Kyung, M.; Lee, P.T.; Knight, T.; Chung, T.P.; Mather, P.T. Polycaprolactone−POSS Chemical/Physical Double Networks. *Macromolecules* **2008**, *41*, 4730–4738. [CrossRef]

32. Frank, K.L.; Exley, S.E.; Thornell, T.L.; Morgan, S.E.; Wiggins, J.S. Investigation of pre-reaction and cure temperature on multiscale dispersion in POSS–epoxy nanocomposites. *Polymer* **2012**, *53*, 4643–4651. [CrossRef]

33. Tan, B.H.; Hussain, H.; Leong, Y.W.; Lin, T.T.; Tjiu, W.W.; He, C. Tuning self-assembly of hybrid PLA-P(MA-POSS) block copolymers in solution via stereocomplexation. *Polym. Chem.* **2013**, *4*, 1250–1259. [CrossRef]

34. Wang, Z.; Leng, S.; Wang, Z.; Li, G.; Yu, H. Nanostructured Organic−Inorganic Copolymer Networks Based on Polymethacrylate-Functionalized Octaphenylsilsesquioxane and Methyl Methacrylate: Synthesis and Characterization. *Macromolecules* **2011**, *44*, 566–574. [CrossRef]

35. Markovic, E.; Clarke, S.; Matisons, J.; Simon, G.P. Synthesis of POSS−Methyl Methacrylate-Based Cross-Linked Hybrid Materials. *Macromolecules* **2008**, *41*, 1685–1692. [CrossRef]

36. Mya, K.Y.; Gose, H.B.; Pretsch, T.; Bothe, M.; He, C. Star-shaped POSS-polycaprolactone polyurethanes and their shape memory performance. *J. Mater. Chem.* **2011**, *21*, 4827–4836. [CrossRef]

37. Wei, K.; Wang, L.; Zheng, S. Organic-inorganic polyurethanes with 3,13-dihydroxypropyloctaphenyl double-decker silsesquioxane chain extender. *Polym. Chem.* **2013**, *4*, 1491–1501. [CrossRef]

38. Turri, S.; Levi, M. Structure, Dynamic Properties, and Surface Behavior of Nanostructured Ionomeric Polyurethanes from Reactive Polyhedral Oligomeric Silsesquioxanes. *Macromolecules* **2005**, *38*, 5569–5574. [CrossRef]

39. Turri, S.; Levi, M. Wettability of Polyhedral Oligomeric Silsesquioxane Nanostructured Polymer Surfaces. *Macromol. Rapid Commun.* **2005**, *26*, 1233–1236. [CrossRef]

40. Oaten, M.; Choudhury, N.R. Silsesquioxane−Urethane Hybrid for Thin Film Applications. *Macromolecules* **2005**, *38*, 6392–6401. [CrossRef]

41. Li, Z.; Tan, B.H. Towards the development of polycaprolactone based amphiphilic block copolymers: Molecular design, self-assembly and biomedical applications. *Mater. Sci. Eng. C* **2014**, *45*, 620–634. [CrossRef] [PubMed]

42. Blanco, I.; Bottino, F.A.; Abate, L. Influence of n-alkyl substituents on the thermal behaviour of Polyhedral Oligomeric Silsesquioxanes (POSSs) with different cage's periphery. *Thermochim. Acta* **2016**, *623*, 50–57. [CrossRef]

43. Wang, W.; Guo, Y.-L.; Otaigbe, J.U. The synthesis, characterization and biocompatibility of poly(ester urethane)/polyhedral oligomeric silsesquioxane nanocomposites. *Polymer* **2009**, *50*, 5749–5757. [CrossRef]

44. Waltman, R.J.; Deng, H.; Wang, G.J.; Zhu, H.; Tyndall, G.W. The Effect of PFPE Film Thickness and Molecular Polarity on the Pick-Up of Disk Lubricant by a Low-Flying Slider. *J. Appl. Polym. Sci.* **2010**, *39*, 211–219. [CrossRef]

45. Sasi, K.R.; Muthukaruppan, A. Dielectric and thermal behaviors of POSS reinforced polyurethane based polybenzoxazine nanocomposites. *RSC Adv.* **2015**, *5*, 33008–33015. [CrossRef]

46. Payne, A.R. The dynamic properties of carbon black-loaded natural rubber vulcanizates Part II. *J. Appl. Polym. Sci.* **1962**, *6*, 57–63. [CrossRef]

47. Cassagnau, P. Melt rheology of organoclay and fumed silica nanocomposites. *Polymer* **2008**, *49*, 2183–2196. [CrossRef]

48. Quinlan, E.; López-Noriega, A.; Thompson, E.; Kelly, H.M.; Cryan, S.A.; O'Brien, F.J. Development of collagen–hydroxyapatite scaffolds incorporating PLGA and alginate microparticles for the controlled delivery of rhBMP-2 for bone tissue engineering. *J. Control. Release* **2015**, *198*, 71–79. [CrossRef]

49. Caracciolo, P.C.; Pita, C.S.; Abraham, G.A.; Me'ndez, J.A.; Molera, J.G. Synthesis, characterization and applications of amphiphilic elastomeric polyurethane networks in drug delivery. *Polym. J.* **2013**, *45*, 331–338. [CrossRef]

50. Gorna, K.; Gogolewski, S. Biodegradable porous polyurethane scaffolds for tissue repair and regeneration. *J. Biomed. Mater. Res. A* **2006**, *79*, 128–138. [CrossRef]

51. Bonzani, I.C.; Adhikari, R.; Houshyar, H.; Mayadunne, R.; Gunatillake, P.; Stevens, M.M. Synthesis of two-component injectable polyurethanes for bone tissue engineering. *Biomaterials* **2007**, *28*, 423–433. [CrossRef] [PubMed]

![polymers logo] *polymers*

MDPI

Article

Composites of Rigid Polyurethane Foams Reinforced with POSS

Sylwia Członka [1,*], Anna Strąkowska [1], Krzysztof Strzelec [1], Agnieszka Adamus-Włodarczyk [2], Agnė Kairytė [3] and Saulius Vaitkus [3]

[1] Institute of Polymer & Dye Technology, Lodz University of Technology, 90-924 Lodz, Poland; anna.strakowska@p.lodz.pl (A.S.); krzysztof.strzelec@p.lodz.pl (K.S.)

[2] Institute of Applied Radiation Chemistry, Faculty of Chemistry, Lodz University of Technology, 90-924 Lodz, Poland; agnieszka.adamus@p.lodz.pl

[3] Vilnius Gediminas Technical University, Faculty of Civil Engineering, Institute of Building Materials, Laboratory of Thermal Insulating Materials and Acoustics, Linkmenu st. 28, LT-08217 Vilnius, Lithuania; agne.kairyte@vgtu.lt (A.K.); saulius.vaitkus@vgtu.lt (S.V.)

* Correspondence: sylwia.czlonka@edu.p.lodz.pl

Received: 23 January 2019; Accepted: 13 February 2019; Published: 14 February 2019

Abstract: Rigid polyurethane foams (RPUFs) were successfully modified with different weight ratios (0.5 wt%, 1.5 wt% and 5 wt%) of APIB-POSS and AEAPIB-POSS. The resulting foams were evaluated by their processing parameters, morphology (Scanning Electron Microscopy analysis, SEM), mechanical properties (compressive test, three-point bending test and impact strength), viscoelastic behavior (Dynamic Mechanical Analysis, DMA), thermal properties (Thermogravimetric Analysis, TGA, and thermal conductivity) and application properties (contact angle, water absorption and dimensional analysis). The results showed that the morphology of modified foams is significantly affected by the type of the filler and filler content, which resulted in inhomogeneous, irregular, large cell shapes and further affected the physical and mechanical properties of resulting materials. RPUFs modified with APIB-POSS represent better mechanical and thermal properties compared to the RPUFs modified with AEAPIB-POSS. The results showed that the best results were obtained for RPUFs modified with 0.5 wt% of APIB-POSS. For example, in comparison with unfilled foam, compositions modified with 0.5 wt% of APIB-POSS provide greater compression strength, better flexural strength and lower water absorption.

Keywords: POSS; ridgid polyurethane foams; mechanical properties; cellular structure

1. Introduction

Rigid polyurethane foams (RPUFs) are highly cross-linked, three dimensional polymers with closed-cell structures which account for about 23% of all polyurethane (PU) production [1]. Due to their exceptional thermal-insulating properties, high resistance to weather conditions, good mechanical properties, and low apparent density [1], they are commonly used in industries such as furniture, automotive construction or in the production of thermal insulation materials [1–4]. RPUFs have become one of the most diverse and widely-used plastics with a continuously increasing global market. The total value of the global RPUF market amounts to 401 billion dollars and is expected to grow to 619 billion dollars in 2018 [5].

Beside the favorable properties of resulting materials, improving the mechanical properties of RPUFs is the foundation for their further potential applications. To successfully employ them in building application, it is critical that foams are combined with other materials or elements that provide mechanical strength and low thermal conductivity. A review of the related literature indicates that applying nanomaterials contributes a better improvement to mechanical properties and heat resistance [6–16]. Cao et al. [13] used an organically-modified montmorillonite to enhance several properties of RPUFs. The presence of nanoclay resulted in a reduction in cell size compared to a pristine RPUF sample. In the nanocomposites RPUF which used high molecular weight polyols, compressive strength dropped by 650%, but T_g increased by 6 °C. Widya and Macasko [7] incorporated montmorillonite-based organonanoclay into RPUFs at a clay loading of 1 wt%. They observed a reduction of mean cell size and a decrease in the diffusion of the blowing agent. Dolomanova et al. [11] investigated PU foam reinforced by single-walled carbon nanotube (SWNT) and multiwalled carbon nanotube (MWNT). The observation showed that both nanotubes improved compressive strength and compressive modulus, but, compared to SWNTs, MWNTs had a more remarkable improvement in the morphology and mechanical properties of the nanocomposite RPUF. This was because the MWNTs could interact with the matrix material and be better dispersed, which in turn affects the nucleation process and leads to a finer cell structure.

Alongside traditional substances, an interesting group of reactive nanofillers are polyhedral oligomeric silsesquioxanes (POSSs). POSSs are unique organic–inorganic nanobuilding blocks, with a rigid siliceous core with silicon (Si) atoms on the vertices and oxygen (O) atoms on the edges [17,18]. Organic groups are attached on the Si atoms, which allows the cage-like cores to be directly bound to polymer chains, via covalent linkage [19]. The cage POSS structures can be easily functionalized with various organic substituents, thanks to which they can participate in polymerization or grafting processes [20–22]. They show a versatile chemical compatibility owing to the diversity of organic substituents [23,24]. Due to the main features, such as hybrid nature and nanometer-sized structures with high surface area and controlled porosity, POSS compounds can be extremely attractive materials to be used as polyurethanes modifier [25]. Studies have shown that the incorporation of POSS molecules into various polymers can influence their thermal stability and degradation behaviour at elevated temperatures, change the glass transition temperature, increase the resistance to composite oxidation, and improve mechanical strength [22,25–27]. Among different POSS-containing polymer materials, hybrid PU-POSS elastomers of various architectures have been widely studied in the literature [17–19,28–31], but PU foam chemically reinforced by POSS constitutes a new class of materials.

The effects of open-cage nanostructure disilanoisobutyl POSS (DSI-POSS) on the properties of rigid PU foams were studied by Hebda et al. [32]. The hybrid composite foams containing 1.5 and 2.0 wt% DSI-POSS showed a reduced number of cells and an increased average area of foam cells in comparison with the unmodified PU. Thermogravimetric analysis results have shown that incorporation of POSS nanoparticles into PU foam does not significantly change the degradation process, however the POSS-modified samples, showed greater mechanical properties. Polyurethane foams reinforced by 1,2-propanediolizobutyl POSS (PHI-POSS) in the amount of 0.25, 0.5 and 0.75 wt% have been synthesized and investigated by Michałowski et al. [33]. The obtained results have shown that the use of PHI-POSS with phosphorus additive flame retardants leads to the reduction of the rigid polyurethane foam's flammability, without significant changes of the foams' crucial mechanical and thermal conductivity properties. Analogue tendency has been observed in the case of polyurethane foams modified with octa(3-hydroxy-3-methylbutyldimethylsiloxy) POSS (OCTA-POSS) [34].

The aim of this research is to develop and test PU foams modified with closed-cage nanostructure aminopropyl isobutyl-POSS (APIB-POSS) and aminoethylaminopropylisobutyl-POSS (AEAPIB-POSS). The influence of different amounts of POSSs on thermal properties (Thermogravimetric Analysis, TGA), dynamic mechanical properties (Dynamic Mechanical Analysis, DMA), physico-mechanical properties (compression strength, three-point bending test, impact strength, apparent density, and water absorption), and morphology of obtained PU composites was examined in the current work. The results obtained in the present work indicate that the addition of APIB-POSS and AEAPIB-POSS in the range of 0.5–5 wt% influences the morphology of analyzed foams and consequently their further mechanical and thermal properties.

Polymers **2019**, *11*, 336

2. Experimental

2.1. Materials and Manufacturing

RPUFs were synthesized from the same polyurethane-modified polyisocyanurate rigid foam formulation based on petrochemical components provided by Purinova Sp. z o.o. at Bydgoszcz, Poland (Izopianol 30/10/C and Purocyn B). Izopianol 30/10/C used in the reaction is a fully formulated mixture containing a mixture of polyester polyol (hydroxyl number ca. 230−250 mg KOH/g, functionality of 2), catalyst (N,N-Dimethylcyclohexylamine), flame retardant (Tris(2-chloro-1-methylethyl)phosphate), and chain extender (1,2-propanediol) [29]. Purocyn B used in the synthesis is a polymeric diphenylmethane 4,4'-diisocyanate (pMDI) [29]. Aminopropylisobutyl-POSS (APIB-POSS) and aminoethylaminopropylisobutyl-POSS (AEAPIB-POSS) were provided by Hybrid Plastics Inc., Fountain Valley, CA, USA. The chemical structures of APIB-POSS and AEAPIB-POSS are presented in Figure 1a,b, respectively. Figure 2a,b shows the optical micrographs obtained for the APIB-POSS and AEAPIB-POSS, respectively. Since POSS molecules have polar groups and hydrophilic properties, strong interfacial interaction, such as hydrogen bonding, can be formed between the POSS molecules and isocyanate leading to the formation of a cross-linked structure. Amine groups present in POSS molecules can react with isocyanates even in the absence of catalyst [35–38]. It has been reported in previous work, that isocyanate has a higher reactivity with amines than with hydroxyl and carboxyl groups. The scheme of the amine reaction with isocyanate is shown in Equation (1). A urea bond is formed with this reaction. The generalized reaction scheme of isocyanate and POSS reaction is shown in Equation (1).

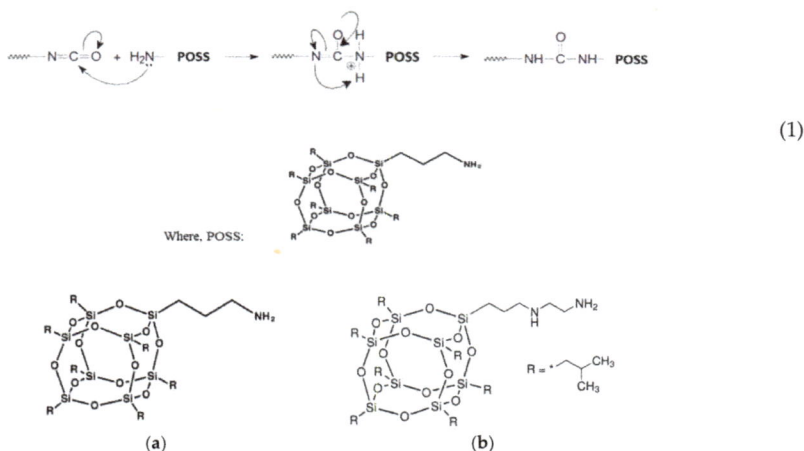

$$\text{(1)}$$

Where, POSS:

(a) (b)

Figure 1. Chemical structure of (a) APIB-POSS and (b) AEAPIB-POSS.

RPUFs were manufactured using a two-step method. In the first step, the component A, containing a polyol (Izopianol 30/10/C) and selected POSS, was prepared by mechanical stirring. The POSS reactive additive was introduced into the polyol in an amount of 0.5, 1.5 and 5 mass% of the polyol weight (wt%) and dispersed by the mechanical stirrer. In the second step, pMDI (Purocyn B) as component B was added to component A and the PU system was mixed using a mechanical stirrer for 10 s. After this time, the mixture was poured into an open mould where free foaming occurred in the vertical direction. RPUFs were conditioned at room temperature for 24 h. After this time, samples were cut with a band saw into appropriate shapes (determined by obligatory standards) and their physico-mechanical properties were investigated.

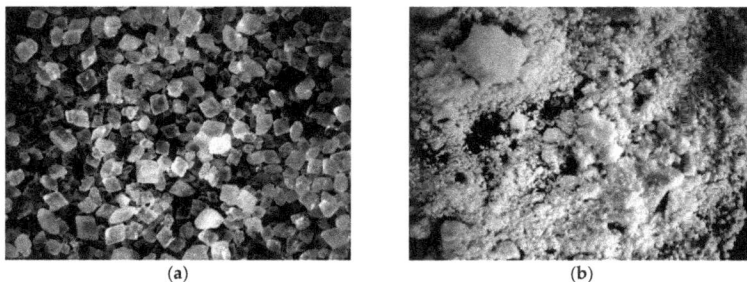

(a)   (b)

Figure 2. APIB-POSS and AEAPIB-POSS observed by a magnification of 50.

2.2. Characterization Techniques

The absolute viscosities of polyol and isocyanate were determined corresponding to ASTM D2930 (equivalent to ISO 2555) using a rotary Viscometer DVII+ (Brookfield, Germany). The torque of samples was measured as a range of shear rate from 0.5 to 100 s^{-1} in ambient temperature.

The apparent density of foams was determined accordingly to ASTM D1622 (equivalent to ISO 845). The densities of five specimens per sample were measured and averaged.

The morphology and cell size distribution of foams were examined from the cellular structure images of foam which were taken using JEOL JSM-5500 LV scanning electron microscopy (JEOL Ltd., Peabody, MA, USA). All microscopic observations were made in the high-vacuum mode and at the accelerating voltage of 10 kV. The samples were scanned in the free rising direction. The average pore diameters, walls thickness and pore size distribution were calculated using ImageJ software (Media Cybernetics Inc., Rockville, MD, USA).

The thermal properties of the synthesized composites were evaluated by TGA measurements performed using the STA 449 F1 Jupiter Analyzer (Netzsch Group, Selb, Germany). About 10 mg of the sample was placed in the TG pan and heated in argon atmosphere at a rate of 10 K min^{-1} up to 600 °C with the sample mass of about 10 mg. The initial decomposition temperatures, $T_{10\%}$, $T_{50\%}$ and $T_{80\%}$ of mass loss were determined.

The compressive strength ($\sigma_{10\%}$) of foams was determined accordingly to the ASTM D1621 (equivalent to ISO 844) using Zwick Z100 Testing Machine (Zwick/Roell Group, Ulm, Germany) with a load cell of 2 kN and a speed of 2 mm min^{-1}. Samples of the specified sizes were cut with a band saw in the direction perpendicular to the foam growth direction. Then, the analyzed sample was placed between two plates and the compression strength was measured as a ratio of load causing 10% deformation of the sample cross-section in the parallel and perpendicular direction to the square surface. The result was the average of five measurements per each sample.

The impact test was conducted in accordance with ASTM D4812 on the pendulum 0.4 kg hammer impact velocity at 2.9 m s^{-1}, with a sample dimension of 10 mm × 10 mm × 100 mm. All tests were conducted at room temperature, 25 °C. At least five samples were prepared for the tests.

A three-point bending test was carried out using Zwick Z100 Testing Machine (Zwick/Roell Group, Ulm, Germany) at room temperature, according to ASTM D7264 (equivalent to ISO 178). The tested samples were bent with a testing speed of 2 mm min^{-1}. Obtained flexural stresses at break (ε_f) results for each sample were expressed as a mean value. The average of five measurements per each type of composition was accepted.

Dynamic mechanical analysis (DMA) was determined using an ARES Rheometer (TA Instruments, New Castle, DE, USA). Torsion geometry was used with samples of that had a thickness of 2 mm. Measurements were examined in the temperature range 20−250 °C at a heating rate of 10 °C min^{-1}, using a frequency of 1 Hz and applied deformation at 0.1%.

Surface hydrophobicity was analyzed by contact angle measurements using the sessile-drop method with a manual contact angle goniometer with an optical system OS-45D (Oscar, Taichung City,

Taiwan) to capture the profile of a pure liquid on a solid substrate. A water drop of 1 μL was deposited onto the surface using a micrometer syringe fitted with a stainless steel needle. The contact angles reported are the average of at least 10 tests on the same sample.

Water absorption of the RPUFs was measured according to ASTM D2842 (equivalent to ISO 2896). Samples were dried for 1 h at 80 °C and then weighed. The samples were immersed in distilled water to a depth of 1 cm for 24 h. Afterwards, the samples were removed from the water, held vertically for 10 s, the pendant drop was removed and then blotted between dry filter paper (Fisher Scientific, Waltham, MA, USA) at 10 s and weighed again. The average of five specimens was used.

Changes in the linear dimensions were looked into with accordance to the ASTM D2126 (equivalent to ISO 2796). The samples were conditioned at temperatures of 70 °C and −20 °C for 14 days. The change in linear dimensions was calculated in % from Equation (2).

$$\Delta l = ((l - l_0)/l_0) \times 100 \tag{2}$$

where l_0 is the length of sample before thermostating and l is the length of sample after thermostating. The average of five measurements per each type of composition was reported.

3. Results and Discussion

3.1. Characterization of POSS-Modified Polyol Premixes

One of the most important parameters for industrial processing of polyols is their rheological behavior. The dispersion of the POSS's particles in the PU matrix was evaluated by optical microscopy. Figure 3 shows the optical micrographs obtained for the samples with 0.5, 1.5 and 5 wt% of APIB-POSS and AEAPIB-POSS. A comparison of the optical images for 0.5 (Figure 3a) and 1.5 wt% of APIB-POSS (Figure 3b) with that of 5 wt% of APIB-POSS (Figure 3c) reveals that the higher concentration of particles is more difficult to disperse since the particles are closer to each other and will readily aggregate due to their high surface area and energy. An analogous trend is observed for premixes with AEAPIB-POSS (Figure 3c–e); however, the premixes show the accumulation of higher aspect ratio particles compared to the premixes with APIB-POSS.

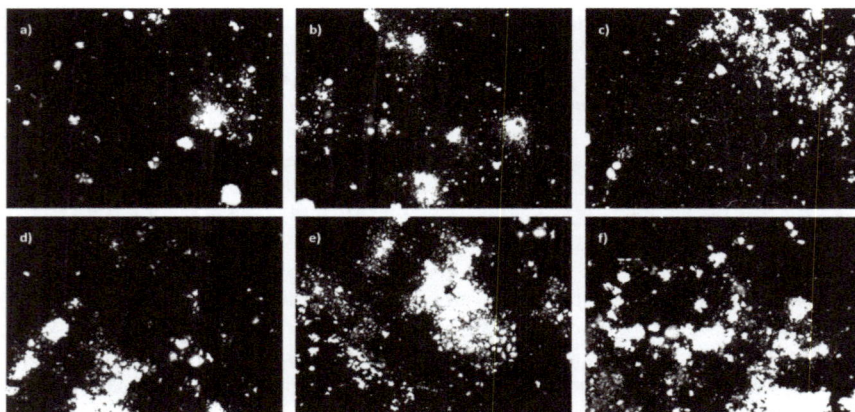

Figure 3. Polyol premixes with (**a**) 0.5 wt%, (**b**) 1.5 wt% and (**c**) 5 wt% of AEAPIB-POSS and (**d**) 0.5 wt%, (**e**) 1.5 wt% and (**f**) 5 wt% of APIB-POSS.

The viscosity of the reactive mixture was measured first, since it is a critical parameter affecting the foaming process [39]. Increased viscosity hinders bubble growth, yielding foams with a lower cell size. Table 1 shows the viscosity of the polyol mixture plus the additives and fillers used in each formulation.

The polyol premixes that contained APIB-POSS and AEAPIB-POSS are characterized by an increase in their viscosity with an increasing filler content, as a result of the presence of POSS particles interacting with the polyether polyol through hydrogen bonding and van der Waal's interaction [40]. The effect of this higher-viscosity formulation is increased for RPUFs with AEAPIB-POSS.

Table 1. Dynamic viscosity and logarithmic plot of the fitting equations for polyol premixes.

Sample Codes	Dynamic Viscosity η [mPa·s]			Fitting Equation	Power Law Index (*n*)	R^2
	0.5 RPM	5 RPM	10 RPM			
PU-0	852	423	379	y = −0.058 + 0.335	0.335	0.970
PU-APIB-0.5	1887	878	765	y = −0.061 + 0.320	0.320	0.968
PU-APIB-1.5	3285	1542	821	y = −0.059 + 0.315	0.315	0.971
PU-APIB-5	6875	4250	1358	y = −0.059 + 0.295	0.295	0.972
PU-AEAPIB-0.5	2831	1317	1148	y = −0.058 + 0.325	0.325	0.979
PU-AEAPIB-1.5	4523	2015	1105	y = −0.060 + 0.315	0.315	0.971
PU-AEAPIB-5	7540	5250	1955	y = −0.062 + 0.310	0.310	0.976

The rheological properties of polyol premixes are shown as the viscosity versus shear rate in Figure 4a. In all systems, the viscosity is generally reduced at increased shear rates. The viscosity of the samples initially decreases sharply and then significantly slows to reach a relatively stable value, due to the fact that particles of liquids reach the best possible arrangement. Such a phenomenon is typical for non-Newtonian fluids with a pseudoplastic nature and is found in many previous works [6,41]. To further analyze the data, the graph of viscosity versus shear rate is converted to log viscosity versus log shear rate form as shown in Fig 4b. From this graph, it can be seen that the curvatures of viscosity versus shear rate can be made close to linear using this log-log format with regression of 0.969–0.978. The power law index (*n*) was calculated from the slopes. All results are presented in Table 1. For the systems containing APIB-POSS, the power law index is lower than that of their AEAPIB-POSS-modified system counterparts. It indicates that the effect of the filler on the pseudoplasticity behavior becomes more significant for systems modified with APIB-POSS, leading to the highly non-Newtonian behavior.

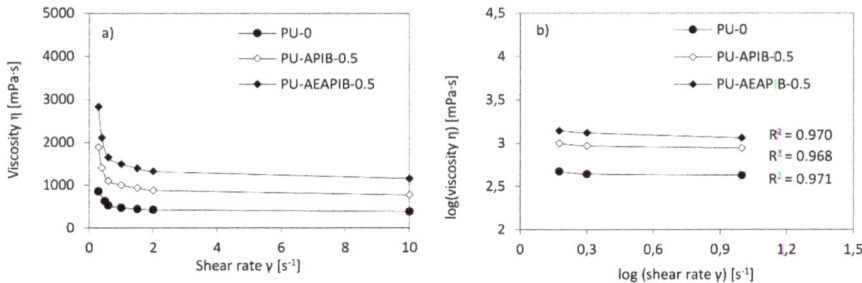

Figure 4. (**a**) Viscosity as a function of shear rate and (**b**) log-log plot of the viscosity vs. shear rate for the selected polyol premixes.

The results presented in Table 2 indicate that with the incorporation of POSS into a polyol mixture, the maximum temperature of reaction (T_{max}) slightly increases. It is believed that a higher temperature of the reaction of the modified compositions is connected with a higher reactivity of the components of the reaction mixture [42]. The presence of the additional groups as a result of the incorporation of POSS filler can lead to the exothermic reaction (see Equation (*1*)) providing more heat evaporation to the system, and consequently a higher temperature of the modified system compared with the PU-0 foam. It should also be pointed out that amine groups of the POSS can act as a catalyst and promote the polymerization reaction between polyol and isocyanate, which is an exothermic reaction [43–45].

This effect is most pronounced in the case of PU composites modified with 0.5 wt% of APIB-POSS and AEAPIB-POSS. With further increasing the concertation of POSS, the maximum temperature during the foaming process decreases. It has been shown that during the synthesis of the foams, the viscosity of the system increased with the addition of the POSS, and thus, a reduction in the efficiency of the reaction may have occured. Basically, analogue tendency has been observed by Kairytė et al. [46], who tested waste ash as a filler and flame retardant in PU foams. The authors have stated that the temperature decrease of PU composites can be attributed to the assumption that fillers absorb part of the heat generated during the foaming reaction.

Table 2. Selected properties of RPUFs.

Sample Codes	T [°C]	Cream Time [s]	Extension Time [s]	Tack-Free Time [s]	Cell Size [μm]	Wall Thickness [μm]	Apparent Density [kg m^{-3}]
PU-0	110	43 ± 4	277 ± 10	341 ± 14	472 ± 10	62 ± 4	38 ± 1
PU-APIB-0.5	135	41 ± 2	312 ± 11	320 ± 12	390 ± 8	66 ± 2	40 ± 2
PU-APIB-1.5	128	46 ± 1	358 ± 12	265 ± 10	274 ± 7	68 ± 3	41 ± 1
PU-APIB-5	120	48 ± 2	370 ± 10	260 ± 12	209 ± 8	70 ± 2	43 ± 1
PU-AEAPIB-0.5	130	50 ± 2	332 ± 12	335 ± 15	364 ± 9	68 ± 5	42 ± 2
PU-AEAPIB-1.5	126	57 ± 4	360 ± 12	277 ± 14	320 ± 8	69 ± 4	44 ± 2
PU-AEAPIB-5	112	69 ± 3	384 ± 18	270 ± 16	184 ± 10	74 ± 4	40 ± 3

3.2. Apparent Density of RPUFs

Apparent density is an important parameter that influences the properties and performances of RPUFs. The values of density of prepared foams are presented in Table 2. In general terms, the apparent density tends to increase when the APIB-POSS and AEAPIB-POSS are added. The reference foam is characterized by an apparent density of 38 kg m^{-3}. The apparent density for RPUFs increases from 40 to 43 kg m^{-3} and from 40 to 44 kg m^{-3} with an increase of APIB-POSS and AEAPIB-POSS content of 0.5 and 5 wt%, respectively. This effect can be explained by the analysis of the role of filler particles on nucleation and cell growth. The POSS particles act as nucleation sites promoting the formation of bubbles, and this is an increasing trend with nanoparticles content, but, at the same time, the growth process of the resulting cells is hindered by an increase of the gelling reaction speed, which is revealed in the greater viscosity. This results in bubble collapse and higher density foams. Moreover, it should be pointed out that another factor affecting the density of RPUFs is higher density of POSS (*ca.* 1.2 g cm^{-3}) compared to the PU foam matrix. This resulted in the increase of the apparent density of studied composites, wich is also in agreement with the results reported in the literature [47,48].

3.3. Foaming Kinetic

The foaming process was determined by measuring the characteristic processing times, such as cream, extension and tack-free time. The results presented in Table 2 indicate a slight increase in cream and extension time for the RPUFs containing APIB-POSS and AEAPIB-POSS fillers in each amount. This dependence is mostly related to the fact that well-dispersed filler in the reaction mixture acts as a nucleating agent in the nucleation process, leading to a greater bubble formation and prolonged cream time [49]. Moreover, further growth of the resulting cells seems to be hindered by the increase in viscosity of modified systems (see Table 1), leading to prolonged cream and extension times, which has also been noted by other researchers [50]. The polymerization kinetics are also affected by the presence of amine groups incorporated with the POSS molecules. Amine groups present in POSS molecules can react with isocyanates even in the absence of a catalyst [35–38]. It has been reported in previous work, that isocyanate has a higher reactivity with amines than with hydroxyl and carboxyl groups [35–38]. Because of this, most NCO groups are consumed in the formation of polyurea, and the number of NCO groups remaining to react with water is quite limited. This leads to less CO_2 gases escaping from the foam structure and thus, to a prolonged start and gel time. Compositions modified with the addition of POSS are also characterized by a shorter tack-free time, indicating that POSS particles act as a curing accelerator. The total characteristic times measured for the POSS-modified compositions are higher than those measured for the PU-0, but still in the range of operation conditions for preparing RPUFs [51,52]. Contrary results were obtained by Liu et al. [17] who determined that compositions modified with waste ash were characterized by a longer tack-free time as well, indicating that the waste ash did not act as a curing accelerator. The authors have stated that this might be related to the fact that not all fillers determine the same reaction manner that is related to the chemical composition and particle size distribution leading to different foaming kinetics.

3.4. Cell. Structure of RPUFs

The morphology of the RPUFs is shown in Figure 5, showing polygon closed-cell structures with many windows. The values of the cell sizes of the foams were statistically analyzed by means of *ImageJ* software from SEM images and the median values are summarized in Table 2.

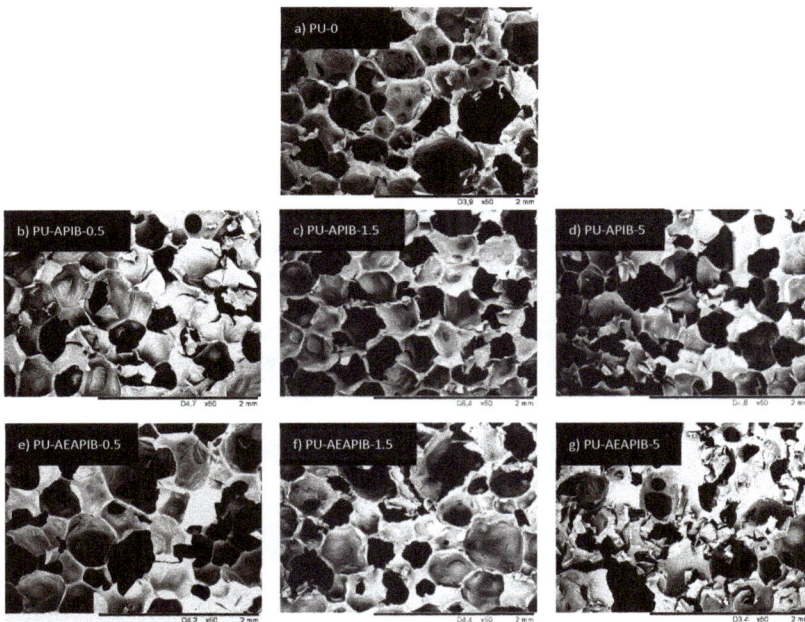

Figure 5. Morphology of (**a**) PU-0, (**b**) PU-APIB-0.5, (**c**) PU-APIB-1.5, (**d**) PU-APIB-5, (**e**) PU-AEAPIB-0.5, (**f**) PU-AEAPIB-1.5, and (**g**) PU-AEAPIB-5 observed at the same magnification.

As observed from the micrograph of the PU-0 (Figure 5a), the cell size and cell distribution are nearly uniform and the foam consists of closed cells with a negligible amount of cells with broken walls. Micrographs of the RPUFs with APIB-POSS are shown in Figure 5b–d. With the addition of APIB-POSS, the overall cell structure became less uniform and the amount of broken cells increased. The filled foams do not exhibit pronounced preferential orientation either; however, they are more irregular and a defective shape of cells with many cracks is observed. The morphology of PU-APIB-0.5 (Figure 5b) is mostly comparable to the PU-0, but with a further increase in the POSS content up to 1.5 wt% (Figure 5b), the overall structure becomes less uniform with a greater content of open cells and noticeable voids present in the structure. This trend is further prominent at 5 wt% filler content (Figure 5d), where an inhomogeneous structure and defective shape of cells with many cracks is observed. A similar trend is observed in RPUFs with AEAPIB-POSS, as shown in Figure 5e–g. With the increasing weight percent of AEAPIB-POSS, the cell distribution is transformed into non-uniform shapes and damaged cells appear. The explanation may be found in poor interfacial adhesion between the filler surface and polymer matrix, which promotes earlier cell collapsing phenomena and increases the high probability of generating open pores [53]. Moreover, the possible interphase interactions between POSS particles and polyurethane in cell struts disturbs the formation of a stable, non-defective foam structure [54]. The dispersion of the particles of used fillers in RPUFs is presented in Figure 6. It is clearly visible that for both series of modified foams, POSS particles are attached to the cell wall. Some dots and projections also become detectable in the cell void and a coarse surface can be seen in the cell struts.

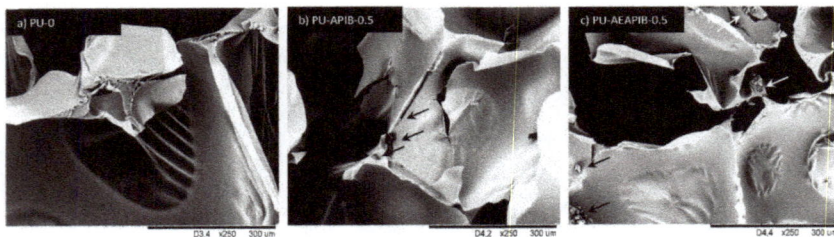

Figure 6. Morphology of (**a**) PU-0, (**b**) PU-APIB-0.5 and (**c**) PU-AEAPIB-0.5 observed at the same magnification.

Table 2 presents the average cell sizes for the prepared PU foams. The addition of POSS into the PU matrix up to 5 wt% leads to a noticeable decrease in cell size, which can be also seen in the SEM images presented in Figure 5. It seems that filler particles can act as gas nucleation sites during the foaming process and assist in the formation of nucleation centers for the gaseous phase [53], thus affecting local rheology surrounding the growing bubbles [7]. The addition of powder filler can change the nucleation mode from homogenous to heterogeneous and reduce the nucleation energy, which in turn promotes the formation of large numbers of small cells [53], increasing the tendency of cell coalescence and leading to higher inhomogeneous cell size distribution. Comparing Figure 7a with Figure 7b, it is obvious that the incorporation of APIB-POSS resulted in smaller cells and a more homogenous structure as compared to their AEAPIB-POSS-modified RPUFs counterparts. In the case of RPUFs modified with APIB-POSS, most pores are located in the range of 300-350 μm. In the case of APIB-POSS-modified RPUFs, two populations of pores can be distinguished: large ones with a diameter of about 800 μm and small ones whose size is about 300 μm. A similar tendency has been reported in other studies [55–57].

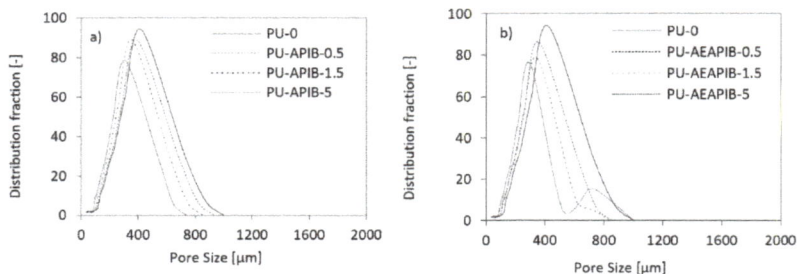

Figure 7. Cell size distributions of RPUFs modified with (**a**) APIB-POSS and (**b**) AEAPIB-POSS.

3.5. Compressive Strength of RPUFs

Another important parameter that impacts performance characteristics is the compressive strength, and the change in its value are presented in Figure 8a,b. The compressive strength of all materials tested in the direction parallel and perpendicular to the direction of foam rise is greater than the strength of the reference foam. In the case of foams with APIB-POSS, the largest increase in compressive strength is observed for the 0.5 wt% load (compared to the PU-0) and it is about 351 kPa in the parallel direction and 159 kPa in the perpendicular direction. In the foams containing 1.5 and 5 wt% of APIB-POSS, there is a decrease in compressive strength compared with RPUFs containing 0.5 wt% of APIB-POSS; however, it is still larger than for the reference foam. A similar trend is observed for foams modified with AEAPIB-POSS. With an increasing concentration of the modifier, the compressive strength decreases from 287 to 255 kPa in the perpendicular direction and from 152 to 140 kPa in the parallel direction. Nonetheless, similar to the APIB-POSS counterparts, the compressive strength of AEAPIB-modified foams is higher compared to the one of reference foams, except for the foam modified with 5 wt% of AEAPIB-POSS.

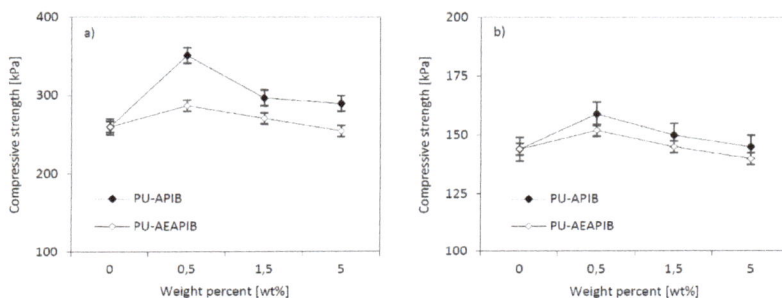

Figure 8. Effect of POSS content on the compressive strength of RPUFs measured (**a**) parallel and (**b**) perpendicular to the foam rise direction.

Such changes in the mechanical properties of composite samples can be explained in terms of characteristic features of their structure. As presented in Figure 5a, reference foam has a mostly spherical and equally distributed cell structure. With increased filler content, it could be observed that foam cell structure becomes more distorted and has a less uniform distribution. At this time, if there is an application of loading, bending and shrinkage of cell walls occurs and results in the development of micro cracks. Therefore, foam strength extremely depends on the initiation of micro cracks and forces on their growth [25]. So it can be explained by the decreasing foam compressive strength with crack initiation and growth.

Moreover, the reason for the decreasing mechanical strength at a high filling rate might also be related to the non-uniform dispersion of particles and polyol mixture. A high tendency to aggregate filler particles, noticeable in the structure, leads to a weakened interfacial adhesion between the filler and effective active surface. In consequence, RPUFs are characterized by a microphase separation of the structure, which leads to the failure of samples in an unexpected manner at random locations in the samples. The non-uniform concentration of the filler in some regions contributed to the embrittlement effect of polymer structures, inhibiting the enhancement of the mechanical properties of RPUFs. By increasing the content of the POSS, the negative effects of the filler such as disruption of the formation of hydrogen bonds and reaction stoichiometry, the probability of the agglomeration of nanoparticles due to the increase of viscosity and inappropriate distribution of nanoparticles is increased. Therefore, the particles interaction with PU macromolecules is decreased and the mechanical properties are weakened. The poor interfacial adhesion between some particles, especially the loose ones as discussed above, the polymer matrix and the uneven dispersion of the filler may lead to the above results, as proven by other authors [58–61].

3.6. Impact Strength of RPUFs

With the increasing concentration of APIB-POSS and AEAPIB-POSS fillers from 0.5 to 5 wt%, the impact strength decreases from 0.46 to 0.31 kJ m^{-2} and from 0.39 to 0.32 kJ m^{-2} as shown in

Figure 9. The impact strength shows a clear increase for RPUFs with APIB-POSS and AEAPIB-POSS by an amount of 0.5 wt% of the fillers. This behavior is related to the good interface reinforcement matrix and the generation of fracture paths through the POSS-reinforced RPUFs. The impact strength decreases with the increasing concentration of APIB-POSS and AEAPIB-POSS from 1.5 to 5 wt%. This result is caused by the POSS particles that serve as points for a localized stress concentration from which the failure begins or generally because of the elasticity reduction of the material with the POSS addition. Thus, the deformability of the RPUF's matrix is reduced, which in turn affects the ductility in the foam surface. With this effect, the foam composite tends to form a weak structure and increase the concentration of POSS, thus reducing the foam's energy absorption, resulting in reduced toughness and impact strength.

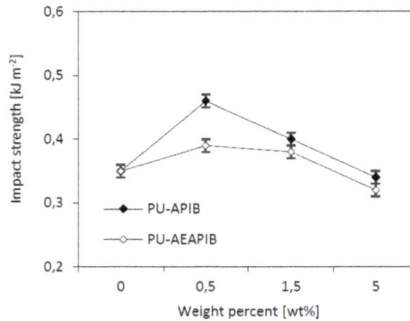

Figure 9. Effect of POSS's content on impact strength of RPUFs.

3.7. Flexural Strength of RPUFs

The comparison between impact strength and tensile strength (σ_f) shows that the same trend can be observed in both properties (Figure 10). Compared to the PU-0, RPUFs modified with 0.5 wt% of APIB-POSS and AEAPIB-POSS exhibits a slight improvement of the σ_f. This behavior is attributed to the good interaction between the filler particles and PU matrix that delays the failure of the material until the mechanical strength of the fillers combined with the PU is reached. The incorporation of both fillers in the amount of 1.5 and 5 wt% leads to a deterioration of σ_f, as a result of greater elasticity, connected with the cellular morphology of PU foams (see Figure 5). Due to an uneven distribution of the fillers in the matrix and many clusters present in the structure of modified materials, the mechanical properties of the resulting composite are reduced. The lack of a reinforcing effect with the incorporation of the filler was also observed in previous studies [62,63].

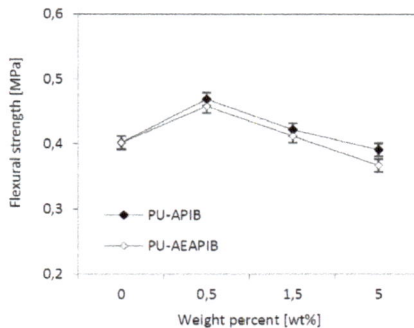

Figure 10. Effect of POSS's content on flexural strength (σ_f) of RPUFs.

3.8. Dynamic Mechanical Analysis (DMA)

RPUFs prepared with different weight percentages of APIB-POSS and AEAPIB-POSS were analyzed by DMA. The most interesting observation from this analysis is the displacement of the peak in the tanδ curves toward higher temperatures for samples modified with APIB-POSS and AEAPIB-POSS. The temperature of the maximum of the peaks associated to the main relaxation of the matrix is related to the PU glass transition temperature (T_g). As shown in Figure 11a,b, the reference and POSS-modified foams exhibit one wide peak in the range of temperature analyzed. The width of the peak becomes broader with the POSS incorporation due to different relaxation mechanisms appearing in the modified materials as a consequence of the added filler. The broadening of the tanδ peak is often assumed to be due to a broader distribution in molecular weight between crosslinking points or heterogeneities in the network structures [17].

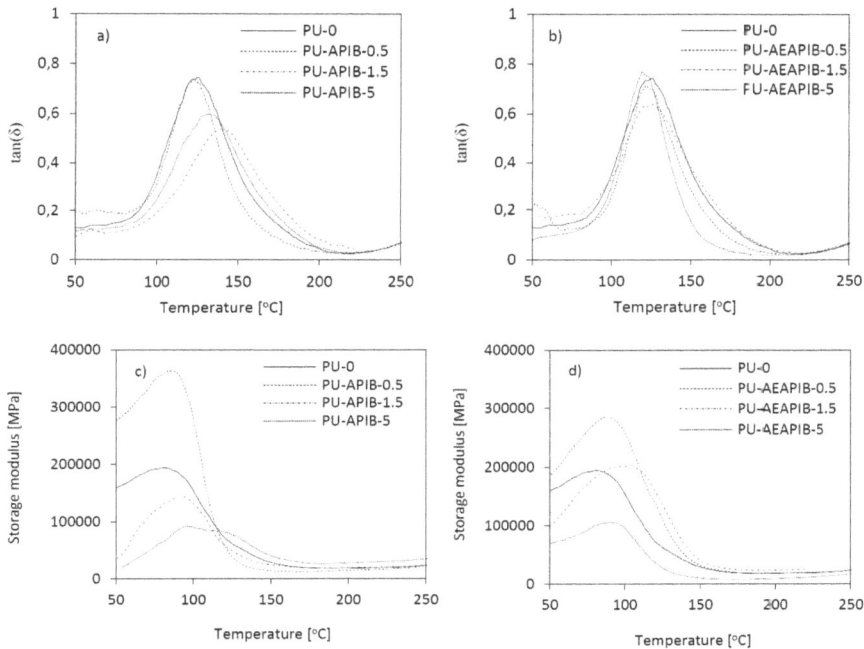

Figure 11. (**a,b**) tanδ and (**c,d**) storage modulus as a function of temperature plotted for RPUFs modified with APIB-POSS and AEAPIB-POSS.

The comparison of PU-0 and POSS-modified composites shows that the addition of 0.5 wt% of APIB-POSS and AEAPIB-POSS results in an increase of storage modulus (E'), however, with the increasing content of the filler in the PU matrix, the value of the storage modulus significantly decreases (Figure 11c,d). This decrement in E' is attributed to the beginning of a thermal transition, which is associated with hard segments phase. The changes observed around 100 °C are attributable to the presence of a high concentration of hydrogen-bonded aromatic urethane groups in the poly(ether-urethane) phase and hard-segment domains which act as macroscopic cross links. The higher E' for RPUFs with 0.5 wt% of APIB-POSS and AEAPIB-POSS indicates more restricted mobility compared to the RPUFs with higher concentrations (1.5 and 5 wt%) of fillers; thus, the introduction of POSS over a certain optimal level, in this kind of RPUFs, produces more flexible materials. This might be connected with the fact that the PU matrix itself is highly crosslinked and an excess of POSS particles may actually disrupt this crosslinking, even if it is multifunctional. It has also been observed, that during the synthesis of the foams the viscosity of the system increased with the addition of the POSS (see Table 1), and thus a reduction in the efficiency of the reaction may have occured, leading to the more mobile polymeric network. It should also be pointed out that the higher viscosity of the system can also pronounce the aggregation of POSS at high loadings, and thus, the interaction with the matrix may be hindered.

3.9. Thermogravimetric Analysis (TGA)

The thermal stability of the RPUFs was characterized by thermogravimetric analysis (TGA) and derivative thermogravimetry (DTG), as illustrated in Figure 12a,b and Figure 12c,d, respectively.

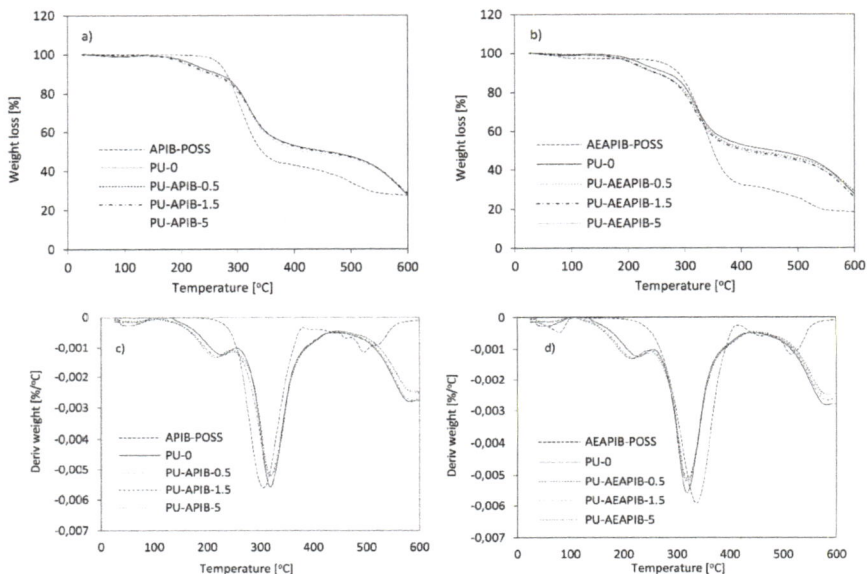

Figure 12. TGA curves for RPUFs modified with (**a**) APIB-POSS, (**b**) AEAPIB-POSS and DTG curves for RPUFs modified with (**c**) APIB-POSS and (**d**) AEAPIB-POSS.

Typically, the thermal degradation of RPUFs consists of three stages. The first step of decomposition is connected with the dissociation of the urethane bond at a temperature between 150 and 330 °C (corresponding to the temperature at 10% of total weight losses) [64,65]. The second step of the degradation of RPUFs occurs in the temperature range between 330 and 400 °C and is ascribed to the decomposition of soft polyol segments (corresponding to the temperature at 50% of total weight losses) [66]. The third step of degradation, ascribed to the degradation of the fragments generated during the second step, which corresponds to the loss of weight of about 80%, occurs at a temperature of 500 °C [64,67].

It has been noticed (Table 3) that there was no significant effect of POSS on the thermal stability of RPUFs, despite the fact that POSS compounds have a significantly different stability from PU-0. However, samples containing 0.5 wt% of POSS are characterized by the highest temperature at 70% weight loss, which indicates the POSS stabilizing properties. On the other hand, increasing the content of POSS causes more NCO groups to be involved in the formation of polyureas, and the number of NCO groups reacted with the polyol is quite limited. This leads to a reduction in the content of more thermally stable polyurethane groups compared to polyurea groups. The degradation of POSS-modified foams in the atmosphere of air runs in three distinct stages with maximum of mass loss at ca. 316–321 °C and 581–586 °C and involves the reaction of oxygen to form hydroperoxides which themselves are unstable and succumbing to decomposition to form more free radicals [68]. The slight increase in the char residues gives rise to the formation of more stable char layers, which may protect the materials from further decomposition and in turn increase the thermal stability. Compared to the foams modified with AEAPIB-POSS, the composites containing APIB-POSS are relatively less thermally stable. The degradation stage of the RPUFs modified with APIB-POSS starts at 245–247 °C, while the composite foams modified with AEAPIB-POSS start to degrade at slightly higher temperatures (248–251 °C). In both cases, for the highest filler content of 5 wt%, one may observe the deterioration of thermal properties. The decrease can be ascribed to the poor dispersion of the fillers and changes of the crosslinking density [6]. This is in agreement with the SEM images, where it is clearly visible that the presence of POSS increases the heterogeneity of RPUF's morphology. The TGA results also suggest that the presence of POSS reduced the weight loss of RPUFs in the initial stage of degradation. This can be attributed to the barrier effect provided by POSS which reduces both the heat and oxygen fluxes toward the polymer surface, which limits the weight loss rate.

Table 3. The results of thermogravimetric analysis of RPUFs.

Sample Code	T_g [°C]	T_5 [°C]	T_{10} [°C]	T_{50} [°C]	T_{70} [°C]	Char Residue [%]
APIB-POSS		267	280	342	531	21.6
AEAPIB-POSS		260	287	350	449	18.4
PU-0	126	220	265	454	591	27.9
PU-APIB-0.5	126	205	245	418	596	29.0
PU-APIB-1.5	129	206	247	406	590	28.6
PU-APIB-5	142	206	247	394	590	27.6
PU-AEAPIB-0.5	126	206	248	418	595	29.2
PU-AEAPIB-1.5	129	212	254	444	593	28.2
PU-AEAPIB-5	122	209	251	415	590	27.9

3.10. Contact Angle and Water Absorption

Polihedral oligomeric silsesquioxanes functionalized by amino groups significantly affected the hydrophobicity of the foams (Figure 13). Regarding water absorption, it is notable that foams modified by POSS absorb less water than the reference sample. This trend decreases with the increase in POSS content in the foam, but still, the affinity for water is lower for the POSS-modified RPUFs system. The use of both APIB-POSS and AEAPIB-POSS allowed the reduction of water absorption by almost one-third compared to the output foam without a modifier.

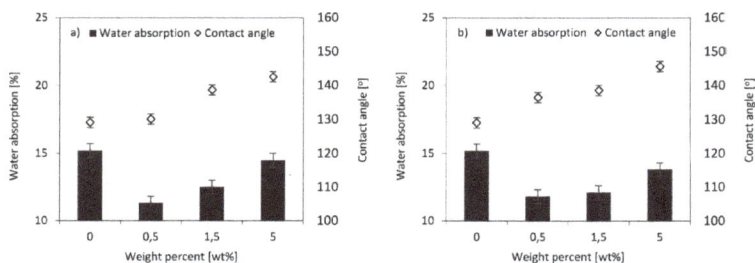

Figure 13. Effect of contact angle on water absorption of RPUFs modified with (**a**) APIB-POSS and (**b**) AEAPIB-POSS.

This tendency is explained by the hydrophobic character of isobutyl-POSS with reactive amino groups [69,70]. The contact angles (θ) of the RPUF composites with various weight percentages of APIB-POSS and AEAPIB-POSS also confirm this trend (Figure 14). It can be observed from the obtained results that by increasing both POSS contents, the water-contact angle of composite foams is increased. With a maximum of 5% of POSS, wettability was significantly reduced. This result implied the improvement of the foam hydrophobicity, leading to a limited absorption of water by the analyzed foams.

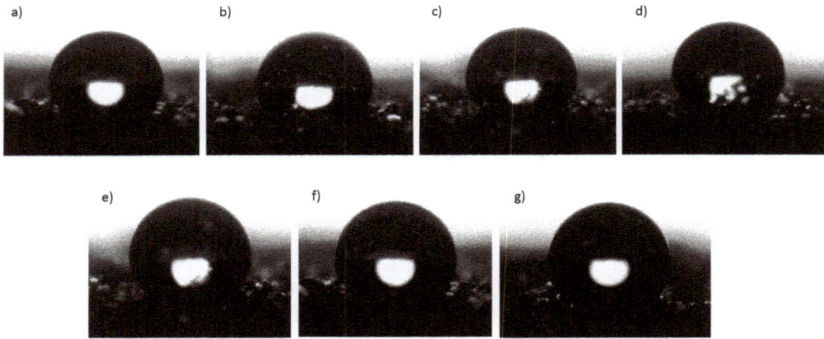

Figure 14. Contact angle of the surface of the (**a**) PU-0, (**b**) PU-APIB-0.5, (**c**) PU-APIB-1.5, (**d**) PU-APIB-5, (**e**) PU-AEAPIB-0.5, (**f**) PU-AEAPIB-1.5, and (**g**) PU-AEAPIB-5.

The % linear changes in length, width and thickness after exposure at 70 °C and −20 °C for up to 14 days for samples modified with APIB-POSS and AEAPIB-POSS are presented in Figure 15. The dimensional stability of RPUFs indicates that the addition of both fillers results in negligible changes of dimensional stability of the modified foams in relation to the PU-0. In all cases, the variations in the sample's dimensions after the special treatment are random and thus, they can be attributed mostly to experimental errors while measuring. According to the industrial standard, the PU panels tested at 70 °C should have less than 3% of the linear change [71]. In each case, the dimensional stability of RPUFs is thus still considered to be mild and within commercially acceptable limits [71].

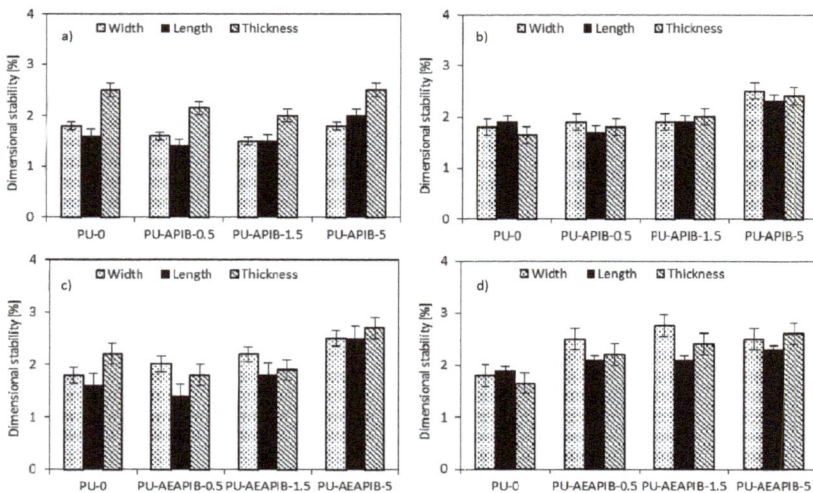

Figure 15. Dimensional stability of RPUFs modified with APIB-POSS (**a,b**) and AEAPIB-POSS (**c,d**) after exposure at 70 °C and −20 °C, respectively.

Polymers **2019**, *11*, 336

4. Conclusions

RPUFs were successfully reinforced using APIB-POSS and AEAPIB-POSS. The impact of POSS on the thermal properties, dynamic mechanical properties, physico-mechanical properties (compressive strength, three-point bending test, impact strength apparent density, and dimensional stability), foaming parameters, and morphology of RPUFs was examined. The presented results indicate that the addition of APIB-POSS and AEAPIB-POSS in the range of 0.5–5 wt% influences the morphology of analyzed foams and consequently, their mechanical and thermal properties. It was noticed that RPUFs modified with APIB-POSS are characterized by smaller and more regular polyurethane cells. This suggests better compatibility between the PU foam matrix and APIB-POSS, compared to the AEAPIB-POSS filler. This results in a significant improvement of the physico-mechanical properties and thermal stability of composites with APIB-POSS. For example, compared to the AEAPIB-POSS-modified RPUFs, composition with 0.5 wt% of the APIB-POSS showed greater compressive strength (351 kPa), higher flexural strength (0.458 MPa) and less water uptake (11% after 24 h). The results obtained in this study confirm that the addition of APIB-POSS and AEAPIB-POSS over a certain optimal level has a negative effect on cell morphology. The addition of both fillers in an amount of 5 wt% led to samples with reduced compression modulus, compressive strength, thermal transitions, and storage modulus with respect to the RPUFs containing 0.5 and 1.5 wt% of the fillers, mainly due to detrimental changes induced by the fillers.

Author Contributions: Each author contributed to the design and implementation of the research, to the analysis of the results and to the writing of the manuscript.

Funding: This research received no external funding.

Conflicts of Interest: The authors declare no conflict of interest.

References

1. Tan, S.; Abraham, T.; Ference, D.; Macosko, C.W. Rigid polyurethane foams from a soybean oil-based Polyol. *Polymer* **2011**, *52*, 2840–2846. [CrossRef]
2. Nikje, M.M.A.; Noruzian, M.; Moghaddam, S.T. Investigation of Fe_3O_4/AEAP supermagnetic nanoparticles on the morphological, thermal and magnetite behavior of polyurethane rigid foam nanocomposites. *Polimery* **2015**, *60*, 26–32. [CrossRef]
3. Xie, H.; Yang, W.; Yuen, A.C.Y.; Xie, C.; Xie, J.; Lu, H.; Yeoh, G.H. Study on flame retarded flexible polyurethane foam/alumina aerogel composites with improved fire safety. *Chem. Eng. J.* **2017**, *311*, 310–317. [CrossRef]
4. Yang, C.; Fischer, L.; Maranda, S.; Worlitschek, J. Rigid polyurethane foams incorporated with phase change materials: A state-of-the-art review and future research pathways. *Energy Build.* **2015**, *87*, 25–36. [CrossRef]
5. Zieleniewska, M.; Leszczyński, M.K.; Kurańska, M.; Prociak, A.; Szczepkowski, L.; Krzyżowska, M.; Ryszkowska, J. Preparation and characterisation of rigid polyurethane foams using a rapeseed oil-based polyol. *Ind. Crop. Prod.* **2015**, *74*, 887–897. [CrossRef]
6. Yan, D.-X.; Xu, L.; Chen, C.; Tang, J.; Ji, X.; Li, Z. Enhanced mechanical and thermal properties of rigid polyurethane foam composites containing graphene nanosheets and carbon nanotubes. *Polym. Int.* **2012**, *61*, 1107–1114. [CrossRef]
7. Widya, T.; Macosko, C.W. Nanoclay-Modified Rigid Polyurethane Foam. *J. Macromol. Sci.* **2005**. *44*, 897–908. [CrossRef]
8. Madaleno, L.; Pyrz, R.; Crosky, A.; Jensen, L.R.; Rauhe, J.C.M.; Dolomanova, V.; de Barros Timmons, A.M.M.V.; Pinto, J.J.C.; Norman, J. Processing and characterization of polyurethane nanocomposite foam reinforced with montmorillonite–carbon nanotube hybrids. *Composites* **2013**, *44*, 1–7. [CrossRef]
9. Thirumal, M.; Khastgir, D.; Singha, N.K.; Manjunath, B.S.; Naik, Y.P. Effect of a Nanoclay on the Mechanical, Thermal and Flame Retardant Properties of Rigid Polyurethane Foam. *J. Macromol. Sci.* **2009**, *46*, 704–712. [CrossRef]
10. Modesti, M.; Lorenzetti, A.; Besco, S. Influence of nanofillers on thermal insulating properties of polyurethane nanocomposites foams. *Polym. Eng. Sci.* **2007**, *47*, 1351–1358. [CrossRef]

11. Dolomanova, V.; Rauhe, J.C.M.; Jensen, L.R.; Pyrz, R.; Timmons, A.B. Mechanical properties and morphology of nano-reinforced rigid PU foam. *J. Cell. Plast.* **2011**, *47*, 81–93. [CrossRef]

12. Zhang, L.; Zhang, M.; Zhou, Y.; Hu, L. The study of mechanical behavior and flame retardancy of castor oil phosphate-based rigid polyurethane foam composites containing expanded graphite and triethyl phosphate. *Poly. Degrad. Stab.* **2013**, *98*, 2784–2794. [CrossRef]

13. Cao, X.; Lee, L.J.; Widya, T.; Macosko, C. Polyurethane/clay nanocomposites foams: Processing, structure and properties. *Polymer* **2005**, *46*, 775–783. [CrossRef]

14. Cai, D.; Jin, J.; Yusoh, K.; Rafiq, R.; Song, M. High performance polyurethane/functionalized graphene nanocomposites with improved mechanical and thermal properties. *Compos. Sci. Technol.* **2012**, *72*, 702–707. [CrossRef]

15. Pauzi, N.N.P.N.; Majid, R.A.; Dzulkifli, M.H.; Yahya, M.Y. Development of rigid bio-based polyurethane foam reinforced with nanoclay. *Composites* **2014**, *67*, 521–526. [CrossRef]

16. Uthaman, N.; Majeed, A. Pandurangan Impact Modification of Polyoxymethylene (POM). *e-Polymers* **2006**, *6*, 1–9. [CrossRef]

17. Liu, H.; Zheng, S. Polyurethane Networks Nanoreinforced by Polyhedral Oligomeric Silsesquioxane. *Macromol. Rapid Commun.* **2005**, *26*, 196–200. [CrossRef]

18. Hernandez, R.; Weksler, J.; Padsalgikar, A.; Runt, J. Microstructural Organization of Three-Phase Polydimethylsiloxane-Based Segmented Polyurethanes. *Macromolecules* **2007**, *40*, 5441–5449. [CrossRef]

19. Raftopoulos, K.N.; Koutsoumpis, S.; Jancia, M.; Lewicki, J.P.; Kyriakos, K.; Mason, H.E.; Harley, S.J.; Hebda, E.; Papadakis, C.M.; Pielichowski, K. Reduced Phase Separation and Slowing of Dynamics in Polyurethanes with Three-Dimensional POSS-Based Cross-Linking Moieties. *Macromolecules* **2015**, *48*, 1429–1441. [CrossRef]

20. Paul, D.; Robeson, L. Polymer nanotechnology: Nanocomposites. *Polymer* **2008**, *49*, 3187–3204. [CrossRef]

21. Song, X.Y.; Geng, H.P.; Li, Q.F. The synthesis and characterization of polystyrene/magnetic polyhedral oligomeric silsesquioxane (POSS) nanocomposites. *Polymer* **2006**, *47*, 3049–3056. [CrossRef]

22. Lee, Y.-J.; Huang, J.-M.; Kuo, S.-W.; Lu, J.-S.; Chang, F.-C. Polyimide and polyhedral oligomeric silsesquioxane nanocomposites for low-dielectric applications. *Polymer* **2005**, *46*, 173–181. [CrossRef]

23. Liu, Y.; Tseng, M.; Fangchiang, M. Polymerization and nanocomposites properties of multifunctional methylmethacrylate POSS. *J. Polym. Sci.* **2008**, *46*, 5157–5166. [CrossRef]

24. Rashid, E.S.A.; Ariffin, K.; Kooi, C.C.; Akil, H.M. Preparation and properties of POSS/epoxy composites for electronic packaging applications. *Mater. Des.* **2009**, *30*, 1–8. [CrossRef]

25. Liu, L.; Tian, M.; Zhang, W.; Zhang, L.; Mark, J.E. Crystallization and morphology study of polyhedral oligomeric silsesquioxane (POSS)/polysiloxane elastomer composites prepared by melt blending. *Polymer* **2007**, *48*, 3201–3212. [CrossRef]

26. Pellice, S.A.; Fasce, D.P.; Williams, R.J.J.; Pellice, S. Properties of epoxy networks derived from the reaction of diglycidyl ether of bisphenol A with polyhedral oligomeric silsesquioxanes bearing OH–functionalized organic substituents. *J. Polym. Sci.* **2003**, *41*, 1451–1461. [CrossRef]

27. Matějka, L.; Strachota, A.; Pleštil, J.; Whelan, P.; Steinhart, M.; Šlouf, M. Epoxy networks reinforced with polyhedral oligomeric silsesquioxanes (POSS). Structure and morphology. *Macromolecules* **2004**, *37*, 9449–9456. [CrossRef]

28. Prządka, D.; Jęczalik, J.; Andrzejewska, E.; Marciniec, B.; Dutkiewicz, M.; Szłapka, M.; Jęczalik, J. Novel hybrid polyurethane/POSS materials via bulk polymerization. *React. Funct. Polym.* **2013**, *73*, 114–121. [CrossRef]

29. Raftopoulos, K.N.; Janowski, B.; Apekis, L.; Pissis, P.; Pielichowski, K. Direct and indirect effects of POSS on the molecular mobility of polyurethanes with varying segment Mw. *Polymer* **2013**, *54*, 2745–2754. [CrossRef]

30. Seifalian, A.; Ghanbari, H. Cardiovascular application of polyhedral oligomeric silsesquioxane nanomaterials: A glimpse into prospective horizons. *Int. J. Nanomed.* **2011**, *6*, 775–786. [CrossRef] [PubMed]

31. Sanchez, C.; Soler-Illia, G.J.D.A.A.; Ribot, F.; Lalot, T.; Mayer, C.R.; Cabuil, V. Designed Hybrid Organic–Inorganic Nanocomposites from Functional Nanobuilding Blocks. *Chem. Mater.* **2001**, *13*, 3061–3083. [CrossRef]

32. Hebda, E.; Ozimek, J.; Raftopoulos, K.N.; Michałowski, S.; Pielichowski, J.; Jancia, M.; Pielichowski, K. Synthesis and morphology of rigid polyurethane foams with POSS as pendant groups or chemical crosslinks. *Polym. Adv. Technol.* **2015**, *26*, 932–940. [CrossRef]

33. Michałowski, S.; Pielichowski, K. 1,2-Propanediolizobutyl POSS as a co-flame retardant for rigid polyurethane foams. *J. Therm. Anal. Calorim.* **2018**, *134*, 1351–1358. [CrossRef]

34. Michałowski, S.; Hebda, E.; Pielichowski, K. Thermal stability and flammability of polyurethane foams chemically reinforced with POSS. *J. Therm. Anal. Calorim.* **2017**, *130*, 155–163. [CrossRef]

35. Mráz, J.; Simek, P.; Chvalová, D.; Nohová, H.; Šmigolová, P. Studies on the methyl isocyanate adducts with globin. *Chem. Biol. Interact.* **2004**, *148*, 1–10. [CrossRef] [PubMed]

36. Schwetlick, K.; Noack, R.; Stebner, F. Three fundamental mechanisms of base-catalysed reactions of isocyanates with hydrogen-acidic compounds. *J. Chem. Soc. Perkin Trans. 2* **1994**, *2*, 599. [CrossRef]

37. Arnold, R.G.; Nelson, J.A.; Verbanc, J.J. Recent Advances in Isocyanate Chemistry. *Chem. Rev.* **1957**, *57*, 47–76. [CrossRef]

38. Guo, C.; Zhou, L.; Lv, J. Effects of Expandable Graphite and Modified Ammonium Polyphosphate on the Flame-Retardant and Mechanical Properties of Wood Flour-Polypropylene Composites. *Polym. Polym. Compos.* **2013**, *21*, 449–456. [CrossRef]

39. Kairytė, A.; Vaitkus, S.; Vėjelis, S.; Girskas, G.; Balčiūnas, G. Rapeseed-based polyols and paper production waste sludge in polyurethane foam: Physical properties and their prediction models. *Ind. Crop Prod.* **2018**, *112*, 119–129. [CrossRef]

40. Amin, M.; Najwa, K. Cellulose Nanocrystals Reinforced Thermoplastic Polyurethane Nanocomposites. Ph.D. Thesis, Australian Institute for Bioengineering and Nanotechnology, The University of Queensland, Brisbane, Australia, 2016. [CrossRef]

41. Michalowski, S.; Cabulis, U.; Kirpluks, M.; Prociak, A.; Kuranska, M. Microcellulose as a natural filler in polyurethane foams based on the biopolyol from rapeseed oil. *Polimery* **2016**, *61*, 625–632.

42. Malewska, E.; Prociak, A. The effect of nanosilica filler on the foaming process and properties of flexible polyurethane foams obtained with rapeseed oil-based polyol. *Polimery* **2015**, *60*, 472–479. [CrossRef]

43. Farkas, A.; Strohm, P.F. Mechanism of Amine-Catalyzed Reaction of Isocyanates with Hydroxyl Compounds. *Ind. Eng. Chem. Fund.* **1965**, *4*, 32–38. [CrossRef]

44. Silva, A.L.; Bordado, J.C. Recent Developments in Polyurethane Catalysis: Catalytic Mechanisms Review. *Catal. Rev.* **2004**, *46*, 31–51. [CrossRef]

45. Ni, H.; Nash, H.A.; Worden, J.G.; Soucek, M.D. Effect of catalysts on the reaction of an aliphatic isocyanate and water. *J. Polym. Sci.* **2002**, *40*, 1677–1688. [CrossRef]

46. Kairytė, A.; Kizinievič, O.; Kizinievič, V.; Kremensas, A. Synthesis of biomass-derived bottom waste ash based rigid biopolyurethane composite foams: Rheological behaviour, structure and performance characteristics. *Composites* **2019**, *117*, 193–201. [CrossRef]

47. Gama, N.V.; Silva, R.; Mohseni, F.; Davarpanah, A.; Amaral, V.; Ferreira, A.; Barros-Timmons, A. Enhancement of physical and reaction to fire properties of crude glycerol polyurethane foams filled with expanded graphite. *Polym. Test.* **2018**, *69*, 199–207. [CrossRef]

48. Kurańska, M.; Prociak, A.; Cabulis, U.; Kirpluks, M.; Ryszkowska, J.; Auguścik, M. Innovative porous polyurethane-polyisocyanurate foams based on rapeseed oil and modified with expandable graphite. *Ind. Crop. Prod.* **2017**, *95*, 316–323. [CrossRef]

49. Takahashi, J.A.; Chaussy, D.; Silva, G.G.; Silva, M.C.; Belgacem, M.N. Composites of rigid polyurethane foam and cellulose fiber residue. *J. Appl. Polym. Sci.* **2010**, *117*, 3665–3672.

50. Song, Z.-L.; Ma, L.-Q.; Wu, Z.-J.; He, D.-P. Effects of viscosity on cellular structure of foamed aluminum in foaming process. *J. Mater. Sci.* **2000**, *35*, 15–20. [CrossRef]

51. John, J.; Bhattacharya, M.; Turner, R.B. Characterization of polyurethane foams from soybean oil. *J. Appl. Polym. Sci.* **2002**, *86*, 3097–3107. [CrossRef]

52. Jin, J.F.; Chen, Y.L.; Wang, D.N.; Hu, C.P.; Zhu, S.; VanOverloop, L.; Randall, D. Structures and physical properties of rigid polyurethane foam prepared with rosin-based polyol. *J. Appl. Polym. Sci.* **2002**, *84*, 598–604. [CrossRef]

53. Sung, G.; Kim, J.H. Influence of filler surface characteristics on morphological, physical, acoustic properties of polyurethane composite foams filled with inorganic fillers. *Compos. Sci. Technol.* **2017**, *146*, 147–154. [CrossRef]

54. Gu, R.; Khazabi, M.; Sain, M. Fiber reinforced soy-based polyurethane spray foam insulation. Part 2: Thermal and mechanical properties. *BioResources* **2011**, *6*, 3775–3790.

55. Tian, H.; Rajulu, A.; Zhang, S.; Xiang, A. Water-Blown Castor Oil-Based Polyurethane Foams with Soy Protein as a Reactive Reinforcing Filler. *J. Polym. Environ.* **2016**, *26*, 15–22.
56. Wolska, A.; Goździkiewicz, M.; Ryszkowska, J. Thermal and mechanical behaviour of flexible polyurethane foams modified with graphite and phosphorous fillers. *J. Mater. Sci.* **2012**, *47*, 5627–5634. [CrossRef]
57. Formela, K.; Hejna, A.; Zedler, Ł.; Przybysz, M.; Ryl, J.; Saeb, M.R.; Piszczyk, Ł. Structural, thermal and physico-mechanical properties of polyurethane/brewers' spent grain composite foams modified with ground tire rubber. *Ind. Crop Prod.* **2017**, *108*, 844–852. [CrossRef]
58. Verdejo, R.; Stämpfli, R.; Alvarez-Lainez, M.; Mourad, S.; Rodriguez-Perez, M.; Brühwiler, P.; Shaffer, M. Enhanced acoustic damping in flexible polyurethane foams filled with carbon nanotubes. *Compos. Sci. Technol.* **2009**, *69*, 1564–1569. [CrossRef]
59. Maharsia, R.R.; Jerro, H.D. Enhancing tensile strength and toughness in syntactic foams through nanoclay reinforcement. *Mater. Sci. Eng.* **2007**, *454*, 416–422. [CrossRef]
60. Saint-Michel, F.; Chazeau, L.; Cavaillé, J.-Y.; Chabert, E. Mechanical properties of high density polyurethane foams: I. Effect of the density. *Compos. Sci. Technol.* **2006**, *66*, 2700–2708. [CrossRef]
61. Nazeran, N.; Moghaddas, J. Synthesis and characterization of silica aerogel reinforced rigid polyurethane foam for thermal insulation application. *J. Non-Cryst. Solids* **2017**, *461*, 1–11. [CrossRef]
62. Gu, R.; Konar, S.; Sain, M. Preparation and Characterization of Sustainable Polyurethane Foams from Soybean Oils. *J. Am. Oil Chem. Soc.* **2012**, *89*, 2103–2111. [CrossRef]
63. Ciecierska, E.; Jurczyk-Kowalska, M.; Bazarnik, P.; Gloc, M.; Kulesza, M.; Krauze, S.; Lewandowska, M.; Kowalski, M. Flammability, mechanical properties and structure of rigid polyurethane foams with different types of carbon reinforcing materials. *Compos. Struct.* **2016**, *140*, 67–76. [CrossRef]
64. Jiao, L.; Xiao, H.; Wang, Q.; Sun, J. Thermal degradation characteristics of rigid polyurethane foam and the volatile products analysis with TG-FTIR-MS. *Polym. Degrad. Stab.* **2013**, *98*, 2687–2696. [CrossRef]
65. Levchik, S.V.; Weil, E.D. Thermal decomposition, combustion and fire-retardancy of polyurethanes—A review of the recent literature. *Polym. Int.* **2004**, *53*, 1585–1610. [CrossRef]
66. Septevani, A.A.; Evans, D.A.; Chaleat, C.; Martin, D.J.; Annamalai, P.K.; Evans, D.A.C. A systematic study substituting polyether polyol with palm kernel oil based polyester polyol in rigid polyurethane foam. *Ind. Crop Prod.* **2015**, *66*, 16–26. [CrossRef]
67. Chattopadhyay, D.; Webster, D.C. Thermal stability and flame retardancy of polyurethanes. *Prog. Polym. Sci.* **2009**, *34*, 1068–1133. [CrossRef]
68. Pagacz, J.; Hebda, E.; Michałowski, S.; Ozimek, J.; Sternik, D.; Pielichowski, K. Polyurethane foams chemically reinforced with POSS—Thermal degradation studies. *Thermochim. Acta* **2016**, *642*, 95–104. [CrossRef]
69. Chen, S.; Guo, L.; Du, D.; Rui, J.; Qiu, T.; Ye, J.; Li, X. Waterborne POSS-silane-urethane hybrid polymer and the fluorinated films. *Polymer* **2016**, *103*, 27–35. [CrossRef]
70. Hojiyev, R.; Ulcay, Y.; Hojamberdiev, M.; Çelik, M.S.; Carty, W.M. Hydrophobicity and polymer compatibility of POSS-modified Wyoming Na-montmorillonite for developing polymer-clay nanocomposites. *J. Colloid Interface Sci.* **2017**, *497*, 393–401. [CrossRef] [PubMed]
71. Badri, K.H.; Ahmad, S.H.; Zakaria, S. Production of a high-functionality RBD palm kernel oil-based polyester polyol. *J. Appl. Polym. Sci.* **2001**, *81*, 384–389. [CrossRef]

![polymers logo] *polymers*

MDPI

Article

Versatile Construction of Single-Tailed Giant Surfactants with Hydrophobic Poly(ε-caprolactone) Tail and Hydrophilic POSS Head

Qiangyu Qian, Jun Xu, Mingzu Zhang, Jinlin He * and Peihong Ni

College of Chemistry, Chemical Engineering and Materials Science, State and Local Joint Engineering Laboratory for Novel Functional Polymeric Materials, Jiangsu Key Laboratory of Advanced Functional Polymer Design and Application, Suzhou Key Laboratory of Macromolecular Design and Precision Synthesis, Soochow University, Suzhou 215123, China; 20164209218@stu.suda.edu.cn (Q.Q.); xujunchemistry@163.com (J.X.); zhangmingzu@suda.edu.cn (M.Z.); phni@suda.edu.cn (P.N.)
* Correspondence: jlhe@suda.edu.cn; Tel.: +86-512-6588-5195

Received: 31 January 2019; Accepted: 9 February 2019; Published: 12 February 2019

Abstract: Giant surfactants refer to a new kind of amphiphile by incorporating functional molecular nanoparticles with polymer tails. As a size-amplified counterpart of small-molecule surfactants, they serve to bridge the gap between small-molecule surfactants and amphiphilic block copolymers. This work reports the design and synthesis of single-tailed giant surfactants carrying a hydrophobic poly(ε-caprolactone) (PCL) as the tail and a hydrophilic cage-like polyhedral oligomeric silsesquioxane (POSS) nanoparticle as the head. The modular synthetic strategy features an efficient "growing-from" and "click-modification" approach. Starting from a monohydroxyl and heptavinyl substituted POSS (VPOSS-OH), a PCL chain with controlled molecular weight and narrow polydispersity was first grown by the ring-opening polymerization (ROP) of ε-CL under the catalysis of stannous octoate, leading to a PCL chain end-capped with heptavinyl substituted POSS (VPOSS-PCL). To endow the POSS head with adjustable polarity and functionality, three kinds of hydrophilic groups, including hydroxyl groups, carboxylic acids, and amine groups, were installed to the periphery of POSS molecule by a high-efficiency thiol-ene "click" reaction. The compounds were fully characterized by NMR, gel permeation chromatography (GPC), MALDI-TOF mass spectrometry, and TGA analysis. In addition, the preliminary self-assembly study of these giant surfactants was also investigated by TEM and dynamic laser light scattering (DLS), which indicated that they can form spherical nanoparticles with different diameters in aqueous solution. This work affords a straightforward and versatile way for synthesizing single-tailed giant surfactants with diverse head surface functionalities.

Keywords: giant surfactant; thiol–ene "click" reaction; polyhedral oligomeric silsesquioxane (POSS); poly(ε-caprolactone); aqueous self-assembly

---

1. Introduction

Amphiphiles refer to a kind of molecules containing chemically distinct segments, such as hydrophobic and hydrophilic parts, linked with chemical or supramolecular bonds. Traditionally, amphiphiles were generally accepted as small-molecule surfactants composed of a hydrophilic polar head and a hydrophobic alkyl chain, which have been broadly used in our daily life, including detergents, dispersants, cosmetics, and pharmaceutical excipients [1]. Afterwards, the region of amphiphiles was extended to amphiphilic block copolymers consisting of hydrophilic and hydrophobic polymeric chains. They have drawn tremendous attention in the past three decades because of their promising applications in various fields [2–6]. It is expected that an integration of both characteristics of small-molecule surfactants and amphiphilic block copolymers would expand the scope of amphiphiles and produce new properties.

In recent years, an emerging category of amphiphiles, named giant surfactants, has been constructed and widely studied through covalently connecting functionalized molecular nanoparticles with polymeric tails [7–14]. They have obtained much attention due to their multivalent nature, promising physical/chemical features, and unique self-assembling behaviors, both in bulk and solution states [10,11,15,16]. As a size-amplified analogue of small-molecule surfactants, giant surfactants serve to fill the niche between small-molecule surfactants and amphiphilic block copolymers. Just like the small-molecule surfactants, the giant surfactants can also be fabricated into various topologies, including single-head/single-tail, single-head/two-tails, or more complex two-head/two-tail (gemini-like), two heads in each terminal of one tail (bolaform-like), and even multiheaded/multitailed style [10]. On the other hand, giant surfactants obviously provide much more space for engineering chemical structures than their small-molecule counterparts. For instance, the polymeric tails can be of different topologies (e.g., linear, cyclic, or branched) or of distinct compositions, while the molecular nanoparticle heads can be modified with various functionalities, such as ionic, non-ionic, bioactive, etc. Therefore, all of these characteristics endow giant surfactants with sophisticated structures and thus result in more complicated self-assembly behaviors, as well as adjustable functional properties [8].

Specifically, the molecular nanoparticles are recognized as unique nanoscale building blocks with fixed shape and volume, precisely-defined chemical structures, as well as versatile surface functionalities, which notably distinguish giant surfactants from the other two typical amphiphiles [9]. To date, representative molecular nanoparticles that have been developed include folded globular proteins [12,17] and cage-like compounds, such as polyhedral oligomeric silsesquioxanes (POSS) [18,19], fullerenes [20], as well as polyoxometalates (POM) [21]. Thereinto, POSS-based molecular nanoparticles have been intensively employed for making giant surfactants, possibly because of the following considerations: (1) POSS molecule is probably the smallest shape-persistent cage-like silsesquioxane nanoparticles with a diameter of about 1.0 nm; (2) the rigid three-dimensional structure of the POSS molecule is relatively stable in common conditions; and (3) it is easy to modify both the cage structures (e.g., T_8 and T_{12}) and the surface functionalities (e.g., mono-, multi-, regio-, homo-, or hetero-functionalized) of POSS molecules. With the aid of living/controlled polymerizations and the powerful "click" reactions [22,23], a variety of POSS-based giant surfactants bearing different kinds of architecture have been constructed [24–32]. It has been shown that, by using such combined methodologies, several crucial molecular parameters of giant surfactants could be manipulated, such as overall molecular weights, surface functionalities, polydispersity, and weight fraction of tails/heads. Recently, we have constructed a library of double-chain giant surfactant region-isomers composed of a hydrophilic hydroxyl-functionalized POSS head and two hydrophobic polystyrene tails with various molecular weights tethered in ortho-, meta-, and para-configurations [33,34]. It was surprising to find that such a minute difference in regio-configuration influences a lot on the self-assembly behaviors of these giant surfactants, which identify regio-chemistry as an additional important factor in tuning the self-assembly of giant surfactants.

Herein, we report the model preparation of single-tailed giant surfactants on the basis of poly(ε-caprolactone) (PCL) end-capped with functional POSS (7R-POSS-PCL, R = OH, COOH, and NH$_3$Cl) using a sequential strategy of polymeric chain growth and POSS head modification. As shown in Scheme 1, PCL carrying seven vinyl groups (VPOSS-PCL) was first grown from VPOSS-OH through stannous octoate-catalyzed ring-opening polymerization (ROP) with controlled molecular weights and narrow polydispersities [35]. As a representative aliphatic polyester, PCL has been widely studied and applied in various biomedical fields due to its excellent biocompatibility, superior synthetic versatility, and flexible mechanical properties. Subsequently, the POSS headgroup was modified with diverse functionalities, including hydroxyl groups, carboxylic acids, and amine groups, via thiol-ene "click" reaction in high efficiency. As anticipated, by means of the controlled ROP reaction and powerful thiol-ene "click" chemistry, this universal method should be readily available to other polymerizable monomers and periphery functionalities.

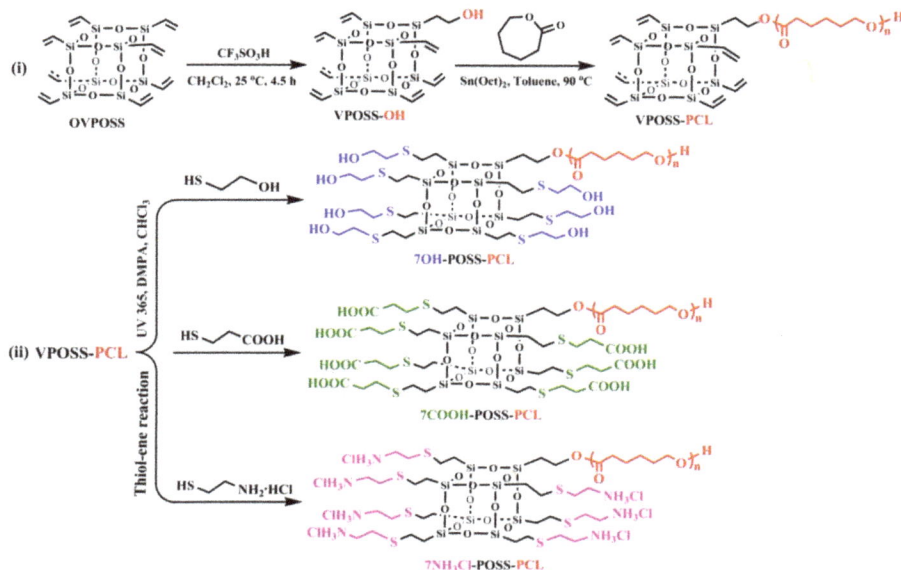

Scheme 1. Synthetic routes of single-tailed giant surfactants 7R-polyhedral oligomeric silsesquioxane (POSS)-poly(ε-caprolactone) (PCL) (R = OH, COOH, and NH₃Cl) via a combination of ring-opening polymerization (ROP) and thiol-ene "click" reaction.

2. Materials and Methods

2.1. Materials

Toluene (A.R., Sinopharm Chemical Reagent, Shanghai, China) and ε-caprolactone (ε-CL, 99%, Acros, Geel, Belgium) were dried by stirring in CaH₂ powder for 24 h at room temperature and distilled under reduced pressure before use. Dichloromethane (CH₂Cl₂, A.R., Sinopharm Chemical Reagent, Shanghai, China) was refluxed with CaH₂ powder and distilled before use. OctavinylPOSS (OVPOSS, 97%, Shanghai Gileader Advanced Material Technology, Shanghai, China) was eluted with CH₂Cl₂ (A.R., Sinopharm Chemical Reagent, Shanghai, China) to remove polar impurities, condensed by rotation distillation, and, finally dried in a vacuum oven to give a white powder. Stannous octoate (Sn(Oct)₂, 95%, Sigma-Aldrich, Saint Louis, MI, USA) was first fractionally distilled under reduced pressure and then diluted with anhydrous toluene to make a solution with a concentration of 0.5 g/mL in a glove box. Triflic acid (99%, J&K Scientific, Shanghai, China), 2,2-dimethoxy-2-phenylacetophenone (DMPA, 98%, TCI, Shanghai, China), 2-mercaptoethanol (98%, Acros, Geel, Belgium), 3-mercaptopropionic acid (99%, Acros, Geel, Belgium), and cysteamine hydrochloride (98%, Acros, Geel, Belgium) were used as received. Milli-Q ultrapure water (18.2 MΩ cm at 25 °C) was generated by a water purification system (Simplicity UV, Millipore, Shanghai, China). All the other chemicals (Sinopharm Chemical Reagent, Shanghai, China) were analytical reagents and used as received, unless otherwise mentioned.

2.2. Synthesis of Monohydroxyl Heptavinyl Substituted POSS (VPOSS-OH)

The monohydroxyl-functionalized heptavinyl POSS (VPOSS-OH) was prepared from commercially available OVPOSS, according to the literature method developed by Feher et al., after some modification [36]. The detailed synthetic procedure is listed as follows: Before reactions, the glassware, including stirring bars, were dried in an oven at 120 °C. OVPOSS (15 g, 23.7 mmol) was added to a round-bottom flask containing 200 mL of dry CH_2Cl_2 and stirred for complete dissolution. After adding triflic acid (4.2 mL, 23.7 mmol) to the above solution, the reaction was conducted at 25 °C for 4.5 h. After that, the mixture was washed three times with saturated aqueous $NaHCO_3$, and the collected organic phase was then mixed with 30 mL of acetone/water (v/v = 4:1) and the hydrolysis was performed at 25 °C for another 12 h. After the reaction, the organic phase was collected and dried with anhydrous Na_2SO_4, and the crude product was obtained after evaporation of the solvent. Column chromatography on silica with CH_2Cl_2/petroleum ether as the eluent afforded VPOSS-OH as a white solid (2.1 g, yield: 13%).

2.3. Preparation of Poly(ε-caprolactone) End-Capped with Heptavinyl Substituted POSS (VPOSS-PCL)

The poly(ε-caprolactone) end-capped with the heptavinyl substituted POSS head (VPOSS-PCL) was obtained by ROP of the ε-CL using VPOSS-OH as the initiator and Sn(Oct)$_2$ as the catalyst [35]. Briefly, VPOSS-OH (0.20 g, 0.31 mmol) was added to a 50 mL of Schlenk flask, which was then heated at 50 °C under a high vacuum to remove the possible residual moisture. After that, 15 mL of dry toluene was transferred to the reactor to dissolve the initiator under stirring. To this solution, ε-CL (1.71 g, 15 mmol) and Sn(Oct)$_2$ (0.3 mL, 0.15 mmol, 0.5 g/mL solution in dry toluene) were added by syringe under a dry nitrogen atmosphere. After degassing the solution by three exhausting–refilling nitrogen cycles, the mixture was kept stirring at 90 °C for 8 h. Afterwards, the viscous solution was concentrated and precipitated in cold methanol thrice. The precipitate was collected and dried at 25 °C under a vacuum to a constant weight, resulting in the product of VPOSS-PCL as a white powder (1.64 g, yield: 86%).

2.4. Typical Procedure for Synthesizing Single-Tailed Giant Surfactants (7R-POSS-PCL)

The single-tailed giant surfactants (7R-POSS-PCL, R = OH, COOH, and NH$_3$Cl) were prepared by the respective reaction between VPOSS-PCL and functional thiols (2-mercaptoethanol, 3-mercaptopropionic acid, and cysteamine hydrochloride) using a UV-irradiated thiol-ene chemistry [37,38]. The representative procedure is described as follows: In a round-bottom quartz flask, VPOSS-PCL (0.15 g, 0.03 mmol), 2-mercaptoethanol (98.3 mg, 1.26 mmol), and DMPA (5.4 mg, 0.021 mmol) were dissolved in 2 mL of CHCl$_3$. After irradiation with UV 365 nm at room temperature for 20 min, the mixture was purified by repeated precipitation in cold methanol. The white solid was collected by centrifugation and dried under a high vacuum for 24 h to give the 7OH-POSS-PCL (0.14 g, yield: 84%). The other two single-tailed giant surfactants, i.e., 7COOH-POSS-PCL and 7NH$_3$Cl-POSS-PCL, were prepared using a similar protocol for synthesizing 7OH-POSS-PCL.

2.5. Self-Assembly of Giant Surfactants in Aqueous Solution

The nanoparticles self-assembled by single-tailed giant surfactants 7R-POSS-PCL in aqueous solution were prepared by a dialysis method. Briefly, 5 mg of polymer sample was dissolved in 1.5 mL dimethylformamide (DMF) in a round-bottom flask, and it was stirred for several hours to achieve complete dissolution. Subsequently, 15 mL of Milli-Q water was added dropwise during a period of 2 h, using an auto-sampling system under moderate stirring. After that, the nanoparticle solution was dialyzed (MWCO 3500) against Milli-Q water for 24 h to remove DMF. The dialysis medium was changed six times during the process. Lastly, the solution was diluted to 25 mL with Milli-Q water to a desired concentration. Dust particles were removed by filtering each solution through an Φ 450 nm microfilter before measurements. The average particle sizes and size distributions of nanoparticles were determined by a Malvern dynamic laser light scattering (DLS, Zetasizer Nano-ZS, Malvern, UK) instrument. The morphologies of the nanoparticles were observed using TEM (HT7700, Hitachi, Tokyo, Japan), operated at an accelerating voltage of 120 kV. The carbon-coated copper grid was placed on the bottom of a glass cell, which was then immediately inserted into liquid nitrogen. After that, 10 μL of the solution was dripped on the grid and the frozen solvent was directly removed in a freeze dryer. The morphologies were then imaged on a normal TEM instrument at room temperature.

2.6. Characterizations

^{1}H NMR and ^{13}C NMR analyses were conducted on a 400 MHz NMR instrument (INOVA-400, Varian, Palo Alto, CA, USA) with CDCl$_3$ or d_6-DMSO as the solvents and tetramethylsilane (TMS) as the internal reference. The number-average molecular weights ($M_{n, GPC}$) and molecular weight distributions (M_w/M_n) of polymers were recorded on a gel permeation chromatography (GPC) instrument (HLC-8320, TOSOH, Tokyo, Japan), which was equipped with a refractive index and UV detectors using two TSKgel SuperMultiporeHZ-N (4.6 × 150 mm, 3.0 μm beads size) columns arranged in a series. It can separate polymers in the molecular weight range of 500–1.9 × 10^5 g/mol. TGA of polymers was performed on a Discovery instrument (TA, New Castle, DE, USA) under a nitrogen atmosphere, and the data were recorded over a temperature range of 30–800 °C at a heating rate of 10 °C/min. MALDI-TOF mass spectra were measured on an UltrafleXtreme MALDI-TOF mass spectrometer (Bruker, Kalsruhe, Germany) equipped with a 1 kHz smart beam-II laser. Before each measurement, the instrument was calibrated by external poly(methyl methacrylate) (PMMA) or polystyrene (PS) standards with desired molecular weights. Trans-2-[3-(4-tert-butylphenyl)-2-methyl-2-propenylidene]-malononitrile (DCTB, >99%, Sigma-Aldrich, Saint Louis, MI, USA) was used as the matrix and prepared in CHCl$_3$ with a concentration of 20 mg/mL. Sodium trifluoroacetate (CF$_3$COONa, >99%, Sigma-Aldrich, Saint Louis, MI, USA) served as the cationizing agent and was dissolved in anhydrous ethanol to make a solution with a concentration of 10 mg/mL. The solutions of matrix and CF$_3$COONa were mixed in a ratio of 10/1 (v/v). The polymers were dissolved in CHCl$_3$ with a concentration of 10 mg/mL. The sample preparation included depositing 0.5 μL of mixture solution of matrix/salt on the wells of a 384-well ground-steel plate, allowing the spots to dry completely, and then adding 0.5 μL of each sample solution on a spot of dry matrix/salt before adding another 0.5 μL of matrix/salt mixture solution on the top of the dry sample. After the solvent was completely evaporated, the plate was inserted in the MALDI mass spectrometer. The attenuation of Nd:YAG laser was adjusted to minimize undesirable polymer fragmentation and to maximize the sensitivity. Data analyses were conducted with Bruker's flexAnalysis software (Bruker Daltonics, Bremen, Germany).

3. Results and Discussion

3.1. Structure Characterization of Functional Initiator VPOSS-OH

The whole synthetic approach shown in Scheme 1 was designed in an effort to achieve giant surfactants with desirable structure and diverse functions using simple reactions and readily available starting chemicals. From the commercially available octavinylPOSS, the monohydroxyl and heptavinyl functionalized VPOSS-OH was easily obtained in an acceptable yield, with all the characterizations consistent with the literature. As shown in Figure 1, the characteristic signals of vinyl protons appeared at δ 5.75–6.25 ppm. After the formation of VPOSS-OH, most of the vinyl protons remained at the same chemical shift and two new peaks could be found at δ 1.1 ppm and δ 3.8 ppm that were ascribed to the methylenes adjacent to hydroxyl group. Moreover, the molecule was also characterized by MALDI-TOF mass spectroscopy, and the result is displayed in Figure 2. It was found that the observed molecular weight (m/z = 672.42 Da) for (M·Na)$^+$ (C$_{16}$H$_{26}$O$_{13}$Si$_8$Na$^+$) was in excellent agreement with the calculated one (m/z = 672.94 Da).

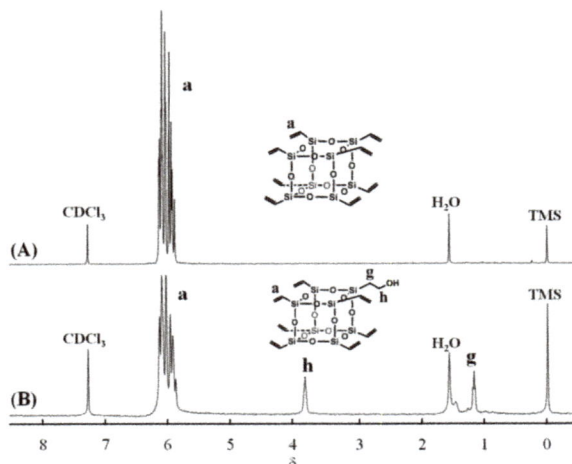

Figure 1. ^{1}H NMR spectra of (**A**) OctavinylPOSS (OVPOSS) and (**B**) VPOSS-OH in CDCl$_3$.

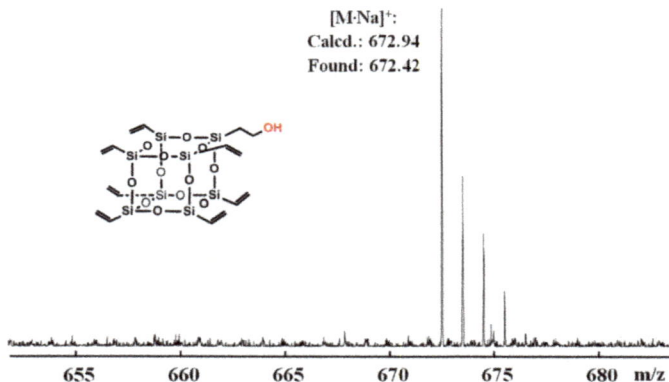

Figure 2. MALDI-TOF mass spectrometry (MS) spectrum of monohydroxyl heptavinyl substituted POSS (VPOSS-OH).

3.2. Synthesis and Characterization of VPOSS-PCL

The versatile hydroxyl group in VPOSS-OH was able to initiate ROP of various cyclic monomers under various conditions. It could also be further transformed to other functional initiating groups, such as tosylate for living cationic polymerization, halide group for atom transfer radical polymerization (ATRP) reaction, or chain transfer agents for reversible addition–fragmentation chain transfer (RAFT) polymerization. Herein, ROP reaction of ε-caprolactone was selected as the model system and this method should also be applicable to other functional cyclic monomers, such as lactides, carbonates, and phosphoesters. The ROP reaction of ε-CL under the catalysis of Sn(Oct)$_2$ has been well studied to prepare PCL with controlled molecular weight and narrow polydispersity. The polymerization was performed at 90 °C in toluene for 8 h using 0.5 equiv. of Sn(Oct)$_2$ to VPOSS-OH, and the polymer was purified by repeated precipitation in methanol to remove the catalyst and unreacted monomer. VPOSS-PCL is obtained as a white powder with a yield of around 86% and readily soluble in most organic solvents. The molecular weights of VPOSS-PCL can be easily tuned by varying the feeding ratio of ε-CL to VPOSS-OH. The polymers were fully characterized by various techniques to confirm the structure and purity.

In the GPC curves shown in Figure 3, three VPOSS-PCL samples with different molecular weights and relatively narrow polydispersities (M_w/M_n around 1.1) were obtained. All the GPC curves showed a unimodal and symmetrical pattern, and the high-molecular-weight samples (VPOSS-PCL-2 and VPOSS-PCL-3) displayed a major distribution that shifts towards the higher molecular weight side. In the typical ^{1}H NMR spectrum (Figure 4A) of VPOSS-PCL-3, the vinyl protons remain at δ 5.75–6.25 ppm (peak a), while the characteristic protons ascribed to methylenes in the PCL backbone showed at peaks b–f, confirming the successful linking of VPOSS with the PCL chain. This was also affirmed by the observation of carbons from both the VPOSS and PCL chain in the ^{13}C NMR spectrum (Figure 4B). Moreover, the integration ratio of peaks a and c in Figure 4A was used to calculate the molecular weight of VPOSS-PCL, and the results are listed in Table 1. The well-defined structure of VPOSS-PCL was also confirmed by the MALDI-TOF mass spectra. It is clear from Figure 5 that all three VPOSS-PCL samples displayed one single molecular weight distribution, and the observed molecular weight was in excellent agreement with the calculated one. For example, for VPOSS-PCL-1, the observed m/z value (2498.49 Da) for $(M_{16} \cdot Na)^+$ with the formula of $C_{112}H_{186}O_{45}Si_8Na^+$ agreed well with the calculated one (m/z 2498.03 Da). The same agreement was also found for the other two VPOSS-PCL samples. In addition, the mass difference between all adjacent two peaks was very close to the caprolactone repeating unit (m/z 114.07 Da). On the other hand, it needs to be pointed out that a minor distribution could be found in VPOSS-PCL-2, shown in Figure 5B$_1$. From the enlarged view shown in Figure 5B$_2$, one could find the minor distribution was probably ascribed to the PCL initiated by residual water during polymerization. Nevertheless, the chemical structure of VPOSS-PCL was thus unambiguously confirmed, and the sample was ready for further modification. Particularly, no fractionation was required in this polymerization process, and this facilitated easy synthesis of gram quantities of polymer samples.

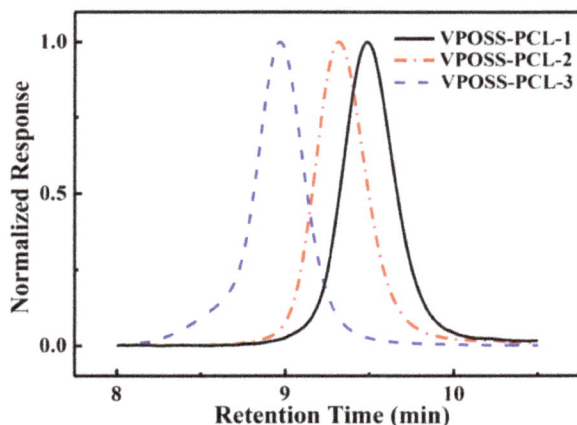

Figure 3. **Gel permeation chromatography** (GPC) traces of VPOSS-PCL samples (VPOSS-PCL-1: $M_{n,\,GPC}$ = 3800 g/mol, M_w/M_n = 1.09; VPOSS-PCL-2: $M_{n,\,GPC}$ = 5000 g/mol, M_w/M_n = 1.07; VPOSS-PCL-3: $M_{n,\,GPC}$ = 7700 g/mol, M_w/M_n = 1.11).

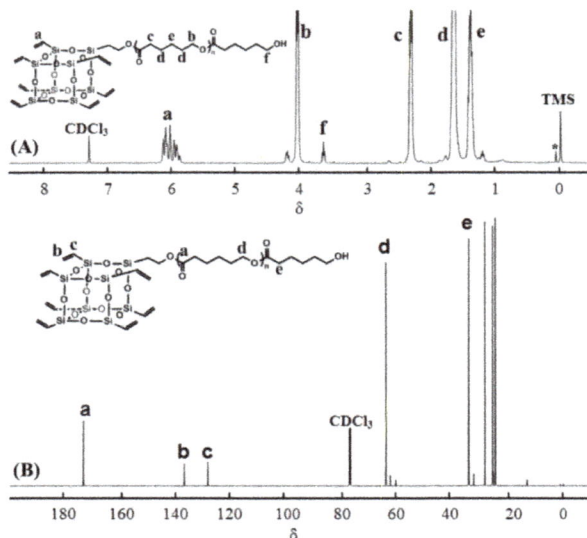

Figure 4. ^1H NMR spectrum (**A**) and ^{13}C NMR spectrum (**B**) of VPOSS-PCL-3 in CDCl$_3$. Asterisk represents resonance from residual silicone grease.

Table 1. Characterization results of VPOSS-PCL.

Entry	Samples	$M_{n, NMR}$ [a]	$M_{n, GPC}$ [b]	M_w/M_n [b]
1	VPOSS-PCL-1	2300	3800	1.09
2	VPOSS-PCL-2	4100	5000	1.07
3	VPOSS-PCL-3	5400	7700	1.11

[a] Calculated on the basis of ^1H NMR analysis in CDCl$_3$; [b] determined by GPC analysis with THF as the eluent.

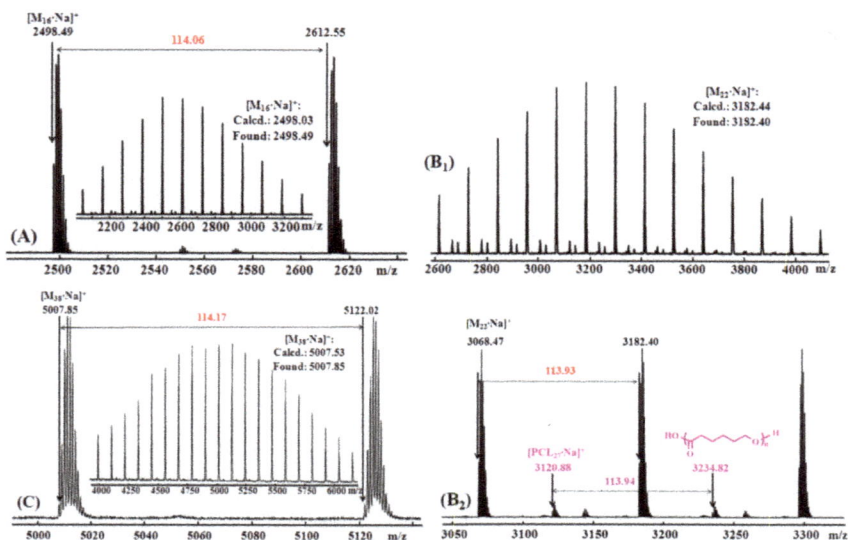

Figure 5. MALDI-TOF MS spectra of (**A**) VPOSS-PCL-1, (**B**) VPOSS-PCL-2 (B$_2$ is the enlarged image of B$_1$), and (**C**) VPOSS-PCL-3.

3.3. Synthesis and Characterization of Single-Tailed Giant Surfactants 7R-POSS-PCL

As a well-established methodology, thiol-ene "click" reaction has been broadly applied for various functionalizations [37]. In particular, it is quite powerful for situations when multiple modifications or sites of poor reactivity are involved in polymers. In this case, in order to tune the properties of the POSS headgroups, a variety of functional groups (-OH, -COOH, -NH$_3$Cl) were successfully introduced to the POSS head. The synthesis was rapid and straightforward from commercially available starting materials, and it was sure that this model functionalization could also be extended to other systems as needed. The functional polymers were characterized by various techniques. In representative GPC curves (Figure 6), the elution profile of 7OH-POSS-PCL-3 was basically the same as that of VPOSS-PCL-3, indicating that the modification with very small functional groups on such a rigid POSS head did not affect the overall hydrodynamic volume a lot. Unfortunately, the other two sets of samples could not be used for GPC measurements since the polar ionic groups have very strong interaction with the separation column.

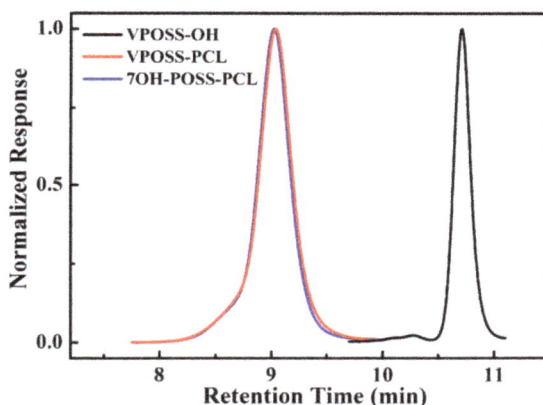

Figure 6. GPC traces of VPOSS-OH ($M_{n, GPC}$ = 900 g/mol, M_w/M_n = 1.14), VPOSS-PCL-3 ($M_{n, GPC}$ = 7700 g/mol, M_w/M_n = 1.11), and 7OH-POSS-PCL-3 ($M_{n, GPC}$ = 7900 g/mol, M_w/M_n = 1.10).

By means of the ^{1}H NMR analysis (Figure 7), the successful ligation of the VPOSS headgroup by different small molecules was proven by the complete disappearance of vinyl proton resonances at δ 5.75–6.25 ppm (peak a in Figure 4A) and the new appearance of thio-ether methylene linkages at δ 2.5–3.7 ppm in the ^{1}H NMR spectra (Figure 7). The chemical shifts agreed well with that of the reported analogues. In particular, the proton resonances of –OH, –COOH, and –NH$_3$Cl were shown at δ 3.8 ppm, 12.1 ppm, and 8.1 ppm, respectively. Moreover, TGA test was performed to further characterize the polymers, and the results are shown in Figure 8. It was clearly observed that the starting degradation temperature of VPOSS-PCL-3 was around 250 °C, while the values for 7OH-POSS-PCL-3, 7COOH-POSS-PCL-3, and 7NH$_3$Cl-POSS-PCL-3 were decreased to lower than 200 °C due to the incorporation of polar small-molecule headgroups. However, the residual weights for these four polymer samples at 800 °C were in the range of about 9%, which could be ascribed to the POSS molecule in polymers. As a result, the model functionalization of VPOSS-PCL with small functional groups was conveniently achieved by thiol-ene ligation, leading to single-tailed giant surfactants with polar heads modified with hydroxyl groups, carboxylic acids, or amine groups.

As a size-amplified counterpart of small-molecule surfactant, the single-tailed giant surfactants were composed of hydrophobic polymeric chain and hydrophilic headgroups. Therefore, the giant surfactants should also self-assemble in aqueous solution to form nanoparticles. The preliminary self-assembly study of the present PCL-based giant surfactants was carried out. The morphology, average particle sizes, and size polydispersity indices (size PDIs) of the nanoparticles self-assembled from various polymers were investigated by DLS and TEM measurements, and the results are displayed in Figure 9. One can find that the giant surfactants mainly formed spherical nanoparticles, and the corresponding size distribution curves of the nanoparticles measured by DLS show monomodal peaks.

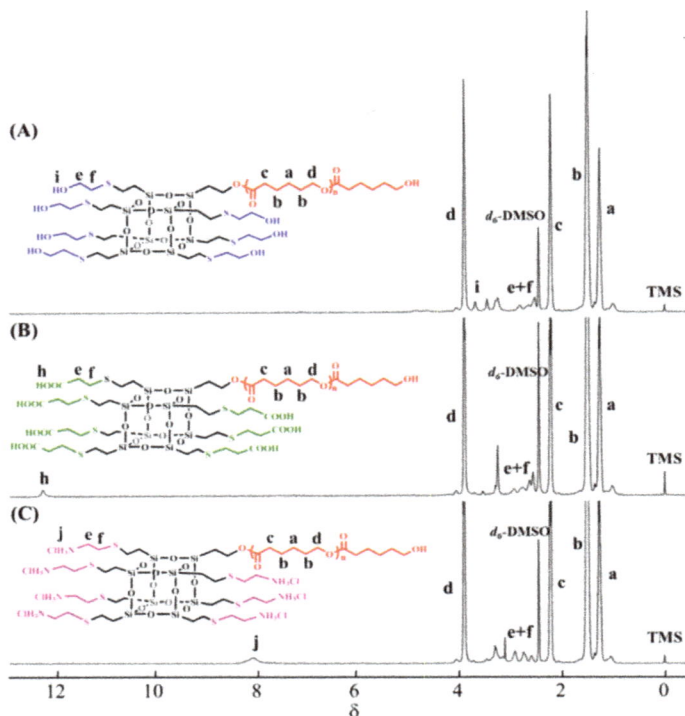

Figure 7. ^{1}H NMR spectra of (**A**) 7OH-POSS-PCL-3, (**B**) 7COOH-POSS-PCL-3, and (**C**) 7NH$_3$Cl-POSS-PCL-3 in d_6-DMSO.

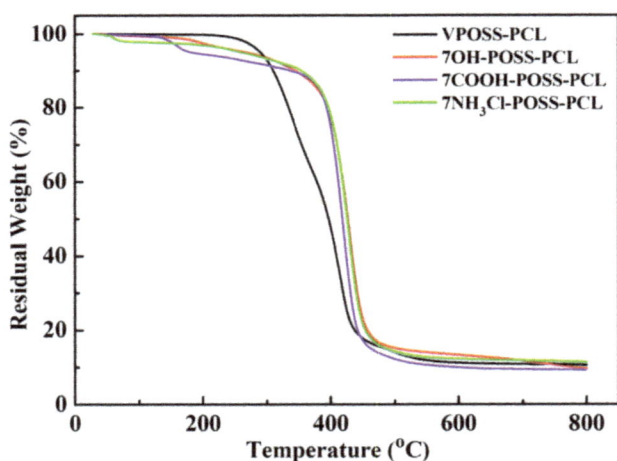

Figure 8. TGA thermograms of VPOSS-PCL-3, as well as the corresponding giant surfactants 7OH-POSS-PCL-3, 7COOH-POSS-PCL-3, and 7NH$_3$Cl-POSS-PCL-3.

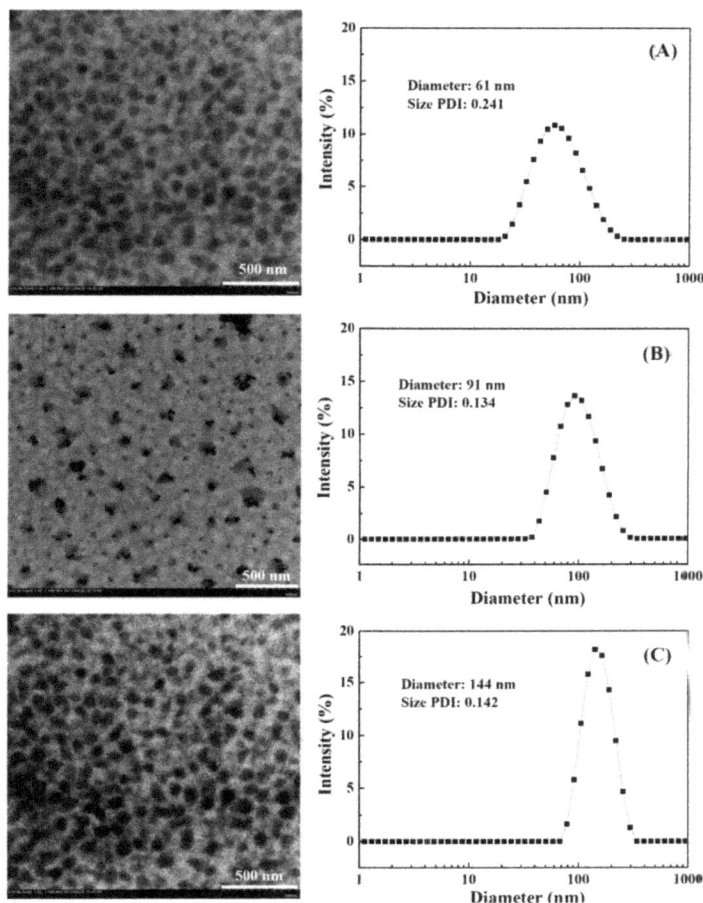

Figure 9. TEM images (scale bars represent 500 nm) and plots of particle size distributions of nanoparticles formed from (**A**) 7OH-POSS-PCL, (**B**) 7COOH-POSS-PCL, and (**C**) 7NH₃Cl-POSS-PCL in aqueous solution with a concentration of 0.2 mg/mL.

4. Conclusions

In summary, we have provided a facile approach to synthesize giant surfactants containing a poly(ε-caprolactone) (PCL) tail and a POSS head with diverse surface functionality by the combination of ROP reaction and thiol-ene "click" chemistry. The polymers were fully characterized by ^{1}H NMR, ^{13}C NMR, GPC, MALDI-TOF mass spectrometry, and TGA. Moreover, the preliminary self-assembly investigation demonstrated that these giant surfactants can form nanospheres with different sizes in aqueous solution. This synthetic way demonstrates a click philosophy by constructing straightforward and diverse structures from an easily available precursor, using a simple set of high-efficiency chemical transformations. It is expected that the methodology should be easily extended to other polymer systems and functional headgroups for fine-tuning the interaction parameters of giant surfactants.

Author Contributions: Data curation, Q.Q. and J.X.; Formal analysis, Q.Q., J.X., and J.H.; Funding acquisition, J.H.; Investigation, Q.Q. and J.X.; Project administration, M.Z., J.H., and P.N.; Supervision, M.Z., J.H., and P.N.; Writing—original draft, J.H., Q.Q. and J.X. contributed equally to this work.

Polymers **2019**, *11*, 311

Acknowledgments: We thank the financial supports from the National Natural Science Foundation of China (21774081), Natural Science Foundation of Jiangsu Province (BK20171212), a Project Funded by the Priority Academic Program Development (PAPD) of Jiangsu Higher Education Institutions, and the foundation of Key Laboratory of Synthetic and Biological Colloids, Ministry of Education, Jiangnan University (No. JDSJ2017-05).

Conflicts of Interest: The authors declare no conflict of interest.

References

1. Karsa, D.R. *Industrial Applications of Surfactants IV*; Royal Society of Chemistry: Cambridge, UK, 1999; pp. 1–23.
2. Schacher, F.H.; Rupar, P.A.; Manners, I. Functional block copolymers: Nanostructured materials with emerging applications. *Angew. Chem. Int. Ed.* **2012**, *51*, 7898–7921. [CrossRef] [PubMed]
3. Feng, H.B.; Lu, X.Y.; Wang, W.Y.; Kang, N.-G.; Mays, J.W. Block copolymers: Synthesis, self-Assembly, and applications. *Polymers* **2017**, *9*, 494. [CrossRef]
4. Cabral, H.; Miyata, K.; Osada, K.; Kataoka, K. Block copolymer micelles in nanomedicine applications. *Chem. Rev.* **2018**, *118*, 6844–6892. [CrossRef] [PubMed]
5. Ge, Z.S.; Liu, S.Y. Functional block copolymer assemblies responsive to tumor and intracellular microenvironments for site-specific drug delivery and enhanced imaging performance. *Chem. Soc. Rev.* **2013**, *42*, 7289–7325. [CrossRef] [PubMed]
6. Mai, Y.Y.; Eisenberg, A. Self-assembly of block copolymers. *Chem. Soc. Rev.* **2012**, *41*, 5969–5985. [CrossRef] [PubMed]
7. Reynhout, I.C.; Cornelissen, J.J.L.M.; Nolte, R.J.M. Synthesis of polymer–biohybrids: From small to giant surfactants. *Acc. Chem. Res.* **2009**, *42*, 681–692. [CrossRef] [PubMed]
8. Zhang, W.-B.; Cheng, S.Z.D. Toward rational and modular molecular design in soft matter engineering. *Chinese J. Polym. Sci.* **2015**, *33*, 797–814. [CrossRef]
9. Zhang, W.-B.; Yu, X.F.; Wang, C.-L.; Sun, H.-J.; Hsieh, I.F.; Li, Y.W.; Dong, X.-H.; Yue, K.; Van Horn, R.; Cheng, S.Z.D. Molecular nanoparticles are unique elements for macromolecular science: From "nanoatoms" to giant molecules. *Macromolecules* **2014**, *47*, 1221–1239. [CrossRef]
10. Yu, X.F.; Li, Y.W.; Dong, X.-H.; Yue, K.; Lin, Z.W.; Feng, X.Y.; Huang, M.J.; Zhang, W.-B.; Cheng, S.Z.D. Giant surfactants based on molecular nanoparticles: Precise synthesis and solution self-assembly. *J. Polym. Sci. Part B: Polym. Phys.* **2014**, *52*, 1309–1325. [CrossRef]
11. Dong, X.-H.; Hsu, C.-H.; Li, Y.W.; Liu, H.; Wang, J.; Huang, M.J.; Yue, K.; Sun, H.-J.; Wang, C.-L.; Yu, X.F.; et al. Supramolecular crystals and crystallization with nanosized motifs of giant molecules. In *Polymer Crystallization I: From Chain Microstructure to Processing*; Auriemma, F., Alfonso, G.C., de Rosa, C., Eds.; Springer International Publishing: Cham, Switzerland, 2016; Volume 276, pp. 183–213.
12. Zhang, W.-B.; Wu, X.-L.; Yin, G.-Z.; Shao, Y.; Cheng, S.Z.D. From protein domains to molecular nanoparticles: What can giant molecules learn from proteins? *Mater. Horiz.* **2017**, *4*, 117–132.
13. Tang, W.; Yue, K.; Cheng, S.Z.D. Molecular topology effects in self-assembly of giant surfactants. *Acta Polym. Sin.* **2018**, *8*, 959–972.
14. Liu, Z.G.; Kong, D.Y.; Dong, X.-H. Two-dimensional assembly of giant molecules. *Sci. China Chem.* **2018**, *61*, 17–24. [CrossRef]
15. Li, Q.X.; Wang, Z.; Yin, Y.H.; Jiang, R.; Li, B.H. Self-assembly of giant amphiphiles based on polymer-tethered nanoparticle in selective solvents. *Macromolecules* **2018**, *51*, 3050–3058. [CrossRef]
16. Wang, Y.Y.; Cui, J.; Han, Y.Y.; Jiang, W. Effect of chain architecture on phase behavior of giant surfactant constructed from nanoparticle monotethered by single diblock copolymer chain. *Langmuir* **2019**, *35*, 468–477. [CrossRef] [PubMed]
17. Xu, L.J.; Zhang, W.-B. The pursuit of precision in macromolecular science: Concepts, trends, and perspectives. *Polymer* **2018**, *155*, 235–247. [CrossRef]
18. Cordes, D.B.; Lickiss, P.D.; Rataboul, F. Recent developments in the chemistry of cubic polyhedral oligosilsesquioxanes. *Chem. Rev.* **2010**, *110*, 2081–2173. [CrossRef]
19. Blanco, I. The rediscovery of POSS: A molecule rather than a filler. *Polymers* **2018**, *10*, 904. [CrossRef]
20. Li, Z.; Liu, Z.; Sun, H.Y.; Gao, C. Superstructured assembly of nanocarbons: Fullerenes, nanotubes, and graphene. *Chem. Rev.* **2015**, *115*, 7046–7117. [CrossRef]

Polymers **2019**, *11*, 311

21. Dolbecq, A.; Dumas, E.; Mayer, C.R.; Mialane, P. Hybrid organic–inorganic polyoxometalate compounds: From structural diversity to applications. *Chem. Rev.* **2010**, *110*, 6009–6048. [CrossRef]

22. Kolb, H.C.; Finn, M.G.; Sharpless, K.B. Click chemistry: Diverse chemical function from a few good reactions. *Angew. Chem. Int. Ed.* **2001**, *40*, 2004–2021. [CrossRef]

23. Li, Y.W.; Dong, X.-H.; Zou, Y.; Wang, Z.; Yue, K.; Huang, M.J.; Liu, H.; Feng, X.Y.; Lin, Z.W.; Zhang, W.; et al. Polyhedral oligomeric silsesquioxane meets "click" chemistry: Rational design and facile preparation of functional hybrid materials. *Polymer* **2017**, *125*, 303–329. [CrossRef]

24. Yue, K.; Liu, C.; Guo, K.; Wu, K.; Dong, X.-H.; Liu, H.; Huang, M.J.; Wesdemiotis, C.; Cheng, S.Z.D.; Zhang, W.-B. Exploring shape amphiphiles beyond giant surfactants: Molecular design and click synthesis. *Polym. Chem.* **2013**, *4*, 1056–1067. [CrossRef]

25. Wang, Z.; Li, Y.W.; Dong, X.-H.; Yu, X.F.; Guo, K.; Su, H.; Yue, K.; Wesdemiotis, C.; Cheng, S.Z.D.; Zhang, W.-B. Giant gemini surfactants based on polystyrene–hydrophilic polyhedral oligomeric silsesquioxane shape amphiphiles: Sequential "click" chemistry and solution self-assembly. *Chem. Sci.* **2013**, *4*, 1345–1352. [CrossRef]

26. Su, H.; Zheng, J.; Wang, Z.; Lin, F.; Feng, X.Y.; Dong, X.-H.; Becker, M.L.; Cheng, S.Z.D.; Zhang, W.-B.; Li, Y.W. Sequential triple "click" approach toward polyhedral oligomeric silsesquioxane-based multiheaded and multitailed giant surfactants. *ACS Macro Lett.* **2013**, *2*, 645–650. [CrossRef]

27. Yu, X.F.; Yue, K.; Hsieh, I.-F.; Li, Y.W.; Dong, X.-H.; Liu, C.; Xin, Y.; Wang, H.-F.; Shi, A.-C.; Newkome, G.R.; et al. Giant surfactants provide a versatile platform for sub-10-nm nanostructure engineering. *Proc. Natl. Acad. Sci., USA* **2013**, *110*, 10078–10083. [CrossRef]

28. Wu, K.; Huang, M.J.; Yue, K.; Liu, C.; Lin, Z.W.; Liu, H.; Zhang, W.; Hsu, C.-H.; Shi, A.-C.; Zhang, W.-B.; et al. Asymmetric giant "bolaform-like" surfactants: Precise synthesis, phase diagram, and crystallization-induced phase separation. *Macromolecules* **2014**, *47*, 4622–4633. [CrossRef]

29. Yue, K.; Huang, M.J.; Marson, R.L.; He, J.L.; Huang, J.; Zhou, Z.; Wang, J.; Liu, C.; Yan, X.S.; Wu, K.; et al. Geometry induced sequence of nanoscale Frank–Kasper and quasicrystal mesophases in giant surfactants. *Proc. Natl. Acad. Sci. USA* **2016**, *113*, 14195–14200. [CrossRef]

30. Yue, K.; Liu, C.; Huang, M.J.; Huang, J.; Zhou, Z.; Wu, K.; Liu, H.; Lin, Z.W.; Shi, A.-C.; Zhang, W.-B.; et al. Self-assembled structures of giant surfactants exhibit a remarkable sensitivity on chemical compositions and topologies for tailoring sub-10 nm nanostructures. *Macromolecules* **2017**, *50*, 303–314. [CrossRef]

31. Li, Z.X.; Fu, Y.; Li, Z.; Nan, N.; Zhu, Y.M.; Li, Y.W. Froth flotation giant surfactants. *Polymer* **2019**, *162*, 58–62. [CrossRef]

32. Huang, M.J.; Yue, K.; Huang, J.; Liu, C.; Zhou, Z.; Wang, J.; Wu, K.; Shan, W.; Shi, A.-C.; Cheng, S.Z.D. Highly asymmetric phase behaviors of polyhedral oligomeric silsesquioxane-based multiheaded giant surfactants. *ACS Nano* **2018**, *12*, 1868–1877. [CrossRef]

33. Wang, X.-M.; Shao, Y.; Xu, J.; Jin, X.; Shen, R.-H.; Jin, P.-F.; Shen, D.-W.; Wang, J.; Li, W.H.; He, J.L.; et al. Precision synthesis and distinct assembly of double-chain giant surfactant regioisomers. *Macromolecules* **2017**, *50*, 3943–3953. [CrossRef]

34. Wang, X.-M.; Shao, Y.; Jin, P.-F.; Jiang, W.; Hu, W.; Yang, S.; Li, W.H.; He, J.L.; Ni, P.H.; Zhang, W.-B. Influence of regio-configuration on the phase diagrams of double-chain giant surfactants. *Macromolecules* **2018**, *51*, 1110–1119. [CrossRef]

35. He, J.L.; Yue, K.; Liu, Y.Q.; Yu, X.F.; Ni, P.H.; Cavicchi, K.A.; Quirk, R.P.; Chen, E.Q.; Cheng, S.Z.D.; Zhang, W.-B. Fluorinated polyhedral oligomeric silsesquioxane-based shape amphiphiles: Molecular design, topological variation, and facile synthesis. *Polym. Chem.* **2012**, *3*, 2112–2120. [CrossRef]

36. Feher, F.J.; Wyndham, K.D.; Baldwin, R.K.; Soulivong, D.; Ziller, J.W.; Lichtenhan, J.D. Methods for effecting monofunctionalization of $(CH_2=CH)_8Si_8O_{12}$. *Chem. Commun.* **1999**, *14*, 1289–1290. [CrossRef]

37. Hoyle, C.E.; Lowe, A.B.; Bowman, C.N. Thiol-click chemistry: A multifaceted toolbox for small molecule and polymer synthesis. *Chem. Soc. Rev.* **2010**, *39*, 1355–1387. [CrossRef] [PubMed]

38. Hu, J.; He, J.L.; Zhang, M.Z.; Ni, P.H. Aapplications of click chemistry in synthesis of tolological polymers. *Acta Polym. Sin.* **2013**, *3*, 300–319.

polymers MDPI

Article

A Novel POSS-Based Copolymer Functionalized Graphene: An Effective Flame Retardant for Reducing the Flammability of Epoxy Resin

Min Li, Hong Zhang, Wenqian Wu, Meng Li, Yiting Xu *, Guorong Chen and Lizong Dai *

Fujian Provincial Key Laboratory of Fire Retardant Materials, College of Materials, Xiamen University, Xiamen, Fujian 361005, China; 20720161150023@stu.xmu.edu.cn (M.L.); 20720171150104@stu.xmu.edu.cn (H.Z.); 20720161150026@stu.xmu.edu.cn (W.W.); 20720171150098@stu.xmu.edu.cn (M.L.); grchen@xmu.edu.cn (G.C.)
* Correspondence: xyting@xmu.edu.cn (Y.X.); lzdai@xmu.edu.cn (L.D.)

Received: 31 December 2018; Accepted: 29 January 2019; Published: 1 February 2019

Abstract: In this study, a novel copolymer, PbisDOPOMA-POSSMA-GMA (PDPG), containing methacryloisobutyl polyhedral oligomeric silsesquioxane (POSSMA), reactive glycidyl methacrylate (GMA), and bis-9,10-dihydro-9-oxa-10-phosphaphenanthrene-10-oxide methacrylate (bisDOPOMA) and derivative functionalized graphene oxide (GO) were synthesized by a one-step grafting reaction to create a hybrid flame retardant (GO-MD-MP). GO-MD-MP was characterized by Fourier transform infrared spectroscopy (FT-IR), X-ray diffraction (XRD), transmission electron microscopy (TEM), Raman spectroscopy, X-ray photoelectron spectroscopy (XPS), and thermogravimetric analysis (TGA). Flame-retardant epoxy resin (EP) composites were prepared by adding various amounts of GO-MD-MP to the thermal-curing epoxy resin of diglycidyl ether of bisphenol A (DGEBA, trade name E-51). The thermal properties of the EP composites were remarkably enhanced by adding the GO-MD-MP, and the residue char of the epoxy resin also increased greatly. With the incorporation of 4 wt % GO-MD-MP, the limiting oxygen index (LOI) value was enhanced to 31.1% and the UL-94 V-0 rating was easily achieved. In addition, the mechanical strength of the epoxy resin was also improved.

Keywords: POSS-based copolymer; graphene oxide; flame retardant; epoxy resin

1. Introduction

Epoxy resins (EP) are widely used in industrial applications, such as surface coatings, castable materials, injection molding materials, and microelectronics packaging materials, due to their excellent mechanical, adhesive, and electrical insulation properties [1–3]. However, the fire hazards of an EP composite significantly limit its applications [4,5]. In order to reduce the risk of fire, many flame retardants have been developed, including halogen flame retardants, halogen-free flame retardants, and organic–inorganic hybrid flame retardants [6–8]. Halogen atoms, especially bromine or chlorine, reduce the exothermic combustion process by blocking gas-phase free-radical chemistry and inhibit combustion in the vapor phase [9]. Unfortunately, the toxicity of the hydrogen halide and brominated furans and dioxins formed during combustion were seriously harmful to human health [10]. Therefore, halogen-free flame retardants have been developed for avoiding toxic gas generation and simultaneously reducing the flammability of EP composites [2].

Over the past few years, 9,10-dihydro-9-oxa-10-phosphaphenanthrene-10-oxide (DOPO) and its derivatives have attracted great attention due to their high reactivity and flame retardancy [11,12]. Technical literature extensively reports that DOPO not only enhances the flame retardancy of epoxy resin but also its thermal stability [13–15]. In addition, the modification of DOPO with polyhedral oligomeric silsesquioxane (POSS), which possesses a cage-like organic–inorganic hybrid structure with a 1–3 nm dimension, can further enhance its flame retardancy [16,17]. During the combustion process, POSS is embedded as a molecular enhancer in the EP matrix, retarding the combustion of the material and reducing the heat release by retarding and restricting the movement of the EP chain. Furthermore, the rich silicon oxide structure in POSS can promote the epoxy resin to form a continuous and anti-oxidation char to reduce its flammability [18–20].

Graphene plays an important role in terms of enhancing the mechanical properties and flame retardancy of EP composites [21–23]. At the same time, the effective dispersion of graphene flakes can achieve high aspect ratios in EP composites, which improves the barrier properties of EP composites during thermal degradation due to the special platelet morphology of graphene flakes [24]. Simultaneously, graphene can promote a dense and uniform carbon layer to form in the condensed phase during the combustion of the EP composites [25]. Our previous work has shown that modified graphene oxide (GO) can be better dispersed in EP composites due to the excellent interface interaction between the modified graphene oxide and the EP composites. As a nanoplate filler capable of being well-dispersed in EP, the modified GO can also be viewed structurally to improve the mechanical properties of the EP composites [26]. Chen et al. [27] synthesized a graphene-based hybrid flame retardant by surface grafting 10-dihydro-9-oxa-10-phosphaphenanthrene-10-oxide-g-(2,3-epoxypropoxy) propyltrimethoxysilane (DPP). This type of flame retardant in EP composites achieved the UL-94 V-0 rating, indicating that it is an effective flame-retardant additive for epoxy resin laminate composites. Hu et al. [28] synthesized a functionalized graphene (PD-rGO) via an in situ reaction. It was incorporated into EP to fabricate flame-retardant EP nanocomposites, and the PHRR and THR values decreased significantly by 43.0% and 30.2%, respectively.

In order to improve the flame retardancy of graphene in EP composites, many studies have reported a different functional graphene flame retardant [29–32]. However, a POSS-based copolymer modified graphene oxide hybrid flame retardant was rarely reported. In this research, a novel POSS-based polymer modified graphene oxide hybrid flame retardant, containing long-chain POSS, DOPO, and reactive glycidyl methacrylate (GMA), was synthesized. To accomplish this, first, a copolymer, PDPG, containing POSS and DOPO, was fabricated via copolymerization using POSSMA, bisDOPOMA, and GMA monomers, then PDPG was grafted to GO to obtain GO-MD-MP. The GO-MD-MP was then used as a reactive functional filler and incorporated in the EP matrix to explore its effect on the comprehensive properties of the EP composites, including flame retardancy, thermal properties, and mechanical strengths.

2. Experiment

2.1. Materials

The methacryloisobutyl POSS (POSSMA) (molecular weight = 943.64 g/mol) was purchased from the Hybrid Plastics Company (Sinoreagent Co., Shanghai, China) and was used as received. The 9,10-Dihydro-oxa-10-phosphaphenanthrene-10-oxide (DOPO) was purchased from Shanghai Eutec Chemical, Shanghai, China. The polyoxymethylene (POM), methacryloyl chloride, 4-aminophenol, 4,4′-diaminodiphenylmethane (DDM), and glycidyl methacrylate (GMA) were purchased from Sinopharm Chemical Reagent Co. Ltd., Shanghai, China. The diglycidyl ether of bisphenol A (DGEBA, trade name E-51), a kind of liquid epoxy resin, was supplied by the Wuxi Resin Factory (Wuxi, China). The triethylamine (TEA) was distilled off before use. The reagents used were of an analytical grade, unless otherwise noted.

2.2. Synthesis of OH-bisDOPO

A volume of 200 mL of ethanol solution of DOPO (6.48 g, 30 mmol), POM (0.9 g, 30 mmol), and 4-aminophenol (1.64 g, 15 mmol) were charged into a 500 mL three-necked glass flask equipped with refluxing under argon protection. After stirring for 30 min at room temperature, the mixture was continuously stirred at 55 °C for 24 h to form a white suspension. The white suspension was then filtered with a Buchner funnel and washed three times with ethanol to obtain an off-white solid. Subsequently, the product, OH-bisDOPO, was dried in a 50 °C vacuum drying oven to give a yield of 78%. The synthetic route is shown in Scheme 1a.

The NMR spectra of OH-bisDOPO are shown in Figure 1a: ^{1}H NMR(DMSO-d6, 400 MHz) δ (ppm)—8.74 (s, 1H), 8.04~8.14 (m, 4H), 7.64~7.75 (m, 4H), 7.48 (m, 2H), 7.35 (m, 2H), 7.26 (m, 2H), 6.98 (m, 2H), 6.41 (m, 2H), 6.32 (m, 2H), and 3.97 (m, 4H); ^{31}P NMR(DMSO-d6, 121MHz) δ (ppm)—31.52 (DOPO).

2.3. Synthesis of bisDOPOMA Monomer

OH-bisDOPO (5.7 g, 10 mmol), TEA (1.03 g, 10 mmol) were dissolved in dichloromethane (100 mL) in a 250 mL round-bottomed glass flask. The mixture was then stirred at 0 °C for 10 min and methacryloyl chloride (1.01 g, 10 mmol) was added dropwise slowly into the mixture. Subsequently, the mixture was stirred continuously at 0 °C for 1 h and then at room temperature for 24 h. After that, the solvent was washed 3 times with deionized water and dried with anhydrous magnesium sulfate. Then, the solvent was added dropwise into 300mL of hexane to obtain a faint yellow precipitate. The bisDOPOMA powders were isolated by filtration followed by drying in a vacuum at 50 °C for 24 h. The synthetic scheme of bisDOPOMA is shown in Scheme 1b.

The NMR spectra of bisDOPOMA are shown in Figure 1b: ^{1}H NMR(DMSO-d6, 400MHz) δ (ppm)—8.03~8.15 (m, 4H), 7.67~7.84 (m, 4H), 7.52 (m, 2H), 7.35 (m, 2H), 7.25 (m, 2H), 7.00 (m, 2H), 6.71 (m, 2H), 6.63 (m, 2H), 6.23 (m, 1H), 5.85 (m, 1H), and 4.03 (m, 4H); ^{31}P NMR (DMSO-d6, 121MHz) δ (ppm)—31.52 (DOPO).

2.4. Synthesis of PDPG

PDPG was synthesized with monomers of bisDOPOMA, POSSMA, and GMA via random free-radical polymerization, as shown in Scheme 1c. GMA (0.57 g, 4 mmol), POSSMA (3.8 g, 4 mmol), bisDOPOMA (10.5 g, 16 mmol), and azodiisobutyronitrile (AIBN) (0.03 g, 0.18 mmol) were added into a 250 mL three-necked glass flask equipped with refluxing under argon with 100 mL of tetrachloroethane, and the three-necked glass flask was placed in an oil bath at 65 °C for 24 h. The product was then precipitated in excess hexane and washed three times with hexane. The resulting white powder of PDPG was dried overnight in a vacuum at 50 °C for 24 h.

The NMR spectra of PDPG are shown in Figure 1c, which give the characteristic peaks of PDPG.

2.5. Synthesis of GO-MD-MP

GO was synthesized from flake graphite using the modified Hummers method. First, flake graphite (3 g) was added into concentrated H_2SO_4/H_3PO_4 (360:40 mL) in a 500ml three-necked glass flask. Then, the three-necked glass flask was placed in an ice bath at 0 °C. $KMnO_4$ (18 g) was added slowly in order to prevent the reaction temperature from exceeding 20 °C and the dispersions were heated to 50 °C and stirred for 12 h. Afterwards, distilled water (400 mL) and 30% hydrogen peroxide (6 mL) were added slowly until the mixed solution turned bright yellow. The mixed solution was then centrifuged, and the obtained solid was washed with deionized water and centrifuged until the pH was neutral. Next, the product, GO, was freeze-dried in a vacuum freeze dryer to obtain dry GO. After that, GO (500 mg) and PDPG (2.5 g) were dissolved with dimethylformamide (DMF) (100 mL) in a 250 mL three-necked glass flask equipped with a condenser and magnetic stirrer under an argon atmosphere. The reaction solution was then heated to 150 °C for 24 h. The obtained GO-MD-MP solution was filtered and then washed with DMF, ethanol, and deionized water three times, respectively. Finally, the product was freeze-dried in a vacuum freeze dryer. The synthetic scheme of GO-MD-MP is shown in Scheme 1d.

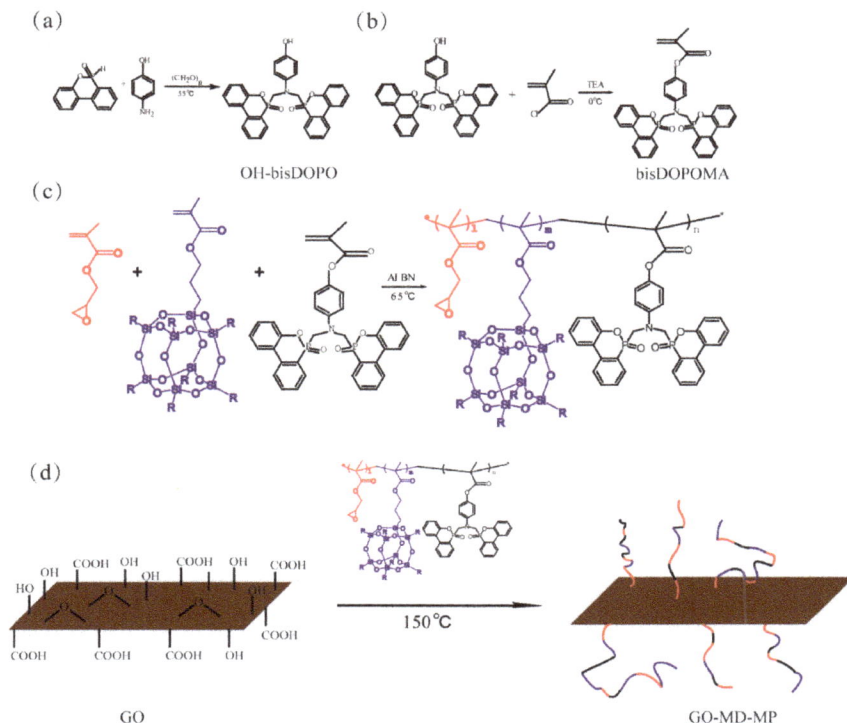

Scheme 1. Synthetic route of OH-bisDOPO (**a**), bisDOPOMA (**b**), PDPG (**c**), and GO-MD-MP (**d**).

2.6. Preparation of the Flame-Retardant EP Composites

We prepared the flame-retardant EP composites with 2 wt % GO-MD-MP. First, GO-MD-MP (0.5 g) was dispersed in 20 mL of chloroform and sonicated for 30 min in an ultrasonic bath. After that, epoxy resins (13.3 g) and the abovementioned GO-MD-MP dispersion in chloroform were mixed and then sonicated for 1 h at room temperature. Subsequently, the chloroform was removed by vacuum evaporation, and then DDM (2.9 g) was added into the mixture at 85 °C in a 50 mL round-bottomed flask equipped with a mechanical stirrer. Finally, the mixture was cast in an aluminum mold, when the mixture became homogeneous. The aluminum mold was then placed using a three-step curing procedure to obtain the thermosetting resins. All the samples were cured at 120 °C for 4 h and then at 140 °C for 2 h. They were then further post-cured at 180 °C for 2 h. Finally, all the samples were cooled naturally to room temperature to prevent stress cracking.

2.7. Characterizations

The nuclear magnetic resonance (^{1}H NMR, ^{31}P NMR) spectra were recorded on a Bruker Advanced II AV400 MHz NMR spectrometer (Bruker, Geneva, Switzerland) with CDCl$_3$ or DMSO-d6 as the solvent. The attenuated total reflectance Fourier transform infrared (ATR-FTIR) spectra were obtained on a Nicolet Avatar 360 spectrophotometer (Thermo Fisher Scientific, Shanghai, China). The X-ray photoelectron spectroscopy (XPS) was conducted using a K-Alpha (Physical Electronics, Inc., Chanhassen, MN, USA) equipped with an Al Kα radiation source. The molecular weight of the PDPG was estimated by gel permeation chromatography (GPC) with a refractive index detector model. Tetrahydrofuran (THF) was used as an eluent at a flow rate of 1 ml/min.

The scanning electron microscopy (SEM) images were obtained using an SU-70 microscope (HITACHI, Tokyo, Japan). The fracture surfaces of the cured EP composites and the char residues of the specimens from the LOI tests were observed. The morphology and structure of the GO and GO-MD-MP were studied by transmission electron microscopy (TEM) using a JEM2100 microscope (JEOL, Tokyo, Japan).

The Raman spectroscopy measurement was performed using a HORIBA Xplora Raman spectrometer (Horiba, Lille, France) with excitation by a 532 nm laser line. The X-ray diffraction (XRD) was performed using a Panalytical X'Pert diffractometer with filtered Cu radiation. Diffractograms (Rigaku, Tokyo, Japan) were obtained in a range from 5° to 55° (2θ) at room temperature.

The thermogravimetric analysis (TGA) was performed with a Netzsch STA 409EP (Netzsch, Selb, Germany) by heating from room temperature to 800 °C at a heating rate of 10 °C/min under a nitrogen atmosphere. The differential scanning calorimeter (DSC) was performed with a Netzsch STA 449C (Netzsch, Selb, Germany) by heating from room temperature to 220 °C at a heating rate of $10\ °C \cdot min^{-1}$ under a nitrogen atmosphere.

The limiting oxygen index (LOI) was measured with a JF-3 instrument (Jiangning, China), and the sample bar dimensions were $100 \times 6 \times 4\ mm^3$. The vertical burning tests were conducted with an instrument from Fire Testing Technology (Jiangning, China), England, based on the UL-94 standard, and the sample bar dimensions were $125 \times 12.5 \times 4\ mm^3$.

The three-point bending experiment was performed on an electronic universal testing machine (AGS-X, Shimadzu, Tokyo, Japan) with a sample size of $120 \times 10 \times 4\ mm^3$ and a crosshead speed of 2 mm/min.

3. Results and Discussion

3.1. Structural and Morphological Characterization of GO-MD-MP and Cured EP Composites

The ^{1}H NMR, ^{31}P NMR spectra of OH-bisDOPO, bisDOPOMA, and PDPG are displayed in Figure 1. All protons can be assigned to the predicted signals. Compared to Figure 1a of OH-bisDOPO, the signal of the phenolic hydroxyl group at 8.74 ppm disappeared in Figure 1b of bisDOPOMA, and new signals at 6.23 and 5.87 ppm for bisDOPOMA were assigned to methylene protons (1H). In addition, the ^{1}H NMR spectra of PDPG is shown in Figure 1c. The related characteristic peaks of PGMA, POSSMA, and bisDOPOPMA were the signals (m) at 3.23 ppm, the signals (g, d, and e) at 0.59 ppm, and the signals (c and j) at 6.2~8.2 ppm. The ratio of GMA, POSSMA, and bisDOPOMA was found to be about 1:1.15:4.11 based on the relative integration of these three kinds of signals above, revealing that the ratio of P and Si was about 0.92. The composition and structure information of PDPG is shown in Table 1. For the ^{31}P NMR spectra, the signal of OH-bisDOPO and bisDOPOMA was in the same position at 31.52 ppm, which indicated that the phosphorus had one chemical environment. Figure 2 shows the GPC of PDPG. The Mn and Mw/Mn of PDPG was 16,018 and $1.55\ g \cdot mol^{-1}$, respectively.

Table 1. Composition and structure information of PDPG calculated by ^{1}HNMR measurement.

Sample	Composition (mol %)			P (at%)	P/Si
	GMA	POSSMA	bisDOPOMA		
PDPG	16	18	66	1.8	0.92

The FTIR spectra of GO and GO-MD-MP are shown in Figure 3a. The characteristic absorption peaks of GO were observed at 3400 cm^{-1} (O–H stretching vibration), 1623.33 cm^{-1} (C=C stretching vibration), 1726.36 cm^{-1} (C=O stretching vibration), 1223.85 cm^{-1} (C–O of epoxy stretching vibration), and 1051.83 cm^{-1} (C–O stretching vibration of alkoxy). For GO-MD-MP, some new bands were observed at 1198.72 cm^{-1} (P=O stretching vibration), 1107.35 cm^{-1} (Si–O stretching vibration), 1378.84 cm^{-1} (C–N stretching vibration), and the typical absorption peaks between 1431 and 1600 cm^{-1}

were assigned to the benzene ring. The appearance of these characteristic peaks indicated that PDPG was successfully grafted onto the GO surface.

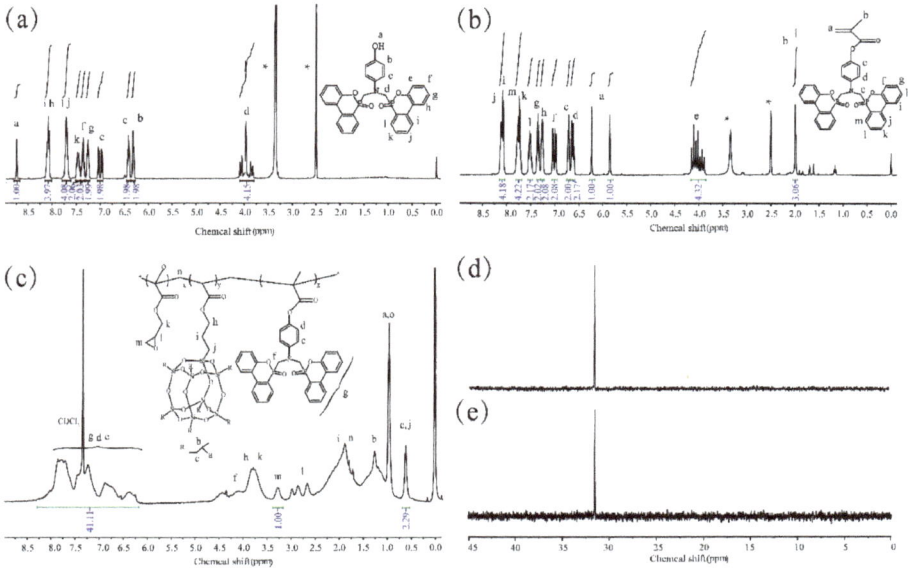

Figure 1. The ^{1}H NMR,^{31}P NMR of OH-bisDOPO (**a,d**), bisDOPOMA (**b,e**), and PDPG (**c**).

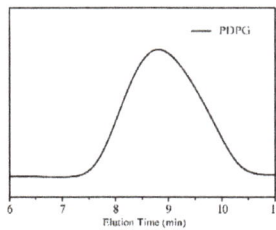

Figure 2. The gel permeation chromatography (GPC) trace recorded for PDPG.

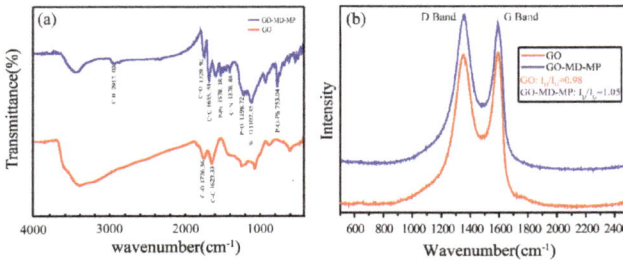

Figure 3. FT-IR (**a**) and Raman spectra (**b**) of graphene oxide (GO) and GO-MD-MP.

Figure 3b shows the Raman spectroscopies of GO and GO-MD-MP. The absorption peaks appearing at 1593 and 1358 cm^{-1} correspond to the G and D bands [33], respectively. The G band represents sp^2 hybrid carbon atoms, while the D band arises from the vibrations of the sp^3 hybrid carbon atoms provided with the defect information for graphene [34]. The ratio of the intensities of the D and G bands (I_D/I_G) is a significant parameter to evaluate the structure of graphene. The I_D/I_G ratio increased from 0.98 for GO to 1.05 for GO-MD-MP, indicating an increase in disordered carbon, compared to the ordered carbon structure, that was due to the PDPG graft.

In order to confirm covalent functionalization, GO and GO-MD-MP were analyzed using XPS. As shown in Figure 4a, two elements, C and O, were detected in the GO spectrum. However, five elements, C, O, P, N, and Si, were observed in GO-MD-MP. Moreover, the C1s, N1s, P2p, and Si2p spectra are elucidated in Figure 4b–e, respectively. In the high-resolution C1s spectrum of GO-MD-MP, six absorbance peaks were distinguished: 284.4 and 284.8 eV were assigned to C=C and C–C in the GO skeleton, 285.5 and 286.7 eV were attributed to C–OH and C–O–C, respectively, 287.2 eV was attributed to C=O, and 288.6 eV was assigned to the C(=O)–O–C band. In the spectrum of N1s, only one peak (N–C) at 400.4 eV was detected. Moreover, in the spectrum of P2p, three peaks at 132.5, 133.7 eV, and 133.1 eV were assigned to P–C, P(=O)–O–C, and P–C–N, respectively. For the Si2p spectrum, two peaks at 102.1 and 103.2 eV were attributed to Si–O–C and Si–O–Si, respectively. In addition, the atom percentages of the various elements in GO-MD-MP were 75.32 at% (C), 20.57 at% (O), 1.01 at% (N), 1.48 at% (P), and 1.62 at% (Si), respectively. The ratio of P and Si was 0.91, which approximately coincided with the value calculated by ^{1}H NMR, indicating that the graft reaction was successful. Thus, according to the measured P (1.48 at%) and Si (1.62 at%) content, the calculated grafting ratio of bisDOPOMA and POSSMA on GO was around 35.38 wt % and 14.41 wt %, respectively.

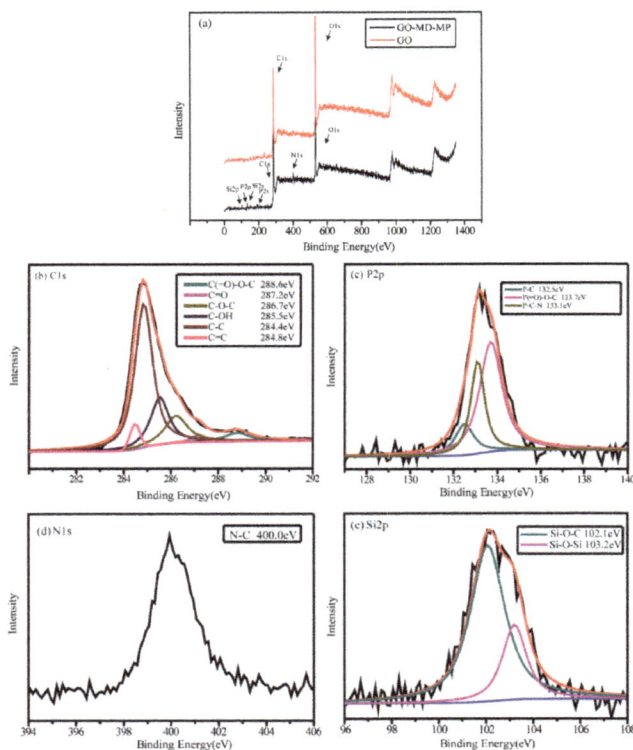

Figure 4. X-ray photoelectron spectroscopy (XPS) survey spectra of GO-MD-MP (**a**), C1s spectrum of GO-MD-MP (**b**), P2p spectrum of GO-MD-MP (**c**), N1s spectrum of GO-MD-MP (**d**), and Si2p spectrum of GO-MD-MP (**e**).

TEM analysis was carried out to study the morphology of the synthesized GO-MD-MP and GO, as shown in Figure 5. GO exhibited folded sheet-like structures, as shown in Figure 5a. Moreover, Figure 5b represents the TEM image of GO-MD-MP where the dark-colored agglomerates are the modified PDPG and the light-colored layers are the graphene sheet.

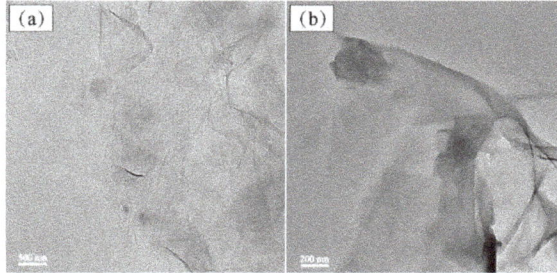

Figure 5. TEM images of GO (**a**) and GO-MD-MP (**b**).

The XRD patterns of GO and GO-MD-MP are represented in Figure 6. The characteristic diffraction peak at $2\theta = 9.5°$ was attributed to GO, corresponding to the (002) reflection of GO, indicating an interlayer distance of 0.93 nm. However, after the chemical modification of GO with PDPG, GO-MD-MP showed no visible diffraction peak in its XRD patterns, indicating that the GO-MD-MP sheet was separated and dispersed randomly, which may contribute to improving the dispersion of GO-MD-MP in epoxy resin.

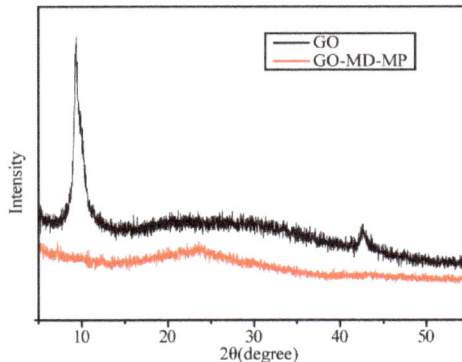

Figure 6. XRD patterns of GO and GO-MD-MP.

The dispersibility of modified graphene was a significant factor in the preparation of graphene-based composites. The enhancement of the nonflammability of graphene in composites also depended on the dispersion of graphene in the matrix. Figure 7 shows the dispersion of GO and GO-MD-MP in different solvents including water, chloroform, THF, and DMF. It shows that GO dispersed well and formed brown homogeneous suspensions in water, THF, and DMF. However, GO settled to the bottom when it was added to chloroform. After GO was modified with PDPG, the dispersions remained homogeneous in chloroform, THF, and DMF due to the good solubility of PDPG in chloroform, THF, and DMF. On the contrary, the hydrophobic group in PDPG on the graphene surface resulted in GO-MD-MP, in water, settling to the bottom of the serum bottle. The results indicate that functionalized graphene with PDPG can improve the solubility and compatibility of GO in organic solvents. It is worth mentioning that the excellent dispersion of GO-MD-MP in chloroform, THF, and DMF is helpful to prepare EP/GO-MD-MP composites, because chloroform, THF, and DMF are also good solvents for epoxy resin.

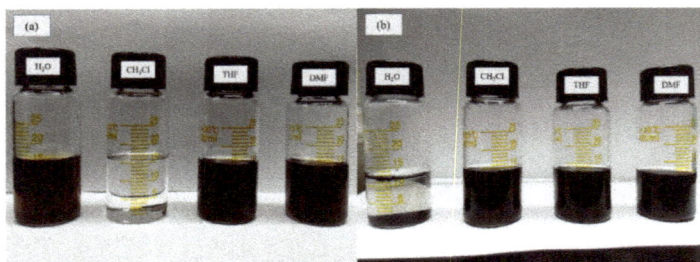

Figure 7. GO (**a**) and GO-MD-MP (**b**) in different solvents.

The thermal degradation curves of GO, PDPG, and GO-MD-MP were investigated using TGA under nitrogen atmospheres. As shown in Figure 8, the mass loss of GO started below 100 °C, and the main weight loss of GO was around 200 °C, which were attributed to the loss of the carboxyl groups, hydroxyl epoxy and hydroxyl, on GO and indicated the insufficient thermal stability of GO. In addition, the char yield ratio of GO was just 43.02% at 700 °C. For PDPG, a mass loss around 120 °C appeared and was attributed to GMA degradation due to the catalytic action of the phosphorus flame retardant. The main mass loss was between 300 and 450 °C, which was due to the degradation of POSS and DOPO. The char yield ratio of PDPG was about 15.70%. When GO was grafted with PDPG, its main weight loss temperature was about 330 °C due to the high bond energy of Si–O and Si-C in the POSS skeleton structure, reflecting that the thermal properties of GO-MD-MP had been significantly improved. On the other hand, the char yield ratio of GO-MD-MP was enhanced to 64.69%, compared to pristine GO, which suggests that when GO is grafted with PDPG its thermal properties are improved.

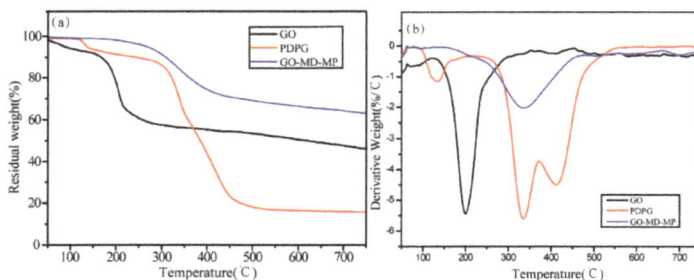

Figure 8. The thermogravimetric analysis (TGA) (**a**) and derivative weight (DTG) (**b**) curves of GO, PDPG, and GO-MD-MP.

The dispersion and interface strength of the graphene-based nanofillers in the polymer matrix are key factors for the properties of functionalized graphene/polymer nanocomposites. The SEM images of the fractured surface of the graphene/polymer nanocomposites is usually used to evaluate the dispersion of functionalized graphene [35]. Figure 9 shows the fractured surface of different cured EP composites. It is commonly known that pure epoxy resin is a brittle material and this was found to be the case in our system, as shown in Figure 9a. When PDPG was added, a lot of curved lines were observed on the fractured surface of the EP composites, which enhanced their toughness. However, some obvious protrusions were on the surface, which resulted in a poor interface strength between PDPG and the EP matrix, which may generate poor mechanical properties. On the contrary, after grafting PDPG to GO, the surface of the EP composites was rough without obvious protrusions, indicating that the interface strength of GO-MD-MP in the EP matrix was enhanced significantly. This may be explained by the fact that PDPG can weaken the intermolecular forces between different nanosheets of graphene. In addition, the covalent bonds connecting GO and PDPG may prevent POSS from aggregating, resulting in a better compatibility between GO-MD-MP and the EP matrix.

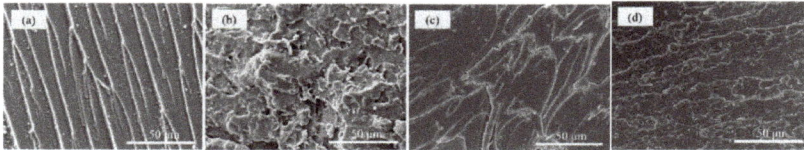

Figure 9. The SEM images of the fractured surface of pure epoxy resin (EP) (**a**), EP-4% PDPG (**b**), EP-2% GO-MD-MP (**c**), and EP-4% GO-MD-MP (**d**) composites.

3.2. Thermal Properties and Flame Retardance of GO-MD-MP/EP Composites

The TGA and DTG curves of the different cured EP composites under nitrogen atmospheres are shown in Figure 10. In Figure 10a, all cured EP composites had a mass loss process that occurred between 300 and 500 °C, which was the decomposition of the epoxy resin curing network. The initial decomposition temperature (T_d) can be adapted to the ambient temperature of epoxy resin. As is shown in Figure 10b, the maximum weight loss rate of GO-MD-MP/EP composites was evidently lower than the pure EP and PDPG/EP composites, indicating that the addition of GO-MD-MP can delay the thermal degradation process of epoxy resin. In addition, the residue chars at 700 °C of the pure EP and EP-4% PDPG were at 17.40% and 18.52%, respectively. However, the residue chars of the EP-2 wt % GO-MD-MP and EP-4 wt % GO-MD-MP were 25.50% and 25.77%, which were higher than that of the pure EP and EP-4 wt % PDPG due to the excellent thermostability of GO-MD-MP.

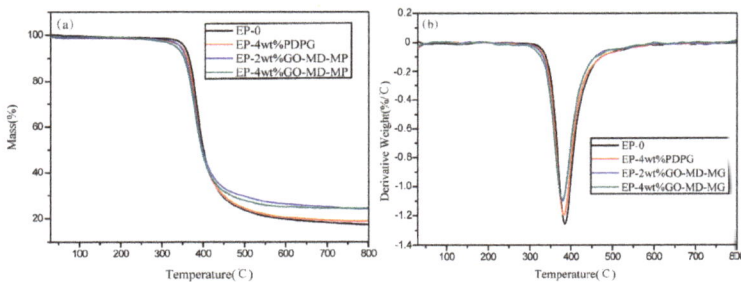

Figure 10. The TGA (**a**) and DTG (**b**) curves of the different cured EP composites.

Glass transition temperature (T_g) is a significant parameter to evaluate the performance and process properties of a polymer. The DSC curves of the pure EP and its different cured EP composites are shown in Figure 11. The T_g value of the pure EP was 157.4 °C. After adding PDPG into EP, the T_g values decreased, which may be attributed to the increasing free volume of the system caused by the large volume of POSS groups. However, when GO-MD-MP was added into EP, the T_g values increased slightly, which was because the effective dispersion of functionalized graphene achieved high aspect ratios in the EP composites, thus resulting in a higher cross-linking density.

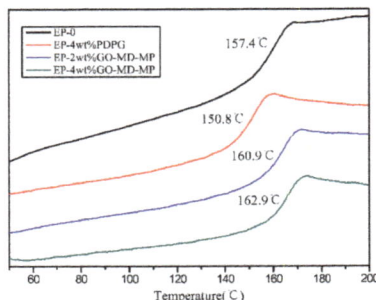

Figure 11. The DSC curves of pure EP and its different cured EP composites.

The flame retardance of the resulting EP composites was tested by the determination of LOI values and the UL-94 test, and the results are presented in Table 2. As is shown in Table 2, the LOI value of pure EP was about 23.4%, indicating that it was very easy to burn. The 4 wt % addition of PDPG in the EP composites resulted in a slight improvement in flame retardancy with a LOI value of 27.9% and a UL-94 rating of just V-1. In contrast, the LOI values of the 4 wt % addition of GO-MD-MP in the EP composites was significantly improved to 31.1% and the UL-94 rating was V-0, revealing that the GO-MD-MP/EP exhibited a great flame retardancy and further indicating that GO-MD-MP played a significant role in the flame retardancy of the epoxy resin. It is worth noting that the same amount of PDPG (4 wt %) and GO-MD-MP (4 wt %) in the EP composites led to a huge difference of flame-retardant EP, and the flame retardancy of EP-4wt%GO-MD-MP was more remarkable than of EP-4 wt % PDPG. Furthermore, the content of the flame-retardant elements, P and Si, in GO-MD-MP is significantly lower than that in PDPG. It is indicated that graphene plays an important role for increasing flame retardancy, which is attributed to not only its excellent barrier properties caused by the unique platelet morphology of graphene, but also its tortuous properties resulting from the effective dispersion of GO-MD-MP in the EP composites [36].

Table 2. Limiting oxygen index (LOI) values and UL-94 data of the cured EP composites.

Samples	LOI (%)	UL-94			Dripping
		Burning Grade	t1(s)	t2(s)	
pure EP	23.4	NR	>30s	-	Yes
PDPG/EP-4%	27.9	V-1	16.2	2.4	No
GO-MD-MP/EP-2%	29.4	V-1	19.6	3.8	No
GO-MD-MP/EP-4%	31.1	V-0	2.7	3.5	No

Figure 12 shows that the SEM images of the outer (a–d) and the inner (e–f) residue chars of the pure EP, the EP-4%PDPG, the EP-2%GO-MD-MP, and the EP-4%GO-MD-MP composites. It is shown that the residue char of the pure EP presented a loose outer layer and a smooth inner layer, which resulted from a lower amount of residue char, further indicating that the pure EP was fully degraded. The introduction of PDPG created a discontinuous outer char layer and a bubbly inner char, since many gaseous products escaped through the outer chars during combustion. In contrast, after adding GO-MD-MP, the outer residue chars of the cured EP composites were much more compact, which prevented the gaseous products escaping through the outer chars. Significant porosity was caused by the bubbles that were produced on the surface of the inner char, indicating that more gaseous products prevented the EP composites from burning. At the same time, these bubbles nucleated below the heated EP surface and could not destroy the outer residue chars. As a result, these residues formed a protective layer to protect the EP composites from burning.

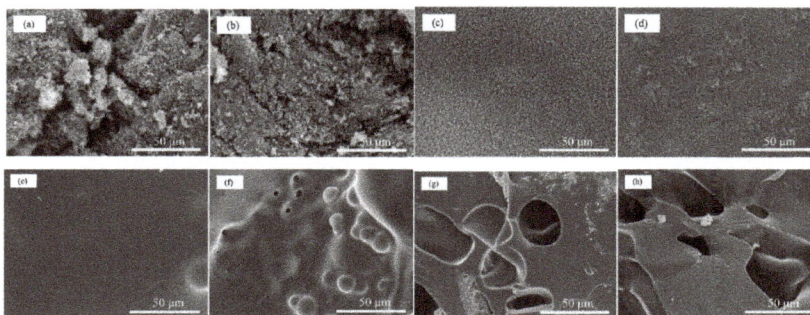

Figure 12. The SEM images of the outer (**a–d**) and the inner (**e–f**) residue chars of the pure EP (**a,e**), the EP-4% PDPG (**b,f**), the EP-2% GO-MD-MP (**c,g**), and the EP-4% GO-MD-MP (**d,h**) composites.

The graphitization degree of the residue char can be used to measure its thermal stability. Raman spectroscopy was conducted to measure the graphitization degree of the pure EP, the EP-4% PDPG, the EP-2% GO-MD-MP, and the EP-4% GO-MD-MP composite residue chars after treating them at 600 °C for 30 min in a muffle furnace. The corresponding Raman spectra are shown in Figure 13. The I_D/I_G values of the four EP composites were 0.936, 0.914, 0.862, and 0.850, respectively. With the I_D/I_G values decreasing in the EP composites, the graphitization degree of the residue char increased, which showed that the thermal stability of the residue chars improved gradually. These results indicated that both PDPG and GO-MD-MP can further improve the flame retardancy of EP composites. However, GO-MD-MP provided more compact residue chars with the same amount of addition.

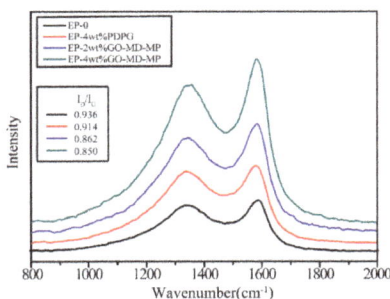

Figure 13. Raman spectra of different cured EP composites residue chars.

3.3. Mechanical Properties

Finally, the mechanical properties of all the cured EP composites were measured via a three-point bending test. The flexural modulus and flexural strength of the EP composites were measured, and the results are shown in Table 3. The flexural modulus of the pure EP, EP-4% PDPG, EP-2% GO-MD-MP, and EP-4% GO-MD-MP was 2308.89, 2145.61, 2658.62, and 2738.46 MPa, respectively. At the same time, the flexural strength of the EP-4% GO-MD-MP composite increased by 15.88% relative to the pure EP, but the EP-4% PDPG decreased. The changes in the mechanical properties of different EP composites can be explained by the interface strength between the flame retardant and the EP matrix. As mentioned above, the poor interface strength of EP-4% PDPG led to the decrease of the flexural modulus. On the contrary, when GO-MD-MP was added into EP, the mechanical properties of the EP composites improved significantly due to the better interface strength between GO-MD-MP and the EP matrix.

Table 3. Flexural modulus (MPa) and flexural strength (N·mm^{-2}) of cured EP and its composites.

Samples	Flexural Modulus (MPa)	Flexural Strength (N·mm^{-2})	Percentage of Increased Flexural Strength (%)
pure EP	2308.89	142.451	0
EP-4% PDPG	2145.61	131.014	−8.03
EP-2% GO-MD-MP	2658.62	154.732	8.62
EP-4% GO-MD-MP	2738.46	165.067	15.88

4. Conclusions

In summary, a novel POSS-based copolymer modified graphene oxide hybrid flame retardant (GO-MD-MP) was successfully synthesized through grafting a POSS-based polymer (PDPG) to the GO backbone. The structure and morphology of GO-MD-MP was investigated by XPS, XRD, FT-IR, TEM, Raman spectroscopy, and TGA. The dispersion and exfoliation of modified graphene oxide was strikingly improved, resulting in a good dispersion of GO-MD-MP in the EP composites. For flame retardancy, the addition of 4 wt % GO-MD-MP to epoxy resin made it pass UL-94 with a V-0 rating. Meanwhile, the LOI value was greatly increased to 31.1%. The enhancement of the flame retardancy was mainly attributed to the excellent barrier function of the char and the tortuous properties that resulted from the effective dispersion of the graphene nanoplatelets, which can effectively prevent heat and oxygen from entering the inner epoxy matrix. As for the mechanical properties, the flexural modulus and flexural strength of the GO-MD-MP/EP composites were improved due to the excellent dispersion and interfacial interaction of modified GO in the EP matrix.

Author Contributions: Y.X. and L.D. designed the experiments, and Min L. performed the experiments. The data processing and analysis were performed by Y.X., Min L., H.Z., W.W., and Meng L. Min L. and Y.X. wrote and revised the paper. G.C., Y.X., and L.D. provided constructive suggestions about this work.

Funding: The work was supported by the joint fund of the Science and Technology Major Project of the Fujian Province (2018HZ0001-1), the Xiamen Science and Technology Major Project (3502Z20171002), and the Graphene Technology Research and Development Project of the Fujian Provincial Development and Reform Commission.

Conflicts of Interest: The authors declare no conflict of interest.

References

1. Feng, Y.; Hu, J.; Xue, Y.; He, C.; Zhou, X.; Xie, X.; Ye, Y.; Mai, Y.W. Simultaneous improvement in the flame resistance and thermal conductivity of epoxy/Al$_2$O$_3$ composites by incorporating polymeric flame retardant-functionalized graphene. *J. Mater. Chem. A* **2017**, *5*, 13544–13556. [CrossRef]
2. Lu, S.Y.; Hamerton, I. Recent developments in the chemistry of halogen-free flame retardant polymers. *Prog. Polym. Sci.* **2002**, *27*, 1661–1712. [CrossRef]
3. Zhang, X.; Alloul, O.; Zhu, J.; He, Q.; Luo, Z.; Colorado, H.A.; Haldolaarachchige, N.; Young, D.P.; Shen, T.D.; Wei, S.; et al. Iron-core carbon-shell nanoparticles reinforced electrically conductive magnetic epoxy resin nanocomposites with reduced flammability. *RSC Adv.* **2013**, *3*, 9453–9464. [CrossRef]
4. Huo, S.; Wang, J.; Yang, S.; Cai, H.; Zhang, B.; Chen, X.; Wu, Q.; Yang, L. Synergistic effect between a novel triazine-based flame retardant and DOPO/HPCP on epoxy resin. *Polym. Adv. Technol.* **2018**, *29*, 2774–2783. [CrossRef]
5. Wang, X.; Hu, Y.; Song, L.; Xing, W.; Lu, H.; Lv, P.; Jie, G. Flame retardancy and thermal degradation mechanism of epoxy resin composites based on a DOPO substituted organophosphorus oligomer. *Polymer* **2010**, *51*, 2435–2445. [CrossRef]
6. Gui, H.; Xu, P.; Hu, Y.; Wang, J.; Yang, X.; Bahader, A.; Ding, Y. Synergistic effect of graphene and an ionic liquid containing phosphonium on the thermal stability and flame retardancy of polylactide. *RSC Adv.* **2015**, *5*, 27814–27822. [CrossRef]
7. Liu, C.; Huang, J.; Zhu, J.; Yuan, C.; Zeng, B.; Chen, G.; Xu, Y.; Dai, L. Synthesis of a novel azaphosphorine flame retardant and its application in epoxy resins. *J. Appl. Polym. Sci.* **2018**, *135*, 45721. [CrossRef]

8. Qian, X.; Song, L.; Yu, B.; Wang, B.; Yuan, B.; Shi, Y.; Hu, Y.; Yuen, R.K. Novel organic-inorganic flame retardants containing exfoliated graphene: Preparation and their performance on the flame retardancy of epoxy resins. *J. Mater. Chem. A* **2013**, *1*, 6822–6830. [CrossRef]

9. Laoutid, F.; Bonnaud, L.; Alexandre, M.; Lopez-Cuesta, J.M.; Dubois, P. New prospects in flame retardant polymer materials: From fundamentals to nanocomposites. *Mater. Sci. Eng. R Rep.* **2009**, *63*, 100–125. [CrossRef]

10. Shaw, S.D.; Blum, A.; Weber, R.; Kannan, K.; Rich, D.; Lucas, D.; Koshland, C.P.; Dobraca, D.; Hanson, S.; Birnbaum, L.S.; et al. Halogenated flame retardants: Do the fire safety benefits justify the risks? *Rev. Environ. Health* **2010**, *25*, 261–306. [CrossRef]

11. Ciesielski, M.; Schafer, A.; Doring, M. Novel efficient DOPO-based flame-retardants for PWB relevant epoxy resins with high glass transition temperatures. *Polym. Adv. Technol.* **2008**, *19*, 507–515. [CrossRef]

12. Shi, Y.Q.; Fu, T.; Xu, Y.J.; Li, D.F.; Wang, X.L.; Wang, Y.Z. Novel phosphorus-containing halogen-free ionic liquid toward fire safety epoxy resin with well-balanced comprehensive performance. *Chem. Eng. J.* **2018**, *354*, 208–219. [CrossRef]

13. Duan, L.; Yang, H.; Shi, Y.; Hou, Y.; Zhu, Y.; Gui, Z.; Hu, Y. A Novel Branched Phosphorus-Containing Flame Retardant: Synthesis and Its Application into Poly(Butylene Terephthalate). *Ind. Eng. Chem. Res.* **2016**, *55*, 10218–10225. [CrossRef]

14. Hoang, D.; Kim, J.; Jang, B.N. Synthesis and performance of cyclic phosphorus-containing flame retardants. *Polym. Degrad. Stab.* **2008**, *93*, 2042–2047. [CrossRef]

15. Wang, C.S.; Shieh, J.Y. Phosphorus-containing epoxy resin for an electronic application. *J. Appl. Polym. Sci.* **1999**, *73*, 353–361. [CrossRef]

16. Haddad, T.S.; Choe, E.; Lichtenhan, J.D. Hybrid styryl-based polyhedral oligomeric silsesquioxane (POSS) polymers. *Mater. Res. Soc. Symp. Proc.* **1996**, *435*, 25–32. [CrossRef]

17. Schwab, J.J.; Lichtenhan, J.D. Polyhedral oligomeric silsesquioxane (POSS)-based polymers. *Appl. Organomet. Chem.* **1998**, *12*, 707–713. [CrossRef]

18. Liu, C.; Chen, T.; Yuan, C.H.; Song, C.F.; Chang, Y.; Chen, G.R.; Xu, Y.T.; Dai, L.Z. Modification of epoxy resin through the self-assembly of a surfactant-like multi-element flame retardant. *J. Mater. Chem. A* **2016**, *4*, 3462–3470. [CrossRef]

19. Song, G.H.; Li, X.S.; Jiang, Q.Y.; Mu, J.X.; Jiang, Z.H. A novel structural polyimide material with synergistic phosphorus and POSS for atomic oxygen resistance. *RSC Adv.* **2015**, *5*, 11980–11988. [CrossRef]

20. Zhang, W.C.; He, X.D.; Song, T.L.; Jiao, Q.J.; Yang, R.J. Comparison of intumescence mechanism and blowing-out effect in flame-retarded epoxy resins. *Polym. Degrad. Stab.* **2015**, *112*, 43–51. [CrossRef]

21. Feng, Y.; He, C.; Wen, Y.; Ye, Y.; Zhou, X.; Xie, X.; Mai, Y.W. Superior flame retardancy and smoke suppression of epoxy-based composites with phosphorus/nitrogen co-doped graphene. *J. Hazard. Mater.* **2018**, *346*, 140–151. [CrossRef] [PubMed]

22. Ming, G.; Li, J.F.; Zhang, X.Q.; Yue, L.N.; Chai, Z.H. The flame retardancy of epoxy resin including the modified graphene oxide and ammonium polyphosphate. *Combust. Sci. Technol.* **2018**, *190*, 1126–1140. [CrossRef]

23. Shi, Y.Q.; Yu, B.; Zheng, Y.Y.; Yang, J.; Duan, Z.P.; Hu, Y. Design of reduced graphene oxide decorated with DOPO-phosphanomidate for enhanced fire safety of epoxy resin. *J. Colloid Interface Sci.* **2018**, *521*, 160–171. [CrossRef] [PubMed]

24. Wang, X.; Song, L.; Yang, H.Y.; Xing, W.Y.; Kandola, B.; Hua, Y. Simultaneous reduction and surface functionalization of graphene oxide with POSS for reducing fire hazards in epoxy composites. *J. Mater. Chem.* **2012**, *22*, 22037–22043. [CrossRef]

25. Wang, X.; Song, L.; Pornwannchai, W.; Hu, Y.; Kandola, B. The effect of graphene presence in flame retarded epoxy resin matrix on the mechanical and flammability properties of glass fiber-reinforced composites. *Compos. Part A Appl. Sci. Manuf.* **2013**, *53*, 88–96. [CrossRef]

26. Huang, G.B.; Gao, J.R.; Wang, X.; Liang, H.D.; Ge, C.H. How can graphene reduce the flammability of polymer nanocomposites? *Mater. Lett.* **2012**, *66*, 187–189. [CrossRef]

27. Chen, W.H.; Liu, Y.S.; Liu, P.J.; Xu, C.G.; Liu, Y.; Wang, Q. The preparation and application of a graphene-based hybrid flame retardant containing a long-chain phosphaphenanthrene. *Sci. Rep.* **2017**, *7*, 8759. [CrossRef]

28. Guo, W.W.; Yu, B.; Yuan, Y.; Song, L.; Hu, Y. In situ preparation of reduced graphene oxide/DOPO-based phosphonamidate hybrids towards high-performance epoxy nanocomposites. *Compos. Part B Eng.* **2017**, *123*, 154–164. [CrossRef]

29. Bao, C.L.; Guo, Y.Q.; Song, L.; Kan, Y.C.; Qian, X.D.; Hu, Y. In situ preparation of functionalized graphene oxide/epoxy nanocomposites with effective reinforcements. *J. Mater. Chem.* **2011**, *21*, 13290–13298. [CrossRef]

30. Guo, Y.Q.; Bao, C.L.; Song, L.; Yuan, B.H.; Hu, Y. In Situ Polymerization of Graphene, Graphite Oxide, and Functionalized Graphite Oxide into Epoxy Resin and Comparison Study of On-the-Flame Behavior. *Ind. Eng. Chem. Res.* **2011**, *50*, 7772–7783. [CrossRef]

31. Liao, S.H.; Liu, P.L.; Hsiao, M.C.; Teng, C.C.; Wang, C.A.; Ger, M.D.; Chiang, C.L. One-Step Reduction and Functionalization of Graphene Oxide with Phosphorus-Based Compound to Produce Flame-Retardant Epoxy Nanocomposite. *Ind. Eng. Chem. Res.* **2012**, *51*, 4573–4581. [CrossRef]

32. Xu, Y.J.; Chen, L.; Rao, W.H.; Qi, M.; Guo, D.M.; Liao, W.; Wang, Y.Z. Latent curing epoxy system with excellent thermal stability, flame retardance and dielectric property. *Chem. Eng. J.* **2018**, *347*, 223–232. [CrossRef]

33. Cao, R.R.; Chen, S.; Liu, H.B.; Liu, H.H.; Zhang, X.X. Fabrication and characterization of thermo-responsive GO nanosheets with controllable grafting of poly(hexadecyl acrylate) chains. *J. Mater. Sci.* **2018**, *53*, 4103–4117.

34. Kudin, K.N.; Ozbas, B.; Schniepp, H.C.; Prud'homme, R.K.; Aksay, I.A.; Car, R. Raman spectra of graphite oxide and functionalized graphene sheets. *Nano Lett.* **2008**, *8*, 36–41. [CrossRef] [PubMed]

35. Cai, D.Y.; Song, M. Recent advance in functionalized graphene/polymer nanocomposites. *J. Mater. Chem.* **2010**, *20*, 7906–7915. [CrossRef]

36. Wang, X.; Kalali, E.N.; Wan, J.T.; Wang, D.Y. Carbon-family materials for flame retardant polymeric materials. *Prog. Polym. Sci.* **2017**, *69*, 22–46. [CrossRef]

polymers

MDPI

Article

Blends of Cyanate Ester and Phthalonitrile–Polyhedral Oligomeric Silsesquioxane Copolymers: Cure Behavior and Properties

Xiaodan Li [1,2,†], Fei Zhou [2,†], Ting Zheng [2,†], Ziqiao Wang [3], Heng Zhou [4], Haoran Chen [3], Lin Xiao [2,*], Dongxing Zhang [2,*] and Guanhui Wang [1,*]

[1] School of Chemistry and Chemical Engineering, Jinggangshan University, No 28, Xueyuan Road, Qingyuan District, Ji'an 343009, China; lixiaodanlixiaodan@126.com
[2] School of Materials Science and Engineering, Harbin Institute of Technology, Harbin 150001, China; angel.flyfly@hotmail.com (F.Z.); zthappy1127@gmail.com (T.Z.)
[3] Harbin FRP Institute, Harbin 150029, China; hit_wzq@126.com (Z.W.); chr0526@163.com (H.C.)
[4] Institute of Chemistry, Chinese Academy of Sciences, No.2 Haidian District, Beijing 100190, China; zhouheng@iccas.ac.cn
*   Correspondence: 14B909032@hit.edu.cn (L.X.); zhangdongxing@hit.edu.cn (D.Z.); wangguanhui@igsu.edu.cn (G.W.);
    Tel.: +86-451-86282455 (L.X.); +86-451-86282455 (D.Z.); +86-796-8100490 (G.W.)
†   These authors contributed equally as first authors to this work.

Received: 15 November 2018; Accepted: 25 December 2018; Published: 1 January 2019

Abstract: Blends of cyanate ester and phthalonitrile–polyhedral oligomeric silsesquioxane copolymers were prepared, and their cure behavior and properties were compared via differential scanning calorimetry (DSC) analysis, thermogravimetric (TG) analysis, dynamic mechanical analysis, Fourier-transform far-infrared (FTIR) spectroscopy, and rheometric studies. The copolymer blends showed high chemical reactivity, low viscosity, and good thermal stability (TG temperatures were above 400 °C). The glass-transition temperature of the blends increased by at least 140 °C compared to cyanate ester resin. The blends are suitable for preparing carbon-fiber-reinforced composite materials via a winding process and a prepreg lay-up process with a molding technique. The FTIR data showed that the polymerization products contained triazine-ring structures that were responsible for the superior thermal properties.

Keywords: phthalonitrile polymers; phthalonitrile-polyhedral oligomeric silsesquioxane copolymers; cyanate ester; blends; thermal properties

1. Introduction

Phthalonitrile-based composites [1–8] with high temperature resistance, ablation resistance, low flammability, and high strength have great potential in the aerospace sector as components for maintaining airframe loads for the next generation of aeronautical and space vehicle systems. They are one of the few to meet the United States Navy's stringent requirements under MIL-STD-2031 for the usage of polymer composites aboard Navy submarines. The model compound of the phthalonitrile monomer, 4,4′-bis(3,4-dicyanophenoxy)biphenyl (BPh), was first discovered at the U.S. Naval Research Laboratory [9–16]. However, the high melting temperature and poor solubility of the monomer limit its application. Many phthalonitrile monomers [17–21] have been synthesized to lower the melting temperature and improve their processing properties and use in resin-infusion fabrication.

We previously [22,23] introduced polyhedral oligomeric silsesquioxane (POSS) into a phthalonitrile system and prepared POSS–phthalonitrile copolymers. The POSS reagents consist of an inorganic silsequioxane cage and have multiple reactive groups that can react with the cyanate group at high temperatures. They provide a unique opportunity to prepare nanocomposites with truly molecular dispersion of the inorganic fillers. Thus, the POSS copolymers are easily prepared at low temperatures with a short curing time. Concurrently, their enhanced properties, such as higher T_g temperature, oxidative resistance, improved mechanical property, and fire resistance, have been shown in many thermoplastics, such as polyethylene, polypropylene, polycarbonate, as well as thermosetting polymers. These include polyimides, epoxy resins, polyurethane, and cyanate ester resins [24–32]. However, POSS–phthalonitrile copolymers are still applicable in resin-infusion fabrication.

To expand the use of resins in the field of composite materials, an urgent problem to overcome is the development of resin systems suitable for the winding process and prepreg lay-up process. The key to solving this problem is to improve the resin fluidity, which can be achieved using two methods. The first method is the use of solvents. However, it is impossible to prepare dense materials with low porosity without efficient removal of the solvent during the molding process. The second method is improvement of the resin fluidity by blending. Polymer blending is a good way to tailor the properties of blended systems that generally have a useful combination of properties derived from each component polymer. Various thermoplastic polymer blends and thermoset polymer blends have been developed and successfully applied. Raj [33] prepared phenolic-urea-epoxy blends that exhibit better mechanical properties and higher thermal stability compared to epoxy resin. Nishimura [34] studied the molecular structure of cyanate ester (NCE)–epoxy blends and discussed the irradiation effect of gamma rays and fast neutrons; they found that the blended resin could survive a design period in a radiation environment. Harvey [35] prepared a homogenous polycarbonate/cyanate ester network and suggested that this kind of blend may have utility in the fabrication of toughed composite structures. Augustine [18] prepared hydroxyl-terminated, polyetheretherketone (PEEK)-toughed epoxy-amino novolac phthalonitrile blends; PEEK reduced the brittleness and improved their shear strength. Dominguez [36] separately blended epoxy resin with biphenyl PN and $n = 4$ PN and studied the cure behavior of the blends. The results showed that phthalonitrile–epoxy blends exhibited good processability and the copolymers had enhanced high-temperature properties (the T_g temperature was 230 °C) compared with cured epoxy resin.

However, while the process of blending phthalonitrile monomers with epoxy resin has been improved, the thermal resistance of the resulting blend is much lower than that of phthalonitrile. Here, we blended the phthalonitrile–polyhedral oligomeric silsesquioxane (POSS)–modified phthalonitrile copolymer with novolac cyanate ester resin (NCE), which has higher thermal resistance. POSS-modified phthalonitrile and highly thermal-resistant NCE resin have similar structures and hence are mutually soluble at high temperatures. Moreover, the active groups of POSS can chemically react with both types of resins to improve their compatibility. The resulting blend has improved processability and resulted in a higher resistance than epoxy resin blends. Therefore, it is necessary to study the blends of phthalonitrile and NCE, which have great potential for engineering applications.

However, research on blends of phthalonitrile and NCE is rare. Therefore, the goal of this study is to further investigate such blends, including their curing behavior, processability, compatibility, and thermal properties by using differential scanning calorimetry (DSC), Fourier-transform infrared (FTIR) spectroscopy, thermogravimetric analysis (TGA), and dynamic mechanical analysis (DMA). These data and rheological properties offer an experimental basis for future applications in carbon-fiber-reinforced composites.

2. Experimental Details

2.1. Materials

BPh (solid, 99%, without any further purification) was synthesized at the Institute of Chemistry of the Chinese Academy of Sciences, Beijing, China. 4,4′-bis(4-aminophenoxy)biphenyl (BAPP, solid, without any further purification) was purchased from Bailingwei, Inc. (J&K Scientific Ltd., Beijing, China). The epoxycyclohexyl polyhedral oligomeric silsesquioxane (EPCHPOSS, semi-solid, without any further purification) was obtained from Hybrid Plastics, Inc. (Hybrid Plastics, Fountain Valley,

CA, USA). NCE resin (liquid, without any further purification) was purchased from Lonza (China) Investments Co., Ltd. (Lonza Ltd., Visp, Switzerland). The structures of the phthalonitrile monomer, curing agent, POSS, and NCE are shown in Figure 1.

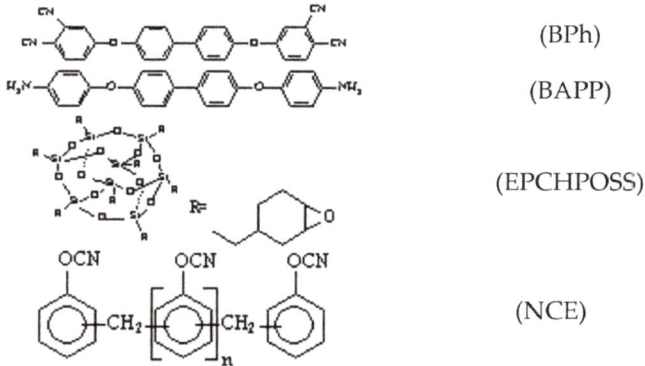

Figure 1. Structures of the phthalonitrile monomer, curing agent, phthalonitrile–polyhedral oligomeric silsesquioxane (POSS), and cyanate ester (NCE) used in this work.

2.2. Preparation of Cyanate and POSS-Phthalonitrile Blends

BPh was melted at 260 °C and BAPP (curing agent, 2 wt.%) was added with continuous stirring. EP0408 was then added at 0.5 wt.% (the concentration shown to have the best thermal stability in our previous work [27]) to the mixture with stirring and then cooled to room temperature; this was named the EPCHPOSS–BPh prepolymer. The EPCHPOSS–BPh prepolymer was pulverized into powder and added to NCE resin at varying compositions (10, 20, 30, and 40 wt.%); the mixtures were named 1090, 2080, 3070, and 4060 prepolymers, respectively. All prepolymers were cured at 280 °C (4 h), 300 °C (8 h), and 325 °C (8 h); these were then subsequently post-cured under an inert N_2 atmosphere at 350 °C (4 h) and 375 °C (2 h). The cured samples were called EPCHPOSS–BPh polymer and 1090, 2080, 3070, and 4060 polymers. The prepolymers were pulverized before performing DSC and rheological tests and studied by TGA, FTIR, and DMA.

2.3. Characterization

DSC experiments were conducted in a flowing N_2 atmosphere on EPCHPOSS–BPh prepolymer and 1090, 2080, 3070, and 4060 prepolymers. The experiments were conducted in a Perkin–Elmer Pyris-6 DSC calorimeter (Perkin–Elmer, Richmond, CA, USA) at a heating rate of 10 °C/min. The DSC curves at different heating rates (10, 20, 30, and 40 °C/min) were measured. The activation energies were calculated using the following equation:

$$\frac{d\left[\ln\left(\beta/T_P^2\right)\right]}{d\left[1/T_P\right]} = -\frac{E}{R} \tag{1}$$

where β is the heating rate, T_p is the peak temperature of each DSC curve at different heating rates, and R is the universal gas constant. Thus, the E value was obtained through the linear dependence of $\ln(\beta/T_p^2)$ on $1/T_p$ at various heating rates.

EPCHPOSS–BPh polymer as well as 1090, 2080, 3070, and 4060 polymers and NCE polymer were pulverized into powder and mixed with spectroscopy grade KBr to make pellets for FTIR. Their chemical structures were studied using a FT-IR spectrometer(Nicolet Avatar 370, ThermoFisher Scientific, Grand Island, NY, USA) with potassium bromide pellets containing a low concentration of sample; the wavenumber range was 500–4000 cm^{-1} at a resolution of 4 cm^{-1}. Thermal analysis was conducted on the polymers using a thermogravimetric analyzer (SDTQ600, TA Instruments, Eden Prairie, MN, USA). TGA tests were performed in air at a scan rate of 10 °C/min and a flow rate of 100 mL/min. The dynamic storage modulus (G') and damping factor (tanδ) of the rectangular phthalonitrile polymer specimens (50 × 10 × 3 mm) were obtained using a DMA instrument (DMS-6100, NSK Ltd., Tokyo, Japan) with a N_2 atmosphere and a temperature of 30–400 °C (rate 4 °C/min, frequency 10 Hz). Thus, the T_g temperature was estimated from the modulus–temperature plots obtained by DMA. Dynamic viscosity measurements were performed on a rheometer (AR-2000, TA Instruments, Eden Prairie, MN, USA). The samples were placed on the platforms and heated from 80 to 280 °C under the fixed strain values of 50% with a fixed frequency of 1 rad/s. The dynamic viscosity and shear storage modulus were obtained.

3. Results and Discussion

The 1090, 2080, 3070, and 4060 resin blends were subjected to DSC scanning from room temperature to 400 °C (Figure 2).

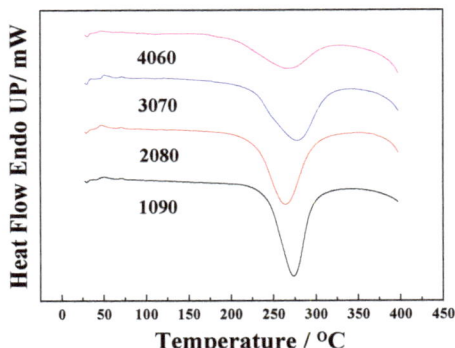

Figure 2. Differential scanning calorimetry (DSC) scanning diagram of 1090, 2080, 3070, and 4060 blend resins.

The DSC curves of the 1090, 2080, 3070, and 4060 resin blends each have a large exothermic reaction peak. This is very different from the DSC curve of the previously studied POSS-modified phthalonitrile [22]. This difference is attributed to the much higher exotherm of NCE [37,38] during the reaction than that of phthalonitrile resin. This masks the endothermic and exothermic peaks in the DSC curves of phthalonitrile resins. As a result, the phthalonitrile content increased. The intensities of the exothermic peaks associated with the curing reaction of NCE became smaller, and the corresponding enthalpy values were −569, −507, −465, −367, and −303 J/g, respectively. The decrease in heat release alleviated the implosion of cyanate resin due to the large amount of released heat. The initial reaction temperatures of the 1090, 2080, 3070, and 4060 resin blends gradually decreased as the phthalonitrile content increased, i.e., 243, 228, 225, 201, and 180 °C, respectively. This indicates that the initial reaction proceeded more easily as the phthalonitrile content increased, but this was accompanied by a decreased reaction intensity, as reflected by the decrease in exothermic energy.

The Kissinger equation [39] was used to calculate the activation energies of the 1090, 2080, 3070, and 4060 resins as 82, 75, 140, and 108 kJ/mol, respectively. Except for the 3070 resin, the other three resins had significantly lower activation energies than phthalonitrile. Activation energy indicates the reactivity of a system. Lower activation energy leads to a higher reactivity, showing that the introduction of the NCE resin significantly improved the reactivity. However, the reactivity began to decrease significantly upon addition of more than 30 wt.% of phthalonitrile. This finding shows that the addition of a small amount of phthalonitrile promoted polymerization reactions; when the contents of the two resins were comparable, their respective polymerization reactions affected each other.

Polymers **2019**, *11*, 54

The rheological properties of the 1090, 2080, 3070, and 4060 resin blends were investigated [40,41], and the results are shown in Figure 3. As the temperature increased, their viscosities all first decreased and then reached a steady state before rapidly increasing. The lowest viscosities of the 1090, 2080, 3070, and 4060 resin blends were 17, 25, 32, and 89 mPa·S, respectively. The viscosities of the 1090, 2080, and 3070 resin blends were very low. However, after the content of EPCHPOSS–BPh prepolymer increased to 40%, the viscosities of the blends increased rapidly to 89 mPa·s, respectively. The temperature at which the viscosity started to increase again from a steady state is termed the gel point of the system; that is, the temperature at which the polymer starts to cure. The gel-point temperatures of the 1090, 2080, 3070, and 4060 resin blends were 278, 278, 271, and 243, respectively, demonstrating that the gel-point temperature decreased as the content of the EPCHPOSS–BPh prepolymer increased. By adjusting its viscosity via temperature control, one can make a blend suitable for the molding of carbon-fiber-reinforced composites in the winding and prepreg lay-up processes. Compared to phthalonitrile, the processability of the blends was greatly improved, leading to more extensive applications of the resin matrix.

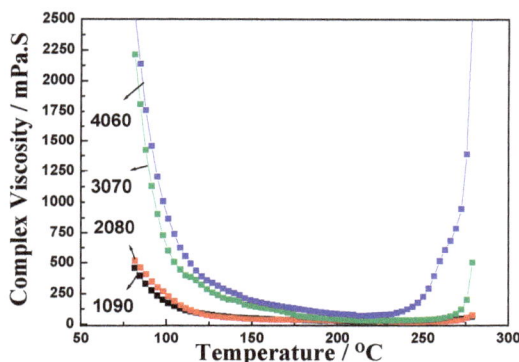

Figure 3. Complex viscosity of blends as a function of temperature. (**a**) 1090, (**b**) 2080, (**c**) 3070, and (**d**) 4060.

The shear storage modulus as a function of temperature curves for 1090, 2080, 3070, and 4060 resins are shown as Figure 4. As the temperature increases, the storage modulus first decreased, then stabilized, and then finally increased quickly. The trend of storage modulus change is consistent with the trend of viscosity change. A lower viscosity leads to greater fluidity. This makes chain movement easier, leading to a smaller modulus. The modulus increased with increasing content of EPCH–BPh prepolymer; the modulus began to increase quickly when the temperature approached the gel temperature.

The structures of the NCE resin polymer, the 1090, 2080, 3070, and 4060 resin blend polymers, and the EPCHPOSS–BPh polymer were studied and compared in Figure 5. The FTIR spectrum of the copolymer is more similar to that of NCE.

In the FTIR spectra of the NCE, 1090, 2080, 3070, and 4060 polymers, the 729-, 744-, 744-, 729-, and 736-cm^{-1} peaks are δCH (out-of-plane deformation) of 1,2,3-trisubstituted benzene; the 809-, 817-, 809-, 817-, and 820-cm^{-1} peaks are δCH (out-of-plane deformation) of the triazine compound; the 1240-, 1240-, 1219-, 1226-, and 1233-cm^{-1} peaks are γC-OC of the ether bond; the 1540-, 1540-, 1533-, 1526-, and 1533-cm^{-1} peaks are γ rings of the triazine ring; and the 1614-, 1599-, 1599-, 1599-, and 1606-cm^{-1} peaks are benzene rings of the aromatic ring [42,43].

The structure of the EPCHPOSS–BPh polymer is mainly composed of triazine rings in which the C atoms are linked by aromatic compounds. In comparison, the structure of the NCE resin is different in that its triazine rings are linked by ether bonds. According to Burchill [44], polymeric products of —CN include the polymer with a triazine ring as the crosslinking point when organic amines and ammonium organic acid salts are used. The mechanism is shown in Figure 6. The structures of the 1090, 2080, 3070, and 4060 copolymers are interpenetrating polymer networks with two types of triazine rings (Figure 7). The structure of the NCE polymer shows that the links of the triazine ring to the ether bond increase the flexibility of the molecular chain, thus lowering the thermal resistance versus the EPCHPOSS–BPh polymer. The introduction of POSS tightly links the network structures of the two polymers.

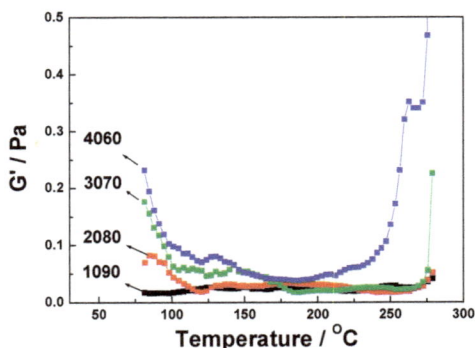

Figure 4. Shear storage modulus (G′) as a function of temperature curves for 1090, 2080, 3070, and 4060 resins.

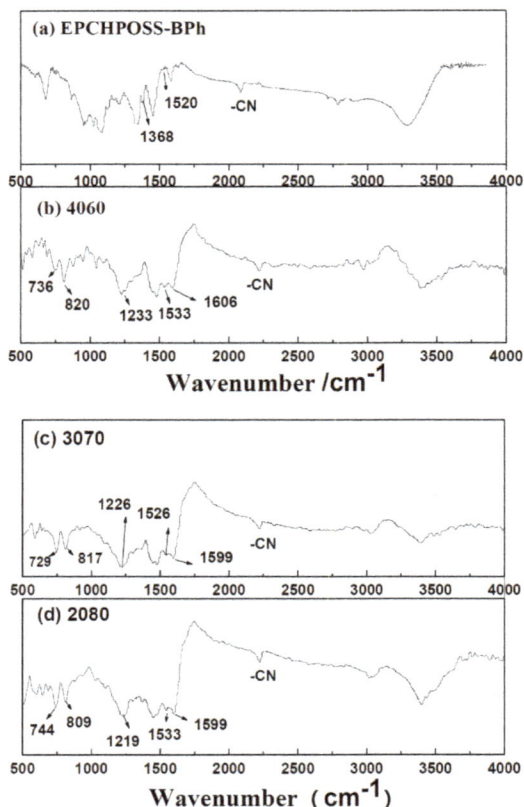

Figure 5. *Cont.*

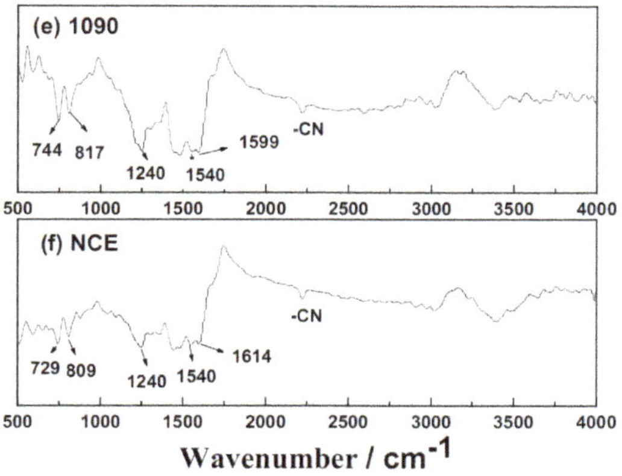

Figure 5. Fourier-transform far-infrared (FTIR) spectra of NCE, EPCHPOSS–BPh, 1090, 2080, 3070, and 4060 copolymers.

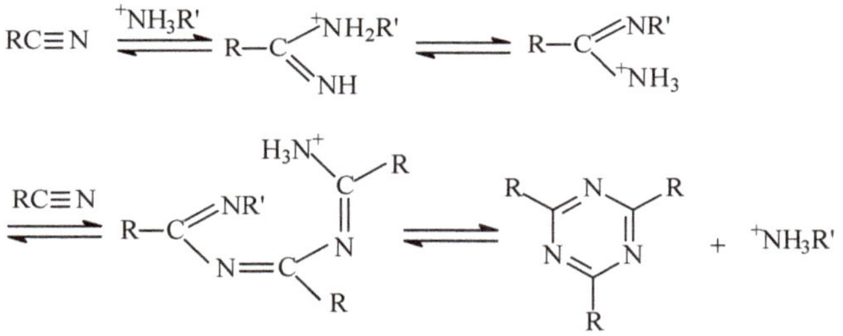

Figure 6. The mechanism of generating a triazine ring [44].

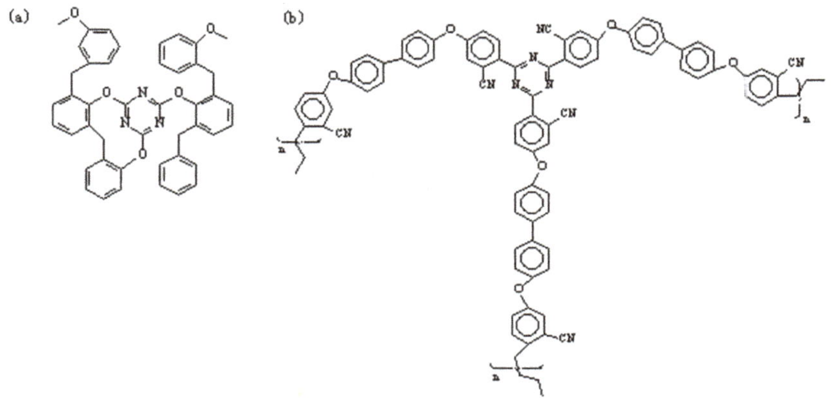

Figure 7. The possible structure of the blends: (**a**) Structure of cured NCE; (**b**) Structure of cured phthalonitrile.

The TG properties of NCE resin polymer, 1090, 2080, 3070, and 4060 resin polymer blends, and EPCHPOSS-BPh polymer were studied and compared [10,45]. The changes in their Thermogravimetry (TG) curves and Derivative Thermogravimetry (DTG) curves as a function of temperature are shown in Figure 8.

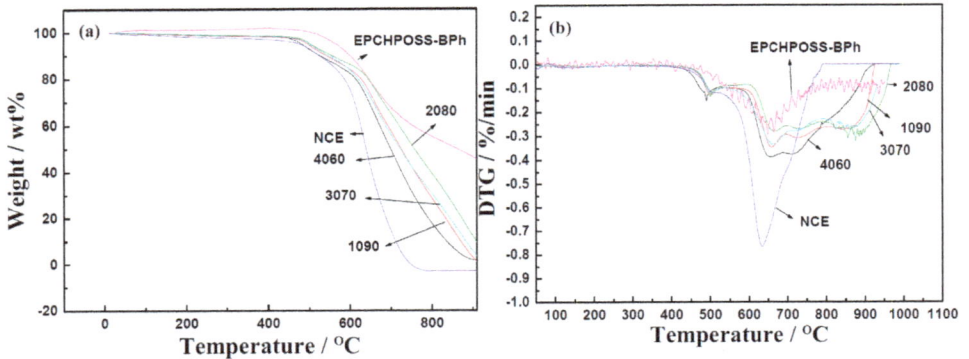

Figure 8. Thermogravimetric analysis (TGA) and Derivative (DTG) plots of NCE, EPCHPOSS–BPh, 1090, 2080, 3070, and 4060 polymers in the air. (**a**) TG curves and (**b**) DTG curves.

The thermal resistance of the NCE polymer is significantly lower than that of other polymers. The rapid decline in mass starting at 593 °C represents a significant difference between the NCE polymer and other polymers. The drastic weight loss causes the TG curve of the NCE polymer to be significantly lower than the other polymers. The temperatures at which the NCE polymer, 1090, 2080, 3070, and 4060 polymer blends, and EPCHPOSS–BPh polymer began to lose weight were 452, 476, 480, 477, 455, and 505 °C, respectively. The temperature at which the NCE polymer, 1090, 2080, 3070, and 4060 polymer blends, and EPCHPOSS–BPh polymer experienced 5% weight loss were 475, 496, 503, 500, and 564 °C, respectively. The resin blend with 20% EPCHPOSS–BPh polymer had the highest thermal resistance and showed a retention of 12% at 900 °C; the NCE polymer had 100% weight loss at this temperature. In addition, the EPCHPOSS–BPh polymer had the highest thermal resistance and showed a retention of 48% at 900 °C (shown in Table 1).

Table 1. Thermal stability of EPCHPOSS–BPh, 1090, 2080, 3070, 4060, and NCE polymers.

Properties	EPCHPOSS–BPh	1090	2080	3070	4060	NCE
$T_{5\%}$ (°C)	564	496	503	500	481	475
Char yield (%)	48	2.7	12	5.5	1.8	0
$T_{lost\ fast}$ (°C)	664	660	669	662	648	630

$T_{5\%}$: temperature of 5% weight loss in a normal air atmosphere. Char yield: percentage of polymer. $T_{lost\ fast}$: temperature of loss fast.

The DTG curves showed that the NCE polymer had the lowest thermal resistance and the largest peak (hence the highest rate) of weight loss. The EPCHPOSS–BPh polymer had the lowest rate of weight loss, followed by the 2080 polymer. There were two stages of the weight-loss rate in the DTG curves. A small weight-loss-rate peak appeared when the decomposition temperature was reached. The temperatures corresponding to the highest weight-loss rates of NCE polymer, 1090, 2080, 3070, and 4060 polymer blends, and EPCHPOSS–BPh polymer were 630, 660, 669, 662, 648, and 664 °C, respectively. As the temperature increased from that corresponding to the highest rate of weight loss, the weight-loss rate of the 2080 polymer gradually decreased; those at 1090, 3070, and 4060 blends showed a plateau, indicating that these blends still continued to lose weight at their highest rates.

The NCE polymer, 1090, 2080, 3070, and 4060 polymer blends, and EPCHPOSS–BPh polymer were tested using DMA. The variations of dynamic storage moduli and damping factors (tanδ) with temperature are shown in Figures 9 and 10, respectively.

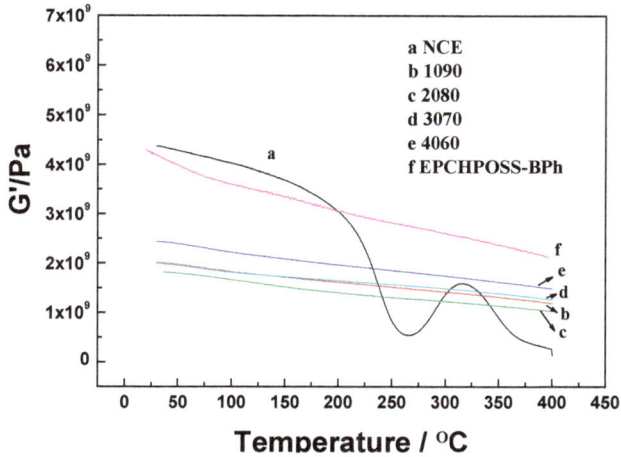

Figure 9. Dynamic storage modulus (G′) as a function of temperature for NCE, EPCHPOSS–BPh, 1090, 2080, 3070, and 4060 polymers under a nitrogenous atmosphere.

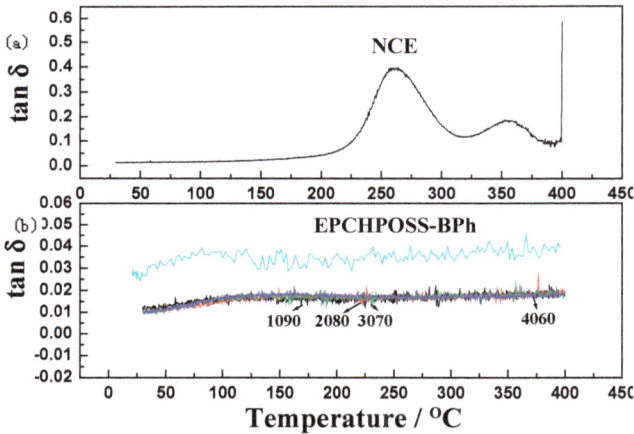

Figure 10. Damping factor (tanδ) as a function of temperature for NCE, EPCHPOSS–BPh, 1090, 2080, 3070, 4060 polymers in the nitrogenous atmosphere: (**a**) NCE, (**b**) EPCHPOSS–BPh, 1090, 2080, 3070, and 4060.

Figure 9 shows that the dynamic storage moduli of the EPCHPOSS–BPh polymer and the 1090, 2080, 3070, and 4060 copolymers gradually decreased with increasing temperatures. This decrease was attributed to the stress relaxation of the polymer network structures. The moduli at room temperature were 4.4, 2.02, 1.89, 1.99, and 2.4 GPa, respectively. When the temperature increased to 400 °C, the moduli decreased to 2.4, 1.19, 1, 1.28, and 1.44 GPa, respectively, amounting to 54%, 58%, 53%, 64%, and 60% of the respective moduli at room temperature. Thus, the 3070 copolymer had the highest modulus retention rate.

The dynamic storage modulus of NCE varied with temperature in a significantly different manner. As the temperature increased, its modulus decreased rapidly from 4.5 GPa at room temperature to the lowest value of 0.52 GPa at 266 °C. From here, the temperature further increased, and its modulus increased moderately. As the temperature increased beyond 316 °C, its modulus decreased rapidly again and reached 0.2 GPa at 400 °C.

Figure 10 shows that the damping factors (tan δ) of EPCHPOSS–BPh polymer and 1090, 2080, 3070, and 4060 copolymers vary smoothly with increasing temperature without the presence of glass-transition peaks, indicating that all three types of polymers formed a stable three-dimensional network structure that prevented the slippage of chain segments as temperature increased. The absence of glass-transition peaks means that the glass-transition temperature was higher than 400 °C. There are two peaks at 259 and 356 °C. In the curve depicting the variation of the damping factor (tanδ) of NCE resin with temperature, these two peaks correspond to the two fluctuations in the variations of dynamic storage modulus with time in Figure 6. The two peaks are associated with the two phases in the curing process, indicating that the polymer chain of the NCE resin underwent a glass transition starting at 259 °C, resulting in the slippage of chain segments. This slipped further at 356 °C. A thorough comparison shows that the thermal resistance of NCE blends significantly improved, and the glass-transition temperature of the blends was at least 140 °C higher than that of NCE [10,11,36].

The T_g temperatures were obtained from DMA for NCE, 1090, 2080, 3070, and 4060 polymers. This was included with DSC data in Figure 11. All curves undergo exothermic processes when the temperature is elevated—especially the NCE polymer, which had the highest heat release. Figure 11 shows that 2080 had no obvious T_g temperature (above 400 °C), while the T_g temperatures of NCE, 1090, 3070, and 4060 were 360 °C, 384 °C, 380 °C, and 377 °C. These were different from the DMA results. The cure extent of the polymer and the aged extent [46] at elevated temperatures will affect the T_g temperature obtained from DSC.

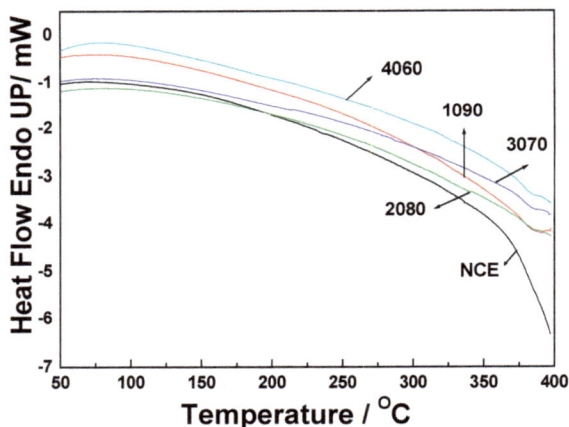

Figure 11. DSC scanning of NCE, 1090, 2080, 3070, and 4060 polymers under nitrogenous atmosphere.

4. Conclusions

The following conclusions can be drawn from the above analysis:

Conclusion 1: The DSC results indicate that as the content of phthalonitrile increased, the exothermic energy of the blend gradually decreased. This alleviates the implosion due to the large amount of heat released from cyanate resin. The activation energies of the 1090, 2080, 3070, and 4060 blends were 81.6, 75.2, 140.3, and 108 kJ/mol, respectively. Except for the 3070 resin, the other three resins had activation energies that were significantly lower than that of phthalonitrile, indicating that the introduction of NCE resin significantly improves the reactivity.

Conclusion 2: The FTIR results indicate that the structures of the 1090, 2080, 3070, and 4060 copolymers were interpenetrating polymer networks with two types of triazine rings. The introduction of POSS tightly linked the network structures of the two types of polymers.

Conclusion 3: The TGA results show that the resin blend with 20% of EPCHPOSS–BPh prepolymer had the highest thermal resistance and showed a retention of 12% at 900 °C.

Conclusion 4: DMA results show that the thermal resistance of the NCE blends were significantly improved, and the glass-transition temperature of the blend was at least 140 °C higher than that of NCE resin.

Conclusion 5: Investigation of the rheological properties shows that the viscosity of the resin blend was still very low and can be adjusted by controlling the temperature. This makes the resin blend suitable for molding the carbon-fiber-reinforced composites in the winding and prepreg lay-up processes. Compared to phthalonitrile, the processability of the resin blend was greatly improved. This suggests extensive applications for this resin matrix.

Author Contributions: X.L., L.X., D.Z. and G.W. conceived and designed the experiments; Z.W. and H.C. performed the experiments; H.Z. synthesized BPh monomer, X.L., F.Z. and T.Z. analyzed the data; X.L. wrote the paper.

Conflicts of Interest: The authors declare no conflict of interest.

References

1. Keller, T.M. Phthalonitrile-based high temperature resin. *J. Polym. Sci. Part. A Polym. Chem.* **1988**, *26*, 3199–3212. [CrossRef]
2. Keller, T.M. Phthalonitrile conductive polymer. *J. Polym. Sci. Part. A Polym. Chem.* **1987**, *25*, 2569–2576. [CrossRef]
3. Keller, T.M. A stable intrinsically conductive polymer. *J. Polym. Sci. Part. C Polym. Lett.* **1986**, *24*, 211–214. [CrossRef]
4. Sastri, S.B.; Keller, T.M. Phthalonitrile polymers: Cure behavior and properties. *J. Polym. Sci. Part. A Polym. Chem.* **1999**, *37*, 2105–2111. [CrossRef]
5. Laskoski, M.; Keller, T.M.; Qadri, S.B. Direct conversion of highly aromatic phthalonitrile thermosetting resins into carbon nanotube containing solids. *Polymer* **2007**, *48*, 7484–7489. [CrossRef]
6. Sastri, S.B.; Armistead, J.P.; Keller, T.M. Phthalonitrile-glass fabric composites. In Proceedings of the 41th International SAMPE Symposium, Anaheim, CA, USA, 24–28 March 1996; pp. 24–28.
7. Sastri, S.B.; Armistead, J.P.; Keller, T.M. Phthalonitrile-carbon fiber composites. *Polym. Compos.* **1996**, *17*, 816–822. [CrossRef]
8. Dominguez, D.D.; Jones, H.N.; Trzaskoma-Paulette, P.P.; Keller, T.M. Mechanical properties of graphite fiber-reinforced phthalonitrile composites. In Proceedings of the 46th International SAMPE Symposium, Long Beach, CA, USA, 6–7 May 2001; pp. 94–107.
9. Laskoski, M.; Dominguez, D.D.; Keller, T.M. Synthesis and properties of a bisphenolA based phthalonitrile resin. *J. Polym. Sci. Part A Polym. Chem.* **2005**, *43*, 4136–4143. [CrossRef]
10. Dominguez, D.D.; Keller, T.M. Low-melting phthalonitrile oligomers: Preparation, polymerization and polymer properties. *High. Perform. Polym.* **2006**, *18*, 283–304. [CrossRef]
11. Dominguez, D.D.; Jones, H.N.; Keller, T.M. The effect of curing Additive on the mechanical properties of phthalonitrile-carbon fiber composites. *Polym. Compos.* **2004**, *25*, 554–561. [CrossRef]
12. Sastri, S.B.; Keller, T.M. Phthalonitrile Cure reaction with aromatic diamines. *J. Polym. Sci. Part. A Polym. Chem.* **1998**, *36*, 1885–1890. [CrossRef]
13. Keller, T.M.; Dominguez, D.D. High temperature resorcinol-based phthalonitrile polymer. *Polymer* **2005**, *46*, 4614–4618. [CrossRef]
14. Keller, T.M. Synthesis and polymerization of multiple aromatic ether phthalonitriles. *Chem. Mater.* **1994**, *6*, 302–305. [CrossRef]
15. Laskoski, M.; Dominguez, D.D.; Keller, T.M. Synthesis and properties of aromatic ether phosphine oxide containing oligomericphthalonitrile resins with improved oxidative stability. *Polymer* **2007**, *48*, 6234–6240. [CrossRef]
16. Laskoski, M.; Schear, M.B.; Neal, A.; Dominguez, D.D.; Ricks-laskoski, H.L.; Hervey, J.; Keller, T.M. Improved synthesis and properties of aryl ether-based oligomericphthalonitrile resins and polymers. *Polymer* **2015**, *67*, 185–191. [CrossRef]
17. Babkin, A.V.; Zodbinov, E.B.; Bulgakov, B.A.; Kepman, A.V.; Avdeev, V.V. Low-melting siloxane-bridged phthalonitriles for heat-resistant matrices. *Eur. Polym. J.* **2015**, *66*, 452–457. [CrossRef]

18. Augustine, D.; Vijayalakshmi, K.P.; Sadhana, R.; Mathew, D.; Nair, C.P.R. Hydroxyl terminated peek-toughed epoxy–amino novolacphthalonitrile blends—Synthesis, cure studies and adhesive properties. *Polymer* **2014**, *55*, 6006–6016. [CrossRef]

19. Augustine, D.; Mathew, D.; Nair, C.P.R. One component phthalonitrilenovolac: Synthesis and characterization. *Eur. Polym. J.* **2015**, *71*, 389–400. [CrossRef]

20. Derradji, M.; Ramdani, N.; Zhang, T.; Wang, J.; Feng, T.T.; Wang, H.; Liu, W.B. Mechanical and thermal properties of phthalonitrile resin reinforced with silicon carbide particles. *Mater. Des.* **2015**, *71*, 48–55. [CrossRef]

21. Zhao, F.H.; Liu, R.J.; Yu, X.Y.; Naito, K.; Tang, C.C.; Qu, X.W.; Zhang, Q.X. Systhesis of a novel naphthyl-based self-catalyzed phthalonitrile polymer. *Chin. Chem. Lett.* **2015**, *26*, 727–729. [CrossRef]

22. Li, X.; Yu, B.; Zhang, D.; Lei, J.; Nan, Z. Cure behavior and thermomechanical properties of phthalonitrile–polyhedral oligomeric silsesquioxane copolymers. *Polymers* **2017**, *9*, 334. [CrossRef]

23. Li, X.; Wang, J.; Sun, Y.; Zhang, D.; Nan, Z.; Lei, J. The effect of thermal ttreatment on the decomposition of phthalonitrile polymer and phthalonitrile-polyhedral oligomeric silsesquioxane (POSS) copolymer. *Polym. Degrad. Stab.* **2018**, *156*, 279–291. [CrossRef]

24. Huang, J.; Wang, W.; Gu, J.; Li, W.; Zhang, Q.; Ding, Y.; Xi, K.; Zhen, Y.; Jia, X. New bead type and high symmetrical diallyl-POSS based emissive conjugated polyfluorene. *Polymer* **2014**, *55*, 6696–6707. [CrossRef]

25. Florea, N.M.; Lungu, A.; Badica, P.; Craciun, L.; Enculescu, M.; Ghita, D.G.; Ionescu, C.; Zgirian, R.G.; Iuvo, H. Novel nanocomposites based on epoxy resin/epoxy-functionalized polydimethylsiloxane reinforced with POSS. *Composites Part. B* **2015**, *75*, 226–234. [CrossRef]

26. Przadka, D.; Jeczalik, J.; Andrzejewska, E.; Marciniec, B.; Dutkiewicz, M.; Szlapka, M. Novel hybrid polyurethane/POSS materials via bulk polymerization. *React. Funct. Polym.* **2013**, *73*, 114–121. [CrossRef]

27. Blanco, I.; Bottino, F.A.; Cicala, G.; Latteri, A.; Recca, A. A Kinetic study of the thermal and the thermal oxidative degradations of new bridge POSS/PS nanocomposites. *Polym. Degrad. Stab.* **2013**, *98*, 2564–2570. [CrossRef]

28. Zhou, Z.; Cui, L.; Zhang, Y.; Zhang, Y.; Yin, N. Preparation and properties of POSS grafted polypropylene by reactive blending. *Eur. Polym. J.* **2008**, *44*, 3057–3066. [CrossRef]

29. Su, C.H.; Chiu, Y.P.; Teng, C.C.; Chiang, C.L. Preparation, characterization and Thermal properties of organic and inorganic composites involving epoxy and polyhedral oligomericSilsequioxane(POSS). *J. Polym. Res.* **2010**, *17*, 673–681. [CrossRef]

30. Blanco, I.; Abate, L.; Bottino, F.A. Synthesis and thermal properties of new dumbbell-shaped isobutyl-substituted POSSs linked by aliphatic bridges. *J. Therm. Anal. Calorim.* **2014**, *116*, 5–13. [CrossRef]

31. Blanco, I.; Abate, L.; Antonelli, M.L.; Bottino, F.A.; Bottino, P. Phenyl heptacyclopentyl–polyhedral oligomericSilsequioxane (ph, hcp-POSS)/polystyrene (PS) nanocomposites: The influence of substituents in the phenyl group on the thermal stability. *eXPRESSPolym. Lett.* **2012**, *6*, 997–1006. [CrossRef]

32. Levicki, J.P.; Pielichowski, K.; De La Croix, P.T.; Janowski, B.; Todd, D.; Liggat, J.J. Thermal degradation studies of polyurethane/POSS nanohybrid elastomers. *Polym. Degrad. Stab.* **2010**, *95*, 1099–1105.

33. Raj, M.M.; Raj, L.M.; Dave, P.N. Glass fiber reinforced composites of phenolic-urea-epoxy resin blends. *J. Saudi Chem. Soc.* **2012**, *16*, 241–246. [CrossRef]

34. Nishimuna, A.; Izumi, Y.; Imaizumi, M.; Nishijima, S.; Hemmi, T.; Shikama, T. Neutron and gamma ray irradiation effects on interlaminar shear strength of insulation materials with cyanate ester-epoxy blended resin. *Fusion Eng. Des.* **2011**, *86*, 1558–1561. [CrossRef]

35. Harvey, B.G.; Guenthner, A.J.; Yandek, G.R.; Gambrea, L.R.; Meylemans, H.A.; Bladwin, L.C.; Reams, J.T. Synthesis and characterization of a renewable cyanate ester/polycarbonate network derived from eugenol. *Polymer* **2014**, *55*, 5073–5079. [CrossRef]

36. Dominguez, D.D.; Keller, T.M. Phthalonitrile-epoxy blends: Cure behavior and copolymer properties. *J. Appl. Polym. Sci.* **2008**, *110*, 2504–2514. [CrossRef]

37. Zhang, S.; Yan, Y.; Li, X.; Fan, H.; Ran, Q.; Fu, Q.; Gu, Y. A novel ultra low-k nanocomposites of benzoxazinyl modified polyhedral oligomeric silsesquioxane and cyanate ester. *Eur. Polym. J.* **2018**, *103*, 124–132. [CrossRef]

38. Lin, Y.; Song, M.; Stone, C.A.; Shaw, S.J. A comprehensive study on the curing kinetics and network formation of cyanate ester resin/clay nanocomposites. *Thermochim. Acta* **2013**, *552*, 77–86. [CrossRef]

39. Wang, C.S.; Lin, C.H. Novel phosphorus-containing epoxy resins. Part II: Curing kinetics. *Polymer* **2000**, *41*, 8579–8586. [CrossRef]

40. Metalwala, Z.; Khoshroo, K.; Rasoulianboroujeni, M.; Tahriri, M.; Johnsonb, A.; Baeten, J.; Fahimipour, F.; Ibrahim, M.; Tayebi, L. Rheological properties of contemporary nanohybrid dental resin composites: The influence of preheating. *Polym. Test.* **2018**, *72*, 157–163. [CrossRef]

41. Dodero, A.; Williams, R.; Gagliardi, S.; Vicini, S.; Alloisio, M.; Castellano, M. A micro-rheological and rheological study of biopolymers solutions: Hyaluronic acid. *Carbohydr. Polym.* **2019**, *203*, 349–355. [CrossRef]

42. Vashchuk, A.; de Anda, A.R.; Starostenko, O.; Grigoryeva, O.; Sotta, P.; Rogalsky, S.; Smertenko, P.; Fainleib, A.; Grande, D. Structure–property relationships in nanocomposites based on cyanate ester resins and 1-heptyl pyridinium tetrafluoroborate ionic liquid. *Polymer* **2018**, *148*, 12–26. [CrossRef]

43. Bartolomeo, P.; Chailan, J.F.; Vernet, J.L. Curing of cyanate ester resin: A novel approach based on FTIR spectroscopy and comparison with other techniques. *Eur. Polym. J.* **2001**, *37*, 659–670. [CrossRef]

44. Burchill, P.J. On the formation and properties of a high-temperature resin from a bisphthalonitrile. *J. Polym. Sci Part A Polym. Chem.* **1994**, *32*, 1–8. [CrossRef]

45. Bershtein, V.; Fainleib, A.; Egorova, L.; Grigoryeva, O.; Kirilenko, D.; Konnikov, S.; Ryzhov, V.; Starostenko, O.; Yakushev, P.; Yagovkina, M.; et al. The impact of ultra-low amounts of introduced reactive POSS nanoparticles on structure, dynamics and properties of densely cross-linked cyanate ester resins. *Eur. Polym. J.* **2015**, *67*, 128–142. [CrossRef]

46. Tao, Q.; Pinter, G.; Krivec, T. Influence of cooling rate and annealing on the DSC T_g of an epoxy resin. *Microelectro. Reliab.* **2017**, *78*, 396–400. [CrossRef]

MDPI

Article

PCL/POSS Nanocomposites: Effect of POSS Derivative and Preparation Method on Morphology and Properties

Mónica Cobos, Johnny R. Ramos, Dailyn J. Guzmán, M. Dolores Fernández and M. Jesús Fernández *

Department of Polymer Science and Technology, Faculty of Chemistry, University of the Basque Country (UPV/EHU), Pº Manuel Lardizábal 3, 20018 San Sebastián, Spain; monica.cobos@ehu.es (M.C.); johnram@hotmail.com (J.R.R.); daylyngm@yahoo.es (D.J.G.); mariadolores.fernandez@ehu.es (M.D.F.)
* Correspondence: mjesus.fernandez@ehu.es; Tel.: +34-943-01-5353

Received: 30 November 2018; Accepted: 23 December 2018; Published: 26 December 2018

Abstract: The incorporation of polyhedral oligomeric silsesquioxanes (POSS) molecules as nanoparticles into polymers can provide improved physico-chemical properties. The enhancement depends on the extent of dispersion of the nanofiller, which is determined by the compatibility with the polymer that is by the POSS type, and the processing method.   In this study, poly(ε-caprolactone)/POSS derivatives nanocomposites (PCL/POSS) were obtained via solution-casting and melt compounding.   Two amino-derivatives containing different alkyl substituents, and ditelechelic POSS-containing hybrid PCL masterbatch were used as nanofillers. The effect of preparation method, POSS content and type on the morphology, thermal, mechanical, and surface properties of nanocomposites were studied.   Morphological analysis evidenced the formation of POSS crystalline aggregates, self-assembled POSS molecules of submicrometer size dispersed in the polymer matrix.   The best dispersion was achieved using the ditelechelic POSS-containing hybrid PCL masterbatch, and comparing the two amino-POSS derivatives, the one with longer alkyl chain of substituents exhibited better degree of dispersion independent of preparation method.   DSC analysis showed the role of POSS derivatives as nucleating agents for PCL. The incorporation of POSS derivatives into the PCL matrix improved thermal stability. The preparation method, POSS type and content had influence on mechanical properties of nanocomposites. POSS nanoparticles enhanced the surface hydrophobicity of PCL.

Keywords: poly(ε-caprolactone) nanocomposite; POSS nanoparticles; dispersion; morphology; thermal properties; mechanical properties; surface properties

---

1. Introduction

In recent years, the growing problem of waste disposal has sparked a great interest in biodegradable polymers [1–4]. Polycaprolactone (PCL) is one of the most attractive biodegradable synthetic polymers due to its commercial availability, biodegradability, and compatibility with different forms of waste disposal and excellent physico-mechanical properties [5]. PCL, a semicrystalline linear aliphatic thermoplastic polyester synthetized by the ring-opening polymerization of caprolactone, exhibits good flexibility, non-toxicity, hydrophobicity, low melting point (58–60 °C) and glass-transition temperature (−60 °C), ease of processing, and compatibility with a range of other polymers [6–8]. PCL has found extensive applications as commodity and biomedical materials, and for agricultural uses [7,9–11], as a substitute material for non-biodegradable polymers. However, modification is highly necessary when it is applied to different requirements. Using nanoparticles significantly improved the properties of PCL [12,13].

Among the variety of nanoscale fillers, polyhedral oligomeric silsesquioxanes (POSS) nanoparticles are considered as one of the most interesting nanofillers that have been used in the preparation of nanocomposites. POSS are a new class of organic/inorganic hybrid materials exhibiting a specific three dimensional cage structure made up of Si–O, with size from 1.5 to 3 nm in diameter. The tetravalent Si atoms of POSS are bound to organic groups, one or more of which may contain reactive functional groups. POSS additives can be incorporated into organic polymers via copolymerization, grafting, or blending [14–17]. Many studies have demonstrated the enhancement in the properties of polymers (thermal and oxidation resistance, mechanical properties, surface hardening, as well as reduced flammability) by incorporating POSS. The dispersion and distribution of nanoparticles in the polymer as well as the interfacial adhesion are key factors for improving the polymer properties. A nanometric dispersion and good distribution is necessary. The morphology depends on the processing method, the chemical structure and concentration of the nanoparticles [18,19]. To achieve the best morphology several methodologies can be used, functional groups can be introduced on the POSS molecules, polymer chains can be grafted on the POSS surface, as well as the optimization of the preparation method.

In previous works, different POSS derivatives have been incorporated into PCL [20–25]. Goffin et al. [20] studied the addition of aminopropylheptakis(isobutyl)-POSS and POSS-*g*-PCL nanohybrid by melt blending into PCL. Individualized POSS nanoparticles dispersed in the nanocomposites were only found when POSS-*g*-PCL nanohybrid was used as a masterbatch. Nanocomposites with enhanced crystallinity and thermo-mechanical properties were obtained. Miltner et al. [21] reported the preparation of POSS nanocomposites based on PCL by in situ polymerization of ε-caprolactone in the presence of aminopropylheptakis(isobutyl)-POSS or by melt mixing techniques, and using the nanohybrid prepared by in situ polymerization as masterbatch for melt mixing with PCL. They found amino-POSS agglomerates arranged in crystalline structure in the nanocomposite prepared by melt mixing, whereas enhanced dispersion quality was observed in the case of the nanocomposite prepared by in situ polymerization. As for the nanocomposite prepared by the masterbatch approach, they found that a higher grafted chain length was more efficient in improving the compatibility between POSS and the PCL matrix. Pan et al. [22] studied the morphology and crystallization of PCL/octaisobutyl-POSS nanocomposites prepared via solution casting method. Aggregates of submicron sized POSS particles were found and enhanced crystallization of PCL. Guan and Qiu [23] studied the crystallization, morphology, and dynamic mechanical properties of PCL in the PCL/octavinyl-POSS nanocomposites obtained by solution blending. They observed fine dispersion, enhanced overall isothermal melt crystallization rates of PCL, and improved storage modulus of the nanocomposites. Miltner et al. [24] investigated the influence of addition of aminopropylheptakis(isobutyl)-POSS into PCL via melt mixing on the thermal properties. A fraction of aggregated POSS structures were observed that did not exerted nucleating effect on PCL crystallization. Lee and Chang [25] reported the effect of trisilanolphenyl-POSS on the thermal and mechanical properties of PCL/POSS nanocomposites prepared by solution mixing. They found that POSS molecules were able to crystallize in the PCL matrix, a decreased degree of crystallinity of PCL, and increased tensile properties.

The purpose of the present study is to investigate the effects of POSS type, POSS concentration, and preparation method (solution-casting and melt compounding) on the morphology, thermal, mechanical, and surface properties of the PCL/POSS nanocomposites. Two amino-derivative POSS containing different alkyl group, aminopropylheptaisobutyl-POSS (APIBPOSS) and aminopropylheptaisooctyl-POSS (APIOPOSS), and ditelechelic POSS-containing hybrid poly(ε-caprolactone) masterbatch, propyl-heptaisooctyl-POSS-*g*-PCL-*g*-propyl-heptaisooctyl-POSS (PIOPOSS–PCL–PIOPOSS), were used as fillers to prepare PCL/POSS nanocomposites. X-ray diffraction (XRD) and transmission electron microscopy (TEM) were used to examine the state of dispersion of POSS nanoparticles in the nanocomposites. Thermal properties were determined using differential scanning calorimetry (DSC) and thermogravimetric analysis

(TGA). The mechanical properties were investigated through tensile tests. Surface properties of nanocomposites were studied by contact angle measurements.

Aminopropyl-heptaisobutyl-POSS is the only amino-POSS derivative used in the so far reported studies focused on PCL/POSS nanocomposites, and neither the effect of preparation method (solution casting and melt mixing), nor the effect of POSS substituent group have been studied [20–25]. The compatibility with the polymer and the dispersion state depends on the nature of the organic groups (R) of POSS molecules, and hence the properties of the new material. Thus, two amine-functionalized POSS were used in this study to investigate the effect of the alkyl chain length of the nonreactive organic substituents attached to the corner silicon atoms on morphology, thermal, and mechanical properties, as well as surface hydrophobicity of PCL. In addition, the masterbatch POSS-PCL-POSS was used to prepared PCL nanocomposites to make a comparison with ungrafted POSS nanoparticles.

2. Materials and Methods

2.1. Materials

Poly(ε-caprolactone) (PCL) (number-average molar mass = 45,000 g/mol), was supplied by Sigma-Aldrich (Munich, Germany). APIBPOSS ($C_{31}H_{71}NO_{12}Si_8$, Fw = 874.58), a white crystalline powder, and APIOPOSS ($C_{59}H_{127}NO_{12}Si_8$, Fw = 1267.32), a pale yellow viscous liquid, were purchased from Hybrid plastics (Hattiesburg, MS, USA). PIOPOSS-PCL-PIOPOSS hybrid was synthesized by grafting APIOPOSS to PCL chains using "click chemistry" as described elsewhere [26]. The alkyne-functionalized POSS (N-(3-(heptaisooctyl POSS) propyl) propiolamide) and bis-azide end-functionalized PCL were used for the click chemistry approach. Figure 1 shows the chemical structure of each POSS. Trichloromethane ($CHCl_3$) was obtained from Scharlau (Barcelona, Spain).

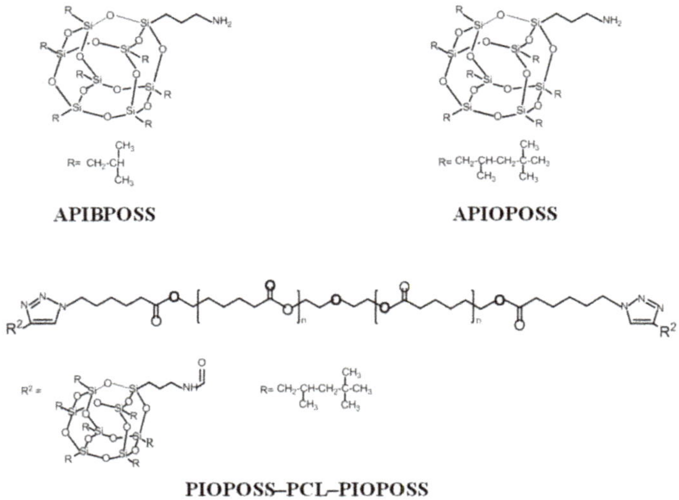

APIBPOSS

R= CH_2-CH
CH_3
CH_3

APIOPOSS

R= CH_2-CH-CH_2-C-CH_3
CH_3 CH_3

PIOPOSS–PCL–PIOPOSS

R= CH_2-CH-CH_2-C-CH_3
CH_3 CH_3

Figure 1. The chemical structure of polyhedral oligomeric silsesquioxanes (POSS) types used in this work.

2.2. Preparation of PCL/POSS Nanocomposites

PCL/APIBPOSS, PCL/APIOPOSS, and PCL/PIOPOSS-PCL-PIOPOSS masterbatch nanocomposite films were prepared by solution blending using trichloromethane as solvent. PCL and the appropriate amounts of POSS were dissolved separately. These two solutions were mixed together and stirred with sonication for 1 h, then cast into glass plates, and after solvent evaporation at room temperature the films were dried in a vacuum oven at 60 °C for 3 days. The nanocomposites were named PCL/APIBPOSS-x-S, PCL/APIOPOSS-x-S, and PCL/PIOPOSS-PCL-PIOPOSS-x-S, where x denotes the weight percentage of POSS derivative.

Melt compounded PCL/APIBPOSS and PCL/APIOPOSS nanocomposites at 2, 5, and 10 wt.% of POSS concentration were prepared in a Minilab II, Haake Rheomix CTW5 mini twin-screw extruder (Waltham, MA, USA) (90 °C, 10 min, 50 rpm). For characterization purposes, the extruded mixture was compression-molded at 60 °C and 300 bars of pressure for 5 min. The nanocomposites were named PCL/APIBPOSS-x-M and PCL/APIOPOSS-x-M where x denotes the weight percentage of POSS derivative.

2.3. Characterization of PCL/POSS Nanocomposites

XRD patterns were recorded on a Bruker D8 Advance X-ray diffractometer (Karlsruhe, Germany) with a graphite monochromator, Cu Kα generator ($\lambda = 0.154$ nm), and operating at 40 kV/30 mA.

TEM micrographs were obtained using a Philips Tecnai G2 20 TWIN TEM (Eindhoven, the Netherlands) at 200 kV accelerated voltage. The samples were cryo-ultramicrotomed using a Leica EM UC6 ultramicrotome apparatus and placed onto Cu grids.

Molecular weight of the PCL matrix (before and after nanocomposites preparation) was determined by gel permeation chromatography (GPC) analysis, using a Waters 2410 HPLC instrument (Milford, MA, USA) equipped with Styragel columns calibrated with polystyrene standards, with tetrahydrofuran as eluent with the flow rate of 1 mL min^{-1}.

DSC analyses were carried out using a TA instruments TA-DSC Q2000 (New Castle, DE, USA) under nitrogen atmosphere at a heating rate of 10 °C min^{-1}. The samples, about 6 mg, were heated from −100 to 100 °C and held in the molten state for 5 min (to erase the thermal history), then cooled to −60 °C at 10 °C min^{-1} and held 5 min, and again heated to 100 °C at 10 °C min^{-1}.

Thermogravimetric analyses were performed on a TA instruments His Res TG-Q-500 instrument (New Castle, DE, USA) under nitrogen or air atmosphere from 50 to 800 °C at a heating rate of 10 °C min^{-1}. The thermal degradation temperature was taken as the onset temperature at which 5% of weight loss occurs.

Tensile tests were carried out with a Universal mechanical testing machine Instron model 5569 (Norwood, MA, USA) according to ASTM D 638, at room temperature, gauge length of 35 mm and speed of 5 mm min^{-1}. Testing was carried out on at least five identical samples of each composition and the average values were reported.

Static contact angles were measured by using the sessile drop technique with a digital goniometer (Filderstadt, Germany) equipped with a dispensing needle (Dataphysics Contact Angle System OCA), using deionized water as liquid. At least 10 measurements were taken at different locations of the each sample surface and average values of static contact angles were reported.

3. Results and Discussion

3.1. Effect of POSS Type and Processing Method on Morphology and Dispersion of POSS in PCL Matrix

The morphology and the state of the POSS dispersion in the PCL matrix were observed with the complementary techniques of XRD and TEM. Figure 2A shows XRD patterns of neat PCL and APIBPOSS, and solution blended PCL/APIBPOSS nanocomposites. A number of strong peaks were observed in the diffractogram of APIBPOSS, revealing a highly crystalline structure.

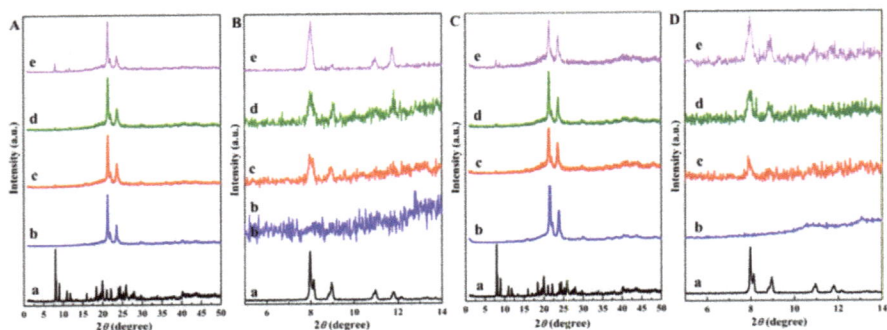

Figure 2. XRD patterns (**A**,**B**) solution blended, (**C**,**D**) melt mixed: (**a**) aminopropylheptaisobutyl-POSS (APIBPOSS), (**b**) Polycaprolactone (PCL), (**c**) PCL/APIBPOSS-2, (**d**) PCL/APIBPOSS-5, (**e**) PCL/APIBPOSS-10; (**B**,**D**): enlarged XRD patterns in the 2θ range 5–14 degrees.

The XRD pattern of PCL showed the characteristic diffraction peaks at 2θ = 21.34°, 21.96°, and 23.56°, corresponding to the (110), (111), and (200) planes, respectively [27]. In the XRD diffrattograms of PCL/APIBPOSS nanocomposites, those peaks remained invariant in comparison with the PCL homopolymer, indicating that the crystalline structure of PCL was unaffected by the presence of POSS. In the enlarged XRD patterns in the 2θ range of 5–14° of PCL/APIBPOSS containing 2 and 5% of nanofiller (Figure 2Bc,d) a weak peak is observed at 2θ = 8° corresponding to the APIBPOSS crystals (Figure 2Ba). This peak centered at 2θ = 8° increased in intensity as the POSS content increased (Figure 2Bd–e), and more crystalline peaks corresponding to POSS were also observed. However, the peak at 2θ = 8.96° almost disappeared at 10 wt.% POSS content. Moreover, from the diffractograms it is observed that crystalline peaks related to POSS crystals are not as sharp as the ones for the neat APIBPOSS, indicating that the dispersed POSS crystals are not as perfect as neat POSS crystals. These observations indicate that PCL/APIBPOSS blends exhibit the characteristic features of the structures of the two separate components. These results suggest that APIBPOSS crystallizes and agglomerates in PCL matrix when it is incorporated in the polymer by solution-casting blending. The interactions between the POSS nanoparticles explain the formation of aggregates.

Figure 2C shows XRD patterns of neat extruded PCL, pure APIBPOSS and melt mixed PCL/APIBPOSS nanocomposites. As with the solution blended nanocomposites, the crystalline peak in the low angle range (2θ = 8°) was also detected (Figure 2D), and became more intense as the APIBPOSS concentration in the blend increased. When comparing these results with those obtained for the solution blended nanocomposites, it is clear that the melt mixed nanocomposite containing 10 wt.% of APIBPOSS exhibited a much less intense APIBPOSS crystalline peak than the solution blended counterpart, indicating that fewer crystalline aggregates of POSS molecules exist in the melt mixed sample. These results allow for the conclusion that in the solution blended nanocomposites the dispersed APIBPOSS particles aggregate together to form a crystalline structure much easier than in the melt blended samples.

Unlike APIBPOSS, APIOPOSS is not crystalline, three amorphous halos were observed in the X-ray diffraction pattern (Figure 3Aa). Figure 3Ac–e presents XRD patterns of solution blended PCL/APIOPOSS nanocomposites. In the enlarged XRD patterns it can be observed that the signals for APIOPOSS are absent in all PCL nanocomposites containing APIOPOSS (Figure 3Bc–e). Similar results were obtained for the melt compounded PCL/APIOPOSS nanocomposites (Figure 3C,D).

Figure 4 shows the XRD patterns of neat APIOPOSS and PCL diol (the precursors of the nanohybrid), telechelic hybrid PCL containing POSS, and PCL. The diffraction pattern of telechelic hybrid displays the characteristic amorphous halo of APIOPOSS at 7.7° and the characteristic diffraction peaks of PCL-diol (Figure 4c). As for solution blended PCL/PIOPOSS-PCL-PIOPOSS masterbatch nanocomposites, XRD patterns (Figure 4Ae–g) were similar to that of neat PCL

(Figure 4Ad). In the enlarged XRD patterns (Figure 4Be–g) the amorphous halos of the telechelic hybrid are absent.

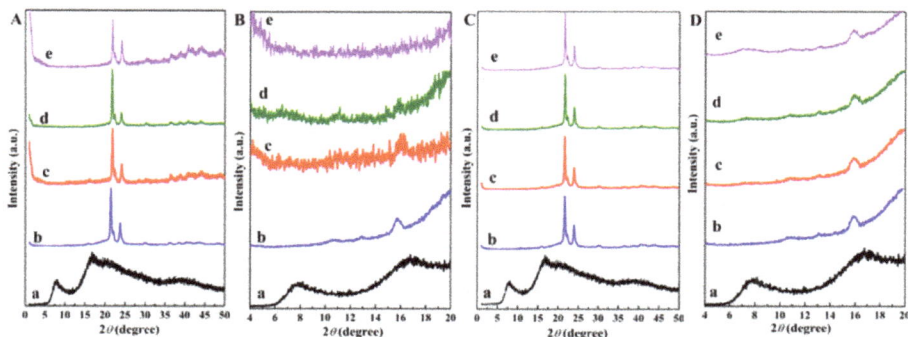

Figure 3. XRD patterns (**A,B**) solution blended, (**C,D**) melt mixed: (**a**) APIOPOSS, (**b**) PCL, (**c**) PCL/APIOPOSS-2, (**d**) PCL/APIOPOSS-5, (**e**) PCL/APIOPOSS-10; (**B,D**): enlarged XRD patterns in the 2θ range 5–20 degrees.

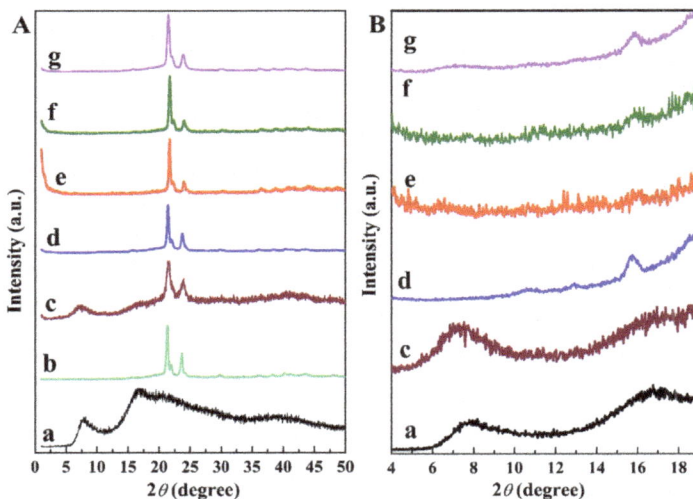

Figure 4. XRD patterns (**A**) solution blended: (**a**) APIOPOSS, (**b**) PCL-diol, (**c**) PIOPOSS-PCL-PIOPOSS, (**d**) PCL, (**e**) PCL/PIOPOSS-PCL-PIOPOSS-2, (**f**) PCL/PIOPOSS-PCL-PIOPOSS-5, (**g**) PCL/PIOPOSS-PCL-PIOPOSS-10; (**B**): enlarged XRD patterns in the 2θ range 5–20 degrees.

The nanocomposite samples were examined with TEM to elucidate the microstructure, the dispersion state of POSS into the matrix, and the interaction level between them. Figures 5 and 6 display TEM images of PCL/POSS blends and neat PCL as reference. TEM images revealed the presence of POSS aggregates. Spherical shaped particles were observed in the micrographs of solution blended PCL/POSS nanocomposites (Figure 5). The size of the POSS aggregates in nanocomposites containing APIBPOSS (Figure 5b–d) was between 130 and 380 nm, while smaller aggregates (110–270 nm) were observed for APIOPOSS nanocomposites (Figure 5e–g) and for the samples prepared by the masterbatch approach (45–135 nm at 5 and 10 wt.% nanohybrid concentrations) (Figure 5h–j). The size of the aggregates depended on the POSS type and concentration. The greatest extent of aggregation and the largest aggregates were observed at the highest POSS concentration. Amino functional-POSS molecules are dispersed in the form of aggregates within the PCL matrix. Comparing the micrographs of solution blended PCL/POSS nanocomposites, it was observed that the largest aggregates were formed in the presence of APIBPOSS, and the best dispersion was achieved when using the masterbatch.

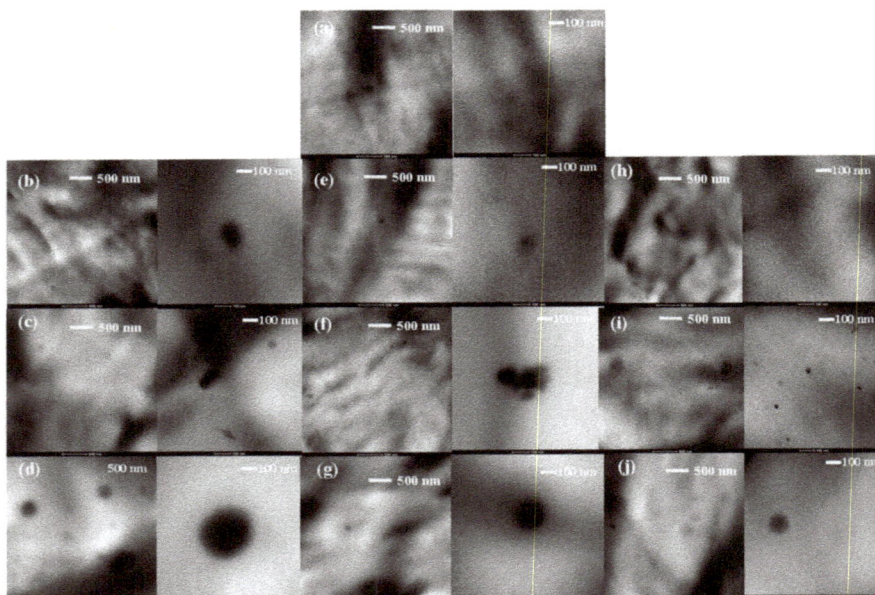

Figure 5. TEM micrographs of solution blended PCL/POSS nanocomposites: (a) PCL; (b) PCL/APIBPOSS-2; (c) PCL/APIBPOSS-5; (d) PCL/APIBPOSS-10; (e) PCL/APIOPOSS-2; (f) PCL/APIOPOSS-5; (g) PCL/APIOPOSS-10; (h) PCL/PIOPOSS-PCL-PIOPOSS-2; (i) PCL/PIOPOSS-PCL-PIOPOSS-5; (j) PCL/PIOPOSS-PCL-PIOPOSS-10.

Solubility parameters can be used for predicting the solubility of one material into another, in this case the POSS nanoparticles into the polymer matrix [18,19,28–30]. Furthermore, the homogeneous dispersion of the nanoparticles is largely dependent on its solubility into the polymer. Greater compatibility and better dispersion characteristics can be expected when mixing materials with similar solubility parameters when compared to those with very different solubility parameters. The literature value of the solubility parameter of the PCL, APIBPOSS, and APIOPOSS are 19.7 $(J/cm^3)^{1/2}$ [31], 17.5 $(J/cm^3)^{1/2}$ [30], and 18.6 $(J/cm^3)^{1/2}$ [30], respectively. APIOPOSS showed the lowest solubility parameter difference with PCL, 1.1$(J/cm^3)^{1/2}$, versus 2.2 $(J/cm^3)^{1/2}$ for APIBPOSS. Therefore, on the basis of this difference a higher solubility of PIOPOSS in the PCL matrix, that is greater compatibility, and less tendency to aggregate in comparison with APIBPOSS is expected. Considering the Flory–Huggins theory for polymer solutions and blends, the Flory–Huggins parameter χ_{12}, included in the definition of the mixing enthalpy, can be related to the solubility parameters of two substances [32] by the relation;

$$\chi_{12} = \frac{V_M}{RT}(\delta_1 - \delta_2)^2 \tag{1}$$

where δ_1 and δ_2 in our case are the solubility parameters of the PCL 1 and POSS 2, respectively, R and T are the gas constant and temperature, respectively, and V_M is a reference volume which is the molar volume of the smallest repeat unit. It is expected that the interaction between PCL and APIOPOSS would be more thermodynamically favorable than the PCL/APIBPOSS, since the values for solubility parameter are closer than that of APIBPOSS. Solubility parameter value is not determined in the case of the ditelechelic hybrid, however, an enhancement in the dispersion state would be expected since the PCL chains covalently bonded to the POSS surface could improve the compatibility between PCL and the POSS nanoparticles. This enhancement of compatibility could be due to the solubility of low molecular weight PCL chains of the hybrid into the PCL matrix. The TEM observations are in agreement with the solubility parameters.

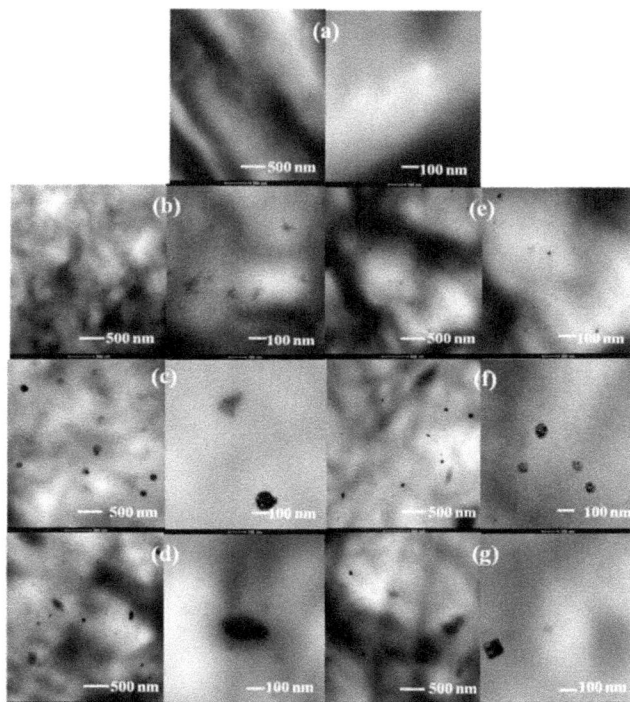

Figure 6. TEM micrographs of melt mixed PCL/POSS nanocomposites: (**a**) PCL; (**b**) PCL/APIBPOSS-2; (**c**) PCL/APIBPOSS-5; (**d**) PCL/APIBPOSS-10; (**e**) PCL/APIOPOSS-2; (**f**) PCL/APIOPOSS-5; (**g**) PCL/APIOPOSS-10.

TEM images of the melt mixed PCL nanocomposites revealed a more uniform dispersion of POSS, although POSS aggregate particles were also observed, in particular at 5 and 10 wt.% amino-POSS loading (Figure 6). The size of the particles was about 60 nm in the composites containing 2 wt.% of amino-POSS, while at higher POSS loading the size was between 110–260 nm for APIBPOSS, and between 70–125 nm for APIOPOSS based nanocomposites. Comparing the micrographs of the melt mixed and solution blended PCL/amino-POSS derivatives composites, it can be observed that the size of the POSS particle aggregates is larger for the solution blended PCL/amino-POSS nanocomposites. In the solution mixing approach, POSS nanoparticles are dispersed in the PCL solution, and nanocomposites are achieved by the evaporation of solvent. Nanoparticles tend to agglomerate due to high surface energy, even though ultrasonic mixing was used to disperse them. In conclusion, the ultrasonication step was not sufficient to properly disperse the POSS nanoparticles within the PCL matrix. In the melt compounding approach, POSS nanoparticles are incorporated into the molten polymer. The better dispersion of amino-POSS nanoparticles achieved by melt mixing as compared with the solution casting process can be due to the high shear force generated during the melt blending. The shear stress can overcome the interactions between the POSS nanoparticles and lead to the breakup of the POSS agglomerates. It has been reported that the shear stresses exerted on the polymer melt during processing helps the dispersion of fillers [33–36].

3.2. Effect of Melt Mixing Processing on PCL Molecular Weight

In order to assess whether degradation of PCL took place at high temperature during melt extrusion by aminolysis reaction between the primary amine function present on POSS nanocages (APIBPOSS and APIOPOSS) and the polymer matrix, molecular weight of PCL was measured. Table S1 shows the M_w, M_n, and the polydispersity index (M_w/M_n) of the neat PCL and nanocomposites extruded at 90 °C and 50 rpm for 10 min. The M_w and M_n values show that PCL did not degrade under the melt mixing conditions of nanocomposites.

3.3. Effect of POSS Type and Processing Method on Thermal Transitions

The determination of thermal transitions of the composites and the study of the effect of POSS type and preparation method on these transitions were conducted by DSC measurements. Figure 7 displays the DSC thermograms (second heating and first cooling scans) of neat PCL, APIBPOSS, and PCL/POSS nanocomposites. From these thermograms, the endothermic and exothermic peaks, corresponding to the melting and crystallization peaks of PCL, respectively, can be clearly seen. All the data as determined from the DSC traces are given in Table 1. The glass transition temperature (T_g) of solution blended and melt processed PCL/POSS nanocomposites is roughly independent on the type and content of POSS.

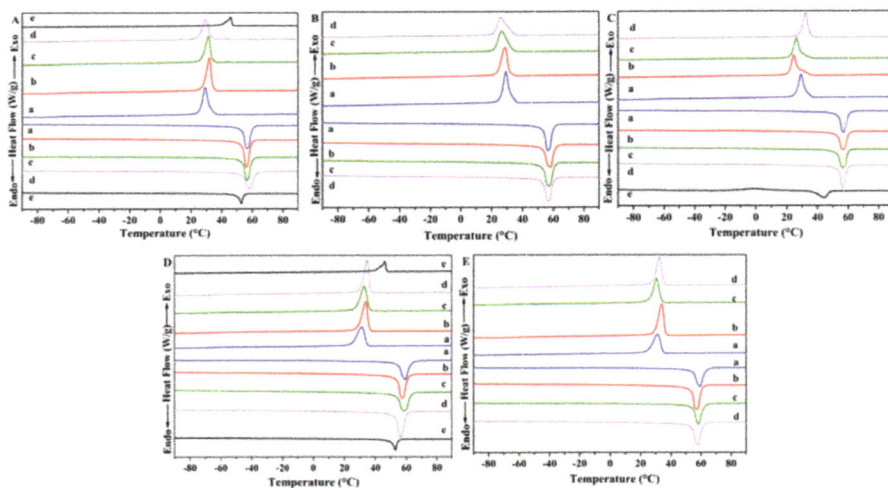

Figure 7. DSC thermograms of: (**A**) solution blended PCL/APIBPOSS; (**B**) solution blended PCL/APIOPOSS; (**C**) solution blended PCL/PIOPOSS-PCL-PIOPOSS; (**D**) melt mixed PCL/APIBPOSS; (**E**) melt mixed PCL/APIOPOSS: (**a**) neat PCL; (**b**) PCL/POSS-2; (**c**) PCL/POSS-5; (**d**) PCL/POSS-10; (**e**) APIBPOSS or PIOPOSS-PCL-PIOPOSS.

Solution blended PCL nanocomposites containing 2 and 5 wt.% APIBPOSS showed similar melting temperatures (T_m) as pristine PCL, while the value of the nanocomposite with 10 wt.% APIBPOSS was slightly higher (2 °C) than that of PCL. No significant differences in T_m value of PCL/APIOPOSS and PCL/POSS nanohybrid masterbatch nanocomposites were observed as compared with neat PCL. PCL/POSS nanocomposites prepared by melt mixing exhibited a slightly lower melting temperature as compared with neat extruded PCL.

Table 1. DSC data for PCL and PCL/POSS composites.

Sample	T_g (°C)	T_c (°C)	ΔH_c (J/g)	T_m (°C)	ΔH_m (J/g)	X_c (%)
APIBPOSS		46.1	18.5	52.9	19.7	
PIOPOSS-PCL-PIOPOSS	−62.0	0	18.5	44.5	28.1	
PCL-S	−61.0	29.3	64.9	56.5	70.2	49.4
PCL/APIBPOSS-2-S	−61.2	32.0	67.9	56.3	72.1	51.8
PCL/APIBPOSS-5-S	−61.8	31.4 44.7	60.5	56.6	66.1	49.0
PCL/APIBPOSS-10-S	−59.6	29.7 44.0	57.2	58.5	57.9	45.3
PCL/APIOPOSS-2-S	−60.7	28.7	65.5	58.0	68.2	49.0
PCL/APIOPOSS-5-S	−62.0	26.7	65.3	57.5	68.5	50.8
PCL/APIOPOSS-10-S	−60.6	26.0	61.4	57.4	63.5	49.7
PCL/PIOPOSS-PCL-PIOPOSS-2-S	−61.3	24.5	67.4	57.0	69.7	50.1
PCL/PIOPOSS-PCL-PIOPOSS-5-S	−60.2	26.3	65.0	57.2	68.5	50.8
PCL/PIOPOSS-PCL-PIOPOSS-10-S	−59.7	32.3	63.4	56.4	69.3	54.2
PCL-M	−58.7	31.2	56.2	59.3	61.2	43.1
PCL/APIBPOSS-2-M	−61.0	33.7	64.5	57.3	70.8	50.9
PCL/APIBPOSS-5-M	−60.6	32.8	62.3	58.5	69.1	51.2
PCL/APIBPOSS-10-M	−60.9	34.7 45.5	61.0	55.9	68.1	53.3
PCL/APIOPOSS-2-M	−60.1	33.9	68.2	57.3	69.7	50.1
PCL/APIOPOSS-5-M	−58.4	30.5	56.0	58.2	56.5	41.9
PCL/APIOPOSS-10-M	−60.5	32.2	62.0	57.9	64.5	50.5

As can be seen in the enlarged DSC cooling scans shown in Figure S1A and from Table 1, the crystallization temperature (T_c) for solution blended nanocomposites containing 2 and 5 wt.% of APIBPOSS were slightly higher than that of neat PCL. Moreover, T_c of PCL first increased and then decreased with increasing the POSS loading in the PCL/APIBPOSS nanocomposites. However, the onset crystallization temperature (data not shown) was almost unaffected by the incorporation of APIBPOSS irrespective of POSS loading. These results indicate that although in the presence of APIBPOSS the crystallization of PCL does not start earlier as compared with neat PCL, it runs much faster, and the degree of enhancement in T_c depends of POSS concentration. This indicates that APIBPOSS nanoparticles act as nucleating agent for the crystallization of PCL. On the other hand, an additional crystallization peak at 44 °C, attributed to APIBPOSS, was also observed. This result is in accordance with that obtained by XRD analysis. T_c value of PCL was unaffected by the addition of 2 wt.% of APIOPOSS, while decreased at higher POSS loading level (Figure 7B and Figure S1B). PCL/APIOPOSS nanocomposites containing 5 and 10 wt.% exhibited a wider crystallization process as compared with that of neat PCL. These results suggest that APIOPOSS did not act as nucleating agent. T_c value of nanocomposites containing 2 and 5 wt.% PIOPOSS-PCL-PIOPOSS hybrid was lower than that of neat PCL, while the value for the blend containing 10 wt.% was higher than that of PCL (Figure 7C and Figure S1C). The onset crystallization temperature (data not shown) was almost unaffected by the incorporation of this nanohybrid irrespective of POSS loading. These results indicate that the crystallization of PCL runs much faster in the presence of 10 wt.% of ditelechelic hybrid, even though the crystallization of PCL does not start earlier as compared with neat PCL.

The T_c values of PCL/APIBPOSS nanocomposites prepared by melt mixing are slightly higher than that of PCL (Figure 7D and Figure S1E), and the crystallization peak at 45 °C attributed to POSS was observed in the blend containing 10 wt.% APIBPOSS. The onset crystallization temperature (data not shown) of melt processed PCL/APIBPOSS composites were shifted toward higher temperatures indicating that those nanoparticles act as nucleating agents. The exothermic peak of the PCL crystallization process was narrower in PCL/APIBPOSS nanocomposites. The blends containing 2 and 10 wt.% APIOPOSS also exhibited a slightly higher T_c value than that of neat PCL (Figure 7E and Figure S1F), a narrower exothermic peak of crystallization, and higher onset crystallization temperatures. These results indicate that APIOPOSS nanoparticles acted as nucleating agents for the crystallization process of PCL.

As a conclusion, the crystallization of PCL is affected by the presence of POSS, the type of POSS derivative, the POSS content, and the preparation method. For the blends prepared by solution-casting,

APIBPOSS acts as a nucleating agent for the crystallization of PCL, while the nucleating effect is inexistent for APIOPOSS, and a high content of the PIOPOSS-PCL-PIOPOSS hybrid is necessary to act as nucleating agent. If nanocomposites are prepared by melt compounding, APIBPOSS acts as nucleating agent, and this effect it is observed only at 2 and 10 wt.% APIOPOSS loading.

The crystallization enthalpy of nanocomposites decreased gradually by increasing the POSS content. For the solution blended nanocomposites, ΔH_c of PCL increased slightly upon addition of 2 wt.% APIBPOSS and then decreased with a further increase in POSS content. The same trend was observed for nanocomposites with the ditelechelic hybrid masterbatch. However, the addition of 2 and 5 wt.% of the APIOPOSS had no effect on ΔH_c value of PCL, and the incorporation of 10 wt.% of POSS led to a reduction in that value. The POSS nanoparticles reduce the crystalline order during crystallization and reduces ΔH_c of PCL. On the other hand, in the melt mixed nanocomposites, the addition of the amino-POSS derivatives led to a slight increase in ΔH_c of PCL.

The percentage crystallinity (% X_c) of PCL was calculated according to the following relation,

$$X_c = \left[\frac{\Delta H_m}{\Delta H_m^0 \times \left(1 - \frac{\%wt_{filler}}{100}\right)} \right] \times 100 \tag{2}$$

where, ΔH_m is the enthalpy of fusion, ΔH^0_m is the enthalpy of fusion of a perfect PCL crystal (142 J g^{-1}) [37], and wt.% filler is the total weight percentage of POSS derivative. For the solution casting samples, the addition of 2 wt.% APIBPOSS led to an increase in the degree of crystallinity of PCL, while it decreased with a further increase in POSS content. This can be due to the formation of aggregates and hence difficulty in transportation of PCL chains to crystal growing surface. The crystallinity of the solution blended PCL/APIOPOSS nanocomposites was almost the same than that of neat PCL. X_c value increased as ditelechelic hybrid content increased, the composite containing 10 wt.% of POSS exhibited a slightly higher degree of crystallinity than PCL. The incorporation of APIBPOSS by melt compounding led to an increase in the degree of crystallinity of PCL that increased with the POSS content. For the APIOPOSS based nanocomposites the addition of 2 and 10 wt.% of POSS led to an increase in the crystallinity of PCL, while the incorporation of 5 wt.% of POSS had no effect.

3.4. Effect of POSS Type and Processing Method on Thermal Stability

Thermogravimetric analysis (TGA) was carried out in an inert (N$_2$) and oxidative (air) atmosphere to study the thermal and thermo-oxidative stability of PCL/POSS nanocomposites. Table 2 shows the thermal decomposition temperatures for 5% weight loss ($T_{5\%}$), the temperature of the maximum loss rate (T_{max}) and the fraction of solid residue at 750° C of the thermograms.

Figure 8 displays TGA curves in N$_2$ atmosphere of neat POSS derivatives, neat PCL and PCL/POSS derivatives nanocomposites, whilst the derivative thermogravimetric (DTG) curves are shown in Figure S2.

APIBPOSS decomposed in a one step process in the 170–260 °C temperature range, the weight loss of APIOPOSS took place in a two-step process in the 200–550 °C temperature range, and PIOPOSS–PCL-PIOPOSS exhibited a single stage degradation process in the temperature range 200–480 °C. As reported in Table 2, $T_{5\%}$ value of PCL was higher than that of neat amino-POSS molecules and slightly lower than that of PIOPOSS-PCL-PIOPOSS hybrid. The solid residue amount at 750 °C was approximately 2.8% for APIBPOSS, 5.5% for APIOPOSS, and 6.4% for PIOPOSS-PCL-PIOPOSS. Thermal stability of neat POSS in nitrogen was observed to decrease in order: PIOPOSS-PCL-PIOPOSS > APIOPOSS > APIBPOSS (Figure 8A). The longer the alkyl chain of substituents in the POSS molecule, the higher the $T_{5\%}$ value (Table 2). Similar results were obtained in the literature [23,38]. On the other hand, PCL decomposed completely in two steps in the temperature range of 280–420 °C. When PCL was blended with APIBPOSS by solution-casting method, all the samples decomposed completely in two steps in the temperature range of 180–450 °C (Figure 8B).

The weight loss of the first stage, in the range of 180–310 °C, was between 2 and 7% and increased with increasing the POSS content, and can be attributed to the decomposition of APIBPOSS, since this filler decomposes in this temperature range (Figure 8A and Figure S2). Improved thermal stability was observed for the nanocomposites with 2 and 5 wt.% of APIBPOSS, being the onset temperature of the degradation more than 50 °C higher than that of the pure PCL, whereas the $T_{5\%}$ value of the nanocomposite containing 10 wt.% was almost 30 °C lower than that of neat polymer matrix, this can be ascribed to the decomposition of APIBPOSS aggregates, larger than those of blends containing 2% and 5% POSS. A significant increase in T_{max} value of the nanocomposites was observed in comparison with neat PCL.

Table 2. Thermogravimetric analysis (TGA) data for PCL, PCL/APIBPOSS, PCL/APIOPOSS, PCL/PIOPOSS-PCL-PIOPOSS composites, neat APIBPOSS, APIOPOSS, and PIOPOSS–PCL-PIOPOSS.

Sample	T_5 (°C)		T_{max} (°C)		Residue (%)	
	N_2	O_2	N_2	O_2	N_2	O_2
APIBPOSS	210	221	224	245	2.8	21.1
APIOPOSS	255	250	283 359	306 320	5.5	30.0
PIOPOSS-PCL-PIOPOSS	301	300	356	349 392	6.4	16.0
PCL-S	292	277	293	320 311	0.6	0.3
PCL/APIBPOSS-2-S	348	279	356	330	0.5	1.1
PCL/APIBPOSS-5-S	346	275	357	350	0.5	1.4
PCL/APIBPOSS-10-S	264	274	357	353	0.8	2.3
PCL/APIOPOSS-2-S	347	269	355	301	1.1	1.1
PCL/APIOPOSS-5-S	336	292	357	338	1.0	1.5
PCL/APIOPOSS-10-S	332	297	355	350	1.7	2.4
PCL/PIOPOSS-PCL-PIOPOSS-2-S	347	339	355	348	0.8	0
PCL/PIOPOSS-PCL-PIOPOSS-5-S	345	335	355	349	0.9	1.3
PCL/PIOPOSS-PCL-PIOPOSS-10-S	343	332	353	349	1.1	2.0
PCL-M	294	267	294	271 342	0.5	0.5
PCL/APIBPOSS-2-M	286	311	289	349	0.8	0.9
PCL/APIBPOSS-5-M	343	318	358	351	0.9	1.7
PCL/APIBPOSS-10-M	277	293	357	351	1.3	1.6
PCL/APIOPOSS-2-M	339	324	355	347	1.0	1.1
PCL/APIOPOSS-5-M	332	326	354	347	2.0	1.5
PCL/APIOPOSS-10-M	307	325	356	349	1.5	2.4

All solution blended PCL/APIOPOSS nanocomposite samples decomposed completely in a single step in the temperature range of 300–450 °C (Figure 8C). The incorporation of APIOPOSS enhanced the thermal stability of PCL, being the values of $T_{5\%}$ improved by more than 40 °C. The $T_{5\%}$ of the nanocomposites decreased continuously with POSS content. A significant increase in T_{max} values of the nanocomposites in comparison with neat PCL was observed. The addition of PIOPOSS-PCL-PIOPOSS led to an enhancement of thermal stability of PCL (Figure 8D). A noticeable increase was observed in the $T_{5\%}$ value (55 °C). T_{max} value was independent of the POSS content.

The improvement in the thermal stability of the PCL/POSS nanocomposites can be explained by the formation of interactions between POSS and PCL matrix. The decline in thermal stability of PCL/amino-POSS nanocomposites as POSS concentration increases can be ascribed to the poorer dispersion level of nanoparticles in the polymer matrix, since larger aggregates were formed. The reduction of the thermal stability in the PCL/APIBPOSS-10-S can be associated to the poorer dispersion of the nanoparticles at this concentration level, as well as to the lower thermal stability of neat APIBPOSS as compared to PCL. The dependence of the thermal stability of POSS containing polymer nanocomposites on the type and content of POSS as well as on dispersion level of POSS nanoparticles in the polymer matrix has been reported in the literature [39].

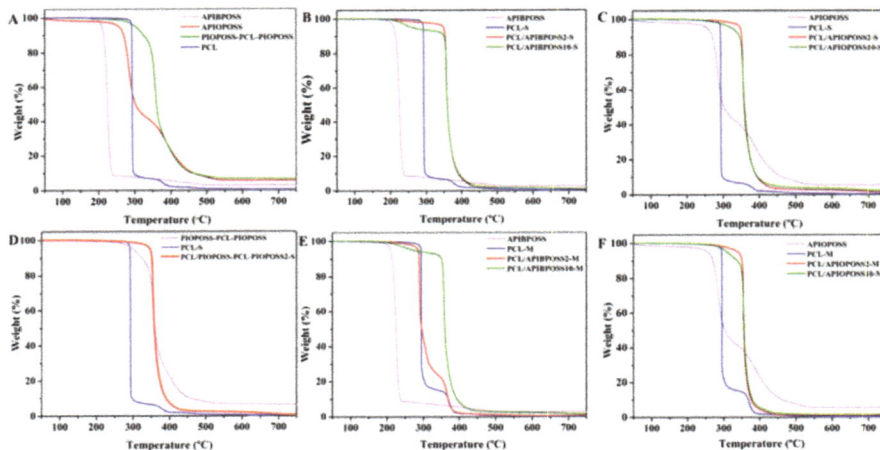

Figure 8. TGA curves of: (**A**) nanofillers and neat PCL; (**B–D**), solution blended nanocomposites; (**E,F**) melt mixed nanocomposites in N_2 atmosphere.

Neat extruded PCL and melt blended PCL/APIBPOSS sample containing 2 wt.% APIBPOSS displayed similar degradation profiles, two decomposition steps in the temperature range of 220–400 °C (Figure 8E). The nanocomposites with 5 and 10 wt.% APIBPOSS decomposed in two steps, in the temperature range 200–325 °C and 325–460 °C, respectively. The weight loss in the first stage was 4% and 8% for the blend with 5 wt.% POSS and 10 wt.% POSS, respectively. This step can be attributed to the decomposition of the amino-POSS filler. For the melt mixed nanocomposites containing 5 wt.% of APIBPOSS, the onset temperature of the degradation is about than 50 °C higher than that of the pure PCL, whereas the $T_{5\%}$ value of the nanocomposites containing 2 and 10 wt.% APIBPOSS are slightly lower than that of neat polymer matrix. The TGA weight loss curves of the melt mixed PCL/APIOPOSS nanocomposites with 2 and 5 wt.% POSS exhibited a single decomposition step between 315 °C and 420 °C, while in the thermogram of the blend with 10% POSS two steps were observed in the 275–330 °C and 330–400 °C temperature range, respectively. The weight loss in the first step was about 8%, and can be attributed to APIOPOSS decomposition. The $T_{5\%}$ value of the PCL/APIOPOSS nanocomposites was higher than that of neat PCL (Table 2 and Figure 8F), and decreased with increasing APIOPOSS content. A significant increase in T_{max} value of the APIBPOSS and APIOPOSS based nanocomposites was observed as compared with neat PCL.

The TG curves in air atmosphere for POSS derivatives, neat PCL and PCL/POSS derivative nanocomposites are displayed in Figure 9, whilst DTG curves are shown in Figure S3. When heated in oxidative atmosphere, APIOPOSS showed a single step process in the 225–550 °C temperature range (Figure 9A), whereas APIBPOSS and PIOPOSS–PCL-PIOPOSS exhibited a two-step process in the 160–400 °C and 200–600 °C temperature range, respectively. Thermal stability in air of neat POSS derivatives decreased in the order: PIOPOSS-PCL-PIOPOSS > APIOPOSS > APIBPOSS. As in case of thermal stability in nitrogen atmosphere, the longer the alkyl chain of substituents in the POSS molecule, the higher the thermo-oxidative stability. The solid residue obtained from the degradations of POSS derivatives appears because of the formation of silica, SiO_2, and higher residue appeared in oxygen than in nitrogen as expected [38,40]. When POSS molecules are added to a polymer matrix they self-segregate to the polymer surface due to the low surface energy of Si atom, upon heating they form a ceramic layer which prevents the heat transfer to the sample and the permeability of volatile products from generating in the degradation process [38,41]. The residue left by APIBPOSS and APIOPOSS after thermo-oxidative decomposition amounted to 21.1% and 30%, respectively. Those values are lower than the theoretical expected value of 54.8% and 38%, respectively, with the total conversion of the inorganic cage to SiO_2. The lower residue for both amino-POSS derivatives is due to the sublimation of a part of the cubic cage of amino-POSS. The amount of residue left by APIOPOSS is 8% lower than expected, versus 33.7% for APIBPOSS, indicating that APIOPOSS is less susceptible to sublimation than APIBPOSS.

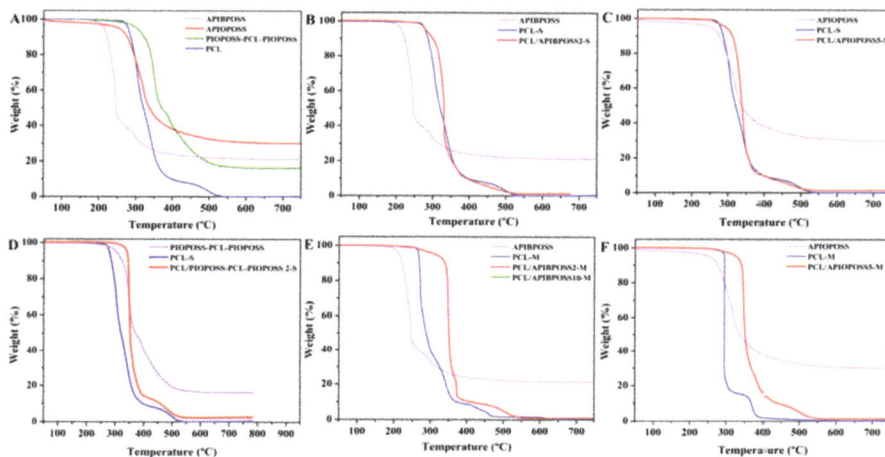

Figure 9. TGA curves of: (**A**) nanofillers and neat PCL; (**B–D**) solution blended nanocomposites; (**E,F**) melt mixed nanocomposites in O_2 atmosphere.

The presence of oxygen led to a faster degradation of the PCL and solution blended PCL/POSS nanocomposites, than in inert conditions (Table 2), the oxidizing atmosphere accelerates the degradation process. The $T_{5\%}$ value of neat PCL was lowered by 15 °C owing to the simultaneous effect of heat and reaction with oxygen. $T_{5\%}$ values of nanocomposites containing 2 wt.% amino-POSS were about 70 °C lower than those under nitrogen atmosphere. When PIOPOSS-PCL-PIOPOSS hybrid was blended with PCL, $T_{5\%}$ value was 10 °C lower than that under nitrogen atmosphere. With regard to the thermo-oxidative behavior of the solution blended PCL/APIBPOSS nanocomposites, the onset decomposition temperatures were similar to that of neat PCL (Figure 9B), while $T_{5\%}$ values of the PCL/ditelechelic hybrid masterbatch nanocomposites (Figure 9D), and that of those containing 5 and 10 wt.% APIOPOSS were higher than that of PCL (Figure 9C). On the other hand, $T_{5\%}$ values of all melt mixed nanocomposites increased by about 40–55 °C, in comparison with neat PCL (Table 2, Figure 9E,F). A noticeable improvement in $T_{5\%}$ values was observed for melt blended amino-POSS based nanocomposites as compared with those prepared by solution mixing, which can be attribute to the more uniform distribution and smaller aggregates of the filler.

On the basis of the above results, it can be concluded that thermo-oxidative stability of PCL improved with the incorporation of the masterbatch (PIOPOSS-PCL-PIOPOSS) and APIOPOSS (at content greater than 2 wt.%) by solution mixing, while it remained unchanged with the incorporation of APIBPOSS. However, the thermo-oxidative stability of PCL was enhanced by the incorporation amino-POSS derivatives by melt mixing. The ditelechelic hybrid masterbatch provided the greatest improvement both in thermal and thermo-oxidative stability of PCL for the entire filler content range studied, that can be explained by the better compatibility with PCL chains, which results in a better extent of dispersion (smaller size aggregates). The enhancement in thermal stability of nanocomposite films can be ascribed to the thermal insulator and mass transport barrier effect of POSS nanoparticles.

3.5. Effect of POSS Type and Processing Method on Mechanical Properties

The Young's modulus, tensile strength and strain at break were determined from experimental tensile stress-strain curves shown in Figure S4. The results as a function of the POSS content and the processing method are shown in Figure 10.

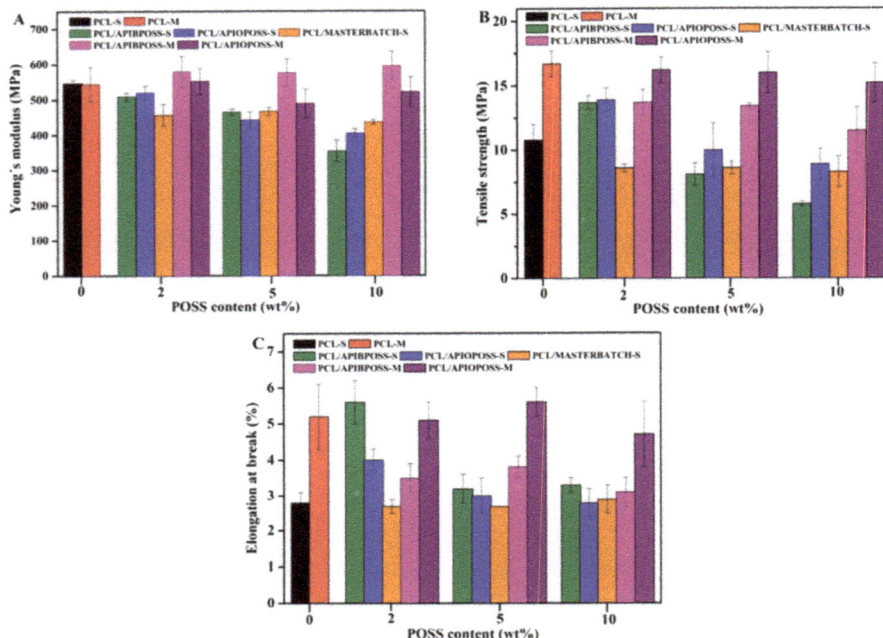

Figure 10. Mechanical properties of PCL/POSS nanocomposite films: (**A**) Young's modulus; (**B**) Tensile strength; and (**C**) Elongation at break.

The results obtained from the solution blended nanocomposites showed a decrease in elastic modulus of the nanocomposite samples as compared to the pure PCL (Figure 10A), and its value was affected by the POSS type and content, decreasing as amino-POSS derivative content increased and remained almost unchanged with PIOPOSS-PCL-PIOPOSS concentration. Comparing the POSS type, the blend containing 10 wt.% APIBPOSS showed the lowest Young's modulus, while that containing 2 wt.% APIOPOSS exhibited the highest one. The decrease in elastic modulus was between 7 and 35% for the PCL/APIBPOSS system, between 5 and 25% for the PCL/APIOPOSS system, and between 15 and 20% for the PCL/PIOPOSS-POSS-PIOPOSS nanohybrid nanocomposites. The Young's modulus in the blend containing 2 wt.% of the PIOPOSS-POSS-PIOPOSS masterbatch is lower than that value for the blends containing 2 wt.% of amino-POSS derivatives. The results indicate that the stiffness of PCL matrix is reduced by the incorporation of APIBPOSS, APIOPOSS. and the PIOPOSS-PCL-PIOPOSS masterbatch via solution blending. This decrease in stiffness can be associated to the decrease in the degree of crystallinity of the PCL matrix with respect to neat PCL only for PCL/APIBPOSS-10-S nanocomposite. Baldi et al. [42] reported the effect of the length of POSS alkyl chain on the mechanical behavior of POSS/PP blends. In their study, they used octamethyl-POSS, octaisobutyl-POSS, and octaisooctyl-POSS. The authors suggested that POSS behave as particles having a siliceous hard-core surrounded by a hydrocarbon soft-shell, which limits the stress transfer from the polymer matrix to the core in dependence on the length of the alkyl groups. The thickness of the shell that is determined by the length of the R side groups of POSS molecules was pointed out as the key factor in permitting the rigid part of the particle to express a reinforcing effect. POSS bearing long substituents behave as rubbery inclusions. Moreover, since the inorganic cage promote the reinforcing action, they stated that the amount of inorganic material dispersed in the polymer matrix must be taken into account, that is, with the same POSS content the inorganic amount depends on the contribution of the organic fraction. The authors found an enhancement in the Young's modulus of PP containing 10 wt.% of octamethyl-POSS, and lack of stiffness enhancement in the case of PP blends containing either octaisobutyl- or isooctyl-POSS.

The difference between APIBPOSS and APIOPOSS is the length of the alkyl chains bounded to the Si atom, four and eight carbon atoms for APIBPOSS and APIOPOSS, respectively. An identical siliceous hard core is enveloped by alkyl chains that constitute a softer outer shell of variable thickness. APIBPOSS nanoparticles have the thinnest soft shell, whereas APIOPOSS has the thickest one. In the PIOPOSS-PCL-PIOPOSS nanohybrid in addition to eight alkyl chains there is a PCL chain grafted. The amount of inorganic material dispersed in the PCL/PIOPOSS-PCL-PIOPOSS blends is lower than that dispersed in the PCL/amino-POSS systems with the same POSS content. Our results confirm the observation of Baldi et al. [42].

The solution blended nanocomposites containing 2 wt.% of APIBPOSS and APIOPOSS displayed an increase in tensile strength at break of about 28%, with respect to PCL (Figure 10B), whilst a higher content of amino-POSS derivatives resulted in a reduction (about 25–45% for APIBPOSS based nanocomposites, and 8–18% for the PCL/APIOPOSS composites). The tensile strength of PCL/PIOPOSS-PCL-PIOPOSS composites were independent of the hybrid content and broke at lower stress than PCL, about 20% reduction was observed. The decline of tensile strength with increasing amino-POSS content could be ascribed to the formation of larger aggregates. At low amino-POSS content, partial tensile strain can be transferred to the hard core of POSS nanoparticles dispersed in the PCL matrix under tensile stress, which leads to the increase of tensile strength. Further addition of filler results in more and larger agglomerates of POSS in PCL and they behave as defects that lead to a decrease of tensile strength. The lower tensile strength attained in the APIBPOSS based nanocomposites as compared with those based on APIOPOSS can be attributed to the larger size of the agglomerates formed in the first case, which act as crack propagation site and lead to the failure under stress. The lack of enhancement in tensile strength in the PCL blends containing PIOPOSS-PCL-PIOPOSS hybrid can be explained by the lower amount of siliceous hard core dispersed in the polymer matrix as compared with the amino-POSS derivatives.

The addition of APIBPOSS nanoparticles induced an increase in the elongation at break of PCL (Figure 10C), 100% with 2 wt.% of APIBPOSS that is an increase in the ductility of PCL. An increase in APIBPOSS concentration above 2% resulted in a decrease in that value. The elongation at break of PCL in the presence of 2 wt.% of APIOPOSS increased by about 40%, about 7% when the APIOPOSS content was 5 wt.%, and a further increase in the nanofiller content had no effect. On the contrary, the addition of PIOPOSS-PCL-PIOPOSS to PCL had almost no effect on the elongation at break of neat polymer matrix. The presence of amino-POSS particles in PCL decreases the brittleness of the material.

The Young's modulus of PCL was almost unaffected by the addition of amino-POSS by melt mixing. On the other hand, the tensile strength of all melt mixed PCL/APIBPOSS nanocomposites was lower than that of PCL, the reduction was between 20% and 30%. The tensile strength of PCL was almost unaffected by the addition of APIOPOSS. The addition of all APIBPOSS nanoparticles induced a reduction between 30 and 40% in the elongation at break of PCL, whilst the incorporation of APIOPOSS induced only a slight change.

The type of POSS derivative and content, and the preparation method have effect on the mechanical properties of PCL/POSS based nanocomposites, due to differences in compatibility and dispersion state. However, the better dispersion state of POSS achieved in the melt mixed PCL/amino-POSS nanocomposites is not enough to achieve good reinforcement, as strong interfacial interaction between the matrix and the filler is also necessary.

3.6. Effect of POSS Type on Surface Properties

Studies reported in the literature indicated that hydrophobicity of polymer matrices increased on incorporation of POSS nanoparticles [43,44]. The hydrophobicity of the nanocomposite surface as a function of POSS type and content was assessed by contact angle measurements with water. The water contact angles (WCA) were measured ten times for each specimen, and the average values are presented in Table S2. The water contact angle images for neat PCL, PIOPOSS-PCL-PIOPOSS hybrid, and PCL/POSS derivatives nanocomposites are displayed in Figure 11.

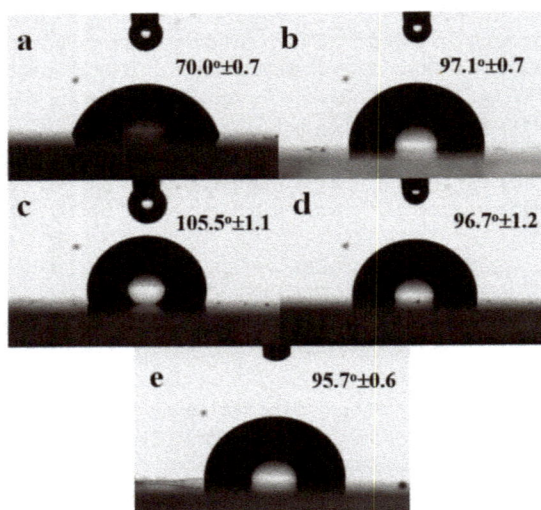

Figure 11. Digital images of water contact angle of (**a**) PCL; (**b**) PIOPOSS-PCL-PIOPOSS; (**c**) PCL/APIBPOSS-5; (**d**) PCL/APIOPOSS-5; and (**e**) PCL/PIOPOSS–PCL-PIOPOSS-5 nanocomposites.

For neat PCL the contact angle was 70.0°, and increased dramatically with loading of POSS derivatives, by 51%, 39%, and 37% for APIBPOSS, APIOPOSS, and ditelechelic hybrid, respectively (Table S2). WCA were almost independent of POSS content. The PCL/APIBPOSS based nanocomposites exhibited the highest value of contact angle. Therefore, from the results it can be concluded that the surface hydrophobicity of the PCL improves by the incorporation of POSS nanoparticles. Surface energy and roughness are some of the factors that affect contact angle [45]. POSS derivatives are organic–inorganic hybrid materials containing a central hydrophobic inorganic core silicon–oxygen–silicon surrounded by organic groups located at the corners of the octahedral siloxane cube. The increase in the contact angle values for PCL/POSS can be explained to the migration of POSS to the external surface of the nanocomposite films due to the low surface energy of the Si atom, thus making the surface more hydrophobic [46,47]. Furthermore the migration of the POSS aggregates to the surface can increase the surface roughness [36,43,47], thereby increasing the water contact angles. The highest water contact angle exhibited by PCL/APIBPOSS nanocomposites can be due to worst dispersion state of POSS and the larger aggregates formed in the presence of this POSS derivative.

4. Conclusions

Three polyhedral oligomeric silsesquioxane (APIBPOSS, APIOPOSS, and PIOPOSS-PCL-PIOPOSS masterbatch) were incorporated into PCL matrix at different contents via solution casting and melt mixing. Morphology, thermal, and mechanical properties depended on the alkyl chain length of the nonreactive organic substituents attached to the corner silicon atoms, the presence of PCL chains covalently attached to POSS nanoparticles, POSS concentration, and preparation method.

Spherical POSS aggregates of submicron size were observed within the PCL matrix, remaining the crystalline nature of APIBPOSS. Melt mixing led to a better dispersion state of amino-POSS nanoparticles than solution blending. The best dispersion in solution blended nanocomposites was achieved when using the masterbatch method. The difference in the dispersion level for different POSS derivative was attributed to their differences in solubility parameters.

A nucleating effect was observed during crystallization process of the melt mixed nanocomposites due to the APIBPOSS and APIOPOSS nanoparticle presence, while only APIBPOSS and high levels of ditelechelic hybrid masterbatch behaved as nucleating agents in the solution blended composites. The nanocomposites obtained by solution blending were more thermally stable than neat PCL. The thermo-oxidative stability of PCL improved with loading of APIOPOSS and ditelechelic hybrid masterbatch by solution mixing, and amino-POSS derivatives by melt blending.

The Young´s modulus of PCL decreased with the incorporation of the three types of POSS by solution-casting method, while remained almost constant for the melt mixed nanocomposites. The incorporation of POSS nanoparticles to PCL by solution blending led to a decrease in the material stiffness and brittleness. However, in the extruded blends the incorporation of APIBPOSS nanoparticles led to more brittle materials. The addition of POSS nanoparticles to PCL render the surface hydrophobic due to presence of filler on the surface.

In this study, we have demonstrated that by controlling the POSS structure and content, and the processing method the properties of POSS/PCL nanocomposites may be tailored.

Supplementary Materials: The following are available online at http://www.mdpi.com/2073-4360/11/1/33/s1; Table S1: Molecular weights of PCL and its nanocomposites; Figure S1: DSC enlarged cooling scans A, B and C: solution blended nanocomposites; D and E: melt mixed nanocomposites: (a) PCL, (b) 2 wt.% POSS, (c) 5 wt.% POSS, and (d) 10 wt.% POSS.; Figure S2: DTG curves of: (A) nanofillers and neat PCL; (B), (C), and (D) solution blended nanocomposites; (E) and (F) melt mixed nanocomposites in N2 atmosphere; Figure S3: DTG curves of: (A) nanofillers and neat PCL; (B), (C), and (D) solution blended nanocomposites; (E) and (F) melt mixed nanocomposites in O2 atmosphere; Figure S4: Stress-strain curves A, B, and C: solution blended nanocomposites; D and E: melt mixed nanocomposites: (a) PCL, (b) 2 wt.% POSS, (c) 5 wt.% POSS, and (d) 10 wt.% POSS; Table S2: Contact angles of neat PCL, PIOPOSS-PCL-PIOPOSS, and PCL/POSS nanocomposites.

Author Contributions: Conceptualization, M.D.F. and M.J.F.; Formal analysis, M.C., J.R.R., D.J.G., M.D.F., and M.J.F.; Funding acquisition, M.D.F. and M.J.F.; Investigation, M.C., J.R.R., and D.J.G.; Methodology, M.D.F. and M.J.F.; Writing—original draft, M.D.F. and M.J.F.; Writing—review & editing, M.D.F. and M.J.F. All authors have given approval to the final version of the manuscript.

Funding: This research was funded by the Basque Government (SAIOTEK 2012 S-PE12UN006) and the University of the Basque Country (UFI11/56).

Acknowledgments: Technical and human support provided by SGIker of UPV/EHU and European funding (ERDF and ESF), is gratefully acknowledged Polymer Technology Group of the Department of Polymer Science and Technology (UPV/EHU) is acknowledged for tensile tests. M. Cobos gratefully thanks the Basque Government for the award of a grant.

Conflicts of Interest: The authors declare that they have no competing interests.

References

1. Tokiwa, Y.; Calabia, B.P.; Ugwu, C.U.; Aiba, S. Biodegradability of plastics. *Int. J. Mol. Sci.* **2009**, *10*, 3722–3742. [CrossRef] [PubMed]

2. Luckachan, G.E.; Pillai, C.K.S. Biodegradable polymers—A review on recent trends and emerging perspectives. *J. Polym. Environ.* **2011**, *19*, 637–676. [CrossRef]

3. Gross, R.A.; Kalra, B. Biodegradable polymers for the environment. *Science* **2002**, *297*, 803–807. [CrossRef] [PubMed]

4. Chandra, R.; Rustgi, R. Biodegradable polymers. *Prog. Polym. Sci.* **1998**, *23*, 1273–1335. [CrossRef]

5. Vert, M.; Feijen, J.; Albertsson, A.C.; Scott, G.; Chiellini, E. *Biodegradable Polymers and Plastics*; Royal Society Chemistry: Cambridge, UK, 1992; ISBN 0-85186-207-1.

6. John, J.; Mani, R.; Battacharya, M. Evaluation compatibility and properties of biodegradable polyester blends. *J. Polym. Sci. Polym. Chem.* **2002**, *40*, 2003–2014. [CrossRef]

7.   Ikada, Y.; Tsuji, H. Biodegradable polyesters for medical and ecological applications. *Macromol. Rapid Commun.* **2000**, *21*, 117–132. [CrossRef]
8.   Mohantya, A.K.; Misra, M.; Hinrichsen, G. Biofibres, biodegradable polymers and biocomposites: An overview. *Macromol. Mater. Eng.* **2000**, *276/277*, 1–24. [CrossRef]
9.   Woodruff, M.A.; Hutmacher, D.W. The return of a forgotten polymer: Polycaprolactone in the 21st century. *Prog. Polym. Sci.* **2010**, *35*, 1217–1256. [CrossRef]
10.   Kweon, H.Y.; Yoo, M.K.; Park, I.K.; Kim, T.H.; Lee, H.C.; Lee, H.S.; Oh, J.S.; Akaike, T.; Cho, C.S. A novel degradable polycaprolactone networks for tissue engineering. *Biomaterials* **2003**, *24*, 801–808. [CrossRef]
11.   Nair, L.S.; Laurencin, C.T. Biodegradable polymers as biomaterials. *Prog. Polym. Sci.* **2007**, *32*, 762–798. [CrossRef]
12.   Yang, K.K.; Wang, X.L.; Wang, Y.Z. Progress in nanocomposite of biodegradable polymer. *J. Ind. Eng. Chem.* **2007**, *13*, 485–500.
13.   Ray, S.S.; Bousmina, M. Biodegradable polymers and their layered silicate nanocomposites: In greening the 21st century materials world. *Prog. Mater. Sci.* **2005**, *50*, 962–1079.
14.   Cordes, D.B.; Lickiss, P.D.; Rataboul, F. Recent developments in the chemistry of cubic polyhedral oligosilsesquioxanes. *Chem. Rev.* **2010**, *110*, 2081–2173. [CrossRef] [PubMed]
15.   Lichtenhan, J.D. Polyhedral oligomeric silsesquioxanes: Building blocks for silsesquioxane-based polymers and hybrid materials. *Comments Inorg. Chem.* **1995**, *17*, 115–130. [CrossRef]
16.   Schwab, J.J.; Lichtenhan, J.D. Polyhedral oligomeric silsesquioxane (POSS)-based polymers. *Appl. Organomet. Chem.* **1998**, *12*, 707–713. [CrossRef]
17.   Pielichowski, K.; Njuguna, J.; Janowski, B.; Pielichowski, J. Polyhedral oligomeric silsesquioxanes (POSS)-containing nanohybrid polymers. *Adv. Polym. Sci.* **2006**, *201*, 225–296.
18.   Misra, R.; Alidedeoglu, A.H.; Jarrett, W.; Morgan, S.E. Molecular miscibility and chain dynamics in POSS/polystyrene blends: Control of POSS preferential dispersion states. *Polymer* **2009**, *50*, 2906–2918. [CrossRef]
19.   Liu, L.; Ming, T.; Liang, G.; Chen, W.; Zhang, L.; Mark, J.E. Polyhedral oligomeric silsesquioxane (POSS) particles in a polysiloxane melt and elastomer. Dependence of the dispersion of the POSS on its dissolution and the constraining effects of a network structure. *J. Macromol. Sci. A* **2007**, *44*, 659–664. [CrossRef]
20.   Goffin, A.L.; Duquesne, E.; Raquez, J.M.; Miltner, H.E.; Ke, X.; Alexandre, M.; Van Tendeloo, G.; Van Mele, B.; Dubois, P. From polyester grafting onto POSS nanocage by ring-opening polymerization to high performance polyester/POSS nanocomposites. *J. Mater. Chem.* **2010**, *209*, 415–9422. [CrossRef]
21.   Miltner, H.E.; Watzeels, N.; Goffin, A.L.; Duquesne, E.; Benali, S.; Dubois, P.; Rahier, H.; Van Mele, B. Quantifying the degree of nanofiller dispersion by advanced thermal analysis: Application to polyester nanocomposites prepared by various elaboration methods. *J. Mater. Chem.* **2010**, *20*, 9531–9542. [CrossRef]
22.   Pan, H.; Yu, J.; Qiu, Z. Crystallization and morphology studies of biodegradable poly(ε-caprolactone)/polyhedral oligomeric silsesquioxanes nanocomposites. *Polym. Eng. Sci.* **2011**, *51*, 2159–2165. [CrossRef]
23.   Guan, W.; Qiu, Z. Isothermal crystallization kinetics, morphology, and dynamic mechanical properties of biodegradable poly(ε-caprolactone) and octavinyl−polyhedral oligomeric silsesquioxanes nanocomposite. *Ind. Eng. Chem. Res.* **2012**, *51*, 3203–3208. [CrossRef]
24.   Miltner, H.E.; Watzeels, N.; Gotzen, N.A.; Goffin, A.L.; Duquesne, E.; Benali, S.; Ruelle, B.; Peeterbroeck, S.; Dubois, P.; Goderis, B.; et al. The effect of nano-sized filler particles on the crystalline-amorphous interphase and thermal properties in polyester nanocomposites. *Polymer* **2012**, *53*, 1494–1506. [CrossRef]
25.   Lee, K.S.; Chang, Y.W. Thermal and mechanical properties of poly(ε-caprolactone)/polyhedral oligomeric silsesquioxane nanocomposites. *Polym. Int.* **2013**, *62*, 64–70. [CrossRef]
26.   Fernández, M.J.; Fernández, M.D.; Cobos, M. Synthesis, characterization and properties of telechelic hybrid biodegradable polymers containing polyhedral oligomeric silsesquioxane (POSS). *RSC Adv.* **2014**, *4*, 21435–21449. [CrossRef]
27.   Hu, H.; Dorset, D.L. Crystal structure of poly(ε-caprolactone). *Macromolecules* **1990**, *23*, 4604–4607. [CrossRef]
28.   Milliman, H.W.; Boris, D.; Schiraldi, D.A. Experimental determination of Hansen solubility parameters for select POSS and polymer compounds as a guide to POSS−polymer interaction potentials. *Macromolecules* **2012**, *45*, 1931–1936. [CrossRef]
29.   Lim, S.K.; Hong, E.P.; Song, Y.H.; Choi, H.J.; Chin, I.J. Poly(ethylene terephthalate) and polyhedral oligomeric silsesquioxane nanohybrids and their physical characteristics. *J. Mater. Sci.* **2010**, *45*, 5984–5987. [CrossRef]

30. Lim, S.K.; Hong, E.P.; Song, Y.H.; Choi, H.J.; Chin, I.J. Thermodynamic interaction and mechanical characteristics of Nylon 6 and polyhedral oligomeric silsesquioxane nanohybrids. *J. Mater. Sci.* **2012**, *47*, 308–314. [CrossRef]

31. Bordes, C.; Fréville, V.; Ruffin, E.; Marote, P.; Gauvrit, J.Y.; Briançon, S.; Lantéria, P. Determination of poly(ε-caprolactone) solubility parameters: Application to solvent substitution in a microencapsulation process. *Int. J. Pharm.* **2010**, *383*, 236–243. [CrossRef]

32. Jang, B.N.; Wang, D.; Wilkie, C.A. Relationship between the solubility parameter of polymers and the clay dispersion in polymer/clay nanocomposites and the role of the surfactant. *Macromolecules* **2005**, *38*, 6533–6543. [CrossRef]

33. Li, Y.; Shimizu, H. Toward a stretchable, elastic, and electrically conductive nanocomposite: Morphology and properties of poly[styrene-b-(ethylene-co-butylene)-b-styrene]/multiwalled carbon nanotube composites fabricated by high-shear processing. *Macromolecules* **2009**, *42*, 2587–2593. [CrossRef]

34. Yongjin, L.; Hiroshi, S. Conductive PVDF/PA6/CNTs nanocomposites fabricated by dual formation of cocontinuous and nanodispersion structures. *Macromolecules* **2008**, *41*, 5339–5344.

35. Cho, J.W.; Paul, D.R. Nylon 6 nanocomposites by melt compounding. *Polymer* **2001**, *42*, 1083–1094. [CrossRef]

36. Misra, R.; Fu, B.X.; Morgan, S.E. Surface energetics, dispersion, and nanotribomechanical behavior of POSS/PP hybrid nanocomposites. *J. Polym. Sci. B Polym. Phys.* **2007**, *45*, 2441–2455. [CrossRef]

37. Crescenzi, V.; Manzini, G.; Calzolari, G.; Borri, C. Thermodynamics of fusion of poly-β-propiolactone and poly-ε-caprolactone. Comparative analysis of the melting of aliphatic polylactone and polyester chains. *Eur. Polym. J.* **1972**, *8*, 449–463. [CrossRef]

38. Fina, A.; Tabuani, D.; Carniato, F.; Frache, A.; Boccaleri, E.; Camino, G. Polyhedral oligomeric silsesquioxanes (POSS) thermal degradation. *Thermochim. Acta* **2006**, *440*, 36–42. [CrossRef]

39. Niemczyk, A.; Dziubek, K.; Sacher-Majewska, B.; Czaja, K.; Dutkiewicz, M.; Marciniec, B. Study of thermal properties of polyethylene and polypropylene nanocomposites with long alkyl chain-substituted POSS fillers. *J. Therm. Anal. Calorim.* **2016**, *125*, 1287–1299. [CrossRef]

40. Blanco, I.; Abate, L.; Bottino, F.A.; Bottino, P. Hepta isobutyl polyhedral oligomeric silsesquioxanes (hib-POSS). *J. Therm. Anal. Calorim.* **2012**, *108*, 807–815. [CrossRef]

41. Zhang, Z.; Gu, A.; Liang, G.; Ren, P.; Xie, J.; Wang, X. Thermo-oxygen degradation mechanisms of POSS/epoxy nanocomposites. *Polym. Degrad. Stab.* **2007**, *92*, 1986–1993. [CrossRef]

42. Baldi, F.; Bignotti, F.; Fina, A.; Tabuani, D.; Ricco, T. Mechanical characterization of polyhedral oligomeric silsesquioxane/polypropylene blends. *J. Appl. Polym. Sci.* **2007**, *105*, 935–943. [CrossRef]

43. Misra, R.; Fu, B.X.; Plagge, A.; Morgan, S.E. POSS-nylon 6 nanocomposites: Influence of POSS structure on surface and bulk properties. *J. Polym. Sci. B Polym. Phys.* **2009**, *47*, 1088–1102. [CrossRef]

44. Shirtcliffe, N.J.; McHale, G.; Newton, M.I. The superhydrophobicity of polymer surfaces: Recent developments. *J. Polym. Sci. Polym. Phys.* **2011**, *49*, 1203–1217. [CrossRef]

45. Bhushan, B.; Jung, Y.C. Wetting study of patterned surfaces for superhydrophobicity. *Ultramicroscopy* **2007**, *107*, 1033–1041. [CrossRef]

46. Fu, H.; Yan, C.; Zhou, W.; Huang, H. Nano-SiO$_2$/fluorinated waterborne polyurethane nanocomposite adhesive for laminated films. *J. Ind. Eng. Chem.* **2014**, *20*, 1623–1632. [CrossRef]

47. Tang, Y.; Lewin, M. Migration and surface modification in polypropylene (PP)/polyhedral oligomeric silsequioxane (POSS) nanocomposites. *Polym. Adv. Technol.* **2009**, *20*, 1–15. [CrossRef]

polymers

Article

Fluoroalkyl POSS with Dual Functional Groups as a Molecular Filler for Lowering Refractive Indices and Improving Thermomechanical Properties of PMMA

Kazunari Ueda [1,2], Kazuo Tanaka [1,]* and Yoshiki Chujo [1]

[1]   Department of Polymer Chemistry, Graduate School of Engineering, Kyoto University, Katsura, Nishikyo-ku, Kyoto 615-8510, Japan; rouge_asteroide_agile_tigre_53@yahoo.co.jp (K.U.); chujo@poly.synchem.kyoto-u.ac.jp (Y.C.)

[2]   Matsumoto Yushi-Seiyaku Co., Ltd., 2-1-3, Shibukawa-cho, Yao-City, Osaka 581-0075, Japan

*   Correspondence: tanaka@poly.synchem.kyoto-u.ac.jp; Tel.: +81-75-383-2604; Fax: +81-75-383-2605

Received: 20 November 2018; Accepted: 30 November 2018; Published: 2 December 2018

Abstract: The dual-functionalized polyhedral oligomeric silsesquioxane (POSS) derivatives, which have the seven fluorinated alkanes and the single acrylate ester on the silica cube, were designed as a filler for lowering the refractive index (RI) and improving thermomechanical properties in poly(methyl methacrylate) (PMMA). The desired dual-functionalized POSS fillers were prepared, and because of its high miscibility, homogeneous films were readily obtained, by the casting method, with the mixture solutions containing the modified POSS and the polymers. From optical measurements, it was found that the larger effects of lowering the RIs of the PMMA matrices were observed from the modified POSS than those of the octa-substituted POSS derivatives with the homogeneous substituents. It should be mentioned that the degradation temperatures and the storage moduli were able to be greatly elevated by loading the present POSS fillers. Finally, it was demonstrated that the methacrylate ester-tethered POSS should be the most effective filler for modulating PMMA (Δn = −0.020, $\Delta Td20$ = +53 °C, $\Delta E'/E'$ = +72%).

Keywords: POSS; filler; low refractive material; thermal stability; fluoropolymer

---

1. Introduction

Organic−inorganic polymer hybrids [1], where both components are homogeneously dispersed in nanometer scales and at molecular level, have been widely known to be a scaffold for producing thermally and mechanically-stable materials with unique functions originating from polymer components. Recently, polyhedral oligomeric silsesquioxane (POSS) has been known to be a promising platform for developing molecular fillers to modulate polymer properties [2–7]. It was shown that the connection of luminescent dyes with POSS induced unique thermal and optical behaviors [8–12]. More recently, simply by loading the POSS derivatives onto polymer matrices, various functions, such as thermal stability, mechanical properties and fire resistance, were improved [13,14]. Therefore, much effort has been directed for exploring syntheses of new POSS derivatives having superior functions [15–17]. We also proposed that POSS is a versatile "element-block", which is a functional building block composed of heteroatoms [18,19], for constructing functional polymer hybrids, according to the preprogrammed design [20–22]. By mixing POSS derivatives with the polymers in the solution, the hybrid materials can be obtained without troublesome sol−gel methods [23–27], and a significant enhancement on the thermal stability was observed [28–32]. In particular, this concept is valid for reconciling trade-off relationships among material properties, such as durability and optical properties [33–35].

Tuning of material properties of low-RI polymers has attracted tremendous attention in the development of modern electronic devices. As an anti-reflection film, low-RI polymers can contribute to the improvement of display resolution by suppressing light scattering, as well as enhancement of device efficiency, by facilitating light extraction [36]. Perfluorinated compounds and multi-pore substances are known to work as a low-RI filler for polymers [37–39]. However, the introduction of these fillers into polymers often induces critical decreases in thermal stabilities and mechanical properties of the polymer matrices. In particular, due to the intrinsically poor compatibility of the fluorine groups with polymers, critical phase separation readily occurs, leading to significant decreases in film-formability and durability. Creation of pores inside materials by employing hollow particles is also valid for efficiently lowering the RI values, whereas the critical losses of mechanical properties were still not overcome. Thus, the development of low-RI fillers, without any loss of thermal and mechanical properties, is strongly needed.

We have previously reported the design strategy for a molecular filler which can simultaneously lower RIs and enhance the durability of conventional polymers, by employing POSS that have dual types of functional groups [34,35]. The cyclopentyl substituents at the seven vertices in the silica cube were expected to reduce the RI values by creating lower density regions in the matrix. Losses of thermal and mechanical properties were compensated by a single substituent, by forming strong interaction with polymer chains. Furthermore, it was presumed that miscibility toward polymers was improved because of the lower symmetry of the dual-functionalized POSS than those of the octa-substituted POSS with the homogeneous substituents [34,35,40]. Indeed, desired filler effects on the optical and thermomechanical properties were observed from the polymer hybrids containing the dual-functionalized POSS fillers. However, the degrees of filler effects were still small, especially in the poly(methyl methacrylate) (PMMA) which is the conventional low-RI polymer although validity of this strategy was able to be demonstrated. Therefore, further sophistication should be required for receiving practical fillers, based on the dual-functionalized POSS.

Herein, we demonstrate molecular fillers by employing POSS as a platform for modulating the optical and thermomechanical properties of the PMMA. Series of the dual-functionalized POSS derivatives that had seven fluorinated alkanes and a single functional group, such as methyl methacrylate ester, cyclopentyl, and octadecyl, on the silica cube, were prepared and loaded onto the PMMA film. Initially, it was confirmed that the present POSS fillers had higher miscibility than the octa-substituted POSS, with the homogeneous substituents. In addition, the larger filler effects on the optical and thermomechanical properties were obtained. The mechanism is discussed in this manuscript.

2. Experimental Section

2.1. General

NMR spectra were measured with a JEOL EX–400 (400 MHz for ^{1}H, 100 MHz for ^{13}C, and 80 MHz for ^{29}Si) spectrometer. Coupling constants (J value) are reported in Hertz. MASS spectra were obtained on a Thermo Fisher Scientific EXACTIVE spectrometer, for atmospheric pressure chemical ionization (APCI). Scanning electron microscopy (SEM) images were obtained using a JEOL JSM-5600 operator with an accelerating voltage of 10 kV. UV-vis transmittance spectra were recorded with a SHIMADZU UV-3600 UV-vis-NIR spectrophotometer. The refractive indices were determined with an Abbe refractometer DR-M4 (accuracy ±0.0002, ATAGO Co., ltd., Tokyo, Japan) at 580 nm at 25 °C. Dynamic mechanical analysis (DMA) was performed on a SDM5600/DMS210 (Seiko Instrument, Inc. (SII), Tokyo, Japan) with a heating rate of 2 °C/min at 1 Hz, with 1% strain, under air. Differential scanning calorimetry (DSC) thermograms were carried out on an SII DSC 6220 instrument, SII, by using ~10 mg of exactly weighed samples, at a heating rate of 10 °C/min, under nitrogen. The glass transition temperatures (T_g) were evaluated from the second monitoring curves, after annealing at 100 °C for 10 min, followed by cooling to 30 °C. Thermogravimetric analysis (TGA) was performed on an EXSTAR TG/DTA 6220, SII, with a heating rate of 10 °C/min up to 500 °C, under nitrogen atmosphere. Residual chloroform was removed by keeping it in a vacuum oven, at 100 °C for 1 h before the TGA measurements. The van der Waals volumes of the POSS derivatives and the monomer unit of the PMMA were calculated after the modeling, using a semi-empirical AM1 method.

2.2. Materials

Trimethoxy(3,3,3-trifluoropropyl)silane (Tokyo Chemical Industry Co., Ltd., Tokyo, Japan), 3-(trichlorosilyl)propyl methacrylate (Sigma-Aldrich Co. LLC, St. Louis, MO, USA), trichlorocyclopentylsilane (Adrich), trichloro(octadecyl)silane (Adrich), and sodium hydroxide (FUJIFILM Wako Pure Chemical Corporation, Osaka, Japan) were purchased and used for the assays, without further purification. PMMA (M_w = 800,000) was purchased from the Nacalai Tesque (Kyoto, Japan). Tetrahydrofuran (THF) and triethylamine were purchased from FUJIFILM Wako Pure Chemical Corporation and Kanto Chemical Co., Inc., Tokyo, Japan and were purified using a two-column solid-state purification system (Glasscontour System, Joerg Meyer, Irvine, CA, USA).

2.3. Syntheses and Preparation of Materials

Synthesis of the Silsesquioxane Partial Cage [$Na_3O_{12}Si_7(C_3H_4F_3)_7$]. $Na_3O_{12}Si_7(C_3H_4F_3)_7$ was synthesized by the following method. Trimethoxy(3,3,3-trifluoropropyl)silane (5.00 g, 22.9 mmol), THF (25 mL), deionized water (0.53 g, 29.4 mmol), and sodium hydroxide (0.40 g, 10.0 mmol) were charged in a round-bottomed flask, equipped with a reflux condenser. After being refluxed for 5 h, the mixture was cooled down to the room temperature and held, with vigorous stirring, for 15 h. The solvent was removed by a rotary evaporator and a white solid was obtained. After being dried in vacuo, 3.6 g of the products were obtained with a 97% yield.

Synthesis of the POSS fillers. $Na_3O_{12}Si_7(C_3H_4F_3)_7$ (2.1 g, 1.8 mmol) and triethylamine (0.27 mL, 2.0 mmol) were dissolved in THF (40 mL) and cooled in an ice bath. Then trichlorosilane (3-methacryloxypropyl-, 0.53 g, 2.0 mmol; cyclopentyl-, 0.41 g, 2.0 mmol; and octadecyl-, 0.79 g, 2.0 mmol) in THF (4 mL) was added slowly to the mixture. The resulting solution was stirred at 0 °C for 4 h and then overnight, at room temperature. After removing the insoluble salts by filtration, the solvent was removed by a rotary evaporator and a white solid was obtained. The white solid was washed with methanol and dried in vacuo to yield the desired product as a white powder (F+MMA POSS, 51%; F+CP POSS, 53%; F+C18 POSS, 54%).

F+MMA POSS: ^{1}H NMR (CDCl$_3$, 400 MHz) δ 6.10 (s, 1 H), 5.57 (s, 1H), 4.13 (t, 2H, J = 6.7 Hz), 2.13 (m, 14H), 1.94 (s, 3H), 1.77 (m, 2H), 0.94 (m, 14H), 0.76 (t, 2H); ^{13}C NMR (CDCl$_3$, 100 MHz) δ 167.4, 136.6, 127.2 (q, J = 276 Hz), 125.4, 66.0, 27.8 (qd, 31 Hz, 4.1Hz), 22.1, 18.2, 7.8, 4.1; ^{29}Si NMR (CDCl$_3$, 80 MHz) δ −66.0, −67.5, −67.7; HRMS (APCI) [(M + Cl)$^-$] calcd. 1256.9853, found 1256.9864.

F+CP POSS: ^{1}H NMR (CDCl$_3$, 400 MHz) δ 2.14 (m, 14H), 1.80 (br, 2H), 1.69−1.37 (br, 6H), 1.14−0.80 (br, 15H); ^{13}C NMR (CDCl$_3$, 100 MHz) δ 127.1 (q, J = 267 Hz), 27.7 (qd, 31 Hz, J = 4.0 Hz), 27.2, 26.9, 21.5, 4.0; ^{29}Si NMR (CDCl$_3$, 80 MHz) δ −65.7, −67.5, −67.7; HRMS (APCI) [(M + Cl)$^-$] calcd. 1198.9798, found 1198.9794.

F+C18 POSS: ^{1}H NMR (CDCl$_3$, 400 MHz) δ 2.11 (m, 14H), 1.57 (br, 32H), 0.89 (br, 17H), 0.63 (br, 2H); ^{13}C NMR (CDCl$_3$, 100 MHz) δ 127.2 (q, J = 276 Hz), 32.9, 32.1, 29.9, 29.5, 27.9, 22.8, 14.1, 4.11; ^{29}Si NMR (CDCl$_3$, 80 MHz) δ −65.3, −67.5, −67.8; HRMS (APCI) [(M+Cl)$^-$] calcd. 1383.1989, found 1383.2002.

Preparation of the Polymer Composites. The mixtures (20 mL) containing 1 g of PMMA and various amounts of POSS fillers in chloroform, were stirred at room temperature for 3 h and then poured into the bottom of the vessels (7 cm × 4.5 cm). After drying at room temperature for 2 days, the film samples were dried again, in a vacuum oven, at 60 °C for 2 h. The resulting films were used for the following measurements.

2.4. Analyses

Determination of the Refractive Indices of the Polymer Composites. According to the Lorentz–Lorenz equation, the refractive index (n) of the polymer (element 1) composites containing fillers (element 2) can be described using the molar fractions (α), molar refractions (R), and the molar volumes (V) [41,42].

$$\frac{n^2-1}{n^2+2} = \alpha_1 \frac{n_1{}^2-1}{n_1{}^2+2} + \alpha_2 \frac{n_2{}^2-1}{n_2{}^2+2} = \alpha_1 \frac{R_1}{V_1} + \alpha_2 \frac{R_2}{V_2} \tag{1}$$

The degree of packing can be described by the molecular packing coefficient K_p that is defined as

$$K_p = \frac{V_{VDW}}{V_{int}} \tag{2}$$

where V_{int} and V_{VDE} are the intrinsic and van der Waals volumes of the molecules, respectively. Therefore, molecules that have significant abilities to lower the interactions between polymer chains would show a smaller K_p value. In addition, according to the Vuks equation, Equation (1) can be transformed in the Equation (3).

$$\frac{n^2-1}{n^2+2} = \alpha_1 \frac{n_1{}^2-1}{n_1{}^2+2} + \alpha_2 \frac{n_2{}^2-1}{n_2{}^2+2} = \alpha_1 \frac{K_{p1}R_1}{V_{VDW,1}} + \alpha_2 \frac{K_{p2}R_2}{V_{VDW,2}} \tag{3}$$

In the case of amorphous polymers, the K_p values (K_{p1}) were determined to be 0.677 [29]. Therefore, we can simply calculate K_{p2} values by measuring the refractive indices of the composite (n), using Equation (3).

3. Results and Discussion

To meet opposite demands, the modified POSS derivatives that had the fluoroalkyl and other functional groups, were designed (Scheme 1). According to previous studies, introduction of the fluoroalkyl groups can efficiently lower the RI values of polymers. However, because of the intrinsic very weak interaction with the polymer chains, critical low miscibility of the polymer matrices was presumed. To compensate for the decrease in miscibility, another functional group, which can be expected to form a strong interaction or a bond with the polymer chains, was introduced into the POSS. It was reported that the methacrylate can form a covalent bond with the PMMA, by heating [40]. As a result, reinforcement of the mechanical properties of the PMMA film was capable. Therefore, the methacrylate ester-tethered POSS (F+MMA POSS) was designed. The octadecyl group in the F+C18 POSS was also expected to form strong interaction with PMMA, by alkyl-chain entanglement [23,35]. The cyclopentyl group was introduced as a comparison to the previous dual-functionalized POSS, which can greatly reduce the RI value of the PMMA (F+CP POSS) [35]. The series of the modified POSS derivatives were conducted, according to Scheme 1. The syntheses of the target POSS derivatives were performed via the preparation of the incomplete cage, followed by the formation of the POSS cage, following the procedures reported elsewhere [35,40]. From the NMR and MS measurements, it was confirmed that the products had the desired structures. All POSS derivatives showed good solubility in chloroform, and film samples for optical and mechanical measurements were prepared through casting with the mixture solutions containing each POSS filler and PMMA, followed by pre-heating at 60 °C for 2 h, to remove the solvent. In the case of the F+MMA POSS, the cross-linking reaction could proceed in this step [40].

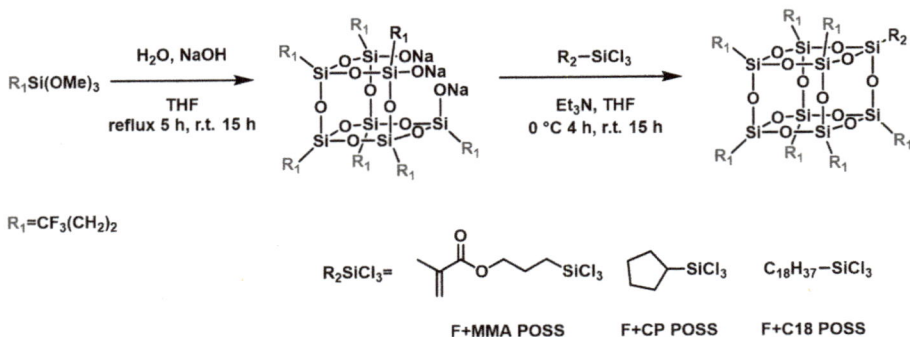

$R_1 = CF_3(CH_2)_2$

$R_2SiCl_3 =$

F+MMA POSS   F+CP POSS   F+C18 POSS

Scheme 1. Synthesis of the polyhedral oligomeric silsesquioxane (POSS) fillers that had a dual type of substituents.

The PMMA films containing 2 mol% (ca. 20 wt%) F+MMA POSS and F+CP POSS mixtures had good transparency, whereas white opaque films were obtained from the F+C18 POSS (Figure 1). From the absorption measurements in the wavelength range from 380 nm to 780 nm, the transparency of the film samples was quantitatively calculated as an averaged value in this wavelength region (Table 1 and Figure S1). The polymer films with F+MMA POSS and F+CP POSS mixtures below 2 mol% presented more than 85% transparencies, while the critical decrease in transparency was detected from the hybrids with the F+C18 POSS. These data indicate that miscibility can be improved by introducing another substituent into the POSS core. The surface morphologies of the hybrid films were investigated using SEM. Critical phase separation or aggregation of the POSS fillers in the size range where the RI values were influenced, was slightly observed in the PMMA hybrids with 2 mol% F+MMA POSS and F+CP POSS (Figure S2). Conversely, by adding the same amounts of F+C18 POSS, SEM images with inhomogeneity were obtained. These results indicate that homogeneous dispersions of the F+MMA POSS and the F+CP POSS mixtures were realized in the PMMA. Especially, the F+PMA POSS mixture should have a higher miscibility with the PMMA than the F+CP POSS. It is likely that 3-methacryloxypropyl-group could interact with the polymer chains, leading to high dispersion in the polymer matrices. In the case of commodity polymers, such as polystyrene and PMMA, the turbidity appeared only by adding 1 mol% (ca. 10 wt%) of the previous molecules, such as the octa-substituted POSS fillers with the homogeneous substituents [23,24]. This fact means that the miscibility of the PMMS could be improved. Decreased molecular symmetry by introducing a single functional group could contribute to miscibility improvement by suppressing crystallization. Most samples showed turbidity at 2 mol%, and the transparency of the films decreased. F+C18 POSS induced white opaque into the film, even at 0.5 mol%. Thereby, we discuss the filler effects with the data sets obtained from samples containing 2 mol% F+MMA POSS and F+CP POSS.

Table 1. Summary of the filler effects on the various properties of the PMMA, by the POSS fillers [a].

POSS filler	Transmittance (%) [b]	Δn [c]	K_{p2}	T_{d20} (°C) [d]	ΔT_{d20} (°C)	T_g (°C)	E' (MPa) [e]
none	92	—	—	215	—	60.6	1780
F+MMA	87	−0.0197	0.093	268	+53	63.8	3060
F+CP	85	−0.0147	0.180	225	+10	61.8	3050
F+C18	45	−[f]	−[f]	272	+57	58.6	2630

[a] 2.0 mol% POSS. [b] Calculated as an averaged value from 380 to 780 nm. [c] Averaged value measured at five points. Calculated from the subtraction of the PMMA films from the refractive index (1.4927). [d] Determined as the decomposition temperature with 20 wt% weight losses. [e] Represented at 30 °C, and the errors are within 5%. [f] Not detectable due to being opaque.

Polymers **2018**, *10*, 1332

Figure 1. Appearances of the poly(methyl methacrylate) (PMMA) hybrid films with 2 mol% POSS fillers.

To evaluate the influence of the POSS fillers on the RI values of the PMMA, optical measurements were performed with the film samples containing variable concentrations of the POSS fillers, using an Abbe refractometer (Figure 2). The averages of the values at five distinct points in the films are shown in Table 1. The RI values of the polymer films were drastically lowered by the fluorinated POSS fillers. By loading the F+MMA POSS with 2 mol%, the degree of the reduction was, approximately, 0.02. F+CP POSS filler could also reduce the RI values (2 mol%, Δn = ca. −0.015). It should be mentioned that degree of the lowering effect of the F+CP POSS was much larger than that of the previous dual-functionalized POSS with cyclopentyl groups (ca. −0.005). These data clearly indicated that the fluorinated POSS can have a superior ability to be an efficient filler, for lowering the refractive indices of the PMMA. It is suggested that the POSS mixtures could be completely dispersed in the polymer matrices, by suppressing the crystallinity of the POSS, as mentioned above. Therefore, a lower density region should be efficiently created around the POSS filler, leading to the lowering effects on the refractive indices. According to previous reports, the degree of decrease in RIs, by the present POSS fillers, is large enough for applications in optical fibers [43,44]. Thus, our concept for the design of molecular fillers might be applicable in the practical light-guiding materials.

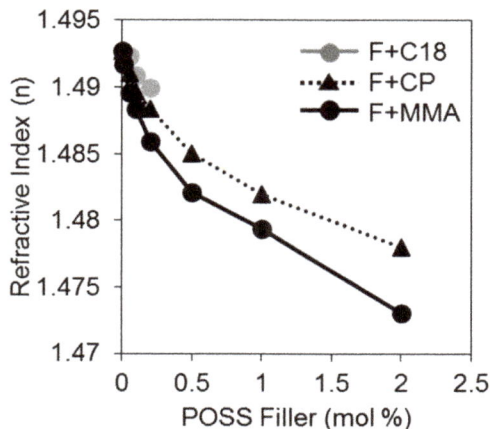

Figure 2. Refractive indices (n) of the PMMA hybrids containing various concentrations of the POSS fillers.

To estimate the degree of packing in polymer chains and filler molecules, the packing coefficients were calculated according to previous reports (see the Supporting Information) [41,42]. In the absence of the POSS filler, the packing coefficient (K_{p1}) of the pristine PMMA film was determined as 0.677 [29]. If the POSS filler could create lower density regions in the matrices, a smaller K_{p2} value than the K_{p1} would be obtained (Table 1) [24]. Apparently, the K_{p2} values were smaller than that of the pure PMMA film. These results clearly indicated that the lowering effect on the RI values of the PMMA, by the POSS filler, should originate from the decreases in local density, around the POSS fillers.

Thermal stabilities of the polymer films containing fluorinated POSS fillers were examined by the TGA. Figure S3 shows the TGA curves from the samples containing the POSS fillers with variable amounts, and Table 1 summarizes the T_{d20}, using the polymer composites containing the POSS fillers. From the curves, it was indicated that degradation occurred through three steps. It was reported that pyrolysis of PMMA proceeds, initially, at the head-to-head linkage, followed by the chain-ends, and the random scission in the main-chains [45]. According to the previous reports, POSS fillers can effectively suppress the second and third steps, which can be evaluated as the degradation temperature with 20% weight losses. Therefore, the thermal stabilizing effects by the POSS fillers were discussed by comparing the T_{d20} values of the polymer matrices [28,29,34,40]. Pure PMMA exhibited a T_{d20} at 215 °C, and significant thermal reinforcement of the PMMA matrices by loading the POSS fillers can be observed. By increasing the amount of the POSS, the T_{d20} values increased. Finally, the T_{d20} value of the PMMA matrices containing 2 mol% of the F+MMA POSS and the F+CP POSS increased by 53 and 11 °C, respectively. Comparing to the previous dual-functionalized POSS (ΔT_{d20} = +2 °C), larger stabilization effects were obtained. According to the previous reports, the methacrylate ester group connected to the POSS reacted with the PMMA matrices, by heating around 60 °C [40]. As a result, significant enhancement of thermal stability was observed. From the analyses, it was shown that cross-linking between the methacrylate ester moiety and PMMA should occur, followed by the formation of the tight binding of the silica cube to polymer chains. In the current case, it has also been suggested that the drastic thermal reinforcement in the hybrid could be induced by the cross-linking reaction at the methacrylate ester moiety, in the F+MMA POSS.

According to the previous DSC result with the octa-substituted POSS with a single kind of alkyl-substituents [28], T_g values were slightly influenced by loading the POSS onto PMMA, with high homogeneity. Meanwhile, significant decreases in T_gs were observed in the heterogeneous films (Table 1). Aggregation of the POSS was obviously observed in the F+C18 POSS-containing film. Due to the low miscibility of the F+C18 POSS, relatively lower T_g should be induced by the filler addition.

Finally, the mechanical properties of the polymer matrices containing the POSS fillers were evaluated by DMA (temperature scan at 1 Hz). The mechanical properties of the polymer composites containing the POSS fillers are listed in Table 1. It was shown that the POSS fillers efficiently improved the rigidity of the PMMA matrices, as represented in the storage modulus (E'), although each POSS molecule created a lower density region. The E' value was enhanced by 72% with the F+PMA POSS and 71% with the F+CP POSS. In the previous reports, loss of E' was induced with the POSS filler possessing the hepta-substituted cyclopentyl and single fluoroalkyl groups [35]. Molecular motions should be effectively suppressed by the rigid POSS core via the single substituent. From these data involving the DSC, the TGA, and the DMA, it can be summarized that the POSS fillers used in this study have significant abilities to simultaneously improve the thermomechanical properties and lower the refractive indices of polymers.

4. Conclusions

By utilizing the molecular fillers based on fluoroalkyl POSS having dual types of functional groups, the polymer hybrids were prepared. From the series of measurements with the hybrid films, it was clearly shown that the POSS fillers can effectively lower the RI values of the PMMA and enhance both the thermal and the mechanical durability. It is proposed that the one functional group should make a strong interaction with the polymer chains, even in the presence of low-miscible hepta-substituents of the fluoroalkyl groups, followed by suppressing the molecular motions. These results indicate that the dual types of functional groups on the POSS core can share roles in simultaneously improving the thermomechanical properties and lowering the refractive indices of the conventional polymers. In other words, it was demonstrated that the POSS played a role only in introducing the positive effects of each functional group, followed by reconciling the trade-off relationship in the polymer functions. Furthermore, it can be said that our concept of designable hybrids, utilizing the preprogrammed-modified POSS could include a huge potential for receiving multi-functional polymeric materials with high stability.

Supplementary Materials: The following are available online at http://www.mdpi.com/2073-4360/10/12/1332/s1.

Author Contributions: Data acquisition, K.U.; Funding acquisition, K.T. and Y.C.; Writing—original draft, K.U. and K.T.; Writing—review & editing, K.T. and Y.C.

Acknowledgments: This work was partially supported by the Magnetic Health Science Foundation (for K.T.) and a Grant-in-Aid for Scientific Research (B) (JP17H03067) and (A) (JP 17H01220), for Scientific Research on Innovative Areas "New Polymeric Materials Based on Element-Blocks (No.2401)" (JP24102013), and for Challenging Research (Pioneering) (JP18H05356).

Conflicts of Interest: The authors declare no conflict of interest.

References

1. Gon, M.; Tanaka, K.; Chujo, Y. Creative Synthesis of Organic–Inorganic Molecular Hybrid Materials. *Bull. Chem. Soc. Jpn.* **2017**, *90*, 463–474. [CrossRef]

2. Gon, M.; Sato, K.; Tanaka, K.; Chujo, Y. Controllable Intramolecular Interaction of 3D Arranged π-Conjugated Luminophores Based on a POSS Scaffold, Leading to Highly Thermostable and Emissive Materials. *RSC Adv.* **2016**, *6*, 78652–78660. [CrossRef]

3. Suenaga, K.; Tanaka, K.; Chujo, Y. Heat-Resistant Mechanoluminescent chromism of the Hybrid Molecule Based on Boron Ketoiminate-Modified Octa-Substituted Polyhedral Oligomeric Silsesquioxane. *Chem. Eur. J.* **2017**, *23*, 1409–1414. [CrossRef] [PubMed]

4. Narikiyo, H.; Gon, M.; Tanaka, K.; Chujo, Y. Control of intramolecular excimer emission in luminophore-integrated ionic POSSs possessing flexible side-chains. *Mater. Chem. Front.* **2018**, *2*, 1449–1455. [CrossRef]

5. Narikiyo, H.; Kakuta, T.; Matsuyama, H.; Gon, M.; Tanaka, K.; Chujo, Y. Development of Optical Sensor for Discriminating Isomers of Fatty Acids Based on Emissive Network Polymers Composed of Polyhedral Oligomeric Silsesquioxane. *Bioorg. Med. Chem.* **2017**, *25*, 3431–3436. [CrossRef] [PubMed]

6. Kakuta, T.; Tanaka, K.; Chujo, Y. Synthesis of Emissive Water-Soluble Network Polymers Based on Polyhedral Oligomeric Silsesquioxane and Their Application as an Optical Sensor for Discriminating the Particle Size. *J. Mater. Chem. C* **2015**, *3*, 12539–12545. [CrossRef]

7. Li, Z.; Kong, J.; Wang, F.; He, C. Polyhedral oligomeric silsesquioxanes (POSSs): An important building block for organic optoelectronic materials. *J. Mater. Chem. C* **2017**, *5*, 5283–5298. [CrossRef]

8. Blanco, I. The Rediscovery of POSS: A Molecule Rather than a Filler. *Polymers* **2018**, *10*, 904. [CrossRef]

9. Zhang, W.; Camino, G.; Yang, R. Polymer/polyhedral oligomeric silsesquioxane (POSS)nanocomposites: An overview of fire retardance. *Prog. Polym. Sci.* **2017**, *67*, 77–125. [CrossRef]

10. Kausar, A. State-of-the-Art Overview on Polymer/POSS Nanocomposite. *Polym. Plast. Technol. Eng.* **2017**, *56*, 1401–1420. [CrossRef]

11. Chruściel, J.J.; Leśniak, E. Modification of epoxy resins with functional silanes, polysiloxanes, silsesquioxanes, silica and silicates. *Prog. Polym. Sci.* **2015**, *14*, 67–121. [CrossRef]

12. Madbouly, S.A.; Otaigbe, J.U. Recent advances in synthesis, characterization and rheological properties of polyurethanes and POSS/polyurethane nanocomposites dispersions and films. *Prog. Polym. Sci.* **2009**, *38*, 1283–1332. [CrossRef]

13. Blanco, I.; Abate, L.; Bottino, F.A. Mono substituted octaphenyl POSSs: The effects of substituents on thermal properties and solubility. *Thermochim. Acta* **2017**, *655*, 117–123. [CrossRef]

14. Zhao, L.; Huang, Y.; Liu, B.; Huang, Y.; Song, A.; Lin, Y.; Wang, M.; Li, X.; Cao, H. Gel polymer electrolyte based on polymethyl methacrylate matrixcomposited with methacrylisobutyl-polyhedral oligomericsilsesquioxane by phase inversion method. *Electrochim. Acta* **2018**, *278*, 1–12. [CrossRef]

15. Blanco, I.; Bottino, F.A.; Bottino, P.; Chiacchio, M.A. A novel three-cages POSS molecule: synthesis and thermal behaviour. *J. Therm. Anal. Calorim.* **2018**, *134*, 1337–1344. [CrossRef]

16. Blanco, I.; Abate, L.; Bottino, F.A. Synthesis and thermal behaviour of phenyl-substituted POSSs linked by aliphatic and aromatic bridges. *J. Therm. Anal. Calorim.* **2018**, *131*, 843–851. [CrossRef]

17. Maegawa, T.; Irie, Y.; Fueno, H.; Tanaka, K.; Naka, K. Synthesis and polymerization of a para-disubstituted T8-caged hexaisobutyl-POSS monomer. *Chem. Lett.* **2014**, *43*, 1532–1534. [CrossRef]

18. Chujo, Y.; Tanaka, K. New polymeric materials based on element-blocks. *Bull. Chem. Soc. Jpn.* **2015**, *88*, 633–643. [CrossRef]

19. Gon, M.; Tanaka, K.; Chujo, Y. Recent Progress in the Development of Advanced Element-Block Materials. *Polym. J.* **2018**, *50*, 109–126. [CrossRef]

20. Tanaka, K.; Chujo, Y. Advanced Functional Materials Based on Polyhedral Oligomeric Silsesquioxane (POSS). *J. Mater. Chem.* **2012**, *22*, 1733–1746. [CrossRef]

21. Tanaka, K.; Chujo, Y. Unique Properties of Amphiphilic POSS and Their Applications. *Polym. J.* **2013**, *45*, 247–254. [CrossRef]

22. Tanaka, K.; Chujo, Y. Chemicals-Inspired Biomaterials; Developing Biomaterials Inspired by Material Science Based on POSS. *Bull. Chem. Soc. Jpn.* **2013**, *86*, 1231–1239. [CrossRef]

23. Okada, H.; Tanaka, K.; Chujo, Y. Preparation of Environmentally Resistant Conductive Silica-Based Polymer Hybrids Containing Tetrathiafulvalen-Tetracyanoquinodimethane Charge-Transfer Complexes. *Polym. J.* **2014**, *46*, 800–805. [CrossRef]

24. Kajiwara, Y.; Nagai, A.; Tanaka, K.; Chujo, Y. Efficient Simultaneous Emission from RGB-Emitting Organoboron Dyes Incorporated into Organic-Inorganic Hybrids and Preparation of White Light-Emitting Materials. *J. Mater. Chem. C* **2013**, *1*, 4437–4444. [CrossRef]

25. Kajiwara, Y.; Tanaka, K.; Chujo, Y. Enhancement of Dye Dispersibility in Silica Hybrids through Local Heating Induced by the Imidazolium Group under Microwave Irradiation. *Polym. J.* **2014**, *46*, 195–199. [CrossRef]

26. Okada, H.; Tanaka, K.; Chujo, Y. Regulation of Responsiveness of Phosphorescence toward Dissolved Oxygen Concentration by Modulating Polymer Contents in Organic−Inorganic Hybrid Materials. *Bioorg. Med. Chem.* **2014**, *22*, 3141–3145. [CrossRef]

27. Okada, H.; Tanaka, K.; Ohashi, W.; Chujo, Y. Photo-Triggered Molecular Release Based on Auto-Degradable Polymer-Containing Organic−Inorganic Hybrids. *Bioorg. Med. Chem.* **2014**, *22*, 3435–3440. [CrossRef]

28. Tanaka, K.; Adachi, S.; Chujo, Y. Structure-Property Relationship of Octa-Substituted POSS in Thermal and Mechanical Reinforcements of Conventional Polymers. *J. Polym. Sci. Part A Polym. Chem.* **2009**, *47*, 5690–5697. [CrossRef]

29. Tanaka, K.; Adachi, S.; Chujo, Y. Side-Chain Effect of Octa-Substituted POSS Fillers on Refraction in Polymer Composites. *J. Polym. Sci. Part A Polym. Chem.* **2010**, *48*, 5712–5717. [CrossRef]

30. Niemczyk, A.; Dziubek, K.; Sacher-Majewska, B.; Czaja, K.; Dutkiewicz, M.; Marciniec, B. Study of thermal properties of polyethylene and polypropylene nanocomposites with long alkyl chain-substituted POSS fillers. *J. Therm. Anal. Calorim.* **2016**, *125*, 1287–1299. [CrossRef]

31. Yuasa, S.; Sato, Y.; Imoto, H.; Naka, K. Fabrication of composite films with poly(methyl methacrylate) and incompletely condensed cage-silsesquioxane fillers. *J. Appl. Polym. Sci.* **2018**, *135*, 46033. [CrossRef]

32. Ueda, K.; Tanaka, K.; Chujo, Y. Remarkably High Miscibility of Octa-Substituted POSS with Commodity Conjugated Polymers and Molecular Fillers for the Improvement of Homogeneities of Polymer Matrices. *Polym. J.* **2016**, *48*, 1133–1139. [CrossRef]

33. Tanaka, K.; Yamane, H.; Mitamura, K.; Watase, S.; Matsukawa, K.; Chujo, Y. Transformation of Sulfur to Organic−Inorganic Hybrids Employed by POSS Networks and Their Application for the Modulation of Refractive Indices. *J. Polym. Sci. Part A Polym. Chem.* **2014**, *52*, 2588–2595. [CrossRef]

34. Ueda, K.; Tanaka, K.; Chujo, Y. Synthesis of POSS Derivatives Having Dual Types of Alkyl Substituents via in situ Sol−Gel Reactions and Their Application as a Molecular Filler for Low-Refractive and Highly-Durable Materials. *Bull. Chem. Soc. Jpn.* **2017**, *90*, 205–209. [CrossRef]

35. Jeon, J.-H.; Tanaka, K.; Chujo, Y. Rational Design of POSS Fillers for Simultaneous Improvements of Thermomechanical Properties and Lowering Refractive Indices of Polymer Films. *J. Polym. Sci. Part A: Polym. Chem.* **2013**, *51*, 3583–3589. [CrossRef]

36. Mont, F.W.; Schubert, E.F. High-refractive-index TiO$_2$-nanoparticle-loaded encapsulants for light-emitting diodes. *J. Appl. Phys.* **2008**, *103*, 083120. [CrossRef]

37. Choi, S.-S.; Lee, H.S.; Kim, E.K.; Baek, K.-Y.; Choi, D.-H.; Hwang, S.S. Synthesis and Characterization of Fluoro-co-Phenyl Silsesquioxane (FCPSQ) for Low Dielectric Constant Materials. *Mol. Cryst. Liq. Cryst.* **2010**, *520*, 231–238. [CrossRef]

38. Ro, H.W.; Kim, K.J.; Theato, P.; Gidley, D.W.; Yoon, D.Y. Novel inorganic-organic hybrid block copolymers as pore generators for nanoporous ultralow-dielectric-constant films. *Macromolecules* **2005**, *38*, 1031–1034. [CrossRef]

39. Hedrick, J.L.; Miller, R.D.; Hawker, C.J.; Carter, K.R.; Volksen, W.; Yoon, D.Y.; Trollsås, M. Templating nanoporosity in thin-film dielectric insulators. *Adv. Mater.* **1998**, *10*, 1049–1053. [CrossRef]

40. Tanaka, K.; Kozuka, H.; Ueda, K.; Jeon, J.-H.; Chujo, Y. POSS-Based Molecular Fillers for Simultaneously Enhancing Thermal and Viscoelasticity of Poly(methyl methacrylate) Films. *Mater. Lett.* **2017**, *203*, 62–67. [CrossRef]

41. Groh, W.; Zimmermann, A. What Is the Lowest Refractive Index of an Organic Polymer? *Macromolecules* **1991**, *24*, 6660–6663. [CrossRef]

42. Tanio, N.; Irie, M. Estimate of Light Scattering Loss of Amorphous Polymer Glass from Its Molecular Structure. *Jpn. J. Appl. Phys. Part I* **1997**, *36*, 743–748. [CrossRef]

43. Evert, A.; James, A.; Hawkins, T.; Foy, P.; Stolen, R.; Dragic, P.; Dong, L.; Rice, R.; Ballato, J. Longitudinally-graded optical fibers. *Opt. Express* **2012**, *20*, 17393–17401. [CrossRef]

44. Hutsel, M.R.; Gaylord, T.K. Concurrent three-dimensional characterization of the refractive-index and residual-stress distributions in optical fibers. *Appl. Opt.* **2012**, *51*, 5442–5452. [CrossRef]

45. Kashiwagi, T.; Inaba, A.; Brown, J.E.; Hatada, K.; Kitayama, T.; Masuda, E. Effects of weak linkages on the thermal and oxidative degradation of poly(methyl methacrylates). *Macromolecules* **1986**, *19*, 2160–2168. [CrossRef]

![polymers logo]

MDPI

Article

Thermal Stability and Flame Retardancy of Polypropylene Composites Containing Siloxane-Silsesquioxane Resins

Arkadiusz Niemczyk [1,*], Katarzyna Dziubek [1], Beata Sacher-Majewska [1], Krystyna Czaja [1], Justyna Czech-Polak [2], Rafał Oliwa [2], Joanna Lenża [3] and Mariusz Szołyga [4]

[1] Faculty of Chemistry, University of Opole, Oleska 48, 45-052 Opole, Poland;
    katarzyna.dziubek@uni.opole.pl (K.D.); beata.sacher@uni.opole.pl (B.S.-M.);
    krystyna.czaja@uni.opole.pl (K.C.)
[2] Faculty of Chemistry, Rzeszow University of Technology, Al. Powstańców Warszawy 6, 35-959 Rzeszów,
    Poland; j.czech@prz.edu.pl (J.C.-P.); oliwa@prz.edu.pl (R.O.)
[3] Central Mining Institute, Plac Gwarków 1, 40–166 Katowice, Poland; jlenza@gig.eu
[4] Centre for Advanced Technologies, Adam Mickiewicz University in Poznań, Umultowska 89C,
    61-614 Poznań, Poland; markussqueeze@wp.pl
* Correspondence: aniemczyk@uni.opole.pl; Tel.: +48-77-452-7146

Received: 30 August 2018; Accepted: 10 September 2018; Published: 13 September 2018

Abstract: A novel group of silsesquioxane derivatives, which are siloxane-silsesquioxane resins (S4SQ), was for the first time examined as possible flame retardants in polypropylene (PP) materials. Thermal stability of the PP/S4SQ composites compared to the S4SQ resins and neat PP was estimated using thermogravimetric (TG) analysis under nitrogen and in air atmosphere. The effects of the non-functionalized and *n*-alkyl-functionalized siloxane-silsesquioxane resins on thermostability and flame retardancy of PP materials were also evaluated by thermogravimetry-Fourier transform infrared spectrometry (TG-FTIR) and by cone calorimeter tests. The results revealed that the functionalized S4SQ resins may form a continuous ceramic layer on the material surface during its combustion, which improves both thermal stability and flame retardancy of the PP materials. This beneficial effect was observed especially when small amounts of the S4SQ fillers were applied. The performed analyses allowed us to propose a possible mechanism for the degradation of the siloxane-silsesquioxane resins, as well as to explain their possible role during the combustion of the PP/S4SQ composites.

Keywords: siloxane-silsesquioxane resins; polypropylene; thermogravimetry; cone calorimeter tests; flame-retardant mechanism

1. Introduction

Polymers (in particular polyolefins) have become an important class of materials in recent years, mostly due to their favorable properties, relatively low cost and easy accessibility. However, the insufficient thermal stability and flammability of these materials still constitute a serious problem which may limit their potential multidirectional applications [1]. Moreover, well-known and effective halogenated flame retardants have been gradually prohibited as they are environmentally toxic [2]. Thus, special efforts are being made to develop a new class of flame retardants for polymeric materials as an alternative for the commonly used halogenated additives. Many studies have already demonstrated that silicones, silanes, silsesquioxanes, silicas and silicates may improve the thermal stability of polymeric materials and that they could be explored as potential environmentally friendly flame retardants [3–6].

Among silicon-based compounds, special attention should be paid to polyhedral oligomeric silsesquioxanes (POSS), which have been successfully applied as fillers for various polymers [7–14], including polyolefins [15–25]. The presence of POSS molecules in these materials contributes generally to improved mechanical [18,24,26–28] and rheological [22,29–31] properties, as well as influenced their melting and crystallization behavior [26,29,31–34]. Moreover, it has already been demonstrated that silsesquioxanes may significantly improve thermal and thermo-oxidative stability [16,18,21,28,30,32,34–37], as well as reduce flammability of polyolefin-based composites [4,28,33]. However, despite the obvious advantages of POSS compounds, their potential commercial applications are limited as they are relatively expensive [38–40].

The search for low-cost alternatives to POSS molecules led to a new class of hybrid silicon compounds: siloxane-silsesquioxane resins [41–53]. They have extended network structures which are composed of silsesquioxane cages connected by siloxane chains that can be functionalized, and which affect their unique properties [51–53].

However, there are only a few papers available to date which report on the application of siloxane-silsesquioxane resins as fillers [38–40,54]. The influence of non-functionalized and *n*-alkyl-functionalized siloxane-silsesquioxane resins on the morphological, structural, thermal and mechanical properties of polypropylene composites was presented in our previous work [54]. It was demonstrated that the presence itself and the length of the alkyl groups in the resin structures play a crucial role in the dispersion of the filler particles in the polypropylene (PP) matrix. The applied resins affected the crystalline structure and improved the crystallization behavior of the polymer matrix. Moreover, the presence of the siloxane-silsesquioxane resin particles enhanced the impact strength and elongation at break of the composites [55]. Subsequently, phenyl-functionalized siloxane-silsesquioxane resin (SiOPh) was applied by Dobrzyńska-Mizera et al. to control the nucleation efficiency of polypropylene composites containing sorbitol derivatives [38–40]. It should, however, be emphasized that there are no literature data available on the thermal stability and fire behavior of polymer composites containing siloxane-silsesquioxane resins as fillers.

In this article, the influence of siloxane-silsesquioxane resins on thermal stability and fire behavior of the polypropylene matrix is discussed. Composites containing 1–10 wt % of non-functionalized (S4SQ-H) as well as *n*-octyl- and *n*-octadecyl-functionalized resins (S4SQ-8 and S4SQ-18, respectively) were prepared by the melt blending method. The thermal stability of neat siloxane-silsesquioxane resins, as well as of the obtained composites was examined by the thermogravimetric (TG) analysis in both inert and oxidative environments. The selected materials were also studied by thermogravimetry-Fourier transform infrared spectrometry (TG-FTIR). Finally, the flame retardancy of the PP composites filled with siloxane-silsesquioxane resins was evaluated on the basis of cone calorimeter tests.

2. Materials and Methods

2.1. Materials

Polypropylene (PP) Moplen HP 400 R (MFR = 23 g/10 min) was provided by Basell Orlen Polyolefins, and was used as the polymer matrix. The non-functionalized (S4SQ-H) and *n*-alkyl-functionalized (S4SQ-8 and S4SQ-18) siloxane-silsesquioxane resins were synthesized in the Centre for Advanced Technologies of the Adam Mickiewicz University in Poznań, according to procedures described elsewhere [51–54]. The S4SQ designation identifies a siloxane-silsesquioxane resin composed of the silsesquioxane cages connected by 4 siloxane groups [54].

2.2. Composites Preparation

Polypropylene composites filled with siloxane-silsesquioxane resins (S4SQ) were prepared by the melt blending method. In the first step, PP masterbatches containing 20 wt % of S4SQ resins were prepared using a HAAKE PolyLab Rheomether (Thermo Fisher Scientific, San Jose, CA, USA) (180 °C, 50 rpm, 15 min). In the second step, the obtained masterbatches were blended with neat PP granulate at suitable weight ratios. The PP/S4SQ composites were obtained with a IM-15 laboratory conical twin screw extruder (ZAMAK, Skawina, Poland) (175–195 °C, 100–200 rpm) coupled with a IMM-15 laboratory injection machine (ZAMAK, Skawina, Poland) (190 °C, 6 MPa). The obtained composites contained 1, 5 and 10 wt % of the appropriate S4SQ resins.

2.3. Characterization

The thermogravimetric analysis of neat siloxane-silsesquioxane resins was carried out using a Q50-TGA thermogravimeter (TA Instruments, Inc., New Castle, DE, USA) under the flow of N_2 or air. The samples (15–25 mg) were loaded on platinum pans and heated from room temperature to 1000 °C at the rate of 10 °C/min. The thermal stability of the obtained polymer materials was evaluated by thermogravimetric (TG) analysis using a TG/DSC1 device (Mettler Toledo, Columbus, OH, USA). The material samples (3–10 mg) were heated under nitrogen or in air atmosphere from room temperature to 500 °C at the rate of 10 °C/min. The values of T_5, T_{25} and T_{50} parameters, identified as the temperatures at which the 5, 25 and 50 wt % weight losses of the samples occurred, respectively, were determined from the TG thermograms. In turn, the T_{max} parameter as the maximum mass loss rate temperature was evaluated on the basis of the derived thermogravimetry (DTG) curves.

Thermogravimetry–Fourier transformed infrared spectroscopy (TG-FTIR) was performed by means of a TGA/DSC1 instrument (Mettler Toledo, Columbus, OH, USA) coupled with an FTIR Nicolet 380 spectrometer (Thermo Fisher Scientific, San Jose, CA, USA). The samples (about 10 mg) were placed in a ceramic crucible and heated from 25 to 900 °C at the rate of 10 °C/min in air atmosphere. The gas carried the decomposition products from the TG through a stainless-steel line into a gas cell with KBr crystal windows for infrared (IR) detection. Both the transfer line and the gas cell were kept at 200 °C to prevent gas condensation. IR spectra were recorded in the spectral range of 4000–500 cm^{-1} with a resolution of 4 cm^{-1} and an average number of scans of 16.

The fire performance was characterized using a mass loss cone microcalorimeter (Fire Testing Technology Ltd., East Grinstead, UK) according to the standard ISO 13927, at the external heat flux of 35 kW/m^2. The dimensions of the samples were 100.0 mm × 100.0 mm × 2.0 mm. The specific data, including the heat release rate (HRR, kW/m^2) and the total heat release (THR, MJ/m^2), were collected or calculated according to the data during combustion. The typical results from the cone microcalorimeter tests were taken as the average from two measurements.

3. Results and Discussion

3.1. TG Studies

Knowledge of the thermal behavior of neat siloxane-silsesquioxane resins was required to understand their role in the degradation of the PP/S4SQ composites. The TG/DTG curves of neat siloxane-silsesquioxane resins under nitrogen and in air atmosphere are shown in Figure 1. The performed analyses and the literature data [55–58] allowed us to propose a possible mechanism for the thermal degradation of the siloxane-silsesquioxane resins, as shown in Scheme 1 and discussed below.

Thermal degradation of the non-functionalized S4SQ-H resin under nitrogen was found to be a two-step process (Figure 1a). In the first step, which started at about 200 °C, detachment of the siloxane chains from the silsesquioxane cages could take place (Scheme 1b). Then, small cyclic oligomers such as hexamethylcyclotrisiloxane (D$_3$) and other higher cyclic oligomers could be obtained (Scheme 1c) as a result of the split of the siloxane chains from the resin structure [55]. The small growth of the sample mass at 300 °C may be explained by the fallout of the silica derivatives from the gas phase oxidation [56]. The second step of the degradation process began at about 500 °C and was most probably associated with thermal decomposition of the silsesquioxane cages to small oligomers and then to the ceramic $Si_xC_yO_z$ structure (Scheme 1d) [55,57].

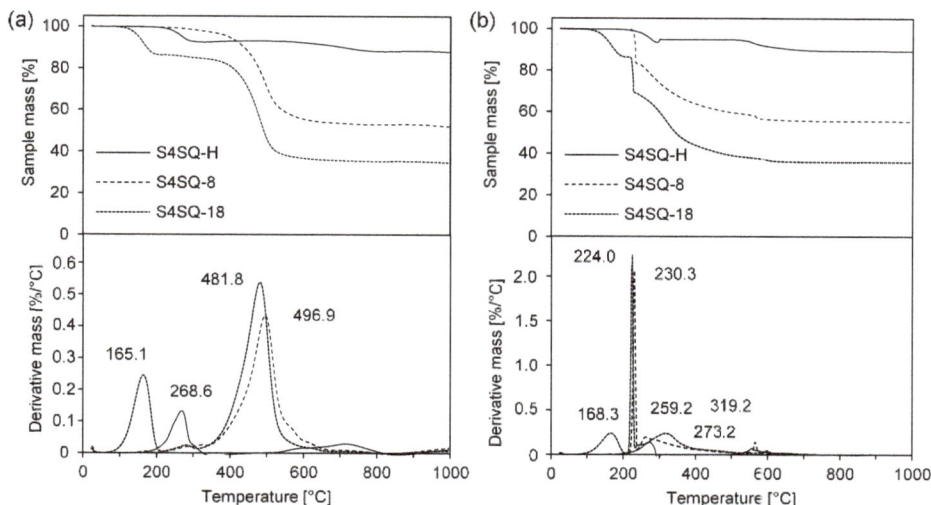

Figure 1. Thermogravimetry (TG)/derived thermogravimetry (DTG) curves of neat siloxane-silsesquioxane (S4SQ) resins (**a**) under nitrogen and (**b**) in air atmosphere.

The *n*-octadecyl-functionalized S4SQ-18 resin also degraded in a two-step process (Figure 1a). However, the first step of degradation of this compound started at much lower temperatures (~100 °C), which could be assigned to the split-off of the alkyl substituents from the siloxane groups (Scheme 1a) [58]. Then, at about 300 °C, the second step began, which might be attributed to the thermal decomposition of the resin structure to $Si_xC_yO_z$ (Scheme 1b–d) [55,57]. Interestingly, degradation of the S4SQ-8 resin which contained the *n*-octyl groups generally occurred in one step (Figure 1a). The initial weight loss started at about 300 °C and could be associated with the loss of the organic groups (Scheme 1a) and formation of the ceramic residue (Scheme 1b–d) at the same time. It should also be noted that the S4SQ-8 resin, which contains slightly shorter *n*-alkyl substituents, exhibited higher thermal stability than the S4SQ-18 compound.

Moreover, based on the amounts of silica residues at 1000 °C (S4SQ-H: 87.9%; S4SQ-8: 52.0%; S4SQ-18: 34.8%), it is possible to assess the approximate weight fraction of the organic substituents in the functionalized resins. It turns out that the *n*-octyl and *n*-octadecyl substituents constituted about 35% and 53% of the S4SQ-8 and S4SQ-18 resins respectively.

The thermo-oxidative degradation of the siloxane-silsesquioxane resins (Figure 1b) gave similar results to those obtained in the analyses performed under nitrogen (Figure 1a). However, the oxidizing action of the oxygen molecules accelerated the degradation processes [59–61], which started at much lower temperatures.

Degradation of the non-functionalized S4SQ-H resin in air atmosphere was also a two-step process (Figure 1b). The siloxane chains started to be detached from the silsesquioxane cages at about 220 °C (Scheme 1b), after which they could be transformed into the oligomers such as D_3 (Scheme 1c) [55]. The silsesquioxane cages, which are the most thermally stable part of the siloxane-silsesquioxane resin structure, were decomposed at about 500 °C, probably to the ceramic residue (Scheme 1d) [55,57]. Once again, a small growth of the sample mass was observed at about 300 °C, which could be assigned (1) to the thermal oxidation of Si-H bonds and formation of Si-O-Si bonds [62] and/or (2) to the fallout of the silica derivatives from the gas phase oxidation [56].

In the case of the siloxane-silsesquioxane resin functionalized with the *n*-octadecyl groups (S4SQ-18), its thermal degradation began at about 100 °C (Figure 1b). The explanation for this may be the split-off of the long *n*-alkyl chain substituents in the resin structure by peroxidation and subsequent fragmentation through classical radical pathways (Scheme 1a) [21,58,61]. Then, at about

220 °C, the organic groups were probably suddenly lost and thermal degradation of the resin structure began (Scheme 1b,c). Next, the sample mass decreased gradually while the temperature rose to 600 °C, at which the ceramic $Si_xC_yO_z$ residue could be formed (Scheme 1d) [55,57]. As regards thermal degradation of the S4SQ-8 resin, it started at about 220 °C with a sharp weight loss, which was assigned to the simultaneous sudden loss of the *n*-octyl groups (Scheme 1a) and formation of the ceramic residue (Scheme 1b–d).

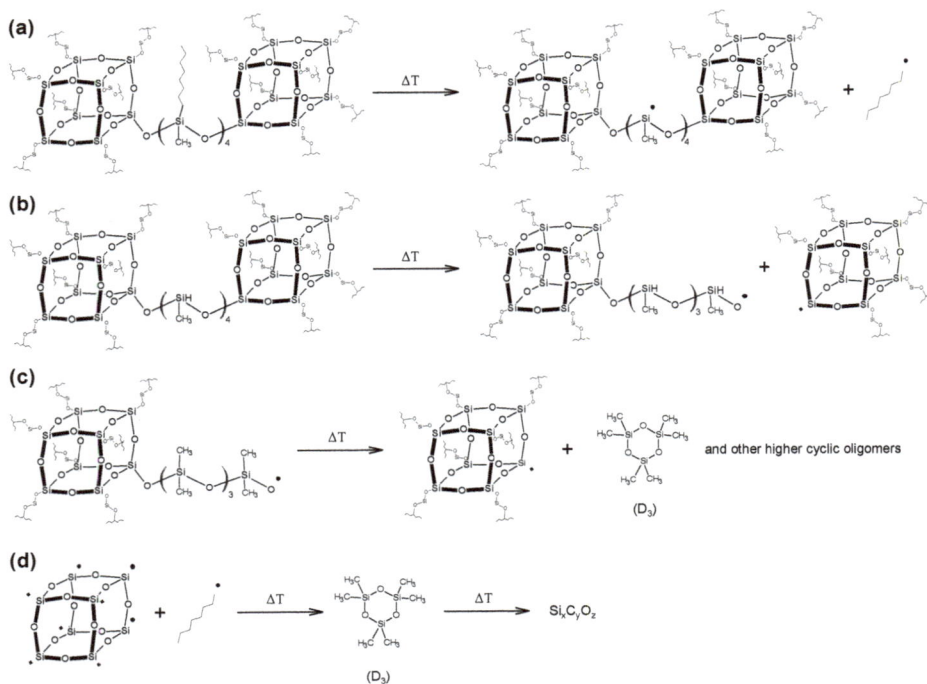

Scheme 1. A possible mechanism for the thermal degradation of siloxane-silsesquioxane resins followed by (**a**) split-off of the alkyl substituents from the siloxane groups, (**b**) detachment of the siloxane chains from the silsesquioxane cages, (**c**) split of the siloxane chains from the resin structure and (**d**) thermal decomposition of the silsesquioxane cages to small cyclic oligomers and then to the ceramic $Si_xC_yO_z$ structure.

The silica residues at 1000 °C (S4SQ-H: 89.3%; S4SQ-8: 55.4%; S4SQ-18: 35.9%) indicate that the approximate weight fractions of the alkyl groups in the S4SQ-8 and S4SQ-18 resins were 34% and 53%, respectively. These values are consistent with the amount of silica residues from the analysis performed under nitrogen. It may, hence, be concluded that the presence of oxygen accelerated thermal decomposition of the siloxane-silsesquioxane resins, especially the split-off of the alkyl substituents from the siloxane groups, but it did not influence the amounts and the composition of the residues.

Figure 2 shows the TG/DTG curves recorded under nitrogen for neat PP and PP composites filled with 1 and 10 wt % of the S4SQ fillers. All these materials presented similar one-step degradation shapes of the TG curves, which means that siloxane-silsesquioxane resins did not change the degradation mechanism of neat PP and that these fillers could only accelerate or delay the onset of this process (Figure 2).

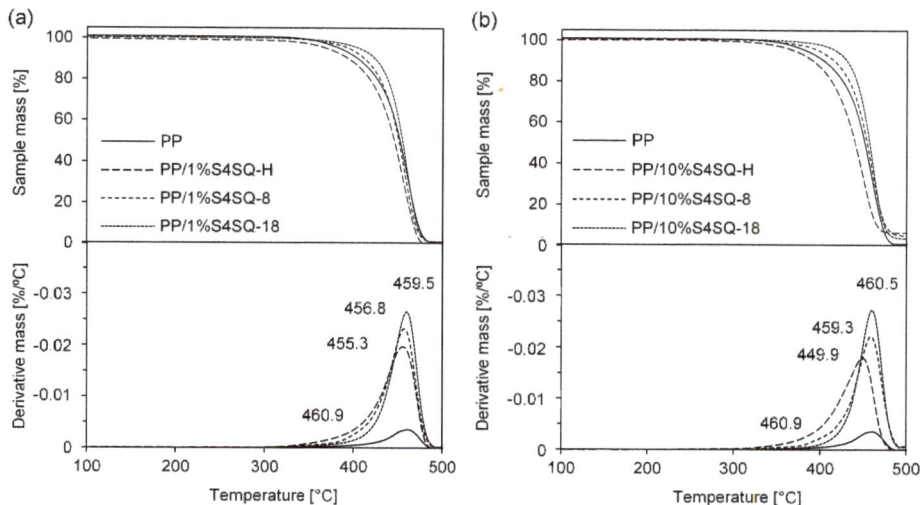

Figure 2. TG/DTG curves of neat PP as well as (**a**) PP/1%S4SQ and (**b**) PP/10%S4SQ composites (under nitrogen atmosphere).

The results of the TG analyses performed under nitrogen and in air atmosphere for neat PP and PP composites containing S4SQ fillers are shown in Table 1.

In the case of the studies carried out under nitrogen, the addition of the siloxane-silsesquioxane resins into the PP matrix caused a slight decrease of the T_{max} values in comparison with neat PP, irrespective of the kind of resin used (Figure 2, Table 1). The filler contents had a negligible influence on the values of this parameter.

Table 1. TG data of neat polypropylene (PP) and PP/S4SQ composites.

Sample	Nitrogen Atmosphere				Air Atmosphere			
	T_{max} (°C)	T_5 (°C)	T_{25} (°C)	T_{50} (°C)	T_{max} (°C)	T_5 (°C)	T_{25} (°C)	T_{50} (°C)
neat PP	460.9	383.7	432.6	451.5	366.9	289.1	325.4	351.8
PP/1%S4SQ-H	455.3	369.1	424.5	445.2	383.3	290.9	336.8	365.3
PP/5%S4SQ-H	-	-	-	-	382.5	292.0	336.5	365.5
PP/10%S4SQ-H	449.9	370.7	419.8	440.7	380.9	282.5	332.6	363.9
PP/1%S4SQ-8	456.8	394.1	434.2	450.0	380.6	291.6	326.8	355.7
PP/5%S4SQ-8	-	-	-	-	400.6	301.5	350.7	380.2
PP/10%S4SQ-8	459.3	398.8	439.0	454.8	388.3	297.8	345.5	372.6
PP/1%S4SQ-18	459.6	405.7	441.5	454.2	404.8	298.4	347.6	379.6
PP/5%S4SQ-18	-	-	-	-	409.0	298.0	346.0	376.7
PP/10%S4SQ-18	460.5	417.7	445.6	457.0	397.7	290.3	336.2	366.2

The T_5, T_{25} and T_{50} values were also found to decrease after the introduction of 1 wt % of S4SQ-H resin into polypropylene (Table 1). Increasing the wt % content of this filler in the material resulted in a slight decrease in the values of these parameters. However, in the presence of 1 wt % of S4SQ-8 and S4SQ-18 fillers, a significant increase in the T_5, T_{25} and T_{50} values was demonstrated (Table 1). Moreover, the values of these parameters increased with the increasing filler wt % content in the PP matrix. This could be explained by good compatibility between the PP chains and resins containing *n*-octyl and *n*-octadecyl substituents in their structures which contributed to a more uniform dispersion of S4SQ-8 and S4SQ-18 particles in the PP matrix. Similar observations were reported in our previous work [32] in the case of polypropylene composites containing POSS nanofillers with *n*-octyl and *n*-octadecyl groups attached to the silicon-oxygen cages.

As regards the results of the TG analyses carried out in air atmosphere, the T_{max} values were increased by 20–40 °C after incorporation of the siloxane-silsesquioxane resins to the PP matrix (Figure 3, Table 1). Increasing the content of these fillers in the materials above 5 wt % resulted in the slight lowering of T_{max}, which is probably associated with the formation of the filler aggregates in the composites. These results are in accordance with the literature data [15,32,33,63,64]. The highest improvement of the T_{max} parameter was observed for the samples containing resin functionalized with the *n*-octadecyl substituents (Table 1).

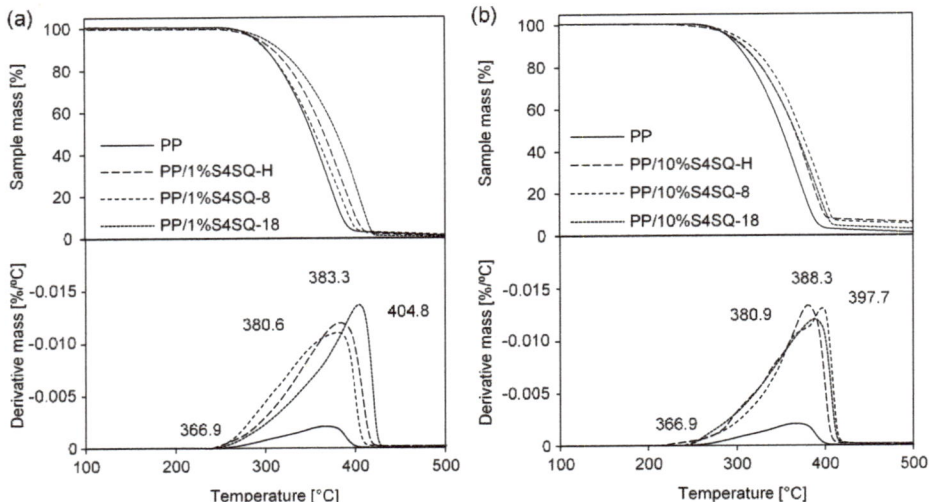

Figure 3. TG/DTG curves of neat PP as well as (a) PP/1%S4SQ and (b) PP/10%S4SQ composites (in air atmosphere).

In the case of the PP/S4SQ-H composites with 1 and 5 wt % filler contents, T_5 values increased by about 2–3 °C, and the T_{25} and T_{50} values by more than 10 °C, in comparison with neat PP (Table 1). However, any further increase of the S4SQ-H content (up to 10 wt %) worsened the thermal stability of the PP/S4SQ-H composites.

A considerable improvement in T_5, T_{25} and T_{50} was observed for composites with the functionalized S4SQ-8 and S4SQ-18 resins, in comparison with neat PP (Table 1), irrespective of the S4SQ filler content. However, it should be noted that the highest thermal stability for the PP/S4SQ-8 and PP/S4SQ-18 composites was observed for the samples containing 5 and 1 wt % of resins, respectively.

The high thermal stability of polypropylene composites with siloxane-silsesquioxane resins as fillers could be explained by the possible formation of a ceramic layer on the surface of the composite material during thermal degradation of S4SQ. As a result, the heat flux to the sample, oxygen diffusion towards the bulk material, and the evolution of the volatile degradation gases from the sample were significantly limited, as was also demonstrated for POSS-containing composites based on polypropylene, polyethylene and polystyrene matrices [4,16,63].

Moreover, some correlation between the thermal stability of the composites and S4SQ filler dispersion in the polymer matrix was observed. The filler aggregates may cause some loosening of the polymer chain packing, as well as some increase in the free volumes which would facilitate the access of oxygen to the material. Thus, uniform dispersion of the siloxane-silsesquioxane resins in PP, which was achieved at lower filler loadings, led to the improved thermal stability of the obtained materials. However, as the siloxane-silsesquioxane resins tend to aggregate at higher wt % contents, the thermal stability of the composites was decreased.

3.2. TG-FTIR Studies

In order to evaluate comprehensively the influence of the siloxane-silsesquioxane resins on the thermal stability of PP/S4SQ composites, the gaseous pyrolysis products of neat PP and composites containing 1 wt % of filler were analyzed by the TG-FTIR method. The detailed characteristics of each TG-FTIR spectrum obtained at different temperatures are shown in Figure 4.

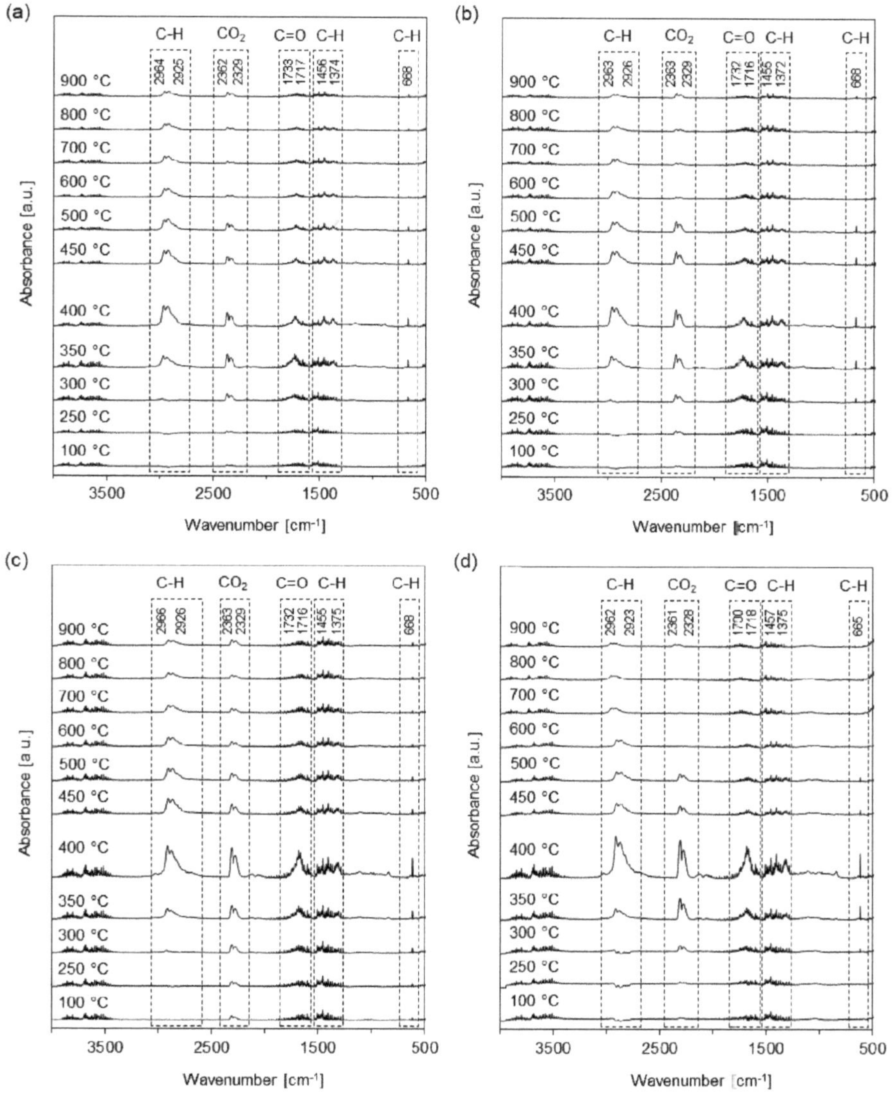

Figure 4. 2D thermogravimetry-Fourier transform infrared spectrometry (TG-FTIR) spectra of pyrolysis products of (**a**) neat PP, (**b**) PP/1%S4SQ-H, (**c**) PP/1%S4SQ-8 and (**d**) PP/1%S4SQ-18 at various temperatures.

Thermal decomposition of both neat PP (Figure 4a) and PP/S4SQ composites (Figure 4b–d) started at about 300 °C. The main degradation products were alkenes, dienes and alkanes, as the absorbances at about 2964 (assigned to the stretching, asymmetric vibrations of -CH$_3$ groups), 2925 (stretching, asymmetric vibrations of -CH$_2$- groups), 1456 (deformation, asymmetric vibrations of -CH$_3$ groups), 1374 (deformation, asymmetric vibrations of -CH$_3$ groups) and 668 cm^{-1} (out of plane deformation vibrations of =CH- groups) were clearly visible at 300–600 °C. Moreover, sharp signals in the range of 1600–1800 cm^{-1} may be attributed to carbonyl-containing compounds such as aldehydes (1733 cm^{-1}), ketones (1717 cm^{-1}) and carboxylic acids (1705 cm^{-1}). Furthermore, the release of nonflammable gases such as H$_2$O, CO$_2$ and CO could also be demonstrated by the signals at 3400–4000, 2300–2390 and 2080–2180 cm^{-1}, respectively.

No additional signals which might be associated with thermal degradation of siloxane-silsesquioxane resins were observed in the spectra of the PP/S4SQ composites (Figure 4b–d), in comparison with the spectrum of neat PP (Figure 4a). It should be noted that the main differences between the spectra of neat PP and the PP/S4SQ composites were visible only at the intensities of the absorbance peaks. A greater increase in the intensity of the peaks associated with the hydrocarbons (in the range of 2800–3000 cm^{-1}), which are the main products of PP decomposition, was found in the case of the PP/1%S4SQ-8 (Figure 4c) and PP/1%S4SQ 18 (Figure 4d) composites. Figure 5 presents the evolution curves of hydrocarbons from the analyzed samples, which are shown as the surface area of the peaks in the range of 2800–3000 cm^{-1} versus temperature.

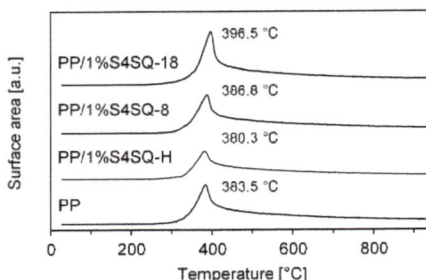

Figure 5. The evolution curves of the hydrocarbons (in the range of 2800–3000 cm^{-1}) evolved from neat PP and PP/S4SQ composites vs. temperature.

The initial release of hydrocarbon gases for neat PP and PP/S4SQ composites occurred at almost the same temperature (at about 260 °C) (Figure 5). However, the maximum of the signal for the PP/1%S4SQ-18 composite appeared at 396.5 °C and it was about 10 °C higher than that for neat PP (383.5 °C), as well as for PP/1%S4SQ-H (380.3 °C) and for PP/1%S4SQ-8 (386.8 °C). Moreover, a greater increase in the intensity of the hydrocarbons in the case of the materials with the functionalized S4SQ-8 and S4SQ-18 resins was probably caused by the presence in the gaseous products of the pyrolysis of both the PP matrix and *n*-alkyl substituents which were split off from the siloxane groups of the filler molecules.

On the basis of the performed analysis, the possible role of the siloxane-silsesquioxane resins during the PP matrix decomposition may be explained. Firstly, decomposition of all the analyzed materials started at almost the same time because the release of the gaseous degradation products occurred at similar temperatures. Thus, it might be assumed that the applied siloxane-silsesquioxane resins did not significantly delay the thermal decomposition of polypropylene. Moreover, it seems that the S4SQ resins do not quench the free radicals generated during decomposition of the PP matrix. Secondly, the highest growth of the intensity of the hydrocarbons occurred in the PP/1%S4SQ-8 and PP/1%S4SQ-18 composites proved that the *n*-alkyl substituents in the siloxane groups of the filler particles did not improve the thermal stability of those composites. However, those groups had a crucial role in the uniform dispersion of the siloxane-silsesquioxane resins in the PP matrix, as it is shown in Scheme 2. Although they are detached from the filler particles at higher temperatures, they provide the formation of a continuous thin ceramic layer on the specimen surface during its thermal decomposition or combustion (Scheme 2b). This layer represents an effective physical barrier which slows down the decomposition of the polymer matrix. On the other hand, non-functionalized siloxane-silsesquioxane resin forms a non-continuous ceramic layer on the sample surface during the composite's decomposition, which is not an effective barrier (Scheme 2a). The formation of the solid ceramic layer on the sample surface could also be proved by the lack of any signals associated with gaseous products of the thermal degradation of the siloxane-silsesquioxane resins (besides the alkyl substituents).

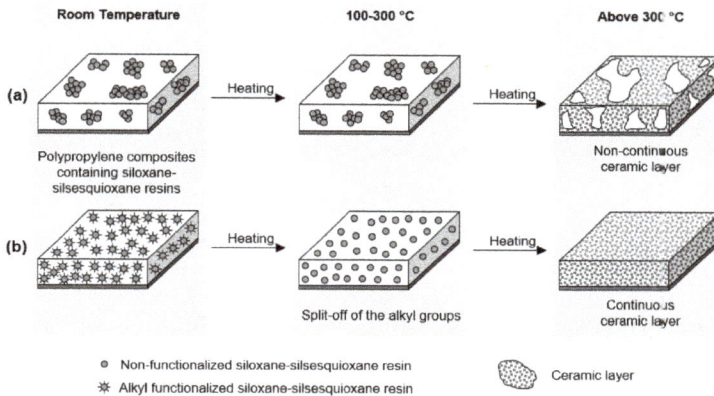

Scheme 2. The possible roles of (**a**) non-functionalized and (**b**) *n*-alkyl-functionalized siloxane-silsesquioxane resins during thermal degradation of PP/S4SQ composites.

3.3. Fire Behavior

Neat PP and the PP/S4SQ composite samples were subjected to cone calorimeter investigations which gave comprehensive information on their performance in a rather well-defined fire test scenario. The main parameters obtained from the cone calorimeter measurements, such as time to ignition (TTI), heat release rate (HRR), fire performance index (FPI), total heat released (THR), effective heat of combustion (EHC), maximum average rate of heat emission (MAHRE) and mass loss during combustion, are reported in Table 2.

Table 2. Main parameters obtained from cone calorimeter measurements for neat PP and PP/S4SQ composites.

Sample	TTI(s)	PkHRR (kW/m^2)	FPI $(m^2 \cdot s/kW)$	MAHRE (kW/m^2)	THR (MJ/m^2)	EHC (MJ/kg)	Mass Loss (g)
PP	40	364	0.110	154	40	76	15.6
PP/1%S4SQ-H	41	354	0.116	185	47	74	16.7
PP/5%S4SQ-H	21	500	0.042	223	44	77	15.9
PP/10%S4SQ-H	19	445	0.043	214	44	78	16.6
PP/1%S4SQ-8	43	227	0.190	115	29	59	14.3
PP/5%S4SQ-8	40	481	0.083	215	48	75	17.0
PP/10%S4SQ-8	-	-	-	-	-	-	-
PP/1%S4SQ-18	40	168	0.247	68	22	66	12.3
PP/5%S4SQ-18	43	328	0.131	160	42	78	15.6
PP/10%S4SQ-18	47	391	0.120	183	47	51	16.2

The addition of just 1 wt % of S4SQ-H resin into the PP matrix did not change the time to ignition (TTI) value in comparison with neat PP (Figure 6). However, increasing the content of the filler in the material resulted in significantly lowered TTI values. On the other hand, the values of that parameter for the PP/S4SQ-8 and PP/S4SQ-18 composites were equal or even higher with respect to neat PP. Thus, it could be concluded that the addition of the S4SQ-H filler into the PP matrix accelerated the combustion process, while the S4SQ-8 and S4SQ-18 particles slowed it down.

The peak of the heat release rate (PkHRR) is a characteristic that displays a very strong and quite complex dependence on the fire scenario [65]. A very sharp HRR peak of neat PP appears at the range of 80–100 s, with a PkHRR of 364 kW/m^2 (Figure 6). The addition of just 1 wt % of the S4SQ resins into the PP matrix caused a significant reduction of PkHRR in comparison with neat PP. It should be noted that all the PP/1%S4SQ composites showed a peak in the heat release rate after the initial strong increase in the heat release rate, as shown in Figure 6, which proves the formation of a protective surface layer [66]. It may be assumed that a char layer is formed during the combustion process on the sample surface and siloxane-silsesquioxane particles undergo a series of oxidation reactions to form $Si_xC_yO_z$ residue. This enhanced char structure protects and insulates the underlying polymer from any further degradation [3]. Similar results were observed by other authors in the case of composites containing POSS fillers [4,66]. Interestingly, increasing the amount of the applied S4SQ fillers in the materials generally resulted in the growth of the PkHRR values. The explanation of this effect may be an aggregation of the siloxane-silsesquioxane resin particles in the polymer which caused the formation of a non-continuous thin ceramic layer on the specimen surface during combustion. This layer does not provide an effective physical barrier for the combustion processes (Scheme 2). Furthermore, a greater deteriorating effect on the flammability properties of PP was found in the case of the composites with S4SQ-8 and S4SQ-18 resins in comparison with S4SQ-H. It may be concluded that the presence of the long *n*-alkyl chain substituents in the structure of the siloxane-silsesquioxane resins had a favorable effect on the filler dispersion in the PP matrix and consequently on the limitation of composite combustion.

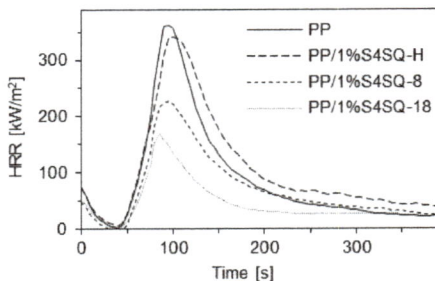

Figure 6. Dynamic curves of HRR versus time for neat PP and PP/S4SQ composites.

The fire performance index (FPI) is used to predict whether a material can easily develop drastic combustion after ignition. FPI is defined as the ratio of TTI to PkHRR, and it indicates the fire resistance of the analyzed material [67]. It can be seen from Table 2 that the PP/1%S4SQ-8 and PP/1%S4SQ-18 composites have the highest FPI values among the all the analyzed samples; hence, they are characterized by the best fire resistance. On the other hand, in the case of the PP/5%S4SQ-H and PP/10%S4SQ-H materials, the FPI values were more than two times lower than for neat PP, which indicates their higher susceptibility to combustion.

The maximum average rate of heat emission (MAHRE) is a good measure of the propensity for fire development under real scale conditions [68]. MAHRE for the PP/1%S4SQ-8 and PP/1%S4SQ-18 composites shows a significant reduction by 25% and 56%, respectively, in comparison with neat PP (Figure 7a). However, increasing the wt % content of S4SQ-8 and S4SQ-18 fillers in the PP matrix or the addition of S4SQ-H filler resulted in a clear growth of the MAHRE values. Therefore, small amounts of the functionalized siloxane-silsesquioxane resins have favorable effects on flame retardancy in the PP/S4SQ composites, whereas a high wt % content of these fillers deteriorates the flame-retardant properties of the obtained materials.

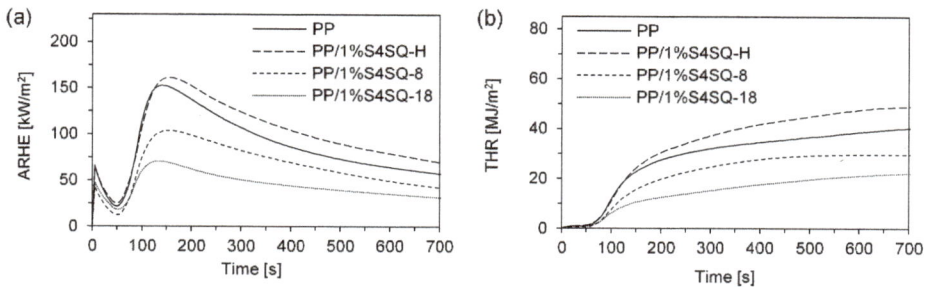

Figure 7. Heat evolving during combustion for neat PP and PP/S4SQ composites: (**a**) average rate of heat emission (AHRE) and (**b**) total heat released (THR).

Furthermore, the amount of the non-combustible fraction in the materials may be determined on the basis of the total heat release (THR) values. Once again, a decrease in THR was observed in the case of the PP/1%S4SQ-8 and PP/1%S4SQ-18 composites compared with neat PP (Figure 7b). A lower THR value may indicate that a part of the polymer is not completely combusted but that it could undergo a carbonization process. On the other hand, in the case of the PP/S4SQ-H composites, as well as the PP/S4SQ-8 and PP/S4SQ-18 with 5 and 10 wt % filler contents, the THR values were higher in comparison with neat PP.

The values of the effective heat of combustion (EHC) of all the analyzed PP/S4SQ composites were equal to or lower than that for neat PP. Hence, the combustion kinetics of these materials were affected by the presence of the siloxane-silsesquioxane resins. This is generally in agreement with the changes in the PkHRR values.

The mass loss values were found to be reduced for the PP/1%S4SQ-8 and PP/1%S4SQ-18 composites in comparison with neat PP (Table 2). The changes in this parameter for the other PP/S4SQ composites were not significant. Moreover, the addition of 1 wt % of the S4SQ-8 and S4SQ-18 resins into the PP matrix resulted in a decrease in the mass loss rate, as shown in Figure 8. The lowest rate of combustion may be attributed to the homogeneous dispersion of these fillers in the matrix. However, increasing the wt % content of the filler in the materials resulted in limiting this effect.

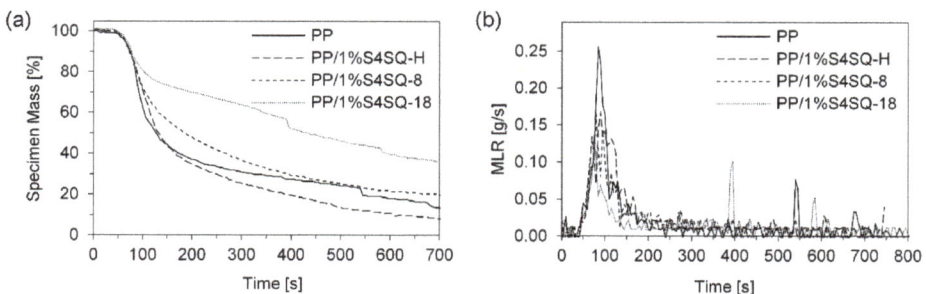

Figure 8. Mass curves during combustion for neat PP and PP/S4SQ composites: (**a**) sample mass (normalized on initial weight) and (**b**) mass loss rate (MLR).

4. Conclusions

The influence of a novel group of silsesquioxane derivatives, siloxane-silsesquioxane resins, on the thermal stability and combustion process of polypropylene composites was examined for the first time. The thermogravimetric studies of neat, non-functionalized, *n*-octyl- and *n*-octadecyl-functionalized S4SQ resins revealed complicated processes in their decomposition which were strongly dependent on the structures of the resins. Based on these results, a possible mechanism for the thermal degradation of the siloxane-silsesquioxane resins was proposed. The TG analyses of the PP materials demonstrated that the S4SQ resins may improve the thermal stability of the composites, especially when they are homogeneously dispersed in the PP matrix. The role of the siloxane-silsesquioxane resins during the thermal decomposition of PP/S4SQ composites was comprehensively evaluated using TG-FTIR. It was found that the presence of the alkyl substituents in the resin structure improved the material's homogeneity, but they were split off at higher temperatures. However, the obtained uniform dispersion of the filler particles in the polymer matrix resulted in the formation of a continuous ceramic layer on the sample surface. Finally, the possible flame retardancy effect of the siloxane-silsesquioxane resins on the PP matrix was examined by cone calorimeter tests. It turns out that small amounts of *n*-alkyl-functionalized siloxane-silsesquioxane resins effectively slow down the combustion process of the material. The values of the PkHRR, MAHRE, THR and EHC parameters were significantly decreased after addition of just 1 wt % of S4SQ-8 and S4SQ-18 filler particles into the PP.

Author Contributions: A.N., K.D. and K.C. conceived and designed the experiments; M.S. synthesized siloxane-silsesquioxane resins and performed their thermal analysis; A.N. prepared composite materials; B.S.-M. performed thermal analysis of composites; J.L. carried out thermogravimetry-Fourier transform infrared spectrometry studies; J.C.-P. and R.O. performed the cone calorimeter tests; A.N. analyzed the data; A.N., K.D. and K.C. wrote the paper.

Funding: This research received no external funding.

Conflicts of Interest: The authors declare no conflict of interest.

References

1. Zhang, W.; Camino, G.; Yang, R. Polymer/polyhedral oligomeric silsesquioxane (POSS) nanocomposites: An overview of fire retardance. *Prog. Polym. Sci.* **2017**, *67*, 77–125. [CrossRef]
2. Georlette, P.; Simons, J.; Costa, L. Halogen-containing fire-retardant compounds. In *Fire Retardancy of Polymeric Materials*; Grand, A.F., Wilkie, C.A., Eds.; Marcel Dekker Inc.: New York, NY, USA, 2000; pp. 245–284, ISBN 0-8247-8879-6.
3. Awad, W.H. Recent Developments in Silicon-Based Flame Retardants. In *Fire Retardancy of Polymeric Materials*; Wilkie, C.A., Morgan, A.B., Eds.; Taylor and Francis Group, LLC: New York, NY, USA, 2010; pp. 187–206, ISBN 978-1-4200-8399-6.
4. Fina, A.; Abbenhuis, H.C.L.; Tabuani, D.; Camino, G. Metal functionalized POSS as fire retardants in polypropylene. *Polym. Degrad. Stab.* **2006**, *91*, 2275–2281. [CrossRef]
5. Blanco, I.; Abate, L.; Bottino, F.A.; Bottino, P. Synthesis, characterization and thermal stability of new dumbbell-shaped isobutyl-substituted POSSs linked by aromatic bridges. *J. Therm. Anal. Calorim.* **2014**, *117*, 243–250. [CrossRef]
6. Yu, L.; Zhou, S.; Zou, H.; Liang, M. Thermal Stability and Ablation Properties Study of Aluminum Silicate Ceramic Fiber and Acicular Wollastonite Filled Silicone Rubber Composite. *J. Appl. Polym. Sci.* **2014**, *131*, 39700. [CrossRef]
7. Hao, N.; Bohning, M.; Schonhals, A. Dielectric Properties of Nanocomposites Based on Polystyrene and Polyhedral Oligomeric Phenethyl-Silsesquioxanes. *Macromolecules* **2007**, *40*, 9672–9679. [CrossRef]
8. Carroll, J.B.; Waddon, A.J.; Nakade, H.; Rotello, V.M. "Plug and Play" Polymers. Thermal and X-ray Characterizations of Noncovalently Grafted Polyhedral Oligomeric Silsesquioxane (POSS)-Polystyrene Nanocomposites. *Macromolecules* **2003**, *36*, 6289–6291. [CrossRef]

9. Li, B.; Zhang, Y.; Wang, S.; Ji, J. Effect of POSS on morphology and properties of poly(2,6-dimethyl-1,4-phenylene oxide)/polyamide 6 blends. *Eur. Polym. J.* **2009**, *45*, 2202–2210. [CrossRef]

10. Jeziórska, R.; Świerz-Motysia, B.; Szadkowska, A.; Marciniec, B.; Maciejewski, H.; Dutkiewicz, M.; Leszczyńska, I. Effect of POSS on morphology, thermal and mechanical properties of polyamide 6. *Polimery* **2011**, *56*, 809–816.

11. Fu, B.X.; Hsiao, B.S.; Pagola, S.; Stephens, P.; White, H.; Rafailovich, M.; Sokolov, J.; Mather, P.T.; Jeon, H.G.; Phillips, S.; et al. Structural development during deformation of polyurethane containing polyhedral oligomeric silsesquioxanes (POSS) molecules. *Polymer* **2001**, *42*, 599–611. [CrossRef]

12. Nanda, A.K.; Wicks, D.A.; Madbouly, S.A.; Otaigbe, J.U. Nanostructured polyurethane/POSS hybrid aqueous dispersions prepared by homogeneous solution polymerization. *Macromolecules* **2006**, *39*, 7037–7043. [CrossRef]

13. Madbouly, S.A.; Otaigbe, J.U. Recent advances in synthesis, characterization and rheological properties of polyurethanes and POSS/polyurethane nanocomposites dispersions and films. *Prog. Polym. Sci.* **2009**, *34*, 1283–1332. [CrossRef]

14. Blanco, I. The Rediscovery of POSS: A Molecule Rather than a Filler. *Polymers* **2018**, *10*, 904. [CrossRef]

15. Fina, A.; Tabuani, D.; Frache, A.; Camino, G. Polypropylene–polyhedral oligomeric silsesquioxanes (POSS) nanocomposites. *Polymer* **2005**, *46*, 7855–7866. [CrossRef]

16. Fina, A.; Abbenhuis, H.C.L.; Tabuani, D.; Frache, A.; Camino, G. Polypropylene metal functionalised POSS nanocomposites: A study by thermogravimetric analysis. *Polym. Degrad. Stab.* **2006**, *91*, 1064–1070. [CrossRef]

17. Zhou, Z.; Cui, L.; Zhang, Y.; Zhang, Y.; Yin, N. Isothermal Crystallization Kinetics of Polypropylene/POSS Composites. *J. Polym. Sci. Polym. Phys.* **2008**, *46*, 1762–1772. [CrossRef]

18. Frone, A.N.; Perrin, F.X.; Radovici, C.; Panaitescu, D.M. Influence of branched or un-branched alkyl substitutes of POSS on morphology, thermal and mechanical properties of polyethylene. *Compos. Part B Eng.* **2013**, *50*, 98–106. [CrossRef]

19. Joshi, M.; Butola, B.S.; Simon, G.; Kukaleva, N. Rheological and Viscoelastic Behavior of HDPE/Octamethyl-POSS Nanocomposites. *Macromolecules* **2006**, *39*, 1839–1849. [CrossRef]

20. Joshi, M.; Butola, B.S. Isothermal Crystallization of HDPE/Octamethyl Polyhedral Oligomeric Silsesquioxane Nanocomposites: Role of POSS as a Nanofiller. *J. Appl. Polym. Sci.* **2007**, *105*, 978–985. [CrossRef]

21. Perrin, F.X.; Panaitescu, D.M.; Frone, F.N.; Radovici, C.; Nicolae, C. The influence of alkyl substituents of POSS in polyethylene nanocomposites. *Polymer* **2013**, *54*, 2347–2354. [CrossRef]

22. Barczewski, M.; Sterzyński, T.; Dutkiewicz, M. Thermo-rheological properties and miscibility of linear low-density polyethylene-silsesquioxane nanocomposites. *J. Appl. Polym. Sci.* **2015**, *132*, 42825. [CrossRef]

23. Niemczyk, A.; Dziubek, K.; Czaja, K.; Szatanik, R.; Szołyga, M.; Dutkiewicz, M.; Marciniec, B. Polypropylene/polyhedral oligomeric silsesquioxane nanocomposites—Study of free volumes, crystallinity degree and mass flow rate. *Polimery* **2016**, *61*, 610–615. [CrossRef]

24. Niemczyk, A.; Adamczyk-Tomiak, K.; Dziubek, K.; Czaja, K.; Rabiej, S.; Szatanik, R.; Dutkiewicz, M. Study of polyethylene nanocomposites with polyhedral oligomeric silsesquioxane nanofillers—From structural characteristics to mechanical properties and processability. *Polym. Compos.* **2017**. [CrossRef]

25. Niemczyk, A.; Dziubek, K.; Czaja, K.; Frączek, D.; Piasecki, R.; Adamczyk-Tomiak, K.; Rabiej, S.; Dutkiewicz, M. Study and evaluation of dispersion of polyhedral oligomeric silsesquioxane and silica filler in polypropylene composites. *Polym. Compos.* **2018**. [CrossRef]

26. Baldi, F.; Bignotti, F.; Fina, A.; Tabuani, D.; Ricco, T. Mechanical Characterization of Polyhedral Oligomeric Silsesquioxane/Polypropylene Blends. *J. Appl. Polym. Sci.* **2007**, *105*, 935–943. [CrossRef]

27. Choi, J.-H.; Jung, C.-H.; Kim, D.-K.; Ganesan, R. Radiation-induced grafting of inorganic particles onto polymer backbone: A new method to design polymer-based nanocomposite. *Nucl. Instrum. Methods B* **2008**, *266*, 203–206. [CrossRef]

28. Fina, A.; Tabuani, D.; Camino, G. Polypropylene–polysilsesquioxane blends. *Eur. Polym. J.* **2010**, *46*, 14–23. [CrossRef]

29. Chen, J.-H.; Chiou, Y.-D. Crystallization Behavior and Morphological Development of Isotactic Polypropylene Blended with Nanostructured Polyhedral Oligomeric Silsesquioxane Molecules. *J. Polym. Sci. Polym. Phys.* **2006**, *44*, 2122–2134. [CrossRef]

30. Fina, A.; Tabuani, D.; Peijs, T.; Camino, G. POSS grafting on PPgMA by one-step reactive blending. *Polymer* **2009**, *50*, 218–226. [CrossRef]

31. Hato, M.J.; Ray, S.S.; Luyt, A.S. Nanocomposites Based on Polyethylene and Polyhedral Oligomeric Silsesquioxanes, 1—Microstructure, Thermal and Thermomechanical Properties Macromol. *Mater. Eng.* **2008**, *293*, 752–762. [CrossRef]

32. Niemczyk, A.; Dziubek, K.; Sacher-Majewska, B.; Czaja, K.; Dutkiewicz, M.; Marciniec, B. Study of thermal properties of polyethylene and polypropylene nanocomposites with long alkyl chain substituted POSS fillers. *J. Therm. Anal. Calorim.* **2016**, *125*, 1287–1299. [CrossRef]

33. Bouza, R.; Barral, L.; Diez, F.J.; Lopez, J.; Montero, B.; Rico, M.; Ramirez, C. Study of thermal and morphological properties of a hybrid system, iPP/POSS. Effect of flame retardance. *Compos. Part B Eng.* **2014**, *58*, 566–572. [CrossRef]

34. Carniato, F.; Fina, A.; Tabuani, D.; Boccaleri, E. Polypropylene containing Ti- and Al-polyhedral oligomeric silsesquioxanes: Crystallization process and thermal properties. *Nanotechnology* **2008**, *19*, 475701. [CrossRef] [PubMed]

35. Butola, B.S.; Joshi, M.; Kumar, S. Hybrid Organic-Inorganic POSS (Polyhedral Oligomeric silsesquioxane)/Polypropylene Nanocomposite Filaments. *Fibers Polym.* **2010**, *11*, 1137–1145. [CrossRef]

36. Carniato, F.; Boccaleri, E.; Marchese, L.; Fina, A.; Tabuani, D.; Camino, G. Synthesis and Characterisation of Metal Isobutylsilsesquioxanes and Their Role as Inorganic–Organic Nanoadditives for Enhancing Polymer Thermal Stability. *Eur. J. Inorg. Chem.* **2007**, *2007*, 585–591. [CrossRef]

37. Blanco, I.; Bottino, F.A.; Cicala, G.; Latteri, A.; Recca, A. A kinetic study of the thermal and thermal oxidative degradations of new bridged POSS/PS nanocomposites. *Polym. Degrad. Stab.* **2013**, *98*, 2564–2570. [CrossRef]

38. Dobrzyńska-Mizera, M.; Dutkiewicz, M.; Sterzyński, T.; Di Lorenzo, M.L. Polypropylene-based composites containing sorbitol-based nucleating agent and siloxane-silsesquioxane resin. *J. Appl. Polym. Sci.* **2016**, *133*, 43476. [CrossRef]

39. Dobrzyńska-Mizera, M.; Dutkiewicz, M.; Sterzyński, T.; Di Lorenzo, M.L. Isotactic polypropylene modified with sorbitol-based derivative and siloxane-silsesquioxane resin. *Eur. Polym. J.* **2016**, *85*, 62–71. [CrossRef]

40. Dobrzyńska-Mizera, M.; Sterzyński, T. Interfacial enhancement of polypropylene composites modified with sorbitol derivatives and siloxane-silsesquioxane resin. *AIP Conf. Proc.* **2015**, *1695*, 020049. [CrossRef]

41. Maciejewski, H.; Dutkiewicz, M.; Byczyński, Ł.; Marciniec, B. Silseskwioksany jako nanonapełniacze. Cz. I. Nanokompozyty z osnową silikonową. *Polimery* **2012**, *57*, 535–544. [CrossRef]

42. Lichtenhan, J.D.; Vu, N.Q.; Carter, J.A. Silsesquioxane-Siloxane Copolymers from Polyhedral Silsesquioxanes. *Macromolecules* **1993**, *26*, 2141–2142. [CrossRef]

43. Liu, Y.R.; Huang, Y.D.; Liu, L. Thermal stability of POSS/methylsilicone nanocomposites. *Compos. Sci. Technol.* **2007**, *67*, 2864–2876. [CrossRef]

44. Gunji, T.; Shioda, T.; Tsuchihira, K.; Seki, H. Preparation and properties of polyhedraloligomeric silsesquioxane–polysiloxane copolymers. *Appl. Organomet. Chem.* **2010**, *24*, 545–550. [CrossRef]

45. Handke, M.; Kowalewska, A. Siloxane and silsesquioxane molecules—Precursors for silicate materials. *Spectrochim. Acta A* **2011**, *79*, 749–757. [CrossRef] [PubMed]

46. Seino, M.; Hayakawa, T.; Ishida, Y.; Kakimoto, M. Synthesis and Characterization of Crystalline Hyperbranched Polysiloxysilane with POSS Groups at the Terminal Position. *Macromolecules* **2006**, *39*, 8892–8894. [CrossRef]

47. Pan, G.; Mark, J.E.; Schaefer, D.W. Synthesis and characterization of fillers of controlled structure based on polyhedral oligomeric silsesquioxane cages and their use in reinforcing siloxane elastomers. *J. Polym. Sci. Polym. Phys.* **2003**, *41*, 3314–3323. [CrossRef]

48. Mantz, R.A.; Jones, P.F.; Chaffee, K.P.; Lichtenhan, J.D.; Gilman, J.W.; Ismail, I.M.K.; Burmeister, M.J. Thermolysis of Polyhedral Oligomeric Silsesquioxane (POSS) Macromers and POSS–Siloxane Copolymers. *Chem. Mater.* **1996**, *8*, 1250–1259. [CrossRef]

49. Ryu, H.S.; Kim, D.G.; Lee, J.C. Polysiloxanes containing polyhedral oligomeric silsesquioxane groups in the side chains; synthesis and properties. *Polymer* **2010**, *51*, 2296–2304. [CrossRef]

50. Zielecka, M.; Bujanowska, E.; Cyruchin, K.; Marciniec, B.; Maciejewski, H.; Dutkiewicz, M. Polysiloxane Composite with Improved Thermal Resistance. PL Patent 393092 A1, 31 May 2013.

51. Szołyga, M.; Dutkiewicz, M.; Maciejewski, H.; Marciniec, B. Siloxane-Silsesquioxane Resins and Method for Obtaining Them. PL Patent 405584 A1, 29 July 2016.

52. Szołyga, M.; Dutkiewicz, M.; Maciejewski, H.; Marciniec, B. Siloxane-Silsesquioxane Resins and Method for Obtaining Them. PL Patent 405587 A1, 29 July 2016.
53. Szołyga, M.; Dutkiewicz, M.; Marciniec, B.; Maciejewski, H. Synthesis of reactive siloxane-silsesquioxane resin. *Polimery* **2013**, *58*, 766–771. [CrossRef]
54. Niemczyk, A.; Czaja, K.; Dziubek, K.; Szołyga, M.; Rabiej, S.; Szatanik, R. Polypropylene composites with functionalized siloxane-silsesquioxane resins as fillers—Comprehensive study of the morphological, structural, thermal and mechanical properties. *Polym. Compos.* **2018**, accepted.
55. Liu, Y.; Shi, Y.; Zhang, D.; Li, J.; Huang, G. Preparation and thermal degradation behavior of room temperature vulcanized silicone rubber-g-polyhedral oligomeric silsesquioxanes. *Polymer* **2013**, *54*, 6140–6149. [CrossRef]
56. Camino, G.; Lomakin, S.M.; Lazzari, M. Polydimethylsiloxane thermal degradation Part 1. Kinetic aspects. *Polymer* **2001**, *42*, 2395–2402. [CrossRef]
57. Colombo, P.; Mera, G.; Riedel, R.; Soraru, G.D. Polymer-derived ceramics: 40 years of research and innovation in advanced ceramics. *J. Am. Ceram. Soc.* **2010**, *93*, 1805–1837. [CrossRef]
58. Tomer, N.S.; Delor-Jestin, F.; Frezet, L.; Lacoste, J. Oxidation, Chain Scission and Cross-Linking Studies of Polysiloxanes upon Ageings. *Open J. Org. Polym. Mater.* **2012**, *2*, 13–22. [CrossRef]
59. Grassie, N.; Scott, G. *Polymer Degradation and Stabilization*; Cambridge University Press: Cambridge, UK, 1985; ISBN 0-521-24961-9.
60. Allen, N.S.; Edge, M. *Fundamentals of Polymer Degradation and Stabilization*; Elsevier Applied Science: London, UK, 1992; pp. 1–21, ISBN 978-1-85166-773-4.
61. Fina, A.; Tabuani, D.; Carniato, F.; Frache, A.; Boccaleri, E.; Camino, G. Polyhedral oligomeric silsesquioxanes (POSS) thermal degradation. *Termochim. Acta* **2006**, *440*, 36–42. [CrossRef]
62. Haruvy, Y.; Webber, S.E. Supported Sol-Gel Thin-Film Glasses Embodying Laser Dyes. 3. Optically Clear SiO$_2$ Glass Thin Films Prepared by the Fast Sol-Gel Method. *Chem. Mater.* **1992**, *4*, 89–94. [CrossRef]
63. Blanco, I.; Bottino, F.A.; Cicala, G.; Latteri, A.; Recca, A. Synthesis and Characterization of Differently Substituted Phenyl Hepta Isobutyl-Polyhedral Oligomeric Silsesquioxane/Polystyrene Nanocomposites. *Polym. Compos.* **2014**, *35*, 151–157. [CrossRef]
64. Palza, H.; Vergara, R.; Zapata, P. Improving the Thermal Behavior of Poly(propylene) by Addition of Spherical Silica Nanoparticles. *Macromol. Mater. Eng.* **2010**, *295*, 899–905. [CrossRef]
65. Schartel, B.; Bartholmai, M.; Knoll, U. Some comments on the use of cone calorimeter data. *Polym. Degrad. Stab.* **2005**, *88*, 540–547. [CrossRef]
66. Song, L.; He, Q.; Hu, Y.; Chen, H.; Liu, L. Study on thermal degradation and combustion behaviors of PC/POSS hybrids. *Polym. Degrad. Stab.* **2008**, *93*, 627–639. [CrossRef]
67. Qu, Y.X.; Chen, Y.; Wang, X.M. *Flame Retarded Polymeric Materials*; National Defense Industry Press: Beijing, China, 2001.
68. Duggan, G.J.; Grayson, S.J.; Kumar, S. New fire classifications and fire test methods for the European railway industry. In *Flame Retardants 2004, Proceedings of the 11th Conference on Interscience Communications, London, UK, 27–28 January 2004*; Interscience Communications Ltd.: London, UK, 2014; pp. 233–240.

MDPI

Article

New Polyhedral Oligomeric Silsesquioxanes-Based Fluorescent Ionic Liquids: Synthesis, Self-Assembly and Application in Sensors for Detecting Nitroaromatic Explosives

Wensi Li [1], Dengxu Wang [1,2], Dongdong Han [1], Ruixue Sun [1,2], Jie Zhang [1,2,*] and Shengyu Feng [1,2,*]

[1] Key Laboratory of Special Functional Aggregated Materials Ministry of Education, School of Chemistry and Chemical Engineering, Shandong University, Jinan 250100, China; anterlina@163.com (W.L.); dxwang@sdu.edu.cn (D.W.); handd1223@163.com (D.H.); sunruixue940622@163.com (R.S.)

[2] National Engineering Technology Research Centre for Colloidal Materials, Shandong University, Jinan 250100, China

* Correspondence: jiezhang@sdu.edu.cn (J.Z.); fsy@sdu.edu.cn (S.F.);
  Tel.: +86-0531-88364628 (J.Z.); +86-0531-88364866 (S.F.)

Received: 8 July 2018; Accepted: 8 August 2018; Published: 15 August 2018

Abstract: In this paper, two different models of hybrid ionic liquids (ILs) based on polyhedral oligomeric silsesquioxanes (POSSs) have been prepared. Additionally, these ILs based on POSSs (ILs-POSSs) exhibited excellent thermal stabilities and low glass transition temperatures. ^{1}H, ^{13}C, and ^{29}Si nuclear magnetic resonance (NMR) spectroscopy, Fourier transform infrared spectroscopy (FT-IR) and X-ray diffraction (XRD) were used to confirm the structures of the IL-POSSs. Furthermore, the spherical vesicle structures of two IL-POSSs were observed and were caused by self-assembly behaviors. In addition, we found it very meaningful that these two ILs showed lower detection limits of 2.57×10^{-6} and 3.98×10^{-6} mol/L for detecting picric acid (PA). Moreover, the experimental data revealed that the products have high sensitivity for detecting a series of nitroaromatic compounds—including 4-nitrophenol, 2,4-dinitrophenol, and PA—and relatively comprehensive explosive detection in all of the tests of IL-POSSs with nitroaromatic compounds thus far. Additionally, the data indicate that these two new ILs have great potential for the detection of explosives. Therefore, our work may provide new materials including ILs as fluorescent sensors in detecting nitroaromatic explosives.

Keywords: fluorescent sensors; nitroaromatic explosives; polyhedral oligomeric silsesquioxane-based ionic liquids; self-assembly behaviors; thiol-ene 'click' reaction

1. Introduction

Ionic liquids (ILs), which have been recognized as a novel class of synthetic materials, have melting points below 100 °C [1–3]. With increasing demand for advanced materials, the development of novel ILs has been increasingly explored due to their unique properties [4], which include high ionic conductivity [5], high thermal stability, non-flammability, negligible vapor pressure, and a wide electrochemical stability window [6,7]. Due to the above advantages, ILs have been applied as functional materials in many chemical and industrial fields including fuel cells and lithium batteries [8], particular as catalysis solvents, capacitors, and electrochemical sensors [3]. Over the past few decades, polyhedral oligomeric silsesquioxane (POSS) materials have attracted much attention due to their unique thermal, chemical, and mechanical stabilities derived from their siloxane (Si–O–Si) frameworks [7,9], which have high bond energies. Furthermore, POSSs exhibit marvellous compatibility [10] with organic materials because of their inorganic core with various functional organic substituents [7,11,12]. Due to their many excellent merits, POSS materials have been extensively used in many fields, including as drug delivery agents [13], liquid crystalline materials [14,15], and light-emitting materials [16]. Given the aforementioned advantages of POSSs, introducing them into ILs is a reasonable approach to creating more novel and functional ILs.

The investigation of IL-POSS materials in different applications has been reported by many researchers. For example, Chujo and Tanaka et al. developed IL-POSSs with higher thermal stabilities than those of ILs linked to the side chains of POSSs because the IL-POSSs are comprised of inorganic frameworks with Si–O–Si bonds [6]. As such, these materials expand the range of applicable temperatures available to ILs. Tan [11] and his coworkers have also prepared IL-POSSs, but the synthetic route is not very convenient and the characterizations are not very comprehensive to illustrate the properties of IL-POSSs. In 2016, Li [17] and his coworkers have synthesized three IL-POSSs via the thiol-ene 'click' reaction, which is a very fast and efficient route to prepare different types of IL-POSSs for various purposes. The thiol-ene 'click' reaction allowed high yields, a variety of functional groups, easy purification, oxygen tolerance, etc. [18]. In addition, this approach has been demonstrated as an efficient and molecular modification procedure [19]. However, the fluorescent properties were not discussed, and there were no more applications of IL-POSSs as novel materials.

Due to the growing concerns about environmental protection and global terrorism activities, the demand for environmentally friendly sensor materials for the detection of nitroaromatic compounds has become a significant goal in the fields of materials chemistry and homeland security [20]. Therefore, fluorescent sensors [21] for the detection of nitroexplosives [22,23] are appealing to many researchers due to their simplicity, high sensitivity, high selectivity [24], cost-effectiveness [25], and fast response times.

Like trinitrotoluene (TNT) [26], picric acid (PA) is an extensively used explosive and is a dangerous environmental pollutant when present in groundwater, soil, etc. [27,28]. Thus, a safe and reliable approach to prevent environmental pollution and terroristic threats from PA is urgently needed. However, many sensors used to detect explosives present problems, such as a high cost [27,29] and difficult sample preparation. Therefore, there is an urgent need for better sensing materials for explosive detection.

IL-POSSs derived from imidazolium [27,30] have shown strong fluorescence and have been applied to the detection of PA in a previous report. However, compounds only functionalized with one or two imidazolium rings may be unable to show excellent fluorescence sensing abilities for detecting explosives compared to multifunctional compounds such as IL-POSSs. Based on this observation, we explored IL-POSS materials synthesized with more imidazolium rings in order to enhance their fluorescence sensitivities and then employed these IL-POSS materials for detecting a series of nitroaromatic compounds, including 4-nitrophenol, 2,4-dinitrophenol and PA. Interestingly, these IL-POSS materials demonstrated fast and efficient fluorescence quenching abilities for the detection of 4-nitrophenol, 2,4-dinitrophenol, and PA. The results presented here indicate that these IL-POSS materials, as new functional and environmentally friendly ionic liquids, could be used as fluorescent sensors in explosive detection.

2. Materials and Methods

2.1. Materials

The starting materials (1-allylimidazole, 1-bromobutane) were purchased from commercial sources (Aladdin Co., Shanghai, China) and used without further purification. 2,2-Dimethoxy-2-phenylacetophenone (DMPA) was purchased from the Aladdin Co. (Beijing, China) and used as received. Tetrahydrofuran (THF) and toluene were purified according to a routine procedure and distilled over sodium before use. Octa(mercaptopropyl)silsesquioxane (denoted as POSS-SH) and octa(chloropropyl)silsesquioxane (denoted as POSS-Cl) were synthesized according to a method described in the literature [11,17].

2.2. Characterization and Measurements

The thiol-ene reaction mixture was irradiated with high-intensity UV light from a Spectroline model SB–100P/FA lamp (365 nm, 100 W, Spectroline Co., Westbury, NY, USA). ^{1}H NMR, ^{13}C NMR, and ^{29}Si NMR spectra were recorded on a Bruker Avance-400 spectrometer (Bruker Co., Rheinstetten, Germany) using CDCl$_3$ or a mixture of CD$_3$OD and acetone-d$_6$ as the solvent and without tetramethylsilane (TMS) as an internal reference. High resolution mass spectra (HRMS) spectra were obtained in the negative mode on an Agilent Technologies 6510 Q-TOF mass spectrometer (Agilent Co., Santa Clara, CA, USA). FT-IR spectra were recorded on a Bruker TENSOR-27 infrared spectrophotometer (Bruker Co., Ettlingen, Germany) via the KBr pellet technique within the wavenumber region from 4000 to 400 cm^{-1}. X-ray diffraction (XRD) patterns were collected on a Bruker-D8 advanced X-ray diffractometer (Bruker Co., Karlsruhe, Germany) with a Cu radiation source ($\lambda = 0.154$ nm) operated at 40 kV and 30 mA using a Ni filter. Data were recorded in the range of $10° \leq 2\theta \leq 80°$ at a scanning rate of 10 °/min. The luminescence (excitation and emission) spectra of the samples were recorded with a Hitachi F-4500 fluorescence spectrophotometer (Rigaku Co., Tokyo, Japan) equipped with a monochromatic Xe lamp as an excitation source. Thermal measurements were carried out using a TA Instruments SDTQ 600 (Mettler Co., Shanghai, China). The IL-POSSs were loaded into aluminium pans, which were then heated from −100 to 25 °C, cooled to −100 °C, and finally reheated to 25 °C. The heating and cooling temperature ramping rates were 10 °C/min. The DSC data are reported in this paper from the second heating cycle. TGA was performed using a Mettler Toledo TGA/DSC1 (Mettler Co., Shanghai, China) at a heating rate of 10 °C/min from room temperature to 700 °C under N$_2$ (10 mL/min) at ambient pressure. To analyze the self-assembly behaviors of the IL-POSSs by TEM observation, samples were prepared by spreading a drop of aggregated solution onto a copper grid, followed by air drying at room temperature before testing on a JEM–1011 (100 kV) electron microscope (JEOL, Tokyo, Japan).

2.3. General Procedures to Synthesize the IL-POSSs

Allyl-min-Br was prepared through a quaternization reaction. IL-POSS-Br was synthesized via a classic procedure, as illustrated in Scheme 1. POSS-SH (1.06 g; 1 mmol), allyl-min-Br (1.96 g; 8 mmol), and DMPA (0.05 g; 2 wt %) were added to a transparent bottle with a 10 mL solvent mixture of CH$_3$OH and CH$_2$Cl$_2$. The starting materials were then irradiated with a UV lamp for 15 min after dissolving completely. Finally, IL-POSS-Br was obtained after solvent evaporation at low pressure and vacuum drying at 60 °C for 24 h. IL-POSS-Cl was prepared through a simple but longer quaternization reaction. POSS-Cl (1.03 g; 1 mmol) and 1-allylimidazole (1.08 g, 10 mmol) were charged to a transparent bottle with 10 mL of toluene. The resultant mixture was heated to 85°C for at least 3 h, and the product was washed with a co-solvent of toluene and hexane several times. Finally, IL-POSS-Cl was obtained by vacuum drying at 60 °C for 24 h.

3. Results

3.1. Synthesis and Characterization

In this paper, we have synthesized two different IL-POSSs: IL-POSS-Br and IL-POSS-Cl. The synthetic routes are described in the following text:

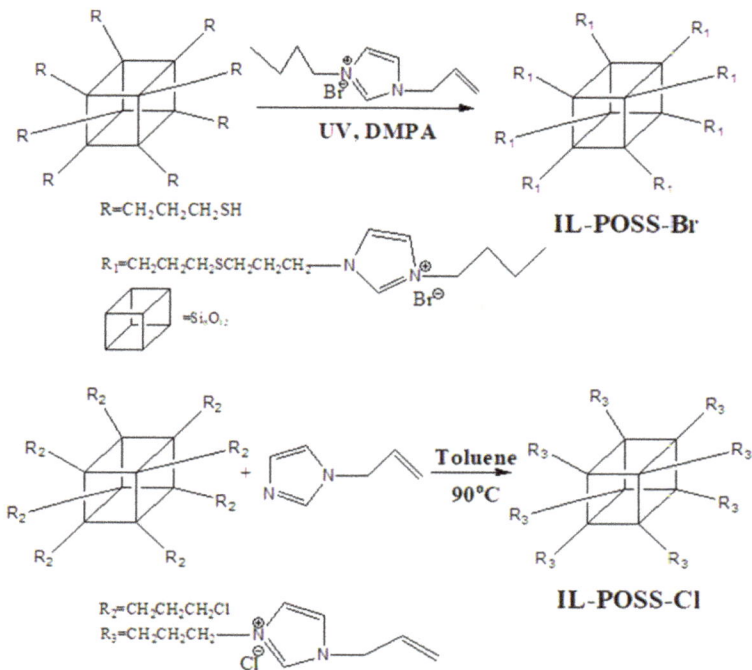

Scheme 1. Synthesis of IL-POSS-Br and IL-POSS-Cl.

First, to synthesize IL-POSS-Br, allyl-3-butylimidazolium bromide (allyl-min-Br) was prepared via the reaction of 1-allylimidazole with n-bromobutane. Afterwards, a thiol-ene reaction was carried out to synthesize IL-POSS-Br according to the route illustrated in Scheme 1 [17]. Fourier transform infrared (FT-IR) spectroscopy was used to characterize the structure and monitor the disappearance of the –SH groups (γ = 2565 cm^{-1}) after 15 min of reaction, coinciding with the appearance of peaks assigned to the imidazole ring at 1580 and 1445 cm^{-1}. Second, with respect to IL-POSS-Cl [11], the peaks corresponding to the imidazole rings were observed after the same amount of time. The peaks corresponding to the silsesquioxane frameworks of these two IL-POSS materials, such as Si–O–Si, appeared at 1050–1130 cm^{-1} (Figure S1). The ^{1}H NMR spectra of the IL-POSSs are shown in Figures S2 and S3. For IL-POSS-Br, whereas the characteristic SH peak at 1.37 ppm and the CH=CH$_2$ peak at approximately 5.5–6.0 ppm vanished, a new peak emerged at 2.25 ppm, and signals corresponding to the imidazole ring appeared at approximately 7.75 and 9.25 ppm. For IL-POSS-Cl, the characteristic imidazole ring peak and CH=CH$_2$ peak are observed at approximately 7.70–9.25 ppm and 5.5–6.0 ppm, respectively. The ^{13}C NMR spectra of IL-POSS-Br and IL-POSS-Cl are shown in Figure S4. These results demonstrate a complete reaction. Moreover, only one peak (δ = −66.8 ppm) was detected in the ^{29}Si NMR spectrum of IL-POSS-Br (Figure S5) [31], illustrating that the POSS cage remains intact during the reaction [7]. Furthermore, we observed the appearance of two peaks at m/z = 749.1942 (IL-POSS-Br) and m/z = 948.2664 (IL-POSS-Cl) in the mass spectra (Figure S6), providing evidence that the targeted products were successfully prepared.

For further study, we employed X-ray diffraction (XRD) to estimate the structures of these two IL-POSSs (shown in Figure S7). Wide peaks appeared near 22.6° in the patterns of both IL-POSSs; such peaks are commonly observed in the XRD patterns of amorphous silica nanocomposites and are caused by the Si–O–Si bonds.

3.2. Thermal Properties

To demonstrate the thermal behaviors of the two IL-POSS materials, thermogravimetric analysis (TGA) and differential scanning calorimetry (DSC) were used to characterize the materials. IL-POSS-Br shows better thermal stability than IL-POSS-Cl, and the thermal decomposition temperatures (T_d at 5 wt % loss) of the materials are approximately 300 °C and 200 °C, respectively (Figure S8). Compared with the IL-POSS prepared by Tan and coworkers [11], the thermal stability of IL-POSS-Cl is slightly lower due to the CH=CH$_2$ group. For IL-POSS-Br, some weight loss occurs at the beginning of the measurement, which is ascribed to the loss of volatile compounds. Consistent with the excellent structures of the IL-POSS materials, these results indicate that strong ionic interactions occur within them, enabling their application of over a wider range of fields.

DSC measurements were conducted under a nitrogen atmosphere and provided evidence that these two IL-POSSs both exhibit a single glass-transition temperature (T_g). IL-POSS-Br exhibits an endotherm near −30 °C (Figure S9), whereas IL-POSS-Cl shows an endotherm near −20 °C (Figure S9). According to the data reported herein, we can deduce that the T_g was strongly affected by the anions and that, as expected, the presence of POSS substantially increased the T_g [32]. Obviously, the size of the anion strongly influenced the resultant T_g, and the data suggest that the T_g decreases with increasing differences in the sizes of the cation and anion. In addition, only one glass transition is observed, which indicates that these anions were homogenously dispersed in the IL-POSSs.

3.3. Self-Assembly Behaviors

We also examined the self-assembly behaviors of the two IL-POSSs via transmission electron microscopy (TEM) to investigate their amphiphilic nature [33]. Using IL-POSS-Br as an example, our study shows that the average diameter of these spherical vesicles is approximately 200 nm when solvated in C$_2$H$_5$OH (shown in Figure 1). To better illustrate these self-assembly behaviors, we suggest a mechanism for vesicle formation (shown in Scheme 2) [17]. The combination of POSS-SH and imidazolium-containing side chains generated a dendritic ionic liquid supported by a POSS cage. These dendritic IL-POSSs can move freely when dissolved in good solvents. Nevertheless, in specific solvents, self-assembly occurred and spherical vesicles formed with POSS cages on the inside and anions on the outside. Obviously, strong electrostatic interactions are present in IL-POSSs [34,35], which causes the cations and anions to aggregate. These data suggest that the spherical vesicle shape is the lowest energy state and is driven by electrostatic interactions. These data suggest that the spherical vesicle shape is the lowest energy state driven by electrostatic interaction [6,36].

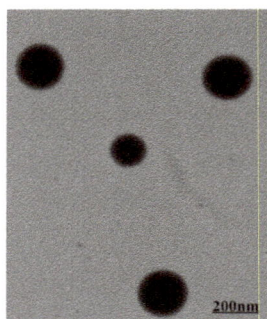

Figure 1. TEM image illustrating the self-assembly behaviors of IL-POSS-Br in ethanol.

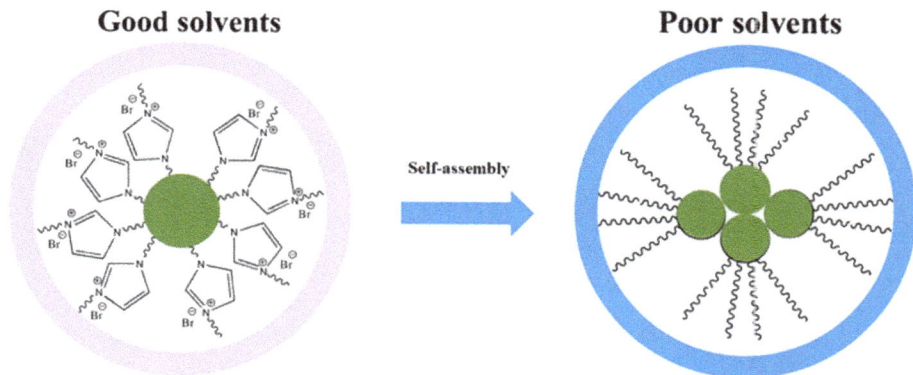

Scheme 2. Proposed working mechanism for IL-POSS aggregation in ethanol. Reproduced with permission from [17]. Copyright Wiley 2016.

With respect to interfacial chemistry, we studied the static contact angle (CA) values of IL-POSS-Br and IL-POSS-Cl with distilled water as the test liquid. The CA of IL-POSS-Br was 69.8°, and the CA of IL-POSS-Cl was 66.1° (Figure S10). These results suggest that anions such as miscible halides strongly influence the miscibility of IL-POSSs and distilled water. In addition, CMC data (Figure S11) are 2.7 and 8.5 mM for IL-POSS-Cl and IL-POSS-Br, respectively. Based on this report, we can predict that ILs can be effectively integrated with amphiphilic molecules for two purposes: to introduce the excellent properties of ILs to the traditional self-assembly and aggregation of amphiphilic molecules and to further expand the development and application of ILs.

3.4. Optical Properties

For further investigation, the fluorescent properties of the IL-POSS materials were characterized. Both IL-POSSs exhibited two emission bands centred at approximately 410 nm and 430 nm when excited at 365 nm, as shown in Figure 2a,b. We propose two causes for the fluorescence of IL-POSSs: the fluorescent properties of the imidazole rings linked to the IL-POSSs and the fluorescence of POSS linked with mercaptopropyl. The splitting of the 3*d* orbital of the Si atom is caused by Si→S coordination bonds [37,38]; therefore, a *d–d* transition occurred due to the rearrangement of electrons in the split orbitals [31], and this transition is beneficial for the fluorescence of the POSS cages. In addition, the emission intensity decreased with decreasing concentration, consistent with the reported data for most organic luminescent compounds.

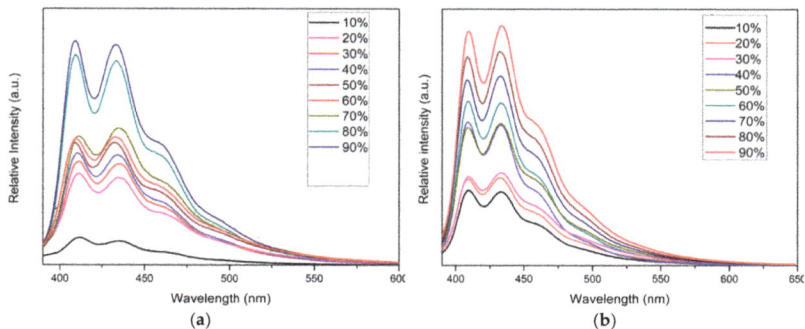

Figure 2. Emission spectra of IL-POSS-Br (a) and IL-POSS-Cl (b) (1×10^{-6} M).

4. Discussion

4.1. IL-POSSs in the Detection of Nitroaromatic Explosives

Based on the aforementioned excellent fluorescence of IL-POSSs, we explored their application in the detection of 4-nitrophenol (NP), 2,4-dinitrophenol (DNP) and picric acid (PA) in ethanol [39]. To examine the sensing abilities and monitor the fluorescent response of IL-POSS-Br and IL-POSS-Cl towards the three aforementioned nitroaromatic compounds, the two IL-POSSs were dissolved in ethanol with gradually increasing NP, DNP, and PA contents, as shown in Figure 3. The spectra clearly show that substantial fluorescence quenching occurred in both IL-POSS materials. The fluorescence intensity gradually weakened with increasing concentrations of the explosives. The sensing studies were performed by recording the changes with excitation at 365 nm. To further investigate the sensitivity for the detection of NP, DNP, and PA, the Stern–Volmer plots were obtained. As shown in Figure 4, the respective Stern–Volmer plots of NP, DNP, and PA in the presence of IL-POSS-Br and IL-POSS-Cl better demonstrate the sensing abilities of the IL-POSSs. The Stern–Volmer plot is linear when the concentration of PA is low. According to the literature [29], the Stern–Volmer constants (K_{sv}) for NP, DNP, and PA can be calculated from the slopes of the Stern–Volmer plots. In Figure 4a–c, the Stern–Volmer constants (K_{sv}) of IL-POSS-Br are 7.925×10^3, 2.974×10^4, and 3.74×10^6 M^{-1} for NP, DNP, and PA, respectively. In Figure 4d–f, the Stern–Volmer constants (K_{sv}) of IL-POSS-Cl are 5×10^3, 2.5×10^4, and 2.41×10^6 M^{-1} for NP, DNP, and PA, respectively. Therefore, the quenching efficiencies of both IL-POSSs in the presence of the studied explosives decrease in the order PA > DNP > NP, which is consistent with the fluorescence intensities observed in Figure 3.

To better illustrate the quenching behaviors of these materials, we calculated their limits of detection (LDs, LD = $3 \times \sigma/K$, σ = standard deviation of blank measurement = 3.20) from the slopes via the approximate linear relationship between the fluorescence intensity of the IL-POSS and the concentration of the relevant explosives; the results are shown in Figure 4. The LDs of IL-POSS-Br are 1.21×10^{-3}, 3.23×10^{-4}, and 2.57×10^{-6} mol/L for NP, DNP, and PA, respectively, and the LDs of IL-POSS-Cl are 1.92×10^{-3}, 3.84×10^{-4}, and 3.98×10^{-6} mol/L for NP, DNP, and PA, respectively.

The aforementioned phenomenon is attributed to two factors [39]: (1) electron transfer from the imidazole rings of IL-POSSs to the electron-deficient nitroaromatic compounds, which enables electron transfer to occur among the functional groups; and (2) competitive absorption or the inner-filter effect (IFE) [40], which is caused by other absorbents or the simultaneous absorption of the excitation and emission lights of the fluorescent materials in both of the detection systems. In Figure 5, the absorption bands of all the explosives show obvious overlap with the excitation spectra of IL-POSS-Cl (the excitation wavelength at 365 nm is included). Thus, the absorption by the explosives filters the light absorbed by the IL-POSSs, resulting in fluorescence quenching. With regards to the fluorescence emission mechanism, after the fluorescent materials absorbed the light or energy, the electrons would be transitioned from the ground state (S_0) to the excited states (S_1, S_2) [41], and due to the unstable state, they would release energy to achieve an absorption competition of the light source energy between the materials and the analytes after a series of vibrations. Consequently, the coexistence of the electron transfer effect and competitive absorption leads to fluorescence quenching behaviors. These data demonstrate that IL-POSS-Br and IL-POSS-Cl show high selectivity for the detection of NP, DNP, and PA.

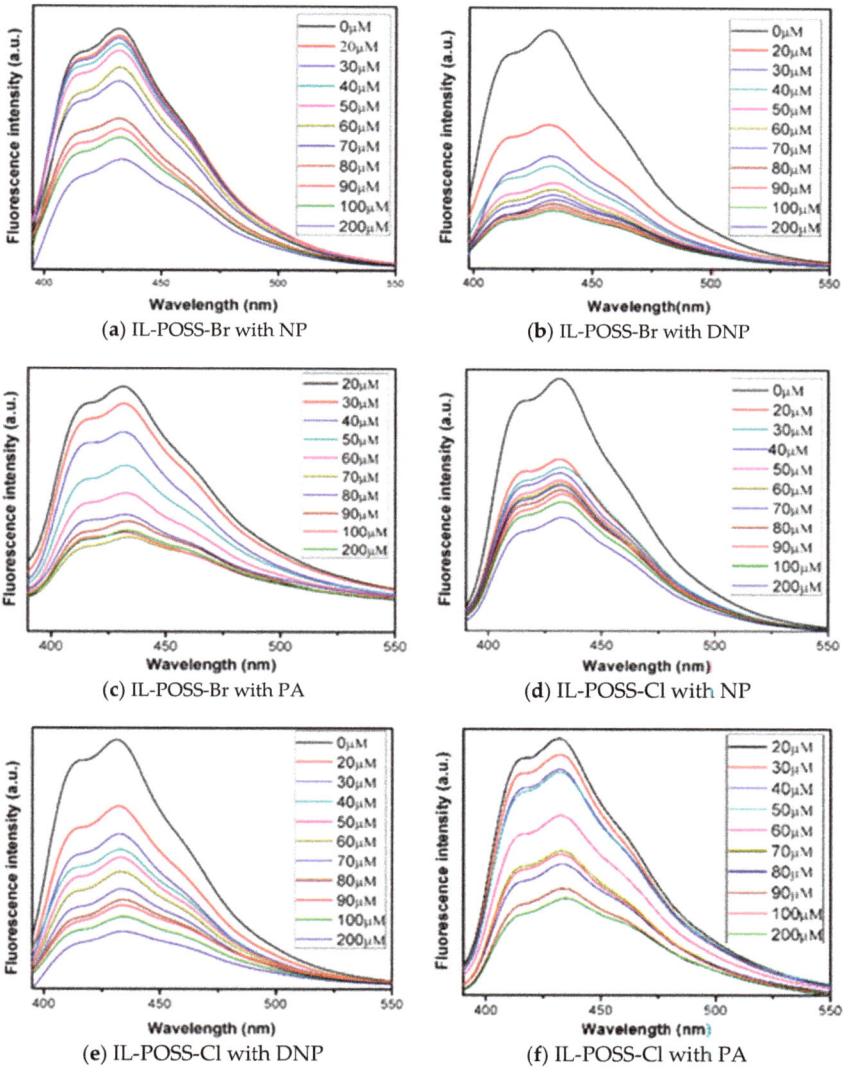

Figure 3. Emission spectra (**a–c**) belong to IL-POSS-Br upon the addition of 4-nitrophenol (NP), 2,4-dinitrophenol (DNP) and picric acid (PA) ethanol solution, respectively. Emission spectra (**d–f**) belong to IL-POSS-Cl upon the addition of NP, DNP, and PA, respectively.

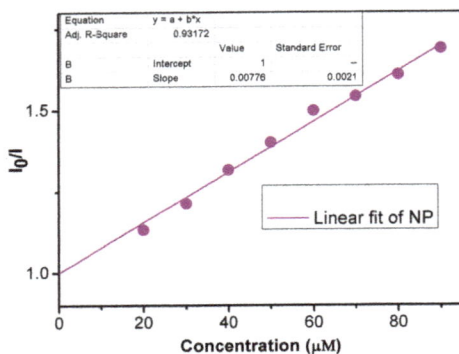

(a) IL-POSS-Br (K_{sv} = 7.925 × 10^3 M^{-1})

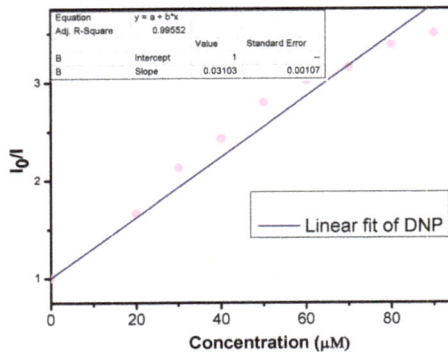

(b) IL-POSS-Br (K_{sv} = 2.974 × 10^4 M^{-1})

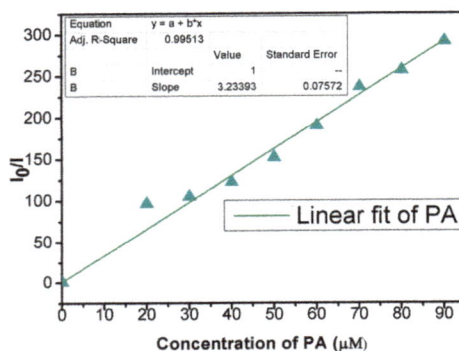

(c) IL-POSS-Br (K_{sv} = 3.74 × 10^6 M^{-1})

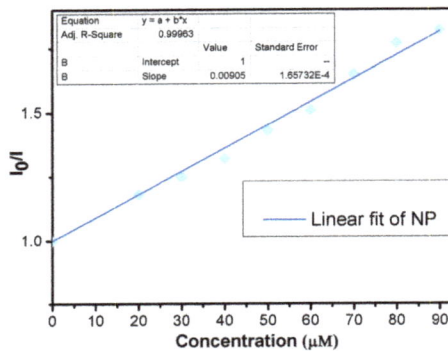

(d) IL-POSS-Cl (K_{sv} = 5 × 10^3 M^{-1})

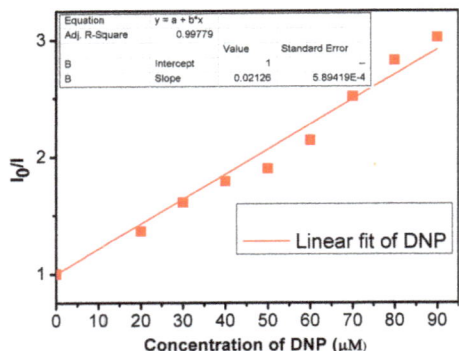

(e) IL-POSS-Cl (K_{sv} = 2 × 10^4 M^{-1})

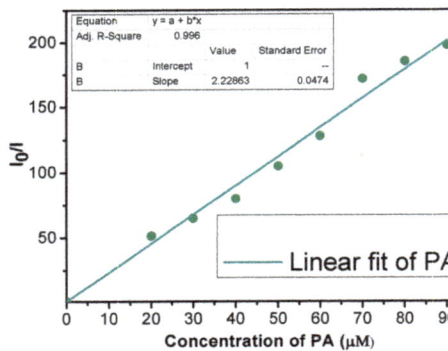

(f) IL-POSS-Cl (K_{sv} = 2.41 × 10^6 M^{-1})

Figure 4. The Stern–Volmer curves (**a–c**) belong to IL-POSS-Br upon the addition of 4-nitrophenol, 2,4-dinitrophenol and picric acid ethanol solution, respectively. Emission spectra (**d–f**) belong to IL-POSS-Cl upon the addition of 4-nitrophenol, 2,4-dinitrophenol, and picric acid ethanol solution, respectively.

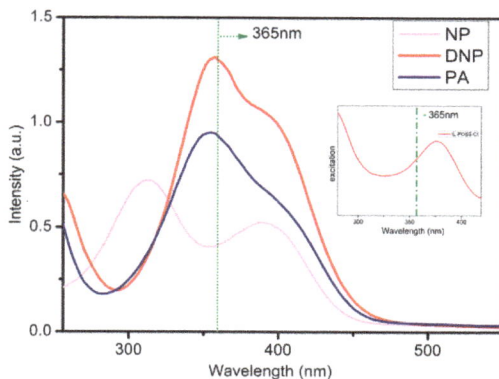

Figure 5. UV–vis absorption spectra of various analytes in ethanol and the excitation spectrum of IL-POSS-Cl in an ethanol suspension.

4.2. Simple and Efficient Testing of IL-POSSs as Sensors for Detecting Nitroaromatic Explosives

For a better visual observation, the IL-POSSs were tested as fluorescent sensors for the detection of PA (shown in Figure 6). Under a UV lamp (365 nm), test papers coated with IL-PCSSs clearly changed colors after being immersed into an aqueous solution of PA [27]. Unlike other fluorescent sensors, IL-POSSs have two advantages: they are more environmentally friendly and more adaptable to 'green' chemistry than certain other sensors because of their basically green synthetic approach, and these IL-POSSs can be expanded to applications in aqueous systems. In comparison, traditional fluorescent sensors are often hydrophobic, which limits their application.

Test paper coated with ILs-POSS   **Then put the test paper into PA solution**

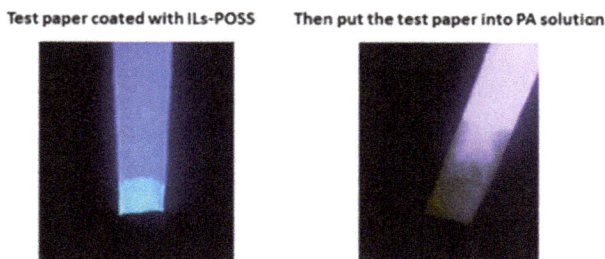

Figure 6. Photographs of test paper coated with IL-POSSs before and after being immersed into the aqueous solution of picric acid.

5. Conclusions

Overall, in this paper, two IL-POSS materials were successfully prepared in high yields via different methods. By comparison of the synthesis time, it has been demonstrated that the thiol-ene 'click' reaction is easily applied in the preparation of IL-POSSs due to its fast and efficient merits. Due to the POSS cage, both IL-POSSs exhibit good thermal stabilities and low glass-transition temperatures, which endows them with a wide range of applicable temperatures. In addition, the unique amphiphilic nature of the two IL-POSSs allows them to aggregate into spherical vesicle structures in select solvents. Furthermore, based on the fluorescent properties of IL-POSSs, we have identified another application, fluorescent sensors that can efficiently detect NP, DNP, and PA, and the comparison of the sensitivities has illustrated that IL-POSS-Br performed better than IL-POSS-Cl with regards to the detection of PA. Additionally, the simple test in Figure 6 is visual proof that the IL-POSSs can be employed as sensors in detecting PA, although this test provides only preliminary evidence. Finally, more functional IL-POSS materials can be fabricated through this reaction and used as effective fluorescent sensors in the future.

Supplementary Materials: The following are available online at http://www.mdpi.com/2073-4360/10/8/917/s1, Figure S1: FTIR spectra of IL-POSS-Br and IL-POSS-Cl, Figure S2: ^{1}H spectra of IL-POSS-Br and HS-POSS. Note: * represents solvent peaks, Figure S3: ^{1}H spectra of IL-POSS-Cl. Note: * represents solvent peaks, Figure S4: ^{13}C spectra of IL-POSS-Br and IL-POSS-Cl. Note: * represents solvent peaks, Figure S5: ^{29}Si spectrum of IL-POSS-Br, Figure S6: HRMS spectra of IL-POSS-Br and IL-POSS-Cl, Figure S7: XRD spectra of IL-POSS-Br and IL-POSS-Cl, Figure S8: TGA curves of IL-POSS-Br and IL-POSS-Cl, Figure S9: DSC curves of IL-POSS-Br and IL-POSS-Cl, Figure S10: Contact angles (CA) of IL-POSS-Br and IL-POSS-Cl.

Author Contributions: W.L., D.W., R.S., and D.H. designed the experiments. W.L. performed the experiments, analyzed the data, and wrote the manuscript. J.Z. and S.F. revised the manuscript.

Funding: This research was funded by the National Natural Science Foundation of China (nos. 21774070 and 21274080) and the Fund for Shandong Province Major Scientific and Technological Innovation Projects (no. 2017CXGC1112).

Acknowledgments: This work was financially supported by the National Natural Science Foundation of China (nos. 21774070 and 21274080) and the Fund for Shandong Province Major Scientific and Technological Innovation Projects (no. 2017CXGC1112).

Conflicts of Interest: The authors declare that they have no conflicts of interest.

References

1. Yao, L.; Zhang, B.; Jiang, H.; Zhang, L.; Zhu, X. Poly(ionic liquid): A new phase in a thermoregulated phase separated catalysis and catalyst recycling system of transition metal-mediated atrp. *Polymers* **2018**, *10*, 347. [CrossRef]

2. Zheng, X.; Lin, Q.; Jiang, P.; Li, Y.; Li, J. Ionic liquids incorporating polyamide 6: Miscibility and physical properties. *Polymers* **2018**, *10*, 562. [CrossRef]

3. Fu, J.; Lu, Q.; Shang, D.; Chen, L.; Jiang, Y.; Xu, Y.; Yin, J.; Dong, X.; Deng, W.; Yuan, S. A novel room temperature POSS ionic liquid-based solid polymer electrolyte. *J. Mater. Sci.* **2018**, *53*, 8420–8435. [CrossRef]

4. Topal, S.Z.; Ertekin, K.; Topkaya, D.; Alp, S.; Yenigul, B. Emission based oxygen sensing approach with tris(2,2'-bipyridyl)ruthenium(ii) chloride in green chemistry reagents: Room temperature ionic liquids. *Microchim. Acta* **2008**, *161*, 209–216. [CrossRef]

5. Shang, D.; Fu, J.; Lu, Q.; Chen, L.; Yin, J.; Dong, X.; Xu, Y.; Jia, R.; Yuan, S.; Chen, Y.; et al. A novel polyhedral oligomeric silsesquioxane based ionic liquids (POSS-ILs) polymer electrolytes for lithium ion batteries. *Solid State Ionics* **2018**, *319*, 247–255. [CrossRef]

6. Tanaka, K.; Ishiguro, F.; Jeon, J.-H.; Hiraoka, T.; Chujo, Y. POSS ionic liquid crystals. *NPG Asia Mater.* **2015**, *7*, 174. [CrossRef]

7. Manickam, S.; Cardiano, P.; Mineo, P.G.; Lo Schiavo, S. Star-shaped quaternary alkylammonium polyhedral oligomeric silsesquioxane ionic liquids. *Eur. J. Inorg. Chem.* **2014**, *2014*, 2704–2710. [CrossRef]

8. Na, W.; Lee, A.S.; Lee, J.H.; Hong, S.M.; Kim, E.; Koo, C.M. Hybrid ionogel electrolytes with POSS epoxy networks for high temperature lithium ion capacitors. *Solid State Ionics* **2017**, *309*, 27–32. [CrossRef]

9. Sun, J.K.; Antonietti, M.; Yuan, J. Nanoporous ionic organic networks: From synthesis to materials applications. *Chem. Soc. Rev.* **2016**, *45*, 6627–6656. [CrossRef] [PubMed]

10. Daver, F.; Kajtaz, M.; Brandt, M.; Shanks, R. Creep and recovery behavior of polyolefin-rubber nanocomposites developed for additive manufacturing. *Polymers* **2016**, *8*, 437. [CrossRef]

11. Tan, J.; Ma, D.; Sun, X.; Feng, S.; Zhang, C. Synthesis and characterization of an octaimidazolium-based polyhedral oligomeric silsesquioxanes ionic liquid by an ion-exchange reaction. *Dalton Trans.* **2013**, *42*, 4337–4339. [CrossRef] [PubMed]

12. Monticelli, O.; Fina, A.; Cavallo, D.; Gioffredi, E.; Delprato, G. On a novel method to synthesize POSS-based hybrids: An example of the preparation of TPU based system. *Express Polym. Lett.* **2013**, *7*, 966–973. [CrossRef]

13. McCusker, C.; Carroll, J.B.; Rotello, V.M. Cationic polyhedral oligomeric silsesquioxane (POSS) units as carriers for drug delivery processes. *Chem. Commun.* **2005**, 996–998. [CrossRef] [PubMed]

14. Saez, I.M.; Goodby, J.W. Chiral nematic octasilsesquioxanes. *J. Mater. Chem.* **2001**, *11*, 2845–2851. [CrossRef]

15. Pan, Q.; Chen, X.; Fan, X.; Shen, Z.; Zhou, Q. Organic–inorganic hybrid bent-core liquid crystals with cubic silsesquioxane cores. *J. Mater. Chem.* **2008**, *18*, 3481–3488. [CrossRef]

16. Ye, S.-H.; Li, L.; Zhang, M.; Zhou, Z.; Quan, M.-H.; Guo, L.-F.; Wang, Y.; Yang, M.; Lai, W.-Y.; Huang, W. Pyridine linked fluorene hybrid bipolar host for blue, green, and orange phosphorescent organic light-emitting diodes toward solution processing. *J. Mater. Chem. C* **2017**, *5*, 11937–11946. [CrossRef]

17. Li, L.; Liu, H. Rapid preparation of silsesquioxane-based ionic liquids. *Chemistry* **2016**, *22*, 4713–4716. [CrossRef] [PubMed]

18. Alves, F.; Nischang, I. Tailor-made hybrid organicinorganic porous materials based on polyhedral oligomeric silsesquioxanes (POSS) by the stepgrowth mechanism of thiolene 'click' chemistry. *Chem. Eur. J.* **2013**, *19*, 17310–17313. [CrossRef] [PubMed]

19. Alves, F.; Scholder, P.; Nischang, I. Conceptual design of large surface area porous polymeric hybrid media based on polyhedral oligomeric silsesquioxane precursors: Preparation, tailoring of porous properties, and internal surface functionalization. *ACS Appl. Mater. Interfaces* **2013**, *5*, 2517–2526. [CrossRef] [PubMed]

20. Shi, G.; Qu, Y.; Zhai, Y.; Liu, Y.; Sun, Z.; Yang, J.; Jin, L. {MSU/PDDA}$_n$ LBL assembled modified sensor for electrochemical detection of ultratrace explosive nitroaromatic compounds. *Electrochem. Commun.* **2007**, *9*, 1719–1724. [CrossRef]

21. Li, Y.; Zhang, W.; Sun, Z.; Sun, T.; Xie, Z.; Huang, Y.; Jing, X. Light-induced synthesis of cross-linked polymers and their application in explosive detection. *Eur. Polym. J.* **2015**, *63*, 149–155. [CrossRef]

22. Yu, R.; Li, Y.; Tao, F.; Cui, Y.; Song, W.; Li, T. A novel double-layer electrospun nanofibrous membrane sensor for detecting nitroaromatic compounds. *J. Mater. Sci.* **2016**, *51*, 10350–10360. [CrossRef]

23. Qin, J.; Chen, L.; Zhao, C.; Lin, Q.; Chen, S. Cellulose nanofiber/cationic conjugated polymer hybrid aerogel sensor for nitroaromatic vapors detection. *J. Mater. Sci.* **2017**, *52*, 8455–8464. [CrossRef]

24. Kumari, S.; Joshi, S.; Cordova-Sintjago, T.C.; Pant, D.D.; Sakhuja, R. Highly sensitive fluorescent imidazolium-based sensors for nanomolar detection of explosive picric acid in aqueous medium. *Sens. Actuators B* **2016**, *229*, 599–608. [CrossRef]

25. Massera, E.; Castaldo, A.; Quercia, L.; Di Francia, G. Fabrication and characterization of polysilsesquioxanes nanocomposites based chemical sensor. *Sens. Actuators B* **2008**, *129*, 487–490. [CrossRef]

26. Karthik, P.; Pandikumar, A.; Preeyanghaa, M.; Kowsalya, M.; Neppolian, B. Amino-functionalized mil-101(Fe) metal-organic framework as a viable fluorescent probe for nitroaromatic compounds *Microchim. Acta* **2017**, *184*, 2265–2273. [CrossRef]

27. Tian, X.; Qi, X.; Liu, X.; Zhang, Q. Selective detection of picric acid by a fluorescent ionic liquid chemosensor. *Sens. Actuators B* **2016**, *229*, 520–527. [CrossRef]

28. Dai, J.; Dong, X.; Fidalgo de Cortalezzi, M. Molecularly imprinted polymers labeled with amino-functionalized carbon dots for fluorescent determination of 2,4-dinitrotoluene. *Microchim. Acta* **2017**, *184*, 1369–1377. [CrossRef]

29. Xiong, J.F.; Li, J.X.; Mo, G.Z.; Huo, J.P.; Liu, J.Y.; Chen, X.Y.; Wang, Z.Y. Benzimidazole derivatives: Selective fluorescent chemosensors for the picogram detection of picric acid. *J. Org. Chem.* **2014**, *79*, 11619–11630. [CrossRef] [PubMed]

30. Peng, R.; Wang, Y.; Tang, W.; Yang, Y.; Xie, X. Progress in imidazolium ionic liquids assisted fabrication of carbon nanotube and graphene polymer composites. *Polymers* **2013**, *5*, 847–872. [CrossRef]

31. Zuo, Y.; Gou, Z.; Li, Z.; Qi, J.; Feng, S. Unexpected self-assembly, photoluminescence behavior, and film-forming properties of polysiloxane-based imidazolium ionic liquids prepared by one-pot thiol–ene reaction. *New J. Chem.* **2017**, *41*, 14545–14550. [CrossRef]

32. Cardiano, P.; Lazzara, G.; Manickam, S.; Mineo, P.; Milioto, S.; Lo Schiavo, S. POSS-tetraalkylammonium salts: A new class of ionic liquids. *Eur. J. Inorg. Chem.* **2012**, *2012*, 5668–5676. [CrossRef]

33. Ye, Q.; Zhou, H.; Xu, J. Cubic polyhedral oligomeric silsesquioxane based functional materials: Synthesis, assembly, and applications. *Chem. Asian J.* **2016**, *11*, 1322–1337. [CrossRef] [PubMed]

34. Cardiano, P.; Fazio, E.; Lazzara, G.; Manickam, S.; Milioto, S.; Neri, F.; Mineo, P.G.; Piperno, A.; Lo Schiavo, S. Highly untangled multiwalled carbon nanotube@polyhedral oligomeric silsesquioxane ionic hybrids: Synthesis, characterization and nonlinear optical properties. *Carbon* **2015**, *86*, 325–337. [CrossRef]

35. Castriciano, M.A.; Leone, N.; Cardiano, P.; Manickam, S.; Scolaro, L.M.; Lo Schiavo, S. A new supramolecular polyhedral oligomeric silsesquioxanes (POSS)–porphyrin nanohybrid: Synthesis and spectroscopic characterization. *J. Mater. Chem. C* **2013**, *1*, 4746–4753. [CrossRef]

36. Jeon, J.-H.; Tanaka, K.; Chujo, Y. Synthesis of sulfonic acid-containing POSS and its filler effects for enhancing thermal stabilities and lowering melting temperatures of ionic liquids. *J. Mater. Chem. A* **2014**, *2*, 624–630. [CrossRef]

37. Zuo, Y.; Lu, H.; Xue, L.; Wang, X.; Wu, L.; Feng, S. Polysiloxane-based luminescent elastomers prepared by thiol-ene 'click' chemistry. *Chemistry* **2014**, *20*, 12924–12932. [CrossRef] [PubMed]

38. Zuo, Y.; Cao, J.; Feng, S. Sunlight-induced cross-linked luminescent films based on polysiloxanes and D-limonene via thiol-ene 'click' chemistry. *Adv. Funct. Mater.* **2015**, *25*, 2754–2762. [CrossRef]

39. Xie, H.; Wang, H.; Xu, Z.; Qiao, R.; Wang, X.; Wang, X.; Wu, L.; Lu, H.; Feng, S. A silicon-cored fluoranthene derivative as a fluorescent probe for detecting nitroaromatic compounds. *J. Mater. Chem. C* **2014**, *2*, 9425–9430. [CrossRef]

40. Sun, R.; Huo, X.; Lu, H.; Feng, S.; Wang, D.; Liu, H. Recyclable fluorescent paper sensor for visual detection of nitroaromatic explosives. *Sens. Actuators B* **2018**, *265*, 476–487. [CrossRef]

41. Guo, L.; Zeng, X.; Lan, J.; Yun, J.; Cao, D. Absorption competition quenching mechanism of porous covalent organic polymer as luminescent sensor for selective sensing Fe^{3+}. *Chem. Select* **2017**, *2*, 1041–1047. [CrossRef]

polymers

MDPI

Article

Preparation and Characterization of Highly Ordered Mercapto-Modified Bridged Silsesquioxane for Removing Ammonia-Nitrogen from Water

Derong Lin [1,*], Yichen Huang [1], Yuanmeng Yang [1], Xiaomei Long [1], Wen Qin [1], Hong Chen [1], Qing Zhang [1], Zhijun Wu [2], Suqing Li [1], Dingtao Wu [1], Lijiang Hu [3] and Xingwen Zhang [3]

[1] College of Food Science, Sichuan Agricultural University, Ya'an 625014, China; 18723091689@163.com (Y.H.); sicau_yym@126.com (Y.Y.); 18428301943@163.com (X.L.); qinwen@sicau.edu.cn (W.Q.); chenhong945@sicau.edu.cn (H.C.); zhangqing@sicau.edu.cn (Q.Z.); lsq03_2001@163.com (S.L.); DT_Wu@sicau.edu.cn (D.W.)
[2] School of Mechanical and Electrical Engineering, Sichuan Agricultural University, Ya'an 625014, China; wzj@sicau.edu.cn
[3] State Key Laboratory of Urban Water Resource and Environment, Harbin Institute of Technology, Harbin 150090, China; hulijiang2008@126.com (L.H.); zhangxinwen@hit.edu.cn (X.Z.)
* Correspondence: lindr2018@sicau.edu.cn; Tel.: +86-835-288-2311

Received: 26 June 2018; Accepted: 19 July 2018; Published: 25 July 2018

Abstract: In acidic conditions, mesoporous molecular sieves SBA-15 and SBA-15-SH were synthesized. Structural characterization was carried out by powder X-ray diffraction (XRD). Fourier Transform infrared spectroscopy (FTIR), scanning electron microscopy (SEM), transmission electron microscopy (TEM), ^{13}C CP MAS-NMR, ^{29}Si CP MAS-NMR and nitrogen adsorption–desorption (BET). The results showed that in SBA-15-SH, the direct synthesis method made the absorption peak intensity weaker than that of SBA-15, while the post-grafted peak intensity did not change. Their spectra were different due to the C-H stretching bands of Si-O-Si and propyl groups. But their structure was still evenly distributed and was still hexangular mesoporous structure. Their pore size increased, and the H-SBA-15-SH had larger pore size. The adsorption of ammonia-nitrogen by molecular sieve was affected by the relative pressure and the concentration of ammonia-nitrogen, in which the adsorption capacity of G-SBA-15-SH was the largest and the adsorption capacity of SBA-15 was the smallest.

Keywords: direct synthesis; grafting synthesis; mercapto-modified; bridged silsesquioxane

1. Introduction

Organic pollutants have gradually increased and have serious environmental impact. Adsorption has become one of the methods of wastewater treatment. Among these, the bridge group and modified silsesquioxane (SSO) have used a variety of optimization methods to promote the development and application of this mesoporous three-dimensional material because of its unique physical and chemical properties [1–8]. For example, the SSO membrane can be used for the removal of anionic compounds, and it can be easily separated from the treated medium [9]. The SSO with a cage structure is called polyhedral oligomeric silsesquioxane (POSS), and POSS was used as a new type of organic-inorganic hybrid material [10]. It was not easy to prepare under alkaline or other harsh conditions. The structure and functional integrity of the POSS nucleus can be not damaged by the "click" reaction of the new method. This was because the effective condensation reaction between the amino oxygen group and aldehyde or ketone produces oxime bonds in order to produce chemical selectivity in the subsequent reaction [11]. The two hydrophilic carboxylic acid functionalized POSS heads and two hydrophobic polystyrene trails which were covalently linked through the rigid septum can synthesize the giant Gemini surfactants through the "click" functionalization [12].

Lin et al. ascertained the relative contributions to adsorption performance from (1) direct competition for sites and (2) pore blockage. A conceptual model was proposed to further explain the phenomena [13]. And we further explained the phenomena suggesting a promising application of cubic mesoporous BPS in wastewater treatment [5]. Hexane, octane, phenyl, and biphenyl-bridged bis (triethoxysilyl) precursors can be used in synthesizing cubic mesoporous (BPS) copolymers. The textural properties of ordered mesoporous hexane bridged polysilsesquioxane (BPS) can be tailored by choosing reaction conditions [2]. And three-dimensional cubic (Pm3n) periodic mesoporous silsequioxanes (PMS) with alkylene bridging groups was prepared by surfactant templating. They are promising candidates for organic pollutant adsorbents from water or air [3]. A vinyl silsesquioxane (VS) was added to a pesticide (citral) to enhance residual, thermal, and anti-ultraviolet properties via three double-bond reactions in the presence of an initiator [14]. After the zeolite molecular sieves were modified by polyhedral oligomeric silsesquioxane (POSS-modified ZMS), the NH_3^+ ino-exchange capacity increased, causing the NH_3–N removal capacity to be enhanced [15].

The mesoporous materials with different properties and characteristics can be prepared by changing the templates and conditions of neutral long chain amines, eighteen alkyl polyoxyethylene ether, and CPBr [16–20]. These mesoporous materials can be widely used in catalysis, semi-conductive nanometer microcrystalline materials, adsorption, and separation fields. The ordered mesoporous silica synthesized by the surfactant template can be used in the fields of catalysis, ion exchange, adsorption, chemical sensors, and nanomaterials with much potentiality [21,22]. At the same time, different ways have also been studied in depth, such as the traditional hydrothermal method, alcohol heating method, two step sol-gel method, morphology control method, skeleton doping method, and surface modification method. Hot water synthesis can be done according to the organic groups, templates, catalysts (acid or alkali), time (agitation, aging, drying). In addition, organic-inorganic grafting was a hot topic in the research of mesoporous molecular sieve in recent years [23,24]. The ordered mesoporous materials prepared with the modified bridge SSO had the advantages of large specific surface area, adjustable pore size, narrow pore distribution, high heat resistance, and good thermal stability, and can show a simple cubic (Pm3n) mesoporous structure. The liquid-crystal-templating mechanism, the generalized liquid-crystal-templating mechanism, and other models were proposed to explain the formation of mesoporous with regular arrangement [25–28]. The selection of bridging groups helped to promote the development and application for these kinds of mesoporous cube materials, which was of great significance for the adsorption of organic pollutants, it can help to promote the development and application of these mesoporous 3D cubic materials [2,3,5,13–15].

Since the 1980s, it has been known that silicon dioxide extracted from rice husks has high activity and performance. Therefore, more attention has been paid to the preparation of high purity silica from rice husks. Rice husk resources are very abundant (>40 million tons/year) in China. The content of silicon in rice husk ash is relatively high, and the SiO_2 content after high temperature calcination is 93.1%. However, most rice husks are burned as garbage, and even become the main source of pollution in some areas. The preparation of silicon organic chemicals from rice husk ash not only broadens the application scope of rice husk ash, but also solves the problem of environmental pollution. Mesoporous molecular sieve SBA-15 had many advantages, such as a large specific surface area, the pore structure rule, convenient pore size regulation, and good thermal stability. It has a broad application prospect in catalytic chemistry and adsorption separation. Therefore, this paper used rice husk as a silicon source and synthesized and studied SBA-15 and SBA-15-SH. It was hoped that the research results of this paper can be used as adsorbents to achieve the purpose of removing ammonia nitrogen from water.

2. Materials and Methods

2.1. Materials

P123 template (analytically pure) was purchased from Sigma Chemical Co. (St. Louis, MO, USA) (3-Mercaptopropyl) trimethoxysilane (MPTMS) (industrial grade) was purchased from Nanjing Fine

Chemical Co., Ltd. (NanJing, China), Anhydrous toluene and ethyl acetate (analytically pure) were purchased from Tianjin BASF Chemical Co., Ltd. (Tianjin, China), Dichloromethane (analytically pure) was purchased from Tianjin Fuyu Fine Chemical Co., Ltd. (Tianjin, China), Hydrochloric acid (37%) and anhydrous ethanol (100%) were obtained from Fisher Scientific (Waltham, MA, USA). Distilled water was obtained from the laboratory, homemade. All reagents were used as received without further purification.

2.2. Direct Synthesis in Hot Water

Two grams of poly (ethylene oxide)-poly (propylene oxide)-poly (ethylene oxide) block copolymer (PEO−PPO−PEO ((EO)20 (PO)70 (EO)20), namely P123) template was dissolved in 30 g of deionized water and stirred for 1 h by constant temperature magnetic stirrer bathed at 40 °C. After adding the treated rice husk ash as silicon source, it was stirred for 30 min violently. An amount of 0.4 g of MPTMS was added and stirred for 45 min. An amount of 10 mL of concentrated hydrochloric acid was added, the mixture was then stirred for 2 h continually. The solution was put in a stainless steel self-pressure autoclave lined with tetrafluoroethylene (PTFE) and then crystaled for 48 h in an oven at 100 °C. It was then poured out and a filtration was taken, it was washed two times repeatedly with 100 mL of ethanol /hydrochloric acid (50:1 v/v) solution for 4 h, the template P123 was removed and then the directly synthesized thiol modified mesoporous SBA-15 (H-SBA-15SH) was obtained. The experiment was performed in triplicate. IR (KBr pellet): v = 3434, 1608, 1078, 956, 807, 456 cm^{-1}; ^{13}C CP MAS-NMR (50.20 MHz) δ = 49.5 (SiCH$_2$CH$_2$C^{3}H$_2$SH), 27.4 (SiCH$_2$C^{2}H$_2$CH$_2$SH), 9.1 (SiC^{1}H$_2$CH$_2$CH$_2$SH); 29Si CP MAS-NMR (39.65 MHz) δ = −48.85 ((SiO)−Si−(OH)$_2$R, T^1 cyclic), −57.11 ((SiO)$_2$−Si−(OH) R, T^2 cyclic), −68.31 ((SiO)$_3$−Si−R, T^3 cyclic), −101.03 ((SiO)$_3$Si−(OH), Q^3 silicon atoms), −110.35 ((SiO)$_4$Si, Q^4 silicon atoms).

2.3. Grafting Synthesis

Two grams of P123 was weighed and dissolved in 30 g of deionized water, stirred for 2 h, bathed at 40 °C to dissolve it, 12.63 g of concentrated hydrochloric acid was added, and it continued to be stirred for 0.5 h. The treated rice husk ash was added and stirred for 3 h. The solution was put into a 50 mL reaction vessel, and crystallized in an oven at 100 °C for 72 h. The mixture was washed with distilled water and filtered, after drying the original powder of SBA-15 sample. Then the sample was heated to 550 °C at a speed of 2 °C/min in air atmosphere and calcined at this temperature for 6 h to remove the template, namely the mesoporous molecular sieves SBA-15. IR (KBr pellet): v = 1642, 1090, 962, 810, 469 cm^{-1}; ^{13}C CP MAS-NMR (50.20 MHz) δ = 76.7 (SiCH$_2$CH$_2$C^{3}H$_2$OH), 73.9 (SiCH$_2$C^{2}H$_2$CH$_2$OH), 18.9 (SiC^{1}H$_2$CH$_2$CH$_2$OH); ^{29}Si CP MAS-NMR (39.65 MHz) δ = −91.74 ((SiO)$_2$Si−(OH)$_2$, Q^2 silicon atoms), −98.72 ((SiO)$_3$Si−(OH), Q^3 silicon atoms), −109.87 ((SiO)$_4$Si, Q^4 silicon atoms).

One gram of SBA-15 and 0.36 g of MPTMS were dissolved in 100 g of anhydrous toluene, stirred, refluxed for 24 h, and then filtered. The filtered solid was refluxed and washed for 24 h with a solution containing 50 g of dichloromethane and 50 g of ethyl acetate. After removing the MPTMS from the surface of the sample, a mercapto group-containing surface modified molecular sieve SBA-15 was obtained, namely G-SBA-15-SH. IR (KBr pellet): v = 1092, 780, 600, 546, 450 cm^{-1}; ^{13}C CP MAS-NMR (50.20 MHz) δ = 49.1 (SiCH$_2$CH$_2$C^{3}H$_2$SH), 27.3 (SiCH$_2$C^{2}H$_2$CH $_2$SH), 9.6 (SiC^{1}H$_2$CH$_2$CH$_2$SH); ^{29}Si CP MAS-NMR (39.65 MHz) δ = −49.21 ((SiO)−Si−(OH)2R, T^1 cyclic), −58.24 ((SiO)$_2$−Si−(OH)R, T^2 cyclic), −69.56 ((SiO)$_3$−Si−R,T^3 cyclic), −102.17 ((SiO)$_3$Si−(OH), Q^3 silicon atoms), −111.20 ((SiO)$_4$Si, Q^4 silicon atoms).(You can see the details in the Supplementary).

2.4. Structure Characterization

X-ray diffraction analysis (XRD): A Philips X' pert instrument A LabX XRD-6000 X-ray diffractometer (Shimadzu Corporation, Kyoto, Japan) using monochromatic Cu Kα radiation was used to record powder X-ray diffraction patterns. A short wavelength particle or a crystalline material was used to reflect the photon surface of an atomic force plane, thus forming a three-dimensional structural form at the atomic force level [29]. Test conditions: CuKa line, λ = 0.15418 nm, tube voltage was 40 kV, tube current was 30 mA, 1°/min of scanning speed.

Fourier Transform Infrared spectrogram (FTIR): Thermo Nicolet Avatar 360 Fourier Transform Infrared Spectrometer (Thermo Fisher Scientific, Waltham, MA, USA) was used to characterize dried samples.

Scanning electron microscopy (SEM): A US-made Quanta 200F electron microscope (FEI, Hillsboro, OR, USA) test sample with an accelerating voltage of 29 KV was used. There was a sample test before the spray treatment.

Transmission electron microscopy (TEM): A Hitachi H-8100 instrument (Hitachi, Tokyo, Japan) at an accelerating voltage of 200 kV recorded the TEM images. The grinded sample was put into the weighing bottle with ethanol as a dispersant and then shocked for 5 min by ultrasonic. After that, the drop solution was taken by a clean drip tube onto the copper, the copper was dried, and the particle morphology of the sample was observed on the machine.

Proton-decoupled ^{13}C and ^{29}Si solid-state NMR spectra were recorded on a Bruker DSX Avance spectrometer.

Nitrogen adsorption–desorption: the nitrogen adsorption–desorption isotherms of the samples were measured at −196 °C and the samples were degassed at 150 °C for 8 h under vacuum. The specific surface area of the sample was calculated by the Brunauer–Emmett–Teller (BET) equation, and the pore size distribution curve was calculated by the Barrett–Joyner–Halenda (BJH) method [30].

The BJH method based on the Kelvin formula has a thermodynamic origin. In the BJH method, the mesopore pore size distribution is usually expressed by graphical form $\triangle V_p/\triangle r_p$ & r_p or d_p (V_p is the volume of mesopore, r_p is the radius of the cylindrical hole, and d_p is the width of the parallel to crack inside). The mesopore volume is completely filled at a relatively high pressure. The pore size distribution depends on whether there is a hysteresis loop. However, the new BJH method is more emphasis on Density Functional Theory [31–33] and Monte Carlo simulation [34]. MCM-41 molecular sieve materials characterized by Density Functional Theory has been reported but used rarely [35].

BET equations are used to determine the specific surface area of mesoporous solids [36]. Under relative low-pressure conditions, the BET equation is the abbreviation of the Langmuir equation and it is relatively good to describe the adsorption process under the condition of relative pressure 0.05–0.35. This specific surface area, as (BET), was obtained from the following equation.

$$a_s\ (BET) = n_m^a \times L \times a_m$$

L is the avogadro constant, n_m^a is the amount of adsorption of a single coating (the surface of the adsorbent in the unit to form a complete single coating), a_m is the average area that accounts for the proportion of adsorbate molecules.

2.5. Effect of the Initial Concentration of Ammonia Nitrogen on the Adsorption of Ammonia Nitrogen

Every modified zeolite (modified by sodium chloride) of a molecular sieve of SBA-15, H-SBA-15-S, G-SBA-15-SH respectively weighed exactly 0.1 g. The number of every weighed molecular sieve should meet the requirements of the following NH$_4$Cl standard solution concentration gradient. Then every one of them was added into a 50 mL triangular flask, at the same time, 50 mL NH$_4$Cl standard working solution with a concentration of 0.1, 0.2, 0.4, 0.8, 1.6 and 2 mg/L were respectively used to test. The solutions were oscillated 120 min by the air flow oscillator, and then static for a period of time, filtering with microporous filter membrane, after that the supernatant was determined by spectrophotometry. Calculating the concentration of ammonia nitrogen in a solution by a standard curve, the test results were as shown in Table 1. The concentration of residual ammonia nitrogen after treatment with three kinds of molecular sieves was calculated by the following formula:

$$q(mg/g) = (C_0 - C) \times V/G$$

q is the adsorption quantity of ammonia nitrogen by zeolite, C_0 is the concentration of ammonia nitrogen in the simulated raw water (mg/L), *C* is the concentration of ammonia nitrogen in solution after treatment (mg/L), *V* is the volume of the added NH_4Cl simulated solution (ML), *G* is the mass of zeolite (g).

Table 1. Textural properties of mesoporous zeolites.

Molecular Sieve Samples	Specific Surface Area (m²/g)	Pore Volume (cm³/g)	Average Pore Size (nm)
SBA-15	310.1	0.43	6.85
H-SBA-15-SH	571.3	0.88	9.26
G-SBA-15-SH	463.8	0.85	6.53

2.6. Statistical Analysis

All experiments were performed in triplicate, as the replicated experimental units and the results were provided with mean ± SD (standard deviation) values. One-way analysis of variance (ANOVA) was performed, and the significance of each mean value was determined ($p < 0.05$) with the Duncan's multiple range test of the statistical analysis system using the SPSS computer program (SPSS, Inc., Chicago, IL, USA).

3. Results and Discussion

3.1. X-Ray Diffraction

According to the Prague formula ($n\lambda = 2d\sin\theta$, d is the distance of the atomic crystal plane in the crystal phase, *n* is an integer, and θ is the Bragg angle), the sample was diffracted in the crystalline phase. The intensity of the X-ray diffraction depends on the 2θ angle diffraction function and the crystal orientation of the sample. The diffraction properties determine the structural properties of the sample (e.g., crystal orientation and size) and the crystalline phase. The orderliness of the pore material was measured by X-ray diffraction with powder. Ordered mesoporous organosilicon material has a very high diffraction peak when 2θ = 2° and a low diffraction peak in the range of 2θ varied from 3° to 8° [23,24]. The cell parameters are calculated by the formula $a_0 = 2d_{100}/ (3)^{-/2}$ in the hexagonal mesopore phase [23]. The difference of the cell parameters and the pore size (measured by nitrogen adsorption–desorption) gives the mesoporous phase pore-thickness. Small angle X-ray diffraction can determine the mesoporous phase of ordered mesoporous materials. A strong diffraction peak d_{100} appears at the diffraction angle 2θ of 0.80° and another two weak diffraction peaks d_{100} and d_{200} appear at 1.4°–1.8°, as shown in Figure 1a, b. The black line represents SBA-15-SH and the red line represents the SBA-15. The three diffraction peaks are typical characteristic peaks of two-dimensional hexagonal structures, and these are attributed to the (100), (110), (200) crystal diffraction peaks of the two-dimensional hexagonal (p6mm) structure respectively, and the d values of the diffraction peaks of different crystal planes satisfy the characteristic relation of hexagonal lattice.

The XRD patterns of the two mesoporous molecular sieves of SBA-15 and H-SBA-15-SH are illustrated schematically in Figure 1a, the three absorption peaks of the chromatogram of H-SBA-15-SH synthesized by direct synthesis method, corresponding to facet absorption peaks of SBA-15 (100, 110, 200) respectively. The intensity of the three absorption peaks is significantly weaker than that of SBA-15, and not only are the main peak of d_{110} and characteristic peaks of d_{200} and d_{100} significantly weakened, but also the d_{100} peak of H-SBA-15-SH is drifted at a high angle compared with *d*100 peak of SBA-15, which related to thiol getting into the molecular sieve skeleton.

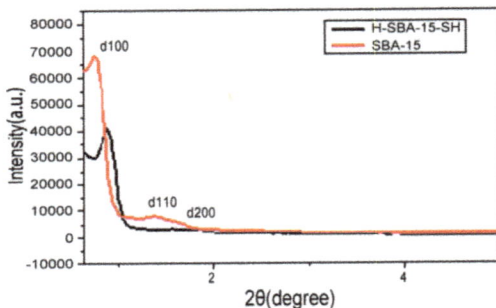

(a)

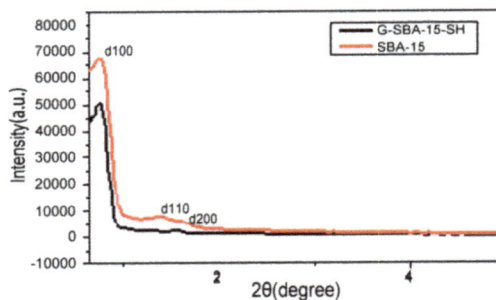

(b)

Figure 1. (**a**) X-ray diffraction patterns of mesoporous SBA-15 and H-SBA-15-SH; (**b**) X-ray diffraction patterns of mesoporous SBA-15 and G-SBA-15-SH.

The XRD patterns of the two mesoporous molecular sieves of SBA-15 and G-SBA-15-SH are illustrated schematically in Figure 1b. The three absorption peaks of the chromatogram of H-SBA-15-SH synthesized by the post-grafting method, also correspond to facet absorption peaks of SBA-15 (100, 110, 200) respectively. Unlike H-SBA-15-SH, the intensity of these peaks hardly changes, and the peak of d100 of G-SBA-15-SH almost has no drift compared with that of SBA-15, which indicates that no effect is produced after introducing the mercapto group into the molecular sieve framework by the post-grafting method, and the G-SBA-15-SH maintain a good pore structure and regular pore arrangement. In a word, the ordered mesoporous materials of G-SBA-15-SH and the synthesized H-SBA-15-SH are consistent with those reports [24].

3.2. Fourier Transform Infrared Spectroscopy

SBA-15 shows the characteristic bands at 1642, 1090, 810, and 469 cm^{-1}, which were attributed to the O–H vibration of the silanol (Si–OH) group or the silanol group of the cross-hydrogen bond [37,38], the symmetric stretching vibration of the Si–OH swing mode [37], the Si–O–Si asymmetric stretching vibration [37,39], the Si–O–Si symmetric vibration [39], and the bending mode of Si–O–Si [40].

At the peak of 1078 cm^{-1}, the peak of H-SBA-15-SH had a tendency to move to the right with respect to the SBA-15, which was caused by the asymmetric Tensile vibration of Si–O–Si [40], while the peak of G-SBA-15-SH was not quite different from the location of SBA-15. It could be found that there was a sharp band between H-SBA-15-SH and G-SBA-15-SH at 3000–2800 cm^{-1}, which was attributed to the C–H stretching belt of propyl group [40].

3.3. Scanning Electron Microscopy

Scanning electron microscope (SEM) photos of SBA-15 and SBA-15-SH mesoporous molecular sieves were shown in Figure 2A. It can be seen that the structure is evenly distributed [41]. The SBA-15 mesoporous molecular sieve had a short cylindrical appearance. After the incorporation of the functional group, it still maintained the hexangular mesoporous structure of SBA-15 [42], but the shape of H-SBA-15-SH was more dispersed and small, and the shape of G-SBA-15-SH was slightly longer. The obvious surface roughness on SEM images may be due to the presence of organic functional groups on the surface and the presence of MPTMS [43]. Overall, the mesoporous structure of SBA-15 was greatly retained in organic functionalization, which was consistent with the XRD results [44].

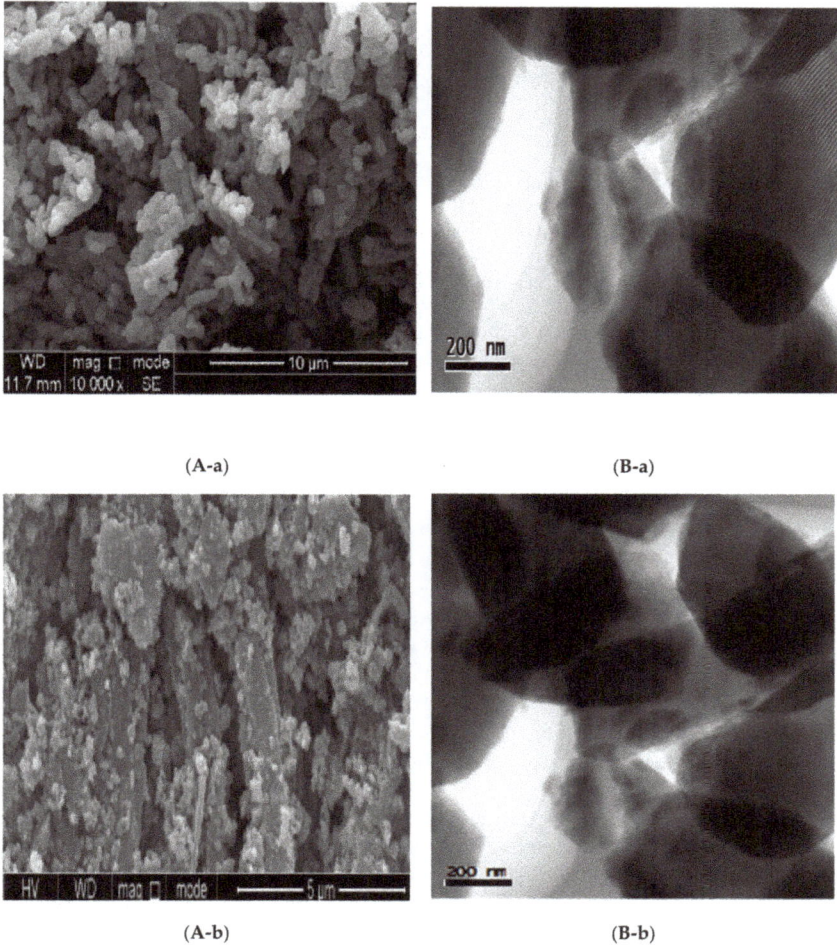

(A-a)

(B-a)

(A-b)

(B-b)

Figure 2. *Cont.*

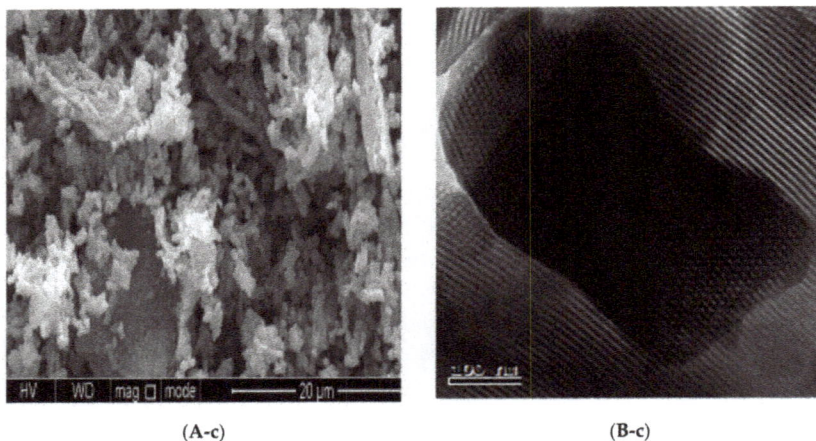

(A-c)   (B-c)

Figure 2. Scanning electron microscopy (SEM (**A**) and transmission electron microscopy (TEM) (**B**) of mesoporous SBA-15 (**a**); H-SBA-15-SH (**b**); G-SBA-15-SH (**c**).

3.4. Transmission Electron Microscopy

The internal morphology of SBA-15 and SBA-15-SH mesoporous molecular sieves were studied by TEM as shown in Figure 2B. Figure 2B showed a better honeycomb structure in the middle of the picture [45]. In Figure 2B-a, it clearly indicated the formation of SBA-15 with a highly ordered hexangular mesoporous structure, where hexangular channels were clearly observed [46]. The SBA-15's tunnel was parallel to the axis and was arranged in an orderly regular cylindrical channel in the direction perpendicular to the axis [46], thus further proving that SBA-15 had a two-dimensional p6mm hexangular structure. After the synthesis of SBA-15-SH, the structure of SBA-15 remained to be maintained, as shown in Figure 2B-b,B-c. The addition of (3-mercapto propyl) trimethoxane did not damage the porous structure of the six party [46].

3.5. Analysis of Pore Size Distribution

Figure 3 shows that the pore size distribution curve of mesoporous molecular sieves SBA-15, which is a typical mesopore pore size distribution with a narrow pore size distribution, varied from 5 to 7 nm (Figure 3). The average pore size was 6.85 nm calculated by the new BJH method. When nitrogen was at 77 K·am, (N_2) = 0.162 nm^2. The specific surface area of the mesoporous SBA-15 was 310.1 m^2/g measured by the BET formula, and the pore volume was 0.43 cm^3/g. However, the BET was lower than that of mesoporous silicone reported in the literature (greater than 1000 m^2/g) [47].

The pore–size distribution curves of mesoporous SBA-15, H-SBA-15-SH and G-SBA-15-SH are shown in Figure 3. As shown in Figure 3, mesoporous H-SBA-15-SH synthesized by direct synthesis and mesoporous G-SBA-15-SH synthesized by post-grafting synthesis both had homogeneous mesoporous structures, but the pore size distribution showed different changes. The pore size distribution of mesoporous H-SBA-15-SH was wider than that of SBA-15. It showed that direct synthesis was a direct introduction of a small amount of mercapto into the skeletal structure of SBA-15, causing pore diameter to increase. The pore size distribution of mesoporous G-SBA-15-SH was narrower than that of mesoporous SBA-15, which indicated that post-grafting method combined a large amount of mercapto groups with siloxanes on the surface of SBA-15 and entered into the inner pores so that the thickness of the hole wall increased and pore size reduced. The conclusion is that the pore size distribution curve was consistent with the XRD analysis.

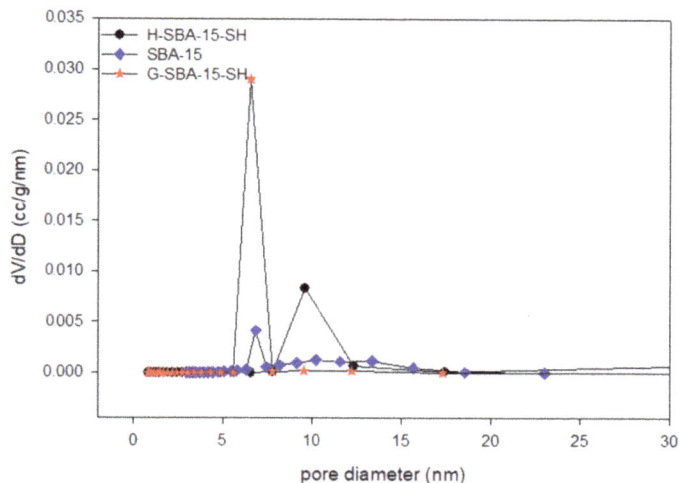

Figure 3. Pore size distributions of mesoporous SBA-15, H-SBA-15-SH, G-SBA-15-SH.

3.6. Analysis of Nitrogen Adsorption–Desorption

The N_2 adsorption–desorption isotherm of the mesoporous molecular sieves SBA-15 is shown in Figure 4b. It can be seen that the N_2 adsorption–desorption isotherms drawn according to the results of the test conform to the typical characteristics of the adsorption isotherm type IV set by IUPAC [36], and meanwhile it showed that the synthetic materials had a mesoporous structure. At the lower relative pressure, the monolayer adsorption occurred, the curves of which showed that the adsorption and desorption lines coincided. Then, multilayer adsorption occurred until the pressure was sufficient to cause capillary condensation. The adsorption isotherm changed greatly, when the adsorption hysteresis occurred, namely the adsorption and desorption curves were separated from each other. In this case, the isotherm had a hysteresis loop. There was an adsorption hysteresis in Type IV due to capillary condensation and a relatively high relative pressure ($0.45 < p/p_0 < 0.8$). The hysteresis loop of type H1 [36,48] existed with parallel and close vertical branches as shown in Figure 4. Type H1 represented adsorption in a narrow mesopore with a small pore size distribution, which was mostly caused by the uniform size and regular shape of the hole. In the range of the small relative pressure, the curve was relatively flat, which may be because nitrogen molecules were adsorbed on the surface of the mesoporous. In the range of the high relative pressure, the adsorption capacity increased rapidly with the increase in relative pressure, which may be related to the nitrogen molecules adsorbed to the mesoporous pores by the monolayer and multilayer adsorption, causing capillary agglomeration. Then the adsorption capacity increased slowly with the increase in the relative pressure and the adsorption gradually saturated. At the same time, there was a large pressure in the middle pressure zone, which was related to the larger pore size of mesoporous molecular sieve SBA-15.

N_2 adsorption isotherms of mesoporous are as shown in Figures 4 and 5. From Figure 4a, the H-SBA-15-SH synthesized by the direct synthesis method has the type IV of adsorption–desorption curve [36,48]. The hysteresis loop shape was similar to that of SBA-15. And with the increase in relative pressure ($p/p_0 > 0.42$), the adsorption curve also has an obvious breakthrough, which was typical capillary adsorption of mesoporous with uniform pore distribution. But unlike Figure 4b, N_2 adsorption capacity increased as shown in Table 1, the pore size increased from 6.85 to 9.26 nm, the specific surface area increased from 310.1 to 571.3 m^2/g, and the pore volume increased from 0.43 to 0.88 cm^3/g. Thus, the H-SBA-15-SH synthesized by direct synthesis has a larger pore size and specific surface area and can further synthesize the carrier of mesoporous catalyst.

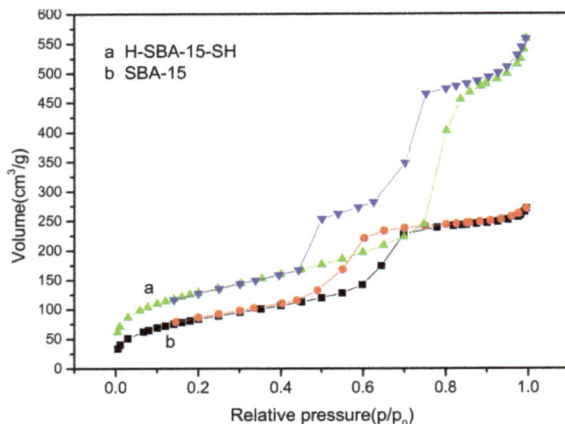

Figure 4. N_2 adsorption isotherms of mesoporous SBA-15 and H-SBA-15-SH.

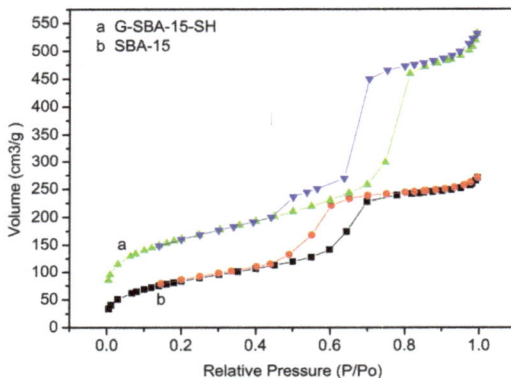

Figure 5. N_2 adsorption isotherms of mesoporous SBA-15 and G-SBA-15-SH.

The N_2 adsorption–desorption isotherm of mesoporous G-SBA-15-SH synthesized by post-grafting was shown in Figure 5a, the isotherm also had type IV of the adsorption–desorption curve [36,48]. The hysteresis loop shape was also similar to that of SBA-15 ordered mesoporous molecular sieves. With the increase in relative pressure ($p/p_0 > 0.42$), the adsorption curve had an obvious breakthrough, which was also a typical capillary adsorption agglomeration of mesoporous with a uniform pore distribution [36,48]. But there was some difference from N_2 adsorption isotherms in ordered mesoporous molecular sieve H-SBA-15-SH synthesized by the direct synthesis method. The hysteresis loop synthesized by the post-grafting method was elongated compared with the SBA-15 ordered mesoporous molecular sieves, the N_2 adsorption capacity decreased and the hysteresis loop shifted to the high p/p_0 value. The hysteresis loop may change from H4 to H3.

The amount of ammonia nitrogen adsorbed was measured by the mass of ammonia nitrogen in water removed by molecular sieves per unit mass. The relationship between the amount of ammonia nitrogen adsorption and the initial concentration of ammonia nitrogen was shown in Tables 2–4. With the increase in the concentration of ammonia and nitrogen, the adsorption amount of ammonia nitrogen increases gradually. That was, with the increase in the ammonia nitrogen initial concentration, the adsorption capacity of the modified molecular sieve to ammonia nitrogen was constantly improved. At the same time, it can be found that the adsorption amount of ammonia nitrogen by G-SBA-15-SH molecular sieve was the largest, the H-SBA-15-SH molecular sieve was the second, while the unmodified SBA-15 molecular sieve had the smallest ammonia nitrogen adsorption. Besides, with the increase in ammonia nitrogen concentration, the ammonia nitrogen adsorption capacity changed more and more. When the concentration was very low, it was difficult to adsorb.

Table 2. Removal of ammonia nitrogen from SBA-15 molecular sieve.

Concentration of Ammonia Nitrogen Solution (mg/L)	Ammonia Nitrogen Content in 50 mL Solution (mg)	The Content of Ammonia Nitrogen in the Solution after Adsorption (mg)	The Additive Amount of Molecular Sieve (g)	Ammonia Nitrogen Adsorption Capacity (mg/g)
0.1	0.005	0.0048	0.1	1.667
0.2	0.01	0.0096	0.1	3.667
0.4	0.02	0.0193	0.1	7.444
0.8	0.04	0.0387	0.1	13.22
1.6	0.08	0.0759	0.1	21.00
2	0.1	0.0971	0.1	29.00

Table 3. Removal of ammonia nitrogen from H-SBA-15-SH molecular sieve.

Concentration of Ammonia Nitrogen Solution (mg/L)	Ammonia Nitrogen Content In 50 mL Solution (mg)	The Content of Ammonia Nitrogen in the Solution after Adsorption (mg)	The Additive Amount of Molecular Sieve (g)	Ammonia Nitrogen Adsorption Capacity (mg/g)
0.1	0.005	0.0050	0.1	2.551
0.2	0.01	0.0096	0.1	6.332
0.4	0.02	0.0198	0.1	11.04
0.8	0.04	0.0391	0.1	17.01
1.6	0.08	0.0796	0.1	24.72
2	0.1	0.0997	0.1	34.33

Table 4. Removal of ammonia nitrogen from G-SBA-15-SH molecular sieve.

Concentration of Ammonia Nitrogen Solution (mg/L)	Ammonia Nitrogen Content in 50 mL Solution (mg)	The Content of Ammonia Nitrogen in the Solution after Adsorption (mg)	The Additive Amount of Molecular Sieve (g)	Ammonia Nitrogen Adsorption Capacity (mg/g)
0.1	0.005	0.0050	0.1	3.312
0.2	0.01	0.0096	0.1	7.898
0.4	0.02	0.0120	0.1	14.17
0.8	0.04	0.0390	0.1	19. 02
1.6	0.08	0.0781	0.1	30. 32
2	0.1	0.0970	0.1	41.31

4. Conclusions

In acidic conditions, pure silicon synthetic SBA-15 with P123 as the template and rice husk ash as the silica source was prepared and was analyzed by FTIR, XRD, SEM, TEM, ^{13}C CP MAS-NMR, ^{29}Si CP MAS-NMR and BET. The result showed that synthetic sample mesoporous SBA-15 had a two-dimensional hexagonal structure with an average pore size of 6.85 nm. The specific surface area of the mesoporous SBA-15 was 310.1 m^2/g and the pore volume was 0.43 cm^3/g. The direct synthesis method of the ordered mesoporous molecular sieves H-SBA-15-SH is the way of introducing the mercapto directly, causing both the pore size, the pore volume, and the specific surface area to increase. The mesoporous molecular sieve can further synthesize the mesoporous catalyst carrier. The ordered mesoporous molecular sieves G-SBA-15-SH synthesized by post-grafting proved that the sulfhydryl group was bonded to the silyl group on the surface of the G-SBA-15-SH, and combined into the interior of the channel, causing the hole wall thickness to increase and the pore size to reduce. The N_2 adsorption capacity decreased. According to the results of the experiment, the ammonia-nitrogen from water can be expected to be removed by the SBA-15-SH.

Supplementary Materials: The following are available online at http://www.mdpi.com/2073-4360/10/8/819/s1.

Author Contributions: D.L. initiated the writing of this manuscript and designed the structure of this manuscript, interpreted results, and drafted the manuscript. Y.H., Y.Y. and X.L. compiled information and made contribution to the revision of the manuscript. L.H., X.Z. designed the structure of this manuscript, interpreted results, and revised the language. W.Q., Q.Z., D.W. and H.C. compiled information and made contribution to the revision of the manuscript. Z.W. and S.L. made certain contribution to the language modification of the manuscript.

Funding: The authors thank the National Science Foundation of China (No. 31340032), "211 Engineering Double Support Plan" (No. 03572081), Sichuan Agricultural University, and the education department of Sichuan Province major project (No. 17ZB0338) for financial support.

Conflicts of Interest: The authors declare no conflict of interest.

References

1. Lin, D.R.; Hu, L.J.; You, H.; Tolbert, S.H.; Loy, D.A. Comparison of new periodic, mesoporous, hexylene-bridged polysilsesquioxanes with Pm3n symmetry versus sol-gel polymerized, hexylene-bridged gels. *J. Non-Cryst. Solids* **2014**, *406*, 139–143. [CrossRef]

2. Lin, D.R.; Hu, L.J.; Tolbert, S.H.; Li, Z.; Loy, D.A. Controlling nanostructure in periodic mesoporous hexylene-bridged polysilsesquioxanes. *J. Non-Cryst. Solids* **2015**, *419*, 6–11. [CrossRef]

3. Lin, D.R.; Hu, L.J.; Li, Z.; Loy, D.A. Influence of alkylene-bridging group length on mesostructure and porosity in cubic (Pm3n) periodic mesoporous bridged polysilsesquioxanes. *J. Porous Mater.* **2014**, *21*, 39–44. [CrossRef]

4. Lin, D.R.; Hu, L.J.; You, H.; Williams, R.J.J. Synthesis and characterization of a nanostructured photoluminescent silsesquioxane containing urea and dodecyl groups that can be patterned on carbon films. *Eur. Polym. J.* **2011**, *47*, 1526–1533. [CrossRef]

5. Lin, D.R.; Zhao, Q.; Hu, L.J.; Xing, B.S. Synthesis and characterization of cubic mesoporous bridged polysilsesquioxane for removing organic pollutants from water. *Chemosphere* **2014**, *103*, 188–196. [CrossRef] [PubMed]

6. Hu, L.C.; Shea, K.J. Organo-silica hybrid functional nanomaterials: How do organic bridging groups and silsesquioxane moieties work hand-in-hand? *Chem. Soc. Rev.* **2011**, *40*, 688–695. [CrossRef] [PubMed]

7. Shea, K.J.; Moreau, J.; Loy, D.A.; Corriu, R.J.P.; Boury, B. Bridged Polysilsesquioxanes. Molecular-engineering nanostructured hybrid organic-inorganic materials. In *Functional Hybrid Materials*; Gómez-Romero, P., Sanchez, C., Eds.; Wiley-VCH: Weinheim, Germany, 2005.

8. Loy, D.A. Sol-gel processing of hybrid organic-inorganic materials based on polysilsesquioxanes. In *Hybrid Materials: Synthesis, Characterization, and Applications*; Kickelbick, G., Ed.; Wiley-VCH: Weinheim, Germany, 2006.

9. Waiman, C.V.; Chesta, C.A.; Gómez, M.L. Hybrid films based on a bridged silsesquioxane doped with goethite and montmorillonite nanoparticles as sorbents of wastewater contaminants. *J. Nanomater.* **2016**, *40*, 6286247. [CrossRef]

10. Zhang, X.; Huang, Y.; Wang, T.; Liu, L. Influence of fibre surface oxidation–reduction followed by silsesquioxane coating treatment on interfacial mechanical properties of carbon fibre/polyarylacetylene composites. *Compos. Part A Appl. Sci. Manuf.* **2007**, *38*, 936–944. [CrossRef]

11. Li, Y.; Dong, X.H.; Zou, Y.; Wang, Z.; Yue, K.; Huang, M.; Zhang, W.B. Polyhedral oligomeric silsesquioxane meets "click" chemistry: Rational design and facile preparation of functional hybrid materials. *Polymer* **2017**, *125*, 303–329. [CrossRef]

12. Wang, Z.; Li, Y.; Dong, X.H.; Yu, X.; Guo, K.; Su, H.; Zhang, W.B. Giant gemini surfactants based on polystyrene–hydrophilic polyhedral oligomeric silsesquioxane shape amphiphiles: Sequential "click" chemistry and solution self-assembly. *Chem. Sci.* **2013**, *4*, 1345–1352. [CrossRef]

13. Lin, D.R.; Hu, L.J.; Xing, B.S.; You, H.; Loy, D.A. Mechanisms of Competitive Adsorption Organic Pollutants on Hexylene-Bridged Polysilsesquioxane. *Materials* **2015**, *8*, 5806–5817. [CrossRef] [PubMed]

14. Lin, D.R.; Kong, M.; Li, L.Y.; Li, X.D.; Zhang, X.W. Enhanced Anti-Ultraviolet and Thermal Stability of a Pesticide via Modification of a Volatile Organic Compound (VOC)-Free Vinyl-Silsesquioxane in Desert Areas. *Polymers* **2016**, *8*, 282. [CrossRef]

15. Lin, D.R.; Hu, L.J.; Zhang, Q.; You, H. Design of a POSS-modified zeolite structure and the study of the enhancement of ammonia-nitrogen removal from drinking water. In *Molecular Environmental Soil Science at the Interfaces in the Earth's Critical Zone*; Springer: Berlin/Heidelberg, Germany, 2010; pp. 118–120.

16. Wu, B.P.; Qi, Y.T.; Yuan, X.D.; Shen, J.; Bi, J.; Li, C. Synthesis of tributyl citrate over 3-SBA-15 mesoprous molecular sieve catalyst. *Ind. Catal.* **2004**, *12*, 32–35. (In Chinese)

17. Liu, T.H. *Preparation of Silcequioxane Hybrid Material from Rice Hush and Properity Study*; Harbin Institute of Technology: Harbin, China, 2007. (In Chinese)

18. Zhai, Q.Z.; Wang, W.; Jiang, T.S.; Wang, Y.; Chen, M.G. Modification for SBA-15 Molecular sieve by Lanthanum(III). *J. Inorg. Mater.* **2004**, *19*, 1212–1216, In Chinese.

19. Coutinho, D.; Aeevedo, A.O.; Dieekmann, G.R.; Balkus, K.J., Jr. Molecular imprinting of mesoporous SBA-15 with chiral ruthenium complexes. *Microporous Mesoporous Mater.* **2002**, *54*, 297–302. [CrossRef]

20. Reddy, S.S.; Raju, B.D.; Kumar, V.S.; Padmasri, A.H.; Narayanan, S.; Rao, K.S.R. Sulfonic acid funetionalized mesoporous SBA-15 for selective synthesis of 4-phenyl-1,3-dioxane. *Catal. Conunun.* **2007**, *8*, 261–266. [CrossRef]

21. Asefa, T.; MacLachlan, M.J.; Grondey, H.; Coombs, N.; Ozin, J.A. Metamorphic channels in periodic mesoporous methylenesilica. *Angew. Chem. Int.* **2000**, *112*, 1878–1881. [CrossRef]

22. Sanchez, C. Design of functional materials: From nanostructured hybrid materials to hierarchical structures. *Abstr. Pap. Am. Chem. Soc.* **2004**, *228*, U490.

23. Beck, J.S.; Vartuli, J.C.; Roth, W.J.; Leonowicz, M.E.; Kresge, C.T.; Schmitt, K.D.; Chu, C.T.W.; Olson, D.H.; Sheppard, E.W.; McCullen, S.B.; et al. A New family of mesoporous molecular-sieves prepared with liquid-crystal templates. *J. Am. Chem. Soc.* **1992**, *114*, 10834–10843. [CrossRef]

24. Kresge, C.T.; Leonowicz, M.E.; Roth, W.J.; Vartuli, J.C.; Beck, J.S. Ordered mesoporous molecular-sieves synthesized by a liquid-crystal template mechanism. *Nature* **1992**, *359*, 710–712. [CrossRef]

25. He, Q.; Shi, J.; Zhao, J.; Chen, Y.; Chen, F. Bottom-up tailoring of nonionic surfactant-templated mesoporous silica nanomaterials by a novel composite liquid crystal templating mechanism. *J. Mater. Chem.* **2009**, *19*, 6498–6503. [CrossRef]

26. Wu, B.; Tong, Z.; Yuan, X. Synthesis, characterization and catalytic application of mesoporous molecular sieves SBA-15 functionalized with phosphoric acid. *J. Porous Mater.* **2012**, *19*, 641–647. [CrossRef]

27. Ziolek, M.; Nowak, I. Synthesis and characterization of niobium-containing MCM-41. *Zeolites* **1997**, *18*, 356–360. [CrossRef]

28. Kruk, M.; Jaroniec, M.; Sayari, A. Structural and surface properties of siliceous and titanium-modified HMS molecular sieves. *Microporous Mater.* **1997**, *9*, 173–182. [CrossRef]

29. Leary, J.J.; Messick, E.B. Constrained calibration curves: A novel application of Lagrange multipliers in analytical chemistry. *Anal. Chem.* **1985**, *57*, 956–957. [CrossRef]

30. Barrett, E.P.; Joyner, L.G.; Halenda, P.P. The determination of pore volume and area distribution of pore volume and area distributions in porous substances 1 computations from nitrogen isotherms. *J. Am. Chem. Soc.* **1951**, *73*, 373–380. [CrossRef]

31. Ciesla, U.; Schuth, F. Ordered mesoporous materials. *Microporous Mesoporous Mater.* **1999**, *27*, 131–149. [CrossRef]

32. Ravikovitch, P.I.; Wei, D.; Chueh, W.T.; Haller, G.L.; Neimark, A.V. Evaluation of pore structure parameters of MCM-41 catalyst supports and catalysts by means of nitrogen and argon adsorption. *J. Phys. Chem. B* **1997**, *101*, 3671–3679. [CrossRef]

33. Ravikovitch, P.I.; Odomhnaill, S.C.; Neimark, A.V.; Schüth, F.; Unger, K.K. Capillary hysteresis in nanopores: Theoretical and experimental studies of nitrogen adsorption on MCM-41. *Langmuir* **1995**, *11*, 4765–4772. [CrossRef]

34. Maddox, M.W.; Olivier, J.P.; Gubbins, K.E. Characterization of MCM-41 using molecular simulation: Heterogeneity effects. *Langmuir* **1997**, *13*, 1737–1745. [CrossRef]

35. Ravikovitch, P.I.; Neimark, A.V. Relations between structural parameters and adsorption characterization of templated nanoporous materials with cubic symmetry. *Langmuir* **2000**, *16*, 2419–2423. [CrossRef]

36. Sing, K.S.W. Reporting physisorption data for gas solid systems with special reference to the determination of surface-area and porosity (recommendations 1984). *Pure Appl. Chem.* **1985**, *57*, 603–619. [CrossRef]

37. Ghosh, B.K.; Hazra, S.; Naik, B.; Ghosh, N.N. Preparation of Cu nanoparticle loaded SBA-15 and their excellent catalytic activity in reduction of variety of dyes. *Powder Technol.* **2015**, *269*, 371–378. [CrossRef]

38. Macina, D.; Piwowarska, Z.; Góra-Marek, K.; Tarach, K.; Rutkowska, M.; Girman, V.; Błachowski, A.; Chmielarz, L. SBA-15 loaded with iron by various methods as catalyst for DeNOx process. *Mater. Res. Bull.* **2016**, *78*, 72–82. [CrossRef]

39. Wang, S.; Wang, K.; Dai, C.; Shi, H.; Li, J. Adsorption of Pb2+ on amino-functionalized core–shell magnetic mesoporous SBA-15 silica composite. *Chem. Eng. J.* **2015**, *262*, 897–903. [CrossRef]

40. Hashemikia, S.; Hemmatinejad, N.; Ahmadi, E.; Montazer, M. Optimization of tetracycline hydrochloride adsorption on amino modified SBA-15 using response surface methodology. *J. Colloid Interface Sci.* **2015**, *443*, 105–114. [CrossRef] [PubMed]

41. Gondal, M.A.; Suliman, M.A.; Dastageer, M.A.; Chuah, G.K.; Basheer, C.; Yang, D.; Suwaiyan, A. Visible light photocatalytic degradation of herbicide (Atrazine) using surface plasmon resonance induced in mesoporous Ag-WO3/SBA-15 composite. *J. Mol. Catal. A Chem.* **2016**, *425*, 208–216. [CrossRef]

42. Pudukudy, M.; Yaakob, Z.; Akmal, Z.S. Direct decomposition of methane over Pd promoted Ni/SBA-15 catalysts. *Appl. Surf. Sci.* **2015**, *353*, 127–136. [CrossRef]

43. Palai, Y.N.; Anjali, K.; Sakthivel, A.; Ahmed, M.; Sharma, D.; Badamali, S.K. Cerium Ions Grafted on Functionalized Meso porous SBA-15 Molecular Sieves: Pre paration and Its Catalytic Activity on p-Cresol Oxidation. *Catal. Lett.* **2018**, *148*, 465–473. [CrossRef]

44. Xie, W.; Fan, M. Biodiesel production by transesterification using tetraalkylammonium hydroxides immobilized onto SBA-15 as a solid catalyst. *Chem. Eng. J.* **2014**, *239*, 60–67. [CrossRef]

45. Cakiryilmaz, N.; Arbag, H.; Oktar, N.; Dogu, G.; Dogu, T. Effect of W incorporation on the product distribution in steam reforming of bio-oil derived acetic acid over Ni based Zr-SBA-15 catalyst. *Int. J. Hydrogen Energy* **2018**, *43*, 3629–3642. [CrossRef]

46. Pudukudy, M.; Yaakob, Z.; Akmal, Z.S. Direct decomposition of methane over SBA-15 supported Ni, Co and Fe based bimetallic catalysts. *Appl. Surf. Sci.* **2015**, *330*, 418–430. [CrossRef]

47. Guan, S.; Inagaki, S.; Ohsuna, T.; Terasaki, O. Cubic hybrid organic-inorganic mesoporous crystal with a decaoctahedral shape. *J. Am. Chem. Soc.* **2000**, *122*, 5660–5661. [CrossRef]

48. Leofanti, G.; Padovan, M.; Tozzola, G.; Venturelli, B. Surface area and pore texture of catalysts. *Catal. Today* **1998**, *41*, 207–219. [CrossRef]

polymers

MDPI

Review

Synthetic Routes to Silsesquioxane-Based Systems as Photoactive Materials and Their Precursors

Beata Dudziec [1,2,*], Patrycja Żak [1,*] and Bogdan Marciniec [1,2]

[1]   Department of Organometallic Chemistry, Faculty of Chemistry, Adam Mickiewicz University in Poznan, Umultowska 89B, 61-614 Poznan, Poland; bogdan.marciniec@amu.edu.pl
[2]   Centre for Advanced Technologies, Adam Mickiewicz University in Poznan, Umultowska 89C, 61-614 Poznan, Poland
*   Correspondence: beata.dudzec@gmail.com (B.D.); pkw@amu.edu.pl (P.Ż.); Tel.: +48-61-829-18-78 (B.D.); +48-61-829-17-31 (P.Ż.)

Received: 5 February 2019; Accepted: 9 March 2019; Published: 16 March 2019

Abstract: Over the past two decades, organic optoelectronic materials have been considered very promising. The attractiveness of this group of compounds, regardless of their undisputable application potential, lies in the possibility of their use in the construction of organic–inorganic hybrid materials. This class of frameworks also considers nanostructural polyhedral oligomeric silsesquioxanes (POSSs) with "organic coronae" and precisely defined organic architectures between dispersed rigid silica cores. A significant number of papers on the design and development of POSS-based organic optoelectronic as well as photoluminescent (PL) materials have been published recently. In view of the scientific literature abounding with numerous examples of their application (i.e., as OLEDs), the aim of this review is to present efficient synthetic pathways leading to the formation of nanocomposite materials based on silsesquioxane systems that contain organic chromophores of complex nature. A summary of stoichiometric and predominantly catalytic methods for these silsesquioxane-based systems to be applied in the construction of photoactive materials or their precursors is given.

Keywords: silsesquioxanes; optoelectronics; OLEDs

---

1. Introduction

Functionalized polyhedral oligomeric silsesquioxanes (POSSs) [1] with a general $(RSiO_{3/2})_n$ formula are inorganic–organic systems composed of a well-defined Si–O–Si core with organic moieties that have become the representative building blocks for hybrid materials (Figure 1).

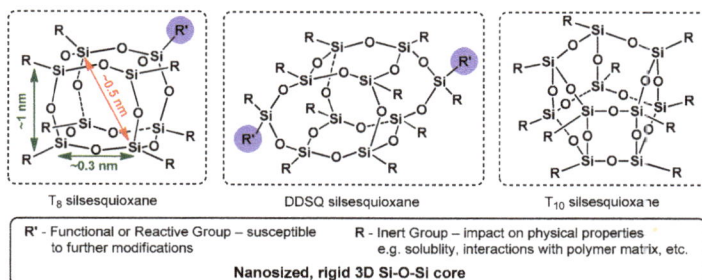

Figure 1. The hybrid (i.e., organic–inorganic) nanostructures of respective cubic T_8, DDSQ and T_{10} silsesquioxanes.

Due to their finely tunable physical and chemical properties, POSSs meet the requirements of both science and industry [2,3]. The silsesquioxane family group contains a small number of well-defined 3D structures. The most recognizable and best known of this is cubic T_8 with either one or eight functional group derivatives [3]. Recently, the so-called double-decker-type silsesquioxane (DDSQ) structure, which differs from the symmetric, cubic one, and features opened (M_4T_8 = DDSQ-4OSi [4]) or closed (D_2T_8 = DDSQ-2Si [5]) frameworks with either two or four reactive moieties, has also gained respective interest [6]. The physicochemical properties of functionalized silsesquioxanes are a particularly important aspect of the research on their synthesis. This is due to their chemical and spatial structure resulting in a hybridic (i.e., organic–inorganic) nature. The presence of a regular element of the Si–O–Si framework corresponds with the silica architecture places silsesquioxanes as fillers (nanofillers) with a perfectly defined structure and nanometric dimensions for polymers modifiers [7]. The well-defined sizes of their molecules, and the presence of different amount and type of functional groups at the POSS core, allows their use in the precise deposition/embedding in the polymer matrix and obtain composite materials with unique properties [2,3,8,9]. The obtained organic–inorganic hybrid materials are characterized by a whole range of interesting physicochemical features (e.g., thermal, mechanical, optical and chemical). This is related to their solubility change, increase in decomposition and glass transition temperatures, improvement of dielectric properties, reduction of heat transfer coefficients, and improvement of oxidation and fire resistance, as well as their impact on the hardness of obtained materials [2,8,10,11]. Due to these aspects, functional silsesquioxanes possess great application potential [2] and could be applied as nanofillers and components of polymers [8,11], building blocks and synthons for a variety of advanced materials with tailored properties [12,13], such as silica models [14], (super)hydrophobic coatings [2], dendrimers and metal carriers. They were applied as immobilizing phase in catalysts [13,15], medicine (e.g., as drugs carriers, components of artificial tissues or in stomatology [16,17]) and opto- and electroluminescent (EL) materials [18].

Organic light emitting diodes (OLEDs) are a highly targeted area of technology because of their expected utility in flat panel displays. There have been discussions over the past decade concerning the family of small molecules and polymers that is best suited for OLEDs [18–23]. Small molecules can be highly purified and vacuum-deposited in multi-layer stacks, which is important both for display lifetime and efficiency. However, vacuum deposition techniques are expensive and limit the practical application, which causes problems achieving full colour displays at high volume. On the other hand, polymers are generally not as pure as small molecules, but can access larger display sizes and full colour at less substantial expense via solution-based deposition techniques.

Nanocomposite materials based on a silsesquioxane architecture that combines the advantages of both small-molecule and organic polymer fragments have been applied to OLEDs. They were introduced by Sellinger et al. in 2003 [24], with the compounds containing a spherical "silica" core with a hole-transporting functionalized periphery (chromophores). The resulting materials have offered numerous advantages for OLEDs, including amorphous properties enhancing thermal resistance (i.e., high glass-transition temperatures (T_g)) and stabilized colour at higher temperatures, as well as low polydispersity, solubility, and high purity. Silsesquioxanes also help to reduce (prevent) the aggregation of chromophores that are susceptible to the π–π intermolecular interaction (aggregation) leading to quenching the fluorescence. In general, the key aspect of the POSS-based OLEDs is to use them as scaffolds for organic chromophores of complex nature, bearing in mind that the characteristic parameters for OLEDs include wavelength (color of light emitted), luminance (a brightness greater than 10,000 cd m^{-2} is desirable) and external quantum efficiency (EQE; greater than 10% is desirable) [2,18]. The most popular are highly π-conjugated arenes (e.g., derivatives of carbazole, fluorene, terfluorene, pyrene, etc.) that have to be anchored properly onto the POSS core via chemical bonding.

The modification methods of silsesquioxanes are based on stoichiometric procedures, such as using prefunctionalized chlorosilane for hydrolytic condensation reaction or nucleophilic substitution. The catalytic transformations of POSS compounds are another synthetic approach and depend on the type and amount of a reactive group attached to the Si–O–Si core (e.g., Si–H or Si–HC=CH$_2$, etc.).

Certain types of transition metal (TM) catalyzed transformations have been applied to modify the abovementioned reactive groups through hydrosilylation (HS), cross-metathesis (CM) and coupling reactions (e.g., silylative coupling (SC), Heck coupling (HC), Sonogashira coupling, etc.). They may be used to attach a specific type of chromophore onto the rigid silsesquioxane core in order to obtain compounds of specific optoelectronic properties that can be applied in the formation of OLEDs devices.

The aim of this paper is to review the synthetic routes that lead to an efficient synthesis of silsesquioxanes with organic coronae of specific optoelectronic properties that constitute a group of electroluminescent (EL) and photoluminescent (PL) materials for the fabrication of electronic and optical devices. The crucial aspect is to designate a specific organic dye of a respective wavelength (color of emitted light) and anchor it to the POSS core using proper reaction procedure, which is dependent on the type of functional group at the Si–O–Si core.

2. Stoichiometric Reactions for the Synthesis of Functionalized Silsesquioxanes as Precursors of Photoactive Compounds

This section concerns stoichiometric reactions leading to the synthesis of silsesquioxane-based systems for the fabrication of photoactive materials and their precursors. It should be noted that the most important class of stoichiometric processes concerns the hydrolytic condensation of respective, prefunctionalized chloro- and alkoxysilanes. This may be part of the greatest problem of the selectivity of obtained products, depending on the kind of solvent, pH, additives, concentration, time, and so on. However, at this stage, silsesquioxanes are obtained with functional groups (e.g., Si–H, Si–HC=CH$_2$, Si–CH$_2$CH$_2$CH$_2$X (X = NH$_2$, Cl), etc.) and that is why it is so crucial. These POSS-based systems are used in further modification, and the form of the modifications is restricted to the nature of new functionality anchored onto the Si–O–Si core.

Hydrolytic condensation may be also used to obtain highly functionalized POSS-based compounds of interesting photophysical properties; however, the reports on this subject are rather rare; the reaction conditions may not be optimized sufficiently, and as a consequence result in rather low products yields. As an example, a paper by Lucenti et al. describes hydrolytic condensation ("corner capping") of hepta(cyclohexyl)silsesquioxane trisilanol with respective tri(methoxy)silylpropyl pyrylene diimide to obtain pyrylene diimide with T$_8$ unit, but only with a 23% yield [25]. This is described in Section 3.1, due to the ease of respective product comparison.

The amidation or esterification processes between respective reactive groups are also worth mentioning. An interesting example of using (3-aminopropyl)hepta(isobutyl)silsesquioxane in the amidation of mono- and bis-anhydrides in microwave radiation was reported by Clarke et al. (Figure 2) [26]. However, irrespective of a fast reaction time, the reaction conditions, and especially the 4-fold excess of silsesquioxane that resulted in only (mostly) moderate to good yields of the respective products (**2a,b**, 15%–54%; **3a–c**, 37%–98%), which may not be satisfying.

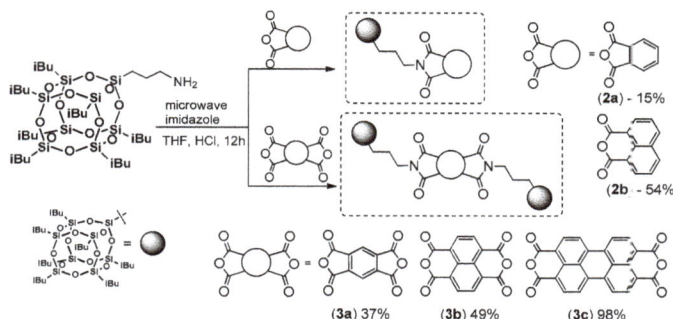

Figure 2. Synthetic path to obtain mono- and di-POSS-based imides and diimides.

In their study, Clarke et al. present the photophysics of the compounds obtained. All products presented absorption bands (in solution, CHCl$_3$) that were similar to and typical of their organic imide counterparts (λ_{ab} = 294 nm (**2a**), λ_{ab} = 309 nm (**2b**), λ_{ab} = 335, 350 nm (**3a**), λ_{ab} = 342, 360, 381 nm (**3b**), λ_{ab} = 459, 490, 526 nm (**3c**)). The emission spectra were much poorer for mono-POSS imide, as **2a** only had one wide band at ca. 330 nm, and for **2b** the strongest three peaks were at 362, 380 and 398 nm, respectively. However, the fluorescence quantum efficiency was very low for these two compounds, ca. Φ = 0.02. On the other hand, for the diimides, the resulting emission spectra were more accurate and reflected the presence of diimide unit (compared with organic counterparts), and three λ_{em} peaks were present (362–390 nm for **3a**, 350–404 nm for **3b** and 533–620 nm for **3c**). Only in the case of **3c** was the λ_{em} red-shifted with Φ that equaled 1.0. The low value of Φ suggests quenching of fluorescence due to an aggregation of organic moieties. In this case, only **3c** had potential for use in optoelectronic device fabrication.

A similar synthetic concept based on the amidation reaction of aminopropylPOSS and pyrylene anhydride, along with its consecutive amidation with methoxypolyethylene glycol amine, was presented by Bai et al. (Figure 3) [27]. However, the isolation yield of the intermediate product after the first amidation reaction with aminopropylPOSS reagents resulted in only a 23% yield, while the sequent amidation reaction that was conducted to obtain the final product—asymmetric pyrylene-based diimide—with 87% yield. As a result, the overall yield of the final product was 20%.

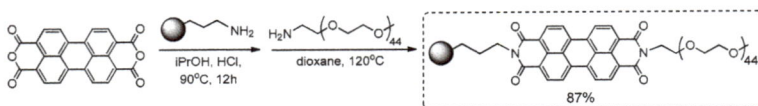

Figure 3. Reaction route for the synthesis of asymmetric pyrylene diimide.

Nevertheless, this compound exhibited interesting photophysical properties. As reported, pyrylene diimides are strongly susceptible to concentration quenching due to intermolecular π–π stacking [28]. In POSS-based pyrylene diimides, the POSS unit prevents the aggregation of pyrylenes, resulting in decreased emission quenching, and these compounds exhibit a strong excimer-like red emission at ca. 620–660 nm (λ_{ex} = 495 nm) [26,29]. The compounds described above were tested as chemosensors for variety of anions (F$^-$, Cl$^-$, Br$^-$, I$^-$, NO$_3^-$, AcO$^-$, ClO$_4^-$ and H$_2$PO$_4^-$), but only in the case of fluorine anions was the red emission found to be quenched, along with its increasing concentration. Interestingly, this was a result of silsesquioxane cage hydrolysis catalyzed by F$^-$, as previously reported in [30–32]. The decomposition of POSS units enables a consecutive and induced aggregation of pyrylene units and results in quenching the fluorescence phenomenon. This, in turn, offers a new optical strategy for toxic F$^-$ ion sensing.

Bai et al. also presented a synthesis of asymmetric pyrylene-based diimine with thePOSS unit as pendant moiety and N-isopropylacrylamide unit that was subjected to atom transfer radical polymerization (ATRP), resulting in a well-defined amphiphilic fluorescent polymer with a 58% yield [29]. Their fluorescence red emission bands were retained at ca. 645 nm (λ_{ex} = 495 nm) and were found to be temperature-dependent (intensity enhanced with an increase in temperature). This could be used to fabricate thermo-responsive materials, in biosensors, and so on.

The imidation reaction was used by Ervithayasuporn et al. to anchor rhodamine B hydrazide units using propyloxy-p-benzaldehyde-functionalized T$_{10}$ silsesquioxane (Figure 4) [33]. The precursor reagent T$_{10}$ was obtained via nucleophilic substitution cage rearrangement of octa(chloropropyl)silsesquioxane, but reported reaction conditions that resulted in a 15% yield of T$_{10}$.

Here, the authors reported on the synthesis of penta-substituted rhodamine B hydrazide T$_{10}$ derivative with very high (98%) yield. The bulkiness of rhodamine B hydrazide moieties enabled the introduction of only five units onto the Si–O–Si core. Nevertheless, the photophysical properties of this compound were found to be interesting in the presence of Hg^{2+} ion, with absorption at λ_{ab} = 520 nm and red emission (λ_{ex} = 520 nm) that increased with the ion concentration rise. The observed active form that exhibited fluorescence was a spiro-lactam form of rhodamine B unit after Hg^{2+} chelation. Analogous tests were performed for others metal ions but none of them gave the same fluorescence response. This was a very interesting work enabling use of the presented compound as a selective dual chemosensor for Hg^{2+} that afforded both fluorescence enhancement and color change from colorless to pink, with a 0.63 ppb detection limit.

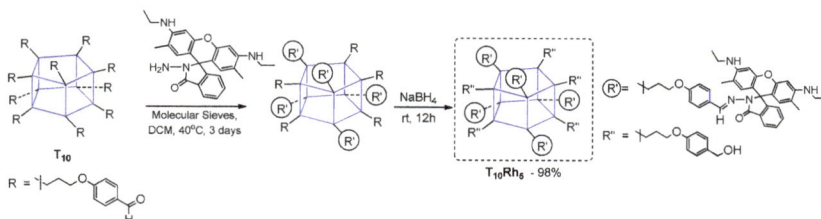

Figure 4. Synthetic path to obtain rhodamine B hydrazidedecorated T_{10} via an imidation reaction.

As reported by Tang, Xu and co-workers [34], an example of the parallel amidation and esterification of 4-(1,2,2-triphenylvinyl)benzoic acid used for the modification of octa(aminopropyl)POSS and octa [hydroxypropylodi(methyl)siloxy]POSS, respectively, resulted in the formation of analogous octafunctionalized silsesquioxanes differing from each other with the spacer (containing amide or ester functionality) between POSS and organic arene coronae. These two systems were obtained with 77% and 58% yields. Although these two compounds exhibited absorption spectra (in THF) at around λ_{ab} = 325 nm, the effective emission band was presented only for the amide derivative (a wide of spectrum ranging from 375 to 600 nm) with a maximum λ_{em} = 458 nm (λ_{em} = 365 nm). This compound was found to be a selective chemosensor for Cu^{2+} ions (other TMs were also tested) by fluorescence quenching in DMSO.

Anchoring the POSS unit on 3,4-ethylenedioxythiophenes was explored by Önal and Cihaner et al. [35–37]. They proposed an amidation reaction of amonipropyl(heptaisobutyl)POSS with anhydride, followed by a palladium-mediated Stille reaction to obtain the respective reagents, with moderate (44%–59%) yields. These systems were reagents in the electrochemical (co-)polymerization, resulting in highly conjugated poly(3,4-ethylenedioxythiophenes) (PEDOTs) known for their interesting electrochemical properties. PEDOTs can be used as elements in electrochromic device fabrication because their color depends on the oxidation state.

Rogach and Choy [38,39] reported the use of mono(3-mercaptopropyl) heptaisobutyloctasilsesquioxane as a dopant in the formation of solid stable perovskite ($CsPbX_3$ (X = Br or I)) colloidal nanocrystals (NCs), with a surface protection of POSS as luminophores exhibiting different emissions colors. This compound may be obtained via the hydrolytic condensation of (3-mercaptopropyl)trimethoxysilane with trisilanol form of not-completely condensed silsesquioxane $(iBu)_7(Si_7O_9)(OH)_3$, but it is also commercially available. The synthetic procedure for obtaining of the final material consists of a simple formation of sulphides bonded (-S-metal) between the nanocrystals and POSS in toluene (added in different stoichiometry). Silsesquioxane, which acts as a hole-blocking layer, is optically transparent and also prevents the NCs from aggregation.

3. Catalytic Reactions Leading to Functionalized Silsesquioxanes as Building Blocks for Photoactive Materials

The synthetic chemistry of organosilicon compounds, including silsesquioxanes, has been profoundly dominated by the TM-mediated catalytic process. This is dictated by high effectiveness and selectivity (regio- and stereo-) of these catalytic transformations. In the case of silsesquioxanes, the proper choice of functional group, as well as the amount anchored to the Si–O–Si core, affects the selection of the respective catalytic reaction to be applied for the modification. As mentioned in Section 1, the silsesquioxanes with Si–H, Si–HC=CH$_2$ or Si–aryl–X (X = I, Br, etc.) have been of the utmost importance due to their easy catalytic modification via hydrosilylation, cross-metathesis or coupling processes.

3.1. Heck Coupling for Obtaining Silylalkenyl Units Anchored at the Si–O–Si Core

Heck coupling (HC), also known as the Mizoroki–Heck reaction, is generally a palladium catalyzed coupling process between unsaturated C=C bonds (here, Si–HC=CH$_2$) and an aryl halide (usually –I, –Br) resulting in the formation of alkenyl-substituted arenes. It occurs in the presence of additional phosphine ligand and equimolar amount of a base, although there are many variations of this process, a wide scope and tolerance to many organic functional groups is observed [40].

Mono-functional silsesquioxanes are generally less frequently applied in obtaining compounds with interesting photophysical properties. Nevertheless, there are a few interesting examples to present that reflect the application of Heck coupling as a method for their synthesis. It is worth mentioning that the compounds are characterized by better thermo- and photostability. Liras et al. presented the synthesis of boron dipyrromethane (BODIPY) derivatives of POSS and their photophysical characterization (Figure 5) [41].

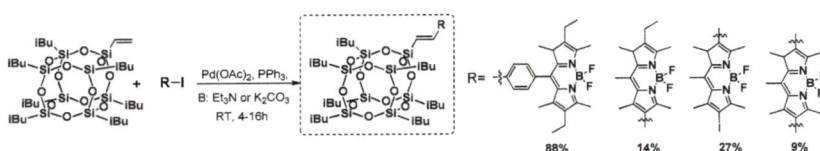

Figure 5. Respective structures for BODIPY derivatives of POSS.

The elaborated synthetic procedure was not so effective, and enabled the respective mono- and disilsesquioxane substituted BODIPY derivatives (BODIPY–POSS) to be obtained with low (9%) to high yields (88%). Their photophysical properties were determined in EtOAc solution and poly(methyl methacrylate) (PMMA) film. Maxima absorption wavelengths were similar in both systems, with λ_{ab} in a range of 521 to 536 nm, and maxima fluorescence emission wavelengths λ_{em} ranging from 532 to 575 nm. Thermal stability was improved for both systems, i.e. mono- and disilsesquioxane substituted BODIPY derivatives. The most photostable hybrid dyes were the systems with a POSS unit linked to the BODIPY by one or two ethylene bonds (π-conjugation). The results obtained indicate a possibility for these systems to be applied where high thermo- and photostability are required.

Monovinylhepta(cyclohexyl)silsesquioxane was reported as an efficient reagent towards Heck coupling (mediated by Pd(P^{t}Bu$_3$)$_2$) with p-iodophenyl pyrylene diimide by Lucenti et al. [25]. This compound synthesized with a moderate 37% yield when compared with an analogous one, obtained via stoichiometric hydrolytic condensation (mediated by MeC$_6$H$_4$SO$_3$H) of hepta(cyclohexyl)silsesquioxane trisilanol with respective tri(methoxy)silylpropyl pyrylene diimide (23% yield). These two systems, where POSS unit and pyrylene diimide moiety were separated either via an aliphatic (propyl) or styryl spacer, were analyzed in terms of their absorption/emission spectra and compared to their pyrylene diimide counterparts. The type of the linker (i.e., aliphatic or arene) did not bear any influence on the photophysics of these. The analyses as expected showed slight red-shift emissions of up to 10 nm in solution (CHCl$_3$), but with very good photoluminescence quantum yields (up to 100%). The analogous analysis for solids resulted in higher red-shifted emission spectra due to the expected aggregation and π–π intermolecular interactions. However, the photoluminescence quantum yields for the POSS-based systems were twice as high as for the respected organic analogue that suggests suppressed aggregation. This favors its possible use in the fabrication of efficient emissive devices. This part is presented in this section for the ease of product comparison, despite the fact that it also concerns condensation reactions.

Heck coupling was also applied by Liu et al. in the synthesis of thermally stable mono- and octa-azobenzene-functionalized T$_8$-based dyes (Figure 6) [42]. The procedure enabled formation of the designed compound with yields of 67% and 80%; however, the reaction conditions were more severe in comparison with the previous example. Both systems were proved to be thermally stable with $T_d^{5\%}$ = 273 and 383 °C for the mono and octa products, respectively. Introduction of these azobenzene moieties reduced the crystallinity of the compound withthe increasing amount of these groups in the molecule. The trans–cis photoisomerisation was examined and supported by DFT (Density Functional Theory) calculations. These systems were fluorescent with a maximum emission wavelength λ_{em} at about 400 nm, which enables their potential application as blue light emission materials.

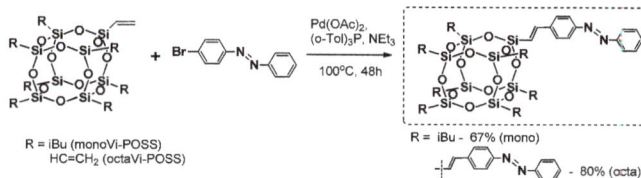

Figure 6. Mono- and octa-azobenzene-functionalized T_8 derivatives obtained in a Heck coupling reaction.

Octa-functional POSS-based OLEDs have attracted far more attention mainly due to the possibility to anchor chromophores moieties (i.e., eight groups) onto the Si–O–Si core. Heck coupling may be conducted in a few combinations and towards the formation of fully substituted products as well as without full substitution.

Pioneering work was presented by Sellinger et al. when they reported on the use octavinyl silsesquioxane (OVS) in a Heck coupling process with a mono-brominated version of a highly π-conjugated arene system in the presence of [Pd(P^{t}Bu$_3$)$_2$], obtaining hexa-substituted OVS with a 75% yield (Figure 7) [43]. Incomplete substitution was found probably due to a steric hindrance of the chromophore used and steric availability of vinyl groups. The obtained compound was verified in terms of its thermal stability, showing very high $T_d^{5\%}$ = 465 °C. Its film and solution (toluene) photoluminescence spectra revealed λ_{em} = 430 and 434 nm respectively. The resulting electroluminescent device showed 18% improvement in external quantum efficiency over their small organic chromophore analogue.

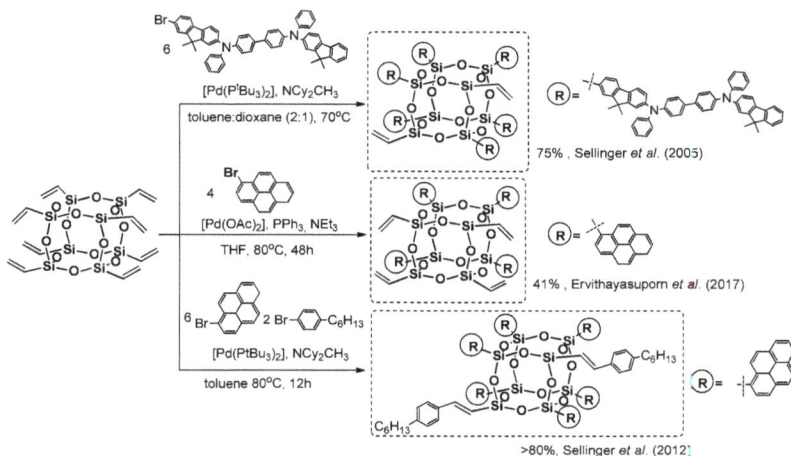

Figure 7. Heck coupling reaction leading to formation of T_8-based systems with two functionalities.

Analogously, not-fully-substituted pyrene-based T_8 cages were obtained by Ervithayasuporn et al. in Heck coupling using a [Pd(OAc)$_2$]/PPh$_3$/NEt$_3$-mediated reaction that, after two days, resulted in the formation of a product with four pyrene substituents (as the most abundant species) with a 41% yield (Figure 7) [44]. The photophysical properties of the product were determined and broad emission bands λ_{em} = 406, 426 and 494 nm of low intensity in DMSO and sharp λ_{em} = 398, 421 and 445 nm of high intensity in THF were revealed. The compound revealed the solvent's polarity dependant fluorescence, the π–π^* emission of sole pyrene moieties was observed in THF, while in DMSO the formation of a pyrene–pyrene excimer through space has been postulated (the authors suggest its confirmation due to a large Stokes shift of $\Delta\lambda$ = 143 nm). This compound was also tested as a sensor for several types of anions and, interestingly, was proved to encapsulate fluoride that results in a π–π^* fluorescence enhancement in DMSO. Recent research by Ervithayasuporn et al. is related, with the synthesis of an analogous compound, but with four anthracene substituents (Figure 7) [45]. The synthetic procedure was based on Heck coupling reaction with the same catalytic system to obtain the desired product with 86% yield. Its photophysical parameters disclosed similar solvent's polarity dependant fluorescence. It was also used to perform additional tests for possibility to encapsulate a variety of anions (e.g., F^-, OH^-, CN^- and PO_4^{3-}) by observation of fluorescence quenching due to from charge-transfer complex formation. This resulted in the possibility to use this compound as a sensor of the abovementioned anions.

Heck coupling to yield pyrene derivatives was also performed by Sellinger et al. They first reported on octavinyl T$_8$ cage functionalization, performing a [Pd(P^{t}Bu$_3$)$_2$]-mediated reaction with bromopyrene [46]. This was performed to obtain complete substitution (8–substituted as most abundant–P8) in comparison to the double Heck reaction product (14–substituted as most abundant–P14) with a >80% yield that was thermally stable up to 495 °C. These two systems were also analyzed in terms of their photoluminescence properties in solution (toluene), λ_{em} = 425 nm (P8) and 431 nm (P14), as well as in thin films, λ_{em} = 494 nm (P8) and 506 nm (P14). This meant a 70–75 nm red-shift effect when compared results in solution vs. in a solid state that suggested possible pyrene unit aggregation. The problem of two high "density" pyrene units was resolved by performing an efficient functionalization of vinyl-substituted T$_8$, T$_{10}$ and T$_{12}$ cages with bromopyrene and a mixture of bromopyrene and a smaller steric reagent (i.e., 4-heptylbenzene) using the same catalytic Pd system (Figure 7) [47].

Obtained products (i.e., monopyrene(heptaisobutyl)silsesquioxane, along with the pyrene/ heptylstyrene- substituted T$_8$, T$_{10}$ and T$_{12}$ cage product mixture) were analyzed with TGA (ThermoGravimetric Analysis), showing their high thermal stability up to 359–469 °C. The films prepared (using solution-processed deposition) from the obtained materials exhibited a sky-blue emission (λ_{em} in the range of 459–486 nm for thin films, and λ_{em} in the range of 400–449 nm for the toluene solution). These results may also suggest chromophore aggregation in a solid state. The obtained compounds were tested in solution-processed OLEDs prepared to obtain materials showing current efficiencies of 3.64% and 9.56 cd A^{-1}, respectively.

Xu et al. presented a Heck coupling-based strategy to obtain anthracene-substituted T$_8$ cage via a [Pd(OAc)$_2$]/PPh$_3$-catalyzed reaction at room temperature in various reagent stoichiometry resulting in differently substituted products (Figure 8) [48]. The substitution was also statistical, but with the majority of the presented forms of products that depended on the possibility to form a complex between palladium with bromoanthracene substrate.

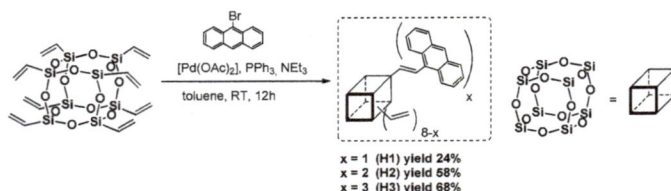

x = 1 (H1) yield 24%
x = 2 (H2) yield 58%
x = 3 (H3) yield 68%

Figure 8. Heck coupling of OVS leading to mono-, di- and tri- anthracene-functional POSS.

Their $T_d^{5\%}$ in the range of 407 °C (H3), 468 °C (H1) and 480 °C (H2), respectively, suggest high thermal stability in comparison with bromoanthracene T_d = 246 °C. The photophysics were verified and the UV-vis spectra revealed the presence of three absorption maxima wavelengths λ_{ab} = 356, 373 and 390 nm for solution in toluene. The emission spectra were recorded in toluene, showing emission maxima wavelengths λ_{em} = 440 (H1), 461 (H3) and 458 (H2) nm, which were compared to the spectra for solids that disclosed red-shifts $\Delta\lambda$ = 42–51 nm, suggesting known phenomena of possible anthracene unit aggregation (the smallest effect for H1). The same authors reported on the further modification of dianthracene(hexavinyl)octasilsesquioxane (H2), and using Heck coupling to introduce one diazophenylnaphtalenediamine moiety followed by an amidation process to introduce a complex structure of a chromophore that possess broad-band absorbing properties [49]. The presence of different chromophores allows the absorbing bands of the dyes to be combined, reduces the possibility of their aggregation and possesses a broad red-emission band at λ_{em} = 638 nm.

Completely octa-substituted T_8 cage obtained via Heck coupling reaction was reported by Liu et al. and used for the synthesis of a water-soluble hybrid nanodot based on POSS and conjugated electrolytes for the Two-Photon Excited Fluorescence (TPEF) analysis that imaginesthe cellular and cancer nuclei (Figure 9) [50,51].

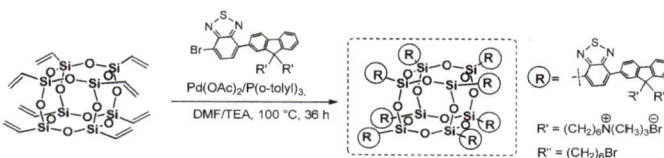

Figure 9. Efficient Heck coupling for the formation of octa-substituted T_8 with conjugated fluorene and benzothiazole units.

The Two-Photon Excited Fluorescence is a microscopic method used to reduce photodamage to a cell, improve resolution and minimize cellular autofluorescence, but it has rarely been exploited, due to the lack of efficient absorption materials that can enter nuclei and in turn selectively stain them. The desired fluorophore was anchored to the POSS core via a Heck coupling procedure of octavinyl silsesquioxane (OVS) and respective monobromo-substituted chromophore in the presence of [Pd(OAc)$_2$/P(o-Tol)$_3$] with a 45% yield. The quaternized silsesquioxane derivative was soluble in water and its average diameter of 3.6 $\pm$ 0.3 nm enabled easy penetration of cellular nuclei. The photophysical properties were verified and absorption maxima λ_{ab} bands at 320 and 465 nm corresponded to the fluorene and benzothiazole units, but they red-shifted by 34 and 75 nm, respectively, due to the –HC=CH–POSS moiety extending the conjugation system. On the other hand, the λ_{em} was located at 617 nm and exhibited a large Stokes effect at 152 nm.

An interesting scientific concept was proposed and performed by Liu et al., who used octavinyl silsesquioxane (OVS) and multibromo-substituted arenes in a Heck Coupling process to yield crosslinked 3D networks of highly hyperbranched structures with a >80% yield (Figure 10) [52,53].

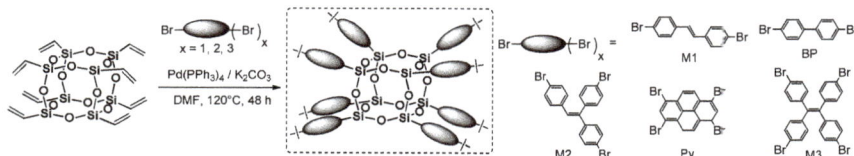

Figure 10. Synthetic route leading to hyperbranched 3D POSS-based materials obtained via Heck coupling.

The reaction procedure was also modified by using mixtures of different chromophores (e.g., BP and Py) in various stoichiometry, which resulted in the formation of macromolecular 3D systems with different chromophore content. The purification procedures of obtained polymeric materials, due to their 3D structure, was necessary due to the presence of non-reacted comonomers, and post-reaction inorganic salts and other residues within the matrix. The isolation was mostly based on filtration and washing with different solvents. For this reason, a Soxhlet apparatus was used with THF and MeOH to obtain pure products. All of the obtained compounds were analyzed in terms of their proven thermal resistance and high $T_d^{5\%}$ at ~390 °C, but the high porosity of these materials was revealed by during BET (Brunauer–Emmett–Teller) analysis. Depending on the type (and amount, if mixture) of respective arene, respective hyperbranched products exhibited various surface areas from 358 to even 848 m^2g^{-1} (when M3 was used as a reagent). The most interesting part of the research was disclosed in defining the photophysical properties of both substrates (M1–M3, BP, Py). In the case of M1, M2 and M3, which showed high blue fluorescence (λ_{em} at ca. 435 nm) in solid state and in solution (DCM = dichloromethane), the respectively obtained hyperbranched products showed strong red shifts (observed at λ_{ex} = 365 nm), and green (λ_{em} = 509 nm for prod.1, derived from M1; λ_{em} = 518 nm for prod.2, derived from M2) or yellow (λ_{em} = 540 nm for prod.3, derived from M3) fluorescence. This may be derived due to the extended π–π conjugation in the 3D network by additional –HC=CH– linkage. The ethanol dispersion of these materials was applied to obtain sensors for various nitroaromatic explosives by fluorescence quenching [53]. On the other hand, modulation of the stoichiometry of BP and Py in Heck coupling with OVS resulted in the formation of nine types of materials that altered the amount of BP and Py units in the frameworks. As a result, they exhibited continuous color change from bluish-green at λ_{em} = 485 nm, to green at λ_{em} = 528 nm, yellow at λ_{em} = 555 nm and finally to red at λ_{em} = 605 nm. High fluorescence quantum yields in the solid state were calculated and ranged from 2.46% to 22.42% [52]. These systems were used along with the thiol-containing polysiloxane matrix to produce a UV-LED device that shows color-transformable properties by simply turning the irradiation on and off.

Heck coupling using the silsesquioxanes may be performed also in the reverse functional group placement (i.e., to applied octa(halogenophenyl)silsesquioxane and styrene derivatives as a source of –HC=CH$_2$ moiety). Respective research on this issue was pioneered by Laine et al. [54]. The primary reagent for the study was cubic octa(phenyl)silsesquioxane (OPS) that was subjected to bromination via a Fe-catalyzed reaction that resulted in [octa(polybromophenyl)]silsesquioxane. The resultant compounds had a different substitution of bromine at the Ph ring, but this could be controlled by reaction stoichiometry to obtain a para-brominated 2,5-dibrominated derivatives. A more-efficient and selective procedure for the synthesis of analogously octa(p-iodophenyl)silsesquioxanes using ICl was also reported [55,56]. These two systems are perfect halogenophenylPOSS reagents for the Heck and other coupling processes (e.g., Suzuki-type) [57]. In 2010, Laine et al. reported on synthesis of a series of octa(stilbene)silsesquioxane derivatives using octa(p-iodophenyl)silsesquioxane and octa(bromophenyl)silsesquioxanes (on average 67% p-substituted, 24% meta, 9% ortho and 3% disubstituted) with respective p-substituted styrenes in Heck coupling reaction conditions (i.e., [Pd$_2$(dba)$_3$] (1 mol %)/[Pd(P^{t}Bu$_3$)$_2$] (2 mol %) that resulted in octa(stilbene)silsesquioxanes with diverse p-phenyl substituents with high 62–86% yields (Figure 11) [58].

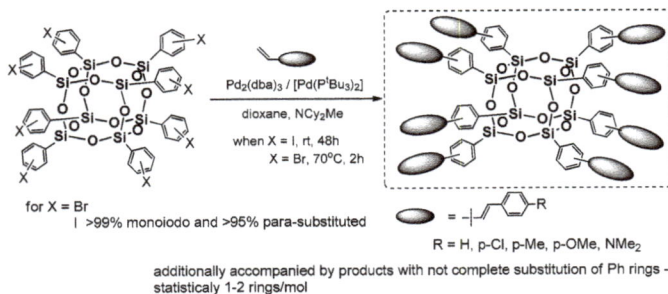

for X = Br
I >99% monoiodo and >95% para-substituted

R = H, p-Cl, p-Me, p-OMe, NMe$_2$

additionally accompanied by products with not complete substitution of Ph rings - statisticaly 1-2 rings/mol

Figure 11. Heck coupling using halogenophenylPOSS reagents.

The desired octa-substituted products were also accompanied by compounds with not-completely substituted phenyl rings, which was a result of not-fully halogenated octa(phenyl)silsesquioxane, especially in the case of Br [54]. The corresponding products were characterized in terms of their thermal resistance ($T_d^{5\%}$ in the range of 318–429 °C). The greatest emphasis was placed on their photophysics assessment and its comparison with trans-stilbene and Me-stilbene-Si(OEt)$_3$. The UV-vis spectra of products (in THF) were similar in shape and red-shifted 5–10 nm (λ_{ab} 298–320 nm) to their molecular counterparts, but their large red shift of 60–100 nm was visible for the emission spectra of these systems when compared with trans-stilbene and Me-stilbene-Si(OEt)$_3$ (λ_{em} 334–436 nm). The greatest difference was noted for stilbene and p-Me-stilbene silsesquioxane derivatives that were also collated with UV-vis and PL spectra for octa(p-Me-stilbene)silsesquioxane and its hepta(p-Me-stilbene)phenylsilsesquioxane. Altogether, the absorption–emission changes that were observed along with the theoretical calculations and additional analysis of stilbene–siloxane and cyclosiloxane, respectively, suggest that the large red-shift may have resulted from interactions of the stilbene π^* orbitals with a LUMO located inside the cage and involving all Si and O atoms, as well as the organic substituents that interact in the excited state. Interestingly, for the NMe$_2$ derivative the solvatochromism phenomenon was observed mainly due to the charge–transfer interaction when the polarity of solvent was changed.

Studies on stilbene derivatives of silsesquioxanes were also conducted. One of the aspects to be studied was the usage of octa(halogenephenyl)silsesquioxane reagents substituted with bromine at phenyl ring in different places (i.e., octa(o-bromophenyl)silsesquioxane, but also octa(2,5-dibromophenyl)silsesquioxane and octa(2,4,5-dibromophenyl)silsesquioxane) [59]. This provided more functionalities per unit volume, exceeding the density when compared to POSS-based dendrimers. These compounds were used for analogous Heck coupling procedure (including catalytic systems, etc.), leading to corresponding octaphenyl silsesquioxane (OPS)-based styrenes with 8, 16 and 24 styryl substituents, with R = 4-Me, R = 4-NBoc and R = 4-Ac groups, respectively [60]. Interestingly, the thermal parameter for these systems is similar to their octa-substituted derivatives. The blue shift of the absorption and the red shift of the emission spectra for the (R-o-styryl)$_8$OPS suggest the interactions of the Si–O–Si cage with the ortho-substituted organic groups. On the other hand, the (Rstyryl)$_{16}$OPS exhibited a longer conjugation length, resulting in higher values of Stoke's shift and (Rstyryl)$_{24}$OPS behavior that revealed the existence of both a regular π–π^* transition and a charge transfer, which could be explained by two excited states involving the LUMO inside the Si–O–Si cage.

The synthetic procedure of Heck coupling was also conducted in a controlled copolymerization of octa(p-iodophenyl)silsesequioxane with 1,4-divinylbenzene (DVB) and 1,4-diethynylbenzene (DEB). Laine et al. reported on a specific formation of copolymeric systems derived from octa(p-iodophenyl)silsesequioxane embedded between DVB or DEB fragments (Figure 12) [61].

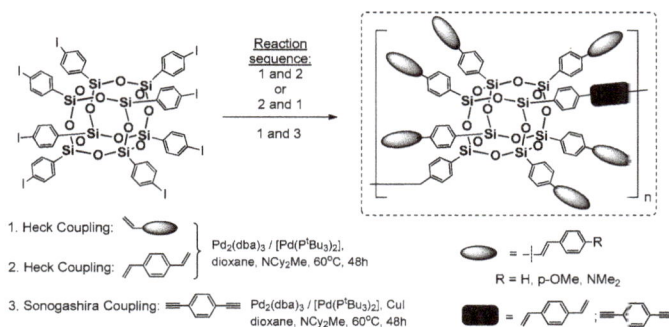

Figure 12. Synthetic approach for linear copolymers obtained via consecutive Heck and Sonogashira coupling.

Polymers **2019**, *11*, 504

The synthetic routes that were tested were based on changing the reaction sequence of octa(p-iodophenyl)silsesquioxane Heck Coupling functionalization (carried out with specific stoichiometry) and linking each silsesquioxane via Heck or Sonogashira cross-coupling with DVB or DEB. The procedure resulted in final copolymeric systems with a moderate 39%–51% yield and respected average molecular weights in the range of 9–30 kDa, as well as suggesting that the formation of di(p-iodostyryl)hexastyrylsilsesquioxanes as the first reaction sequence results in higher molecular weights of the resulting copolymeric systems. The photophysics of the two types of copolymers significantly depended on the organic linker. The DVB copolymers exhibited very little conjugation due to high 1,4-DVB fragments, whereas the DEB showed a strong red-shift at ~40 nm emission. This suggests electron delocalization through the Si–O–Si cage. Again, the p-aminostilbene copolymer derivative showed charge-transfer stabilization.

Laine and co-workers expanded the studies conducted on T_{10} and T_{12} cages with reactive Si–HC=CH, p-iodophenyl and bromophenyl substituents. They previously reported on the possibility of fluoride ion (TBAF = tetra-n-butylammonium fluoride) catalyzed cage rearrangements between polyvinyl- and polyphenylsilsesquioxane, resulting T_{10} and T_{12} with mixed reactive group placement (depending on the reaction condition and stoichometry) [30,31]. The research concerned using vinylT$_{10/12}$ cage mixture and its functionalization using Heck and double Heck coupling with bromostyrene, 1-bromonaphthalene and 9-bromoanthracene, as well as Grubbs first generation catalyzed cross-metathesis of the vinylT$_{10/12}$ with p-substituted styrenes (Figure 13) [62].

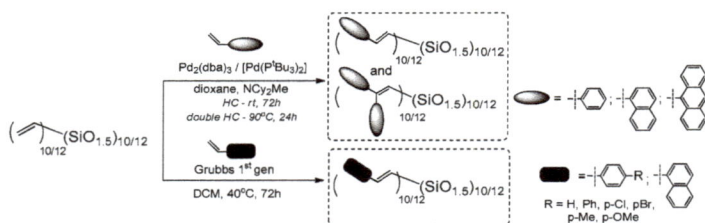

Figure 13. T_{10} and T_{12} arene derivatives obtained by Heck Coupling and Cross-Metathesis.

In parallel, the reverse Heck Coupling (i.e., reaction) of (Br-styrenyl)T$_{10/12}$ cage mixture with p-substituted styrenes was conducted (Figure 14).

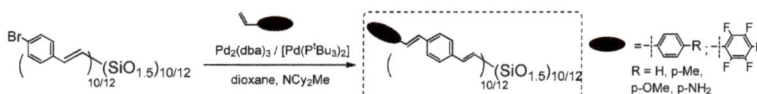

Figure 14. T_{10} and T_{12} stilbene derivatives obtained by Heck coupling and cross-metathesis.

The results for both processes are presented here for the ease of their comparison (CM is described in Section 3.3). The described catalytic processes are parallel routes for analogical T_{10} and T_{12} structure formations, and also similarly efficient. The products were compared, their thermal resistance was verified and they exhibited $T_d{}^{5\%}$ up to 445 °C. The UV-vis and PL spectra in THF were analyzed and, in general, were similar to those obtained for T_8 analogues. Interestingly, the UV-vis absorption maxima of products obtained via Heck and double Heck coupling were almost the same (in the range 254–374 nm), while the emission of double-functionalized compounds was a bit red-shifted (310–465 nm), as should be expected for the slightly extended, excited state of conjugation. The p-substituted stilbene-based T_{10} and T_{12} exhibited emission maxima dependent on the electronic properties on the substituents (e.g., the largest red shift of λ_{em} for the = –NH$_2$ group, that results from a charge transfer). For this derivative, a solvatochromism was also observed for MeCN (acetonitrile), THF and cyclohexane solvents used. Additionally, the dependence of emission spectra shapes (maxima) and their intensity on the solvent used (MeCN:THF, hexane:THF) and the spectra collected for various compound concentration suggested their aggregation. This possible aggregation may be responsible for the exciplex formation and seems to differ significantly from the one known for traditional organic molecules. These findings might suggest smaller band gaps, meaning a possible use of these systems in 3D hole/electron transport devices.

The research connected with the use of T_{10} and T_{12} cages was carried on for copolymeric systems. Again, Laine et al. performed experiments using octa(vinyl)silsesquioxane T_8 (OVS) and octa(p-iodopheyl)silsesquioxane T_8 to obtain a mixture of T_{10} and T_{12} with mixed vinyl- and p-iodophenyl-substituent placement (with a 90% yield) along with an analogous reaction for vinyl and phenyl-substituted T_{10} and T_{12} cages [63]. The vinyl/phenyl-mixed T_{10} and T_{12} systems were substrates for the following:

- Heck Coupling reaction (analogous catalytic conditions described in [61]) with 1,4-dibromobenzene and 4,4'-dibromostilbene, resulting in lightly branched non-linear copolymeric systems with a 83–86% yield. This is the fastest way to obtain the desired copolymeric compounds.Cross-metathesis (described in detail in Section 3.3) reaction, using Grubbs first generation catalyst with 1,4-dibromobenzene to obtain (again) slightly branched non-linear copolymeric system with a 89% yield.The vinyl/(p-iodo)phenyl-mixed T_{10} and T_{12} cages were substrates for the reverse synthetic path (i.e., cross-metathesis reaction, using Grubbs first generation catalyst with 1,4-dibromobenzene to obtain a slightly branched non-linear copolymeric system with a 95% yield).The abovementioned copolymer (i.e., its p-iodophenyl substituents) was post-modified via a Heck coupling reaction using R-styrenes (H-styrene, OMe-styrene) to obtain R-styrene functionalized copolymer with a 67% yield.

All these six types of copolymers were tested in terms of their photophysical properties, with results collated for those with bis(vinyltriethoxysilyl)benzene and 4,4'-bis(vinyltriethoxysilyl)stilbene used as models. All copolymers exhibited a red shifted emission spectra that were strongest for the copolymers obtained via Heck Coupling (described in the first point) at ca. 20–60 nm for the DVB derivative when compared with the model compounds. This is attributed to shorter conjugated linkers, enabling Si–O–Si cage participation in the 3D excited states. The compounds were obtained in particular because of their 3D interaction in the excited state that corresponds with their application in OLEDs devices.

3.2. Sonogashira Coupling for the Introduction of Silylalkynyl Moieties onto the Silsesquioxane Core

Sonogashira coupling (also known as Sonogashira–Hagihara) is a palladium-mediated reaction between an unsaturated –C≡CH terminal alkyne moiety and an aryl halide (usually –Br, –I, –OTf) that results in the formation of substituted alkynes. It is conducted in the presence of copper halide and requires stoichiometric amount as a base. The variations of this process are numerous and concern catalytic systems (including additional ligands), but it is a useful process in organic and organosilicon chemistry, and tolerant of a number of functional groups [64].

This type of coupling requires the presence of –C≡CH or –phenyl-halogen moiety attached to POSS core. The first reports on this process in silsesquioxane chemistry concerned using octa(bromophenyl)POSS and octa(iodophenyl)POSS derivatives with phenylacetylnes in [Pd_2(dba)_3] and [Pd(PtBu_3)_2] catalyzed reaction, conducted at RT and 60 °C for 48 or 24 h, respectively and resulting in 56%–90% yield for –Br product, and 67%–90% yield for –I derivative [65]. In 2016, Chujo et al. applied this procedure using octa(iodophenyl)POSS with π-conjugated phenylacetylenes (Figure 15) [66]. These compounds however, were obtained with rather moderate yields (i.e., Ph3POSS = 20%; tBuPOSS = 19%; and iPrPOSS = 22%).

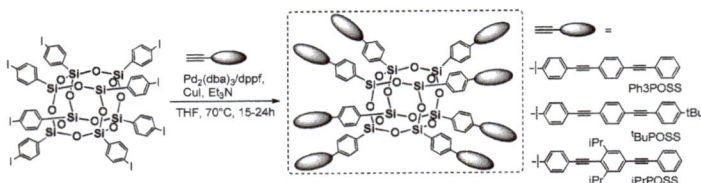

Figure 15. Synthesis route for octa-substituted T$_8$ phenylacetylenes obtained via Sonogashira reaction.

Nonetheless, all of these systems were verified in terms of their thermal resistance which was proved to be high (T_d = 496–526 °C). Their photophysical properties were disclosed via UV-vis and photoluminescence analysis and compared with their organic, trimthylsilyl analogues. The absorption spectra revealed the presence of two maxima wavelengths in the area λ_{ab} = 323–327 and 347–350 nm, and did not present any significant changes regarding the structural differences of substituents. The emission spectra were recorded in solution (CHCl$_3$) and in a solid state. The relatively broader and red-shifted emission was recorded for Ph3POSS along with higher quantum efficiency in comparison to the organic analogue. This was explained by the inorganic core presence as well as the mobility and π-conjugation of phenyl rings (undisturbed for alkyl derivatives). On the other hand, a strong, red-shifted λ_{em} maxima at ca. 405–427 nm was recorded, resulting from unfavorable intermolecular interactions. The authors reported high thermal resistance for the luminescence properties (after UV irradiation λ = 365 nm) that were maintained even at 250 °C.

3.3. Cross-Metathesis and Related Modfications for Silylalkenyl Fragment Formations

Cross-Metathesis (CM) is one of the fundamental catalytic transformations in the chemistry of organosilicon compounds [67]. In this manner, it is a TM-carbene-mediated (TM = Ru, Mo, etc.) process of C=C bond cleavage and formation of unsaturated organosilicon derivatives. Silsesquioxanes, which are promising precursors with Si–HC=CH$_2$ reactive moiety may be placed in this group of reagents [68–71]. CM may be applied as a single synthetic route leading to desired POSS-based compounds, or used as a valuable tool for obtaining precursors of silsesquioxanes with reactive groups for consecutive reactions (e.g., Heck or Sonogashira coupling). In this matter, the described process was used to obtain a series of compounds exhibiting interesting photophysical properties.

Cole-Hamilton et al. contributed to this issue with the use of octavinyl silsesquioxane (OVS) grafted with styrene derivatives using Grubbs first generation catalyst (4 mol %) and 2–3 equiv. of styrene per one Si–HC=CH$_2$ group and resulting in good 67%–78% product yields (Figure 16) [72,73]. The modifications of styryl rings resulted in changes in absorption and emission spectra, especially when aldehyde moiety was introduced and resulted in quenching emission. The POSS derivatives were reported to exhibit red shifts and broader spectral emissions due to the involvement of a silsesquioxane core in electronic delocalisation. It should be noted that CM reaction conditions may be improved so that not as much styrene excess is required to complete OVS conversion (24 h) [74].

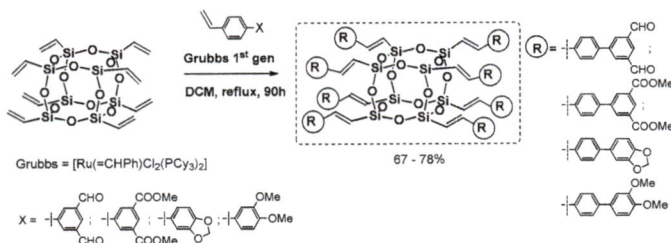

Figure 16. Cross-metathetic reaction path for the synthesis of octa(styryl)silsesquioxane.

Cross-metathesis was employed as precursor synthetic process for obtaining mono(p-bromostyryl)heptaisobutylsilsesquioxane (M1-Br) by Naka et al. [75]. This compound was applied as a reagent in two subsequent Sonogashira coupling reactions, resulting in dumbbell-shaped π-conjugated systems with POSS as pendant groups (Figure 17).

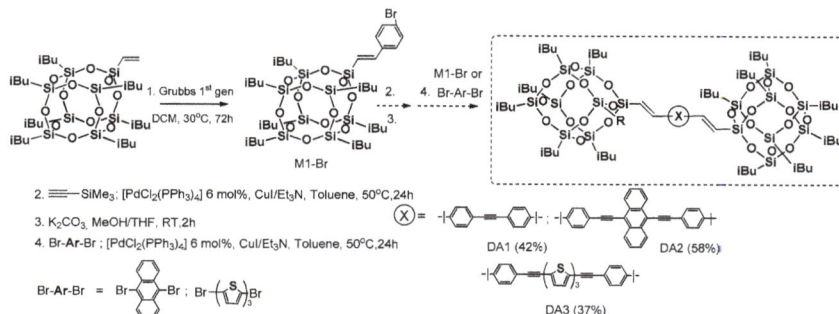

Figure 17. Synthetic reaction route based on sequential cross-metathesis and Sonogashira coupling for dumbbell-shaped π-conjugated systems with POSS as pendant groups.

Unfortunately, the reaction conditions were based on the use of 3 equiv. of para-bromo-styrene, and 3 mol % of Grubbs first generation catalyst (added to the reaction in two portions) enabled the formation of precursor M1-Br with only a 68% yield, contrary to the procedure developed in our group [76]. The next steps of the process were performed with moderate 37%–58% yields of respective final dumbbell products (DA13). These were characterized in terms of their photophysical and thermal properties in comparison with their organic counterparts. Their $T_d^{1\%}$ was respectively high (268–282 °C). The interesting part of the research concerns absorption and emission spectra performed in a solution vs. in a solid. The λ_{em} maxima in the solid state were red-sifted in a range of 34–50 nm and equaled 401, 542 and 520 nm for DA13, respectively. This was noticed after UV irradiation (352 nm) that revealed blue, yellow and greenish-yellow emissions in POSS-based systems contrary to their organic analogues. These results also indicate the participation of the Si–O–Si cage in the delocalization of electrons, and prevention of the π-conjugated parts from interaction.

Laine et al. have profoundly contributed to the development of cross-metathetic transformation using the most useful reagent in this process (i.e., octa(vinyl)silsesquioxane T_8 (OVS), but also decavinylT$_{10}$ and dodecavinylT$_{12}$ cages). The selective introduction of substituted (E)-styrenyl moiety at Si–O–Si cores and for the bromo-derivatives enabled their consecutive modification via Heck coupling (Figure 18) [71,77]. Interestingly, the cross-metathesis for OVS proceeded with 0.5 mol % of the catalyst and at room temperature only in the cases of T_{10} and T_{12} at 40 °C. The purification of these compounds concerned filtration over Celite, precipitation and column chromatography, obtaining final products with 74%–82%yields. Additionally, the p-NH$_2$-vinylStilbene product was transformed into respective benzamide derivatives. All of the compounds were characterized in terms of photophysical properties. For T_8 compounds, the absorption (λ_{ab} 260–358 nm) and emission (λ_{em} 304–482 nm) spectra showed red shifts in comparison with their organic counterparts (styrene and p-vinylstilbene).

The octa(XStyryl)silsesquioxane T_8 showed smaller red-shifts than octa(R'Stilbenevinyl) silsesquioxane T_8compounds (10–40 nm shifts vs. 30–80 nm shifts—i.e., Stokes shifts at ca. 50 nm) as might be expected due to a larger conjugation for the R'StilbenevinylT$_8$ compounds. The red shift is also larger in the presence of more electron-donating groups (–OMe). The NH$_2$StilbeneVinylT$_8$ exhibited effects of solvent polarity on its emission behavior (DCM and MeCN) due to charge-transfer (CT) interactions that were also proved by DFT calculations, as noted previously [58]. The StilbeneVinylT$_{10}$ and T_{12} derivatives were separated effectively due to the proper solvent choice. Their photophysical properties were compared along with the StilbeneVinylT$_8$ to observe the impact of the cage size and symmetry measured in solution (THF) and films. In general, the red-shift emission was observed for all three compounds, due to the π-conjugation, and the additional red-shifted spectra for films revealed the aggregation of chromophores, as expected. Additionally, two-photon spectroscopy analysis was performed to compare the polarization and non-linear absorption properties of the separated T_8, T_{10} and T_{12} cages along with the p-triethoxysilylvinylstilbene and p-vinylstilbene. Due to the differences in symmetry, especially between T_{10} and T_{12} cages, these compounds displayed changes in the dipole moment of excitation and exhibited different emission maxima wavelengths, with excitation at λ_{ex} = 333 nm ($T_{10} - \lambda_{em}$ = 386 nm, high intensity), λ_{ex} = 400 nm ($T_{10} - \lambda_{em}$ = 453 nm) and λ_{ex} = 800 nm ($T_{10} - \lambda_{em}$ = 450 nm, the highest intensity when compared with T_8 and T_{12}). Indeed, the observed impact on fluorescence decreased for bigger T_{12} cages, which may have been derived from the cage symmetry and/or the chromophores' proximity. Moreover, a red-edge effect in these 3D molecular structures was observed (i.e., the red-shifted emission wavelengths upon shifting of the excitation wavelength to the red end of the abs. spectra. The highest effect was observed in the case of deca(Stilbene)silsesquioxane T_{10}, which suggests a possible application of these compounds as a new class of hybrid materials offering switchable emissions.

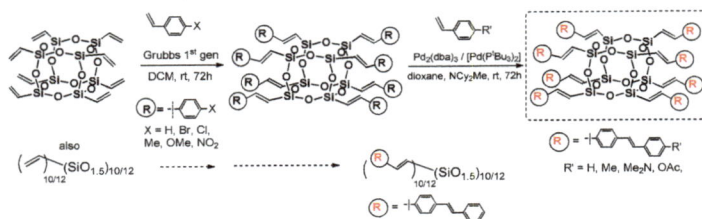

Figure 18. Reaction path for the selective sequence of cross-metathesis and Heck coupling towards (E)-styrenyl-substituted T_8, T_{10} and T_{12}.

The cross-metathesis process, leading to polymeric POSS systems (e.g., ADMET reaction), was described in Section 3.1 due to the ease of comparison of the products, and is analogous in structure to a Heck coupling reaction [63].

The Núñez research group presented an interesting combination of analogous cross-metathesis and Heck coupling sequences leading to octa(carborane-Stilbenevinyl)silsesquioxane T_8 cages (Figure 19) [78]. The cross-metathesis of OVS with p-bromostyrene was perfumed using the synthetic procedure reported by Laine [71]. As a result, octa(Br-styrene)T_8 was applied as a reagent in efficient Heck coupling with four types of carboranyl-substituted vinylstyrene, producing a good isolation yield of 1–4 products (43–65%). The importance of this research lies in the combination of carboranes known from their 3D electron delocalization and the fact that they were additionally coupled with π-conjugated arene fragments along with the T_8 silsesquioxane unit, enhancing thermal stability and also affecting the electronic properties of these systems. All compounds were characterized in terms of their absorption–emission properties (analysis performed in DCM) and supported by DFT calculations. The absorption spectra shapes were similar with λ_{ab} near 338 nm and all emission maxima wavelengths red-shifted to ~392 nm (λ_{ex} = 340 nm). Interestingly, the highest symmetry of product 1 in comparison with products 2–4 (Figure 19) was visible in the broadening of fluorescence in the red part of the spectra that indicates intermolecular interactions of the o- and m-substituted carborane units. This also caused the fluorescence quantum yield to be the highest for 1 (~60%). Additionally, the emission spectra of drop-casted films revealed its strong bathochromic effect (λ_{em} 450–484 nm) due to the aggregation and intermolecular interactions of substituents. However, these systems exhibited remarkable thermal resistance, with up to 87% residue on the initial weight at 1000 °C, making them excellent candidates for luminescent materials.

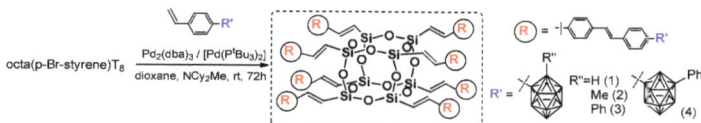

Figure 19. Synthesis of T_8-based ortho- and metha-carboranyl-substituted vinylstilbenes.

3.4. Metallative Coupling (Si, Ge) as a Complemenatry Catalytic Reaction towards the Formation of Alkenylsilyl Moieties

The importance of catalytic transformations in the chemistry of organosilicon compounds is undeniable, and the reactivity of the transition metal–p block element (TM–E) bond often determines the kinetics and selectivity (regio- and stereo-) of these processes. This aspect also concerns silsesquioxanes. A class of catalytic processes attractive in the organosilicon chemistry is silylative coupling (trans-silylation), discovered and developed by the Marciniec group over 20 years ago. This is a process between vinylsilanes and olefins, involving the activation of the =C–H bond at the α and β carbon atom of the vinyl group and the C_{vinyl}–Si bond in the vinylsilane molecule, with simultaneous elimination of the ethylene molecule catalyzed by TM–H/Si (TM = Ru, Ir, Rh, etc.) complexes (the insertion–elimination mechanism that we proved) [79–82]. This leads to unsaturated E/Z and *gem*-alkenylsilanes. Because the selectivity of this reaction is strictly controlled, it results in a selective formation of E-isomers depending on the reaction conditions and catalytic system. The silylative coupling may be considered a complementary process to the abovementioned cross-metathesis. Despite the fact that these two reactions proceed via different mechanisms and catalytic systems, they result (when optimized) in the selective formation of E-1,2-substituted silylalkenes (in contrast to silylative coupling, which may be regioselective towards isomeric 1,1-substituted alkenes).

In the chemistry of vinyl-substituted silsesquioxane modifications using both reactions, silylative coupling, metathesis and also hydrosilylation, our team gained a vast experience. In the scope of this research, it was shown that the studied processes could be used for the synthesis of unsaturated silsesquioxane-based systems using mono- and octavinyl silsesquioxanes (T_8) as well as divinyl-substituted DDSQ reagents. The silylative coupling reaction mediated by the [RuHCl(CO)(PCy₃)₂]/[CuCl] system was found to be regioselective process leading to both mono- and octa[(E)alkenyl]-substituted silsesquioxane T_8 cages (Figure 20) [74,76], as well as containing metalloids (Ge) [83]. The reactions were conducted with strictly controlled stoichiometry and were clean, as the only byproduct was the evolution of ethylene.

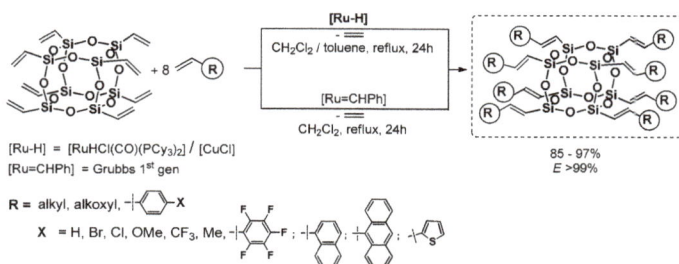

Figure 20. The silylative coupling vs. cross-metathesis leading to octa[(E)alkenyl]-substituted silsesquioxane T_8.

These research projects were expanded to a new member of a silsesquioxane family (i.e., a divinyl-substituted double-decker silsesquioxanes which were used as reactants in the silylative coupling and cross-metathesis (only for E = Ge) processes with a wide range of olefins, including highly π-conjugated arenes (Figure 21) [70,84–86]. It resulted in the development of a stereoselective method for the preparation of new molecular, dialkenyl-substituted DDSQ derivatives with a well-defined and documented structure ((*E*) isomer of –HC=CH–). The applied processes are a complementary tool for the synthesis of this type of systems unique in literature.

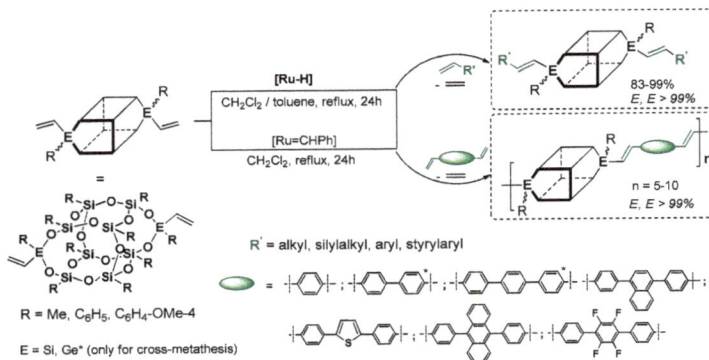

Figure 21. Synthetic protocol for stereoselective silylative coupling and cross-metathesis with olefins/dienes of divinyl-substituted DDSQ systems.

Interestingly, when dienes were applied, they rendered it possible to exploit the methodology for effective and stereoselective formation of new, macromolecular systems with the DDSQ core embedded in the copolymer chain and the preservation of silylene–ethylene–arylene fragments with double-bond (*E*) geometry. This catalytic protocole was the first example in the literature of the synthesis of macromolecular hybrid systems obtained via silylative coupling and cross-metathesis protocols in which the arene unit was stereoselectively connected by the ethenyl bridge with the DDSQ silsesquioxyl fragment.

This fragment concerns the silylative (metallative) coupling process which leads to the highly efficient and stereoselective formation of (*E*)-alkenylsilyl moieties that are anchored to the silsesquioxane core, irrespective of its structure (cubic or double-decker). It also concerns the amount of Si–HC=CH$_2$ reactive groups (one, two or eight). Moreover, it may lead to the formation of both molecular and macromolecular silsesquioxane-based compounds. Due to the presence of the (*E*)-alkenylsilyl units, the resulting functionalized silsesquioxanes are complementary to the architecture of those compounds obtained via cross-metathesis and Heck coupling. Our methodology is an effective, straightforward and clean (only ethylene is a gaseous byproduct) route leading to these systems that are obtained from commercially available reagents. Due to the presence of the (*E*)-alkenylsilyl fragment and its conjugation with π-arenes, these molecules are potentially of high photophysical interest, as was proved in the case of the analogous POSS-based systems described previously in this review in chapter 3.1 and 3.3. As a result, this kind of "bottom up" approach—referring to the synthesis of silsesquioxanes with desired physical and chemical properties—is known, justifiable and used.

3.5. Hydrosilylation of Alkenes and Alkynes as Effcient Procedures for the Introduction of Silylalkanyl and Silylalkenyl Moieties tothe Si–O–Si Core

In the chemistry of organosilicon compounds, the hydrosilylation process is of key importance and, in this case, it has a predominant role as a fundamental industrial tool for the synthesis of the aforementioned systems [87,88]. A catalytic insertion of Si–H into multiple bonds, in particular into carbon–carbon unsaturated bonds (C=C and C≡C) including heteroatom, occurs in the presence of transition metal complexes (TM complexes), especially platinum Karstedt's catalyst = [Pt$_2$(dvds)$_3$], and nickel, rhodium and iridium based systems in homo- and heterogeneous forms, but also with Lewis acids (e.g., AlCl$_3$). The reaction conditions, catalytic systems and reagents' structures influence the stereo-or even regioselectivity of hydrosilylation leading towards saturated and unsaturated molecular, but also macromolecular organosilicon systems [82]. Hydrosilylation is also a very popular catalytic procedure to anchor organic dyes onto a Si–O–Si core of a cubic T_8 cage (POSS) or a double-decker (DDSQ) type.

Imae and Kawakami's group reported a new type of POSS derivative containing carbazole as a photo- and electroactive chromophore [89,90]. The carbazole unit was introduced via hydrosilylation between octa(dimethylsiloxy)silsesquixane (Q_8M_8) and 9-vinylcarbazole in the presence of commercially available Karstedt's ([Pt$_2$(dvds)$_3$]) as well as Speier's (H$_2$PtCl$_6$) catalyst (Figure 22).

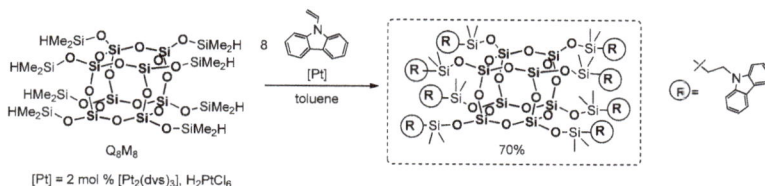

[Pt] = 2 mol % [Pt$_2$(dvs)$_3$], H$_2$PtCl$_6$

Figure 22. Hydrosilylation reaction towards efficient synthesis of octakis [2-carbazol-9-yl]-ethyldimethylsiloxy]silsesquioxane.

The reaction was performed in dry toluene using a small excess of olefin to ensure complete conversion of the reactants, and it led to the formation of the expected product with a 70% yield. Its TGA showed that the product was stable up to ~400 °C in air, and 450 °C in N$_2$ (i.e., it exhibited higher stability than its organic analogoue, poly(9-vinylcarbazole) (PVCz), which was also proved by DSC anaysis and was expected by POSS addition). The T_g = 37 °C was much lower than that of the PVCz (T_g = 190 °C) due to the flexible spacer between POSS and carbazole. The photophysical parameters were analyzed in solution (THF) and in a solid state, and were compared to their organic couterparts (9-ethylcarbazole (EtCz) and poly(9-vinylcarbazole) (PVCz)). The emmision spectrum of the product (λ_{ex} = 250 nm) showed two peaks, λ_{em} = 353 and 370 nm, which is almost the same as that of the EtCz. Similar compatibility was obtained in terms of quantum yields in air (Φ_p = 0.27 and Φ_{EtCz} = 0.30). On the other hand, the PVCz showed weak and broad peaks with a lower quantum yield (Φ_{PVCz} = 0.10) resulting from the formation of excimer. The λ_{em} for solids was more red-shifted towards the wavelength region 350–400 nm and 450 nm, indcating a greater aggregation in crystalline product than that in amorphous material, but the values of the photoluminescence quantum yields were the same. The result suggests that the introduction of the rigid POSS core isolated each of the carbazole units, preventing their aggregation and the formation of excimer, which resulted in a useful concept to apply this new material in a solid-state optelectronic device.

Interesting research on the synthesis of carbazole-substituted double-decker-shaped silsesquioxane (DDSQ-4Cz) also obtained via a hydrosilylation reaction (mediated by Karstedt's catalyst) was presented by Miyashita et al. (Figure 23) [91].

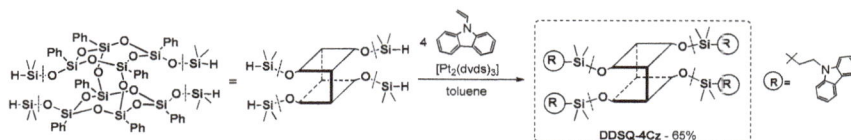

Figure 23. Reaction path towards tetra(carbazole)-substituted DDSQ (DDSQ-4Cz).

The obtained material's thermal stability was higher than 355 °C. Photophysical properties of DDSQ-4Cz were disclosed via UV-vis and photoluminescence analysis and compared with its organic analogue (N-vinylcarbazole). The absorption spectrum (in CHCl$_3$) revealed of two peaks (λ_{ab} between 325–350 nm) atributted to the π–π* absorption of carbazole, which closely resembled that of N-vinylcarbazole. Interestingly, the photoluminescene spectra of DDSQ-4Cz and N-vinylcarbazole were practically the same (λ_{em} = 350–300 nm), indicating that the four carbazole units were isolated by the rigid core, thereby preventing excimer formation. The authors also demonstrate the obtained material's potential application in organic electronic diodes. They fabricated an electroluminescent device that showed a maximum brightness of 320 cd m^{-2} at 20 V drive voltage with 40 mA cm^{-2} current density. This proved that DDSQ are promising compounds useful for OLEDs.

Dudziec and Marciniec's groups reported on dihydro-substituted double-decker silsesquioxane (DDSQ-2SiH) efficient modification using Pt-Karstedt's mediated hydrosilylation reaction protocol with precise reaction time control (FTIR in situ). This facilitated selective formation of a series of molecular as well as macromolecular DDSQ-based systems with ethyl-bridged π-conjugated arenes (Figure 24) [92]. Obtained compounds were verified in terms of their thermal stability (higher than organic couterparts) as well as their photophysics.

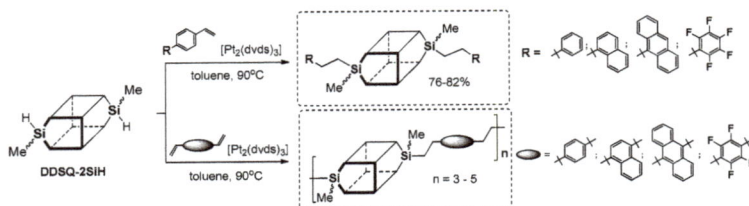

Figure 24. General procedure for the synthesis of molecular and macromolecular aryl-ethyl-double-decker-shaped silsesquioxanes via hydrosilylation.

A UV-vis analysis of the resultng molecular and macromolecular compounds in comparison to the respective arenes was performed. For the DDSQ-based products, the absorption spectra combined the shape of the DDSQ unit as well as the organic counterpart. For the 1-(4-vinylphenyl)naphthalene derivative, the absorption spectrum showed a 21 nm blue-shifted torn peak at λ_{ab} = 257 and 330 nm, which was different than the basic arene (λ_{ab} = 278 nm). This result may suggest the involvement of Si–O–Si core in the excited state, as has already been reported by Laine and Sellinger for mono- and octa-functionalized silsesquioxanes [47,58]. The photophysical and thermal features could enable the application of these compounds in OLEDs devices. Our group studied the hydrosilylation of olefins and alkynes with silsesquioxanes with one, four and eight reactive Si–H moieties, which constitute a valuable group of precursors used in the formation of photoactive materials [93–97].

In 2014 Wang, He and Chin demonstrated a practical route to the synthesis of a series of three hybrid polymers containing a periodically interconnected DDSQ unit and oligofluorenes in the main chain [98]. Hydrosilylation was employed to prepare these materials by sequential addition of dihydro-substituted DDSQ (DDSQ-2SiH) and fluorenes: monomer (n = 1), dimer (n = 2) and trimer (n = 3) in the presence of the Karstedt's catalyst (Figure 25).

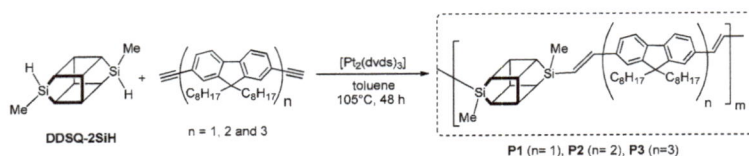

Figure 25. Synthesis of oligomers with the DDSQ unit and oligofluorenes in the main chain via hydrosilylation.

The obtained copolymers (P1–P3) had molecular weights of M_n = 9700, (Polydispersity Index) PDI = 2.55; M_n = 18,700, PDI = 1.45; and M_n = 39,300, PDI = 1.34, respectively, and were soluble in common oragnic solvent. TG analyses indicated a high level of thermal resistance in the obtained systems, up to 446 °C. This means that the block copolymer containing DDSQ units displayed an improvement in their thermal stability when compared to neat poly(fluorene) (PF, 417 °C). The authors also demonstrated that the thermal stability of these products was affected by the ratio of DDSQ in the polymer chain, since the DDSQ can lower thermal conductivity and work as a thermal barrier to isolate heat. All polymers were thermoplastic, with significantly higher T_g (up to 125 °C) than the corresponding PF ($T_g = 55$ °C). The optical properties of the hybrid nanocomposites were determined and compared with the chromophores used. The absorption and emission spectra (in solution and solid) were different. According to the literature, conjugated polymers exhibited significant changes in their solid-state spectra because of interchain interaction (e.g., for PF, λ_{em} changed from 459 to 539 nm in the solid state). Surprisingly, in obtained products, no obvious λ_{em} peak shift was observed for solid- and solution-state spectra, while the λ_{ab} of polymer films (solid) was slightly blue-shifted in comparison to those in a solution. These slight shifts were probably due to the aggregation-induced disorder. The authors of the study observed small 4–13 nm red shifts in the PL λ_{em} for solution compared to the solid state, mainly because of the restriction in conformational motion. It is worth mentioning that λ_{ab} and λ_{em} gradually red-shift along with the increase of chain length (i.e., from P1 to P3). It was demonstrated that all the hybrid polymers exhibited a quantum yield higher than 95%, while for the PF it was only 79%. The reason for such high values of P1–P3 quantum yields could be the confinement of excitons within each oligofluorene unit due to the interconnecting DDSQ cage. Thus, the oligofluorene fragments behaved as isolated, not conjugated chromophores. According to the authors, these results suggested that the optical properties of oligofluorenes were faithfully maintained in the proposed final materials. To demonstrate the potential of synthesized materials in organic electronic applications, the authors successfully prepared two OLED devices that maintained color stability and purity when compared to PF.

In 2006, Shim et al. demonstrated that a hydrosilylation reaction may be used for the efficient synthesis of POSS-based blue-light electroluminescent (EL) nanoparticles containing terfluorene moieties on each of their eight arms [99]. This material was obtained via the Pt-catalyzed hydrosilylation between commercially available octa(dimethylsiloxy)silsesquioxare T8 (Q_8M_8) and an allyl-functionalized terfluorene (Figure 26).

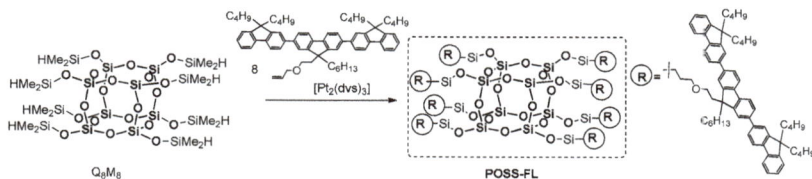

Figure 26. Synthesis of octakis(terfluorene)silsesquioxane via hydrosilylation.

The material displayed high thermal stability ($T_d{}^{5\%} = 379$ °C) and good film-forming properties (tests on a quartz or an indium tin oxide plate). The photophysical analyses were performed in solution (THF) and in solid state. The maximum λ_{ab} in solution was 352 nm (353 nm in a solid state) with PL emission maxima λ_{ab} at 394 and 415 nm (401 and 420 nm in a solid state). These results were compared with those for the organic counterpart poly(dihexylfluorene) and were coincided both in solution and in solid state. This good spectral overlap suggests that POSS–FL can be used as a dopant of blue-light-emissive conjugated polymers, such as polyfluorenes, to increase their quantum efficiencies through energy transference and the isolation of chromophores, as well as in applications requiring electroluminescent nanoparticles. It is worth mentioning that the external quantum efficiency in EL devices of POSS-derivative doped poly(dihexylfluorene) blends were found to be four to eight times higher than that of their organic analogue.

Other examples based on the same procedure for anchoring the chromophore part onto the T8 core were presented by Lee, Mochizuki and Jabbour group [100]. They used a hydrosilylation process for the synthesis of a series of macromolecular materials composed of an inorganic Q_8M_8 core and organic fluorescent emitters covalently attached either monochromatically or in a combination of mixed emitters (POSS with different chromophore systems) (Figure 27).

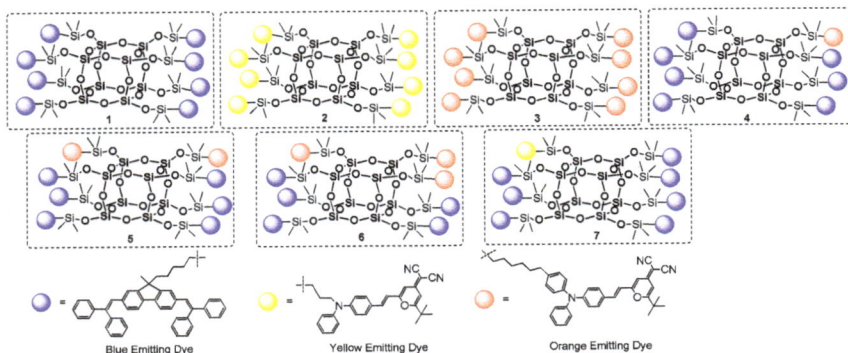

Figure 27. Synthetic hydrosilylation route to T$_8$-based systems with mixed organic fluorescent emitters.

Hydrosilylation was performed using Karstedt's catalyst with fixed and controlled reagents' stoichiometry; however. the final product yields may be not satisfying and may suggest further optimization of the procedure. Monochromatic POSS-emitter products (1–3) were obtained with a 60%, 47% and 61% yield, respectively, while isolated yields of POSS containing a combination of two different emitters (4–7) were not higher than 26%. All compounds were proved to have higher thermal stability than their organic emitters. The most interesting part of the research was disclosed in defining the photophysical properties of obtained compounds. Solution and thin-film photoluminescence spectra were measured and compared for both the free emitters and the modified POSS. It was demonstrated that monochromatic products (1–3) have similar absorbance and emission spectra to their free emitter counterparts both in solution and in a solid state. Thus, using POSS core as a scaffold does not influence the color of the emitters applied. Surprising results were obtained for products bearing more than one type of emitter (4–7). It was found that lower-energy emissions (orange or yellow) dominated in the photoluminescence spectrum for materials 4–7 due to a strong intramolecular energy transference from the neighboring higher-energy blue emitter on the POSS. This high degree of energy transference derives from two major factors: the overlapping of the blue emitter's emission band with its lower-energy absorption band, and the blue and orange emitters' short intramolecular distance. Overall, the authors indicated that these materials were easily processable and could be spin-coated from solution for the fabrication of OLED devices. They were also prepared and tested with dopants of selected POSS-based systems.

In 2009, Hwang et al. described the synthesis of a similar type of material, i.e., POSS-based electroluminescent nanoparticles, containing anthracenenaphthyl fragments at each of their eight arms [101]. The procedure involves the hydrosilylation of Q$_8$M$_8$ with vinyl-functionalized 9-naphtalene-2-yl-10-phenyl anthracene in the presence of 2 mol % of Karstedt's catalyst (Figure 28).

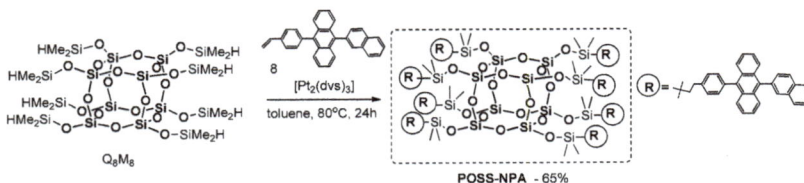

Figure 28. Synthesis of octa(anthracenenaphthyl)silsesquioxane T$_8$ via hydrosilylation.

POSS–NPA was analyzed to investigate its thermal and photophysical properties. It was found to be thermally stable up 450 °C, with high T_g = 154 °C. The UV-vis absorption and PL emission maxima of the final product in solution (chlorobenzene) were found to be 378 and 433 nm, while those characteristics for the product in a solid state were 379 and 464 nm, respectively. These results indicate the aggregation of POSS–NPA molecules and their intermolecular interactions. The authors fabricated an electroluminescent (EL) device that shows blue light emissions.

Xu and Su reported on the synthesis of a new star-type POSS-based molecular hybrid containing stilbene chromophore [102]. These materials were prepared by the hydrosilylation of a series of alkynylstilbenes with octahydridosilsesquioxane (OHS) (Figure 29).

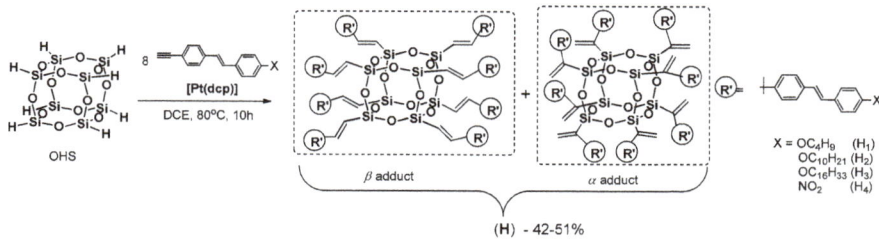

Figure 29. Synthetic pathway for the hydrosilylation of terminal alkynes with OHS.

The reaction was performed in refluxing 1,2-dichloroethane (DCE) in the presence of platinum dicyclopentadiene complex [Pt(dcp)], and led to the formation of expected products with isolated yields ranging from 42 to 51%. However, this catalyst seems to have lower selectivity, as in all cases the formation of α and β isomer mixture was detected (α:β = 39–43:57–61). All products exhibited high thermal stability ($T_d{}^{5\%}$ up to 320 °C, T_g up to 250 °C). The THF solutions of the products were easily cast into films, whereas films of the chromophore could not be obtained in the same conditions. The photophysics were verified and the UV-is spectra (in THF) revealed the presence of strong absorption peaks at λ_{ab} = 336 (H1), 339 (H2), 335 (H3) and 360 (H4) nm. It was also found that alkoxy chain length exerts little influence on the absorption of products H1–3. In the case of H4 the absorption peak was red-shifted, which may be attributable to a stronger intramolecular interaction and formation of the excited state. It was demonstrated that the products (H1–4) and their free emitter counterparts showed nearly the same λ_{ab} and broader absorption bands for the hybrids (H), that may have originated from the σ–π conjugation effect of Si–C=C– in obtained materials. The authors proved that the combination of an inorganic POSS core with stilbene chromophores endowed the resulting materials with novel optical limiting properties and high thermal stability. However, there wano information on how the presence of α and β isomers in H affected these properties.

In 2016, Le et al. developed a new spherosilicate-based compound containing carbazole and pyrene units [103]. This new organic–inorganic amorphous hybrid material was obtained in a two-step reaction involving: (1) synthesis of chromophore using alkylation of N-dipyrenylcarbazole with allyl bromide, and (2) hydrosilylation of obtained olefin with Q_8M_8 in the presence of Karstedt's catalyst (Figure 30).

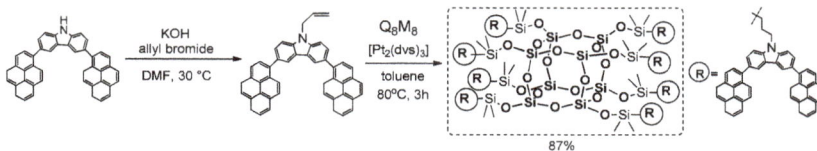

Figure 30. Functionalization of octahydrospherosilicate with N-allyl-carbazole derivative.

The authors demonstrated that this newly-developed POSS derivative exhibits excellent thermal stability, efficient control of POSS dispersion behavior and good film-forming properties. The photophysical properties of the product were determined and compared with chromophore used in solution (CHCl$_3$) and in solid state. The UV-vis spectra of both compounds showed a strong absorption band at λ_{ab} = 281 nm, which can be attributed to the π–π^* local electron transition of the carbazole fragment, and a longer-wavelength absorption band at λ_{ab} ca. 349 nm, which corresponds to the π–π^* electronic transition of the chromophore backbones. The shape and position of the PL emission spectra was different for the product and chromophore, which can be attributed to the fact that bulky POSS cages reduce the extent of substrate aggregation. This conclusion also supports measurement of PL quantum efficiency (PLQE) of the product (90%), which was much higher than that of the separate chromophore (66%). The results of emission in solid state were interesting, and showed ca. 20 nm red-shift in λ_{em} compared to the solution-phase spectra. This suggests the presence of weak van der Waals forces and intermolecular π–π interactions of adjacent chromophore fragments. Although significantly red-shifted behavior was observed in the film state, the final product exhibited an emission band centered at 454 nm, which was substantially blue-shifted compared to that of the substrate film (463 nm). These observations indicate that the POSS cage inhibits the aggregation of chromophore units to improve their color purity and stability both in solution and the solid state. This also promotes its possible applications in solution-processed optoelectronic devices.

Li and Xu et al. reported on the synthesis of two organic–inorganic nanohybrid materials obtained by decorating spherosilicate POSS cores with 2,3,4,5-tetraphenylsilole units through a one-step hydrosilylation process (Figure 31) [104]. As presented in Figure 31, Q$_8$M$_8$ and Q$_8$V$_8$ were used as synthetic scaffolds and two different silole derivatives were chosen as their reagents. Expected products (**1, 2**) were isolated with 57% and 47% yields, respectively. Their thermal and photophysical properties were determined along with their utility as potential sensors for the selective detection of nitroaromatic explosives in aqueous media. As expected, thermal stability of these systems was improved (up to 390 °C). Photophysical properties of both products were solvent-dependent. It was observed that product **1** emits weakly at 486 nm in THF and at 496 nm in a mixture of THF/water (1:9), while in 90% water content the photoluminescence intensity is 68-fold higher than in pure THF.

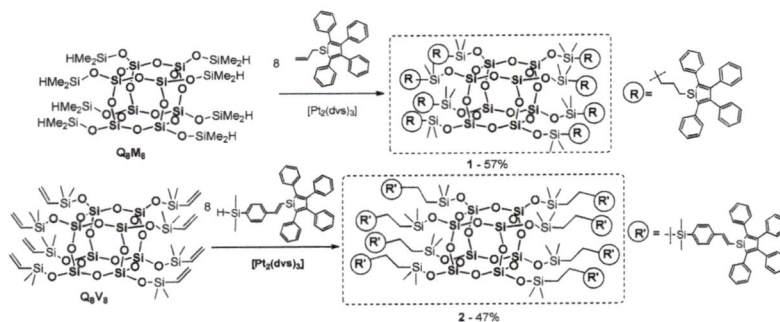

Figure 31. Hydrosilylation-based functionalization of Q$_8$M$_8$ and Q$_8$V$_8$ with allyl-tetraphenylsilole and hydrosilylphenyl-tetraphenylsilole.

These results suggest that the PL of **1** clearly demonstrates the aggregation-enhanced emission (AEE) of this molecule. Similar emissive behavior was also observed for **2**. For a quantitative comparison, measurements of fluorescence quantum were performed. The absolute fluorescence quantum values of the thin solid films of both products reached as high as 51% and 45%, respectively, meaning they were higher than those of their corresponding parent silole molecules in the solid state. Obtained results proved that the POSS core improved the fluorescence quantum yield of the AEE materials in both the solution and in the aggregated state.

Further examples of the use of silsesquioxane derivatives in hydrosilylation with azo-chromophores in order to form functionalizable POSS nanoparticles have been reported by the Xu, Tsukruk and Miniewicz groups independently [105–109]. The procedures, allowing the synthesis of POSS derivatives with azobenzene mesogenes, have been proposed via the hydrosilylation of respective octahydro-substituted T_8 silsesquioxanes. Selected structures of final materials (**1–10**) are presented in Figure 32.

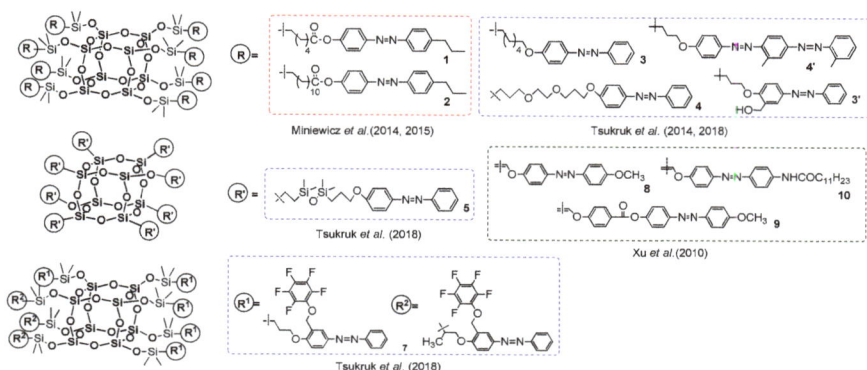

Figure 32. The POSS derivatives with azobenzene-based chromophores obtained by hydrosilylation.

Xu presented a group of azobenzene-containing POSS-based star-like hybrid functional materials obtained via hydrosilylation octahydridosilsesquioxane with three different azobenzene chromophores containing terminal alkynyl moiety as the reactive fragments [109]. The reactions were carried out in the presence of platinum dicyclopentadiene complex [Pt(dcp)] at 80 °C, using an eight-fold molar excess of olefin relative to silsesquioxane. However, the process was not selective, and expected products were obtained as a mixture of α- and β-adducts with yields ranging from 48% to 54%. Analyzed thermal stability was proved to be higher than of the respective chromophores and suggests its enhancement with anchoring chropomhores onto POSS core. Interestingly, the authors did not observe any sign of melting or glass transition of the products up to the decomposition temperatures, though all substrates were crystalline solids. These amorphous characters of compounds may result from a star-like structure with rigid azobenzene chromophore units protruding from a spherical POSS core in 3D thus minimizing any π–π interactions. The photophysical properties of all products were disclosed using UV-vis and PL analyses, and were compared with their organic analogues. Product **10** and its corresponding azobenzene chromophore displayed nearly the same maximum absorption wavelengths and spectral pattern, indicating that Si–O–Si core has no significant effect on the electronic structure of the final material. The UV-vis spectra of products (**8–10**) in THF showed strong absorption bands at λ_{ab} = 356, 347 and 365 nm, respectively, which can be assigned to the π–π* local electron transition of the azobenzene fragments. It was also proven that azobenzene moieties influence the λ_{ab} (e.g., the maximum absorption peak of compound **9** shows a 9 nm blue shift compared to that of **8**, which may have resulted from the rigid spacer –PhCOO– unit twisting the planar azobenzene fragment. Analogous results were achieved for spectra recorded in a solid state (λ_{ab} = 360, 347 and 366 nm). This suggests the star-type structure in products enables avoiding the aggregation for azobenzene units in solid state. The authors showed that obtained materials had good film-forming properties, high thermal stability and novel optical limiting properties that could be potentially applied as opto- and electroluminescent materials.

Miniewicz et al. reported on two types of novel nanoparticles based on a POSS surrounded by eight covalently attached units of azobenzene mesogens differing in the length of their aliphatic chains (**1, 2**) (Figure 32) [105,106]. The compounds were selectively obtained in β isomers using Karsted's catalyst and obtained with 68.5% and 71% yields, respectively. The UV-vis spectra (in CHCl$_3$) of the stable forms of **1** and **2** exhibited features that were characteristic of parent azobenzene molecules in the wavelength range 250–550 nm, as well as a strong π–π* band at λ_{ab} ca. 338 nm (for **1**) and 335 nm (for **2**) and much weaker broad bands with maxima at λ_{ab} ca. 445 nm (for **1**) and 440 nm (for **2**), respectively. These results suggest that, in equilibrium and at room temperature, the azobenzene units are predominantly in their trans configuration. Different absorption spectra were recorded after homogeneously being irradiated by intense broad UV light (360 nm) for 2 min. The disappearance of a band at 335 nm and the increase of intensity of the 440 nm band was observed. This transformation is characteristic of the reversible trans–cis isomerization process in azobenzene units. In further studies, the authors proved that the addition of the POSS compound with mesogenic branches containing azo dye undergo photoisomerization enhanced by UV light.

The Tsukruk group presented a similar type of nanoparticles, but have reported two different approaches to fabricating azo-functionalized POSS compounds via Karstedt's mediated hydrosilylation. These were carried out in 48 and 72 h, with high 83–94% yields (Figure 32) [107,108]. Firstly, they obtained branched azo–POSS structures based on azo dyes possessing flexible spacers of different chemical natures and lengths between the POSS core and the azobenzene units (**3–5**). Secondly, they synthesized the final material by isolating the azobenzene moieties with a constant short spacer between the azo dyes and POSS core (**6, 7**). The product **7** was obtained as mixture of α and β isomer, and the distribution of these isomers at molecular level was statistical. The higher thermal stability of all systems was proved (for products **3** and **5** the melting point temperatures were detected). The photophysical properties of all products were determined and compared with azo dyes (in solution and solid state). The UV-vis spectra of prepared azobenzene-modified POSS compounds showed two characteristic bands: λ_{ab} at 334–347 nm, which was related to the π–π* transition of the trans-azobenzene form, and λ_{ab} at 427–443 nm, originating from a typical n–π* transition. It was also observed that λ_{ab} peak positions (π–π*) of the initial substrates were identical to their respective POSS products. However, the absorption bands corresponding to π–π* transition in the trans isomers of all products were broader. This may have resulted from partial chromophore–chromophore aggregation of the azobenzene moieties. The UV-vis absorption spectra of the prepared films closely resembled solution spectra. Moreover, the products with the isolated azobenzene fragments (**3–5**) revealed a significant decrease in the trans–cis photoisomerization rate in solution as compared to other systems (**6, 7**). In the solid state, the photoisomerization rates of all products were almost the same. These results indicate that the presence of isolation groups effectively prevents aggregation of substituents attached to the POSS core in films.

A different part of POSS-based frameworks exhibiting interesting photophysical properties are TM complexes, mainly Ir and Pt, that are anchored by functional groups playing the role of a ligand to the silsesquioxane core. The phosphorescent TM complexes used as emitters in OLEDs have attracted much attention because they can utilize singlet as well as triplet excitons through spin-orbital coupling with metal ions [110]. Platinum, and particularly iridium, are the most applicable metals, as they possess high quantum efficiency, brightness, a variety of colors and short excited-state lifetimes, and may be used as an alternative to classical OLEDs or in electrochemical devices, etc. On the other hand, silsesquioxane-based systems using conjugated arenes as substituents exhibit interesting photophysical and also thermal properties. Their incorporation into organic-light-emitting materials prevents aggregation of the organic moieties that lead to emission quenching and/or the lowering of quantum efficiency. The POSS core is suggested to act as a hybrid hole-transport material when compared to traditional hole-transport materials. Additionally, rigid Si–O–Si core improves the thermal stability of the resulting material. These silsesquioxane features have encouraged researchers to consider its application as a scaffold for organic substituents that play role of a ligand to anchor the metal ion and constitute a POSS-based complex. In this part, the proper choice of a coordination sphere for a desired complex (i.e., the presence of respective functional groups at the Si–O–Si core acting as donating groups in ligands, as well as modulating metal ancillary ligands), has evolved over the recent years, and interesting reports on TM complexes anchored onto silsesquioxane cores appeared. Their application in the fabrication of light-emitting devices has been observed previously [110]. The literature review suggests a hydrosilylation reaction (mediated by Karstedt's complex) as an easy methodology, since the bridge or spacer between the POSS core and the potential ligand play a minor role [111]. This is why the aliphatic connection resulting from hydrosilylation of alkene moiety by Si–H groupis the predominant synthetic procedure that enables formation of POSS-based compounds and their use as potential ligands for TM-complex synthesis.

Jabbour et al. pioneered this issue and in 2009 reported on the efficient formation of a series of not only iridium-based, but also platinum-based complexes (Figure 33) [112,113]. The synthetic procedure involved a few stages, which is a general approach for this kind of research. First, the proper design of an organic ligand structure is required. Usually the *N*- and *O*-based ligands are the most favorable, (e.g., based on acetylacetone and pyridine derivatives (2-phenylpyridine, 3-phenylisoquinoline), etc.).

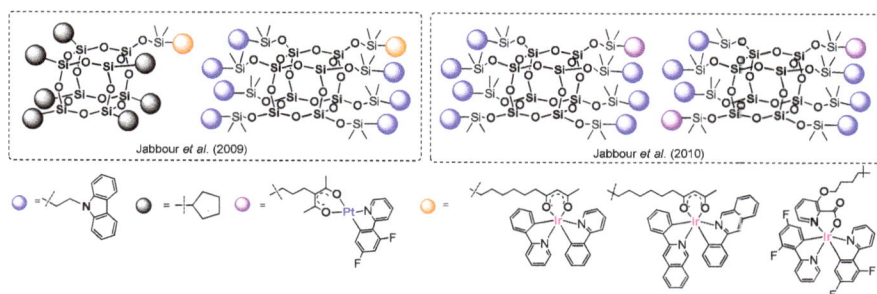

Figure 33. The POSS-based complexes of Ir and Pt obtained by hydrosilylation.

Formation of a respective alkenyl-functionalized linker, placed on the ligand moiety, was used for combining this molecule with the POSS core. Next, the selection and introduction of a proper ancillary ligand to the TM coordination sphere is important. The final part of the research concerns performance of a hydrosilylation reaction to anchor the complex with its functionality onto the POSS core. In these examples, the mono- and octa-substituted POSS was used as a reagent, with the $-OMe_2Si$-H moiety as reactive group. In the case of octa-functional systems, one Ir-based complex fragment and one and two Pt-based complex fragments were attached to the silsesquioxane. The rest of the Si–H moieties were reacted with allyl-carbazole. The authors obtained these systems with diverse yields, from very high (92%), to moderate (54–63%) and rather low (~20%), and their photophysical properties were analyzed ($CHCl_3$, solid film). In general, the absorption properties of these molecules derive from the absorption of metal-to-ligand charge transfers (MLCT) at λ_{ab} ca. 340–350 nm. The carbazole moieties, together with ligand-centered absorption units, result in shorter wavelengths. New emission peaks at $\lambda_{em} = 465$ and 610 nm (green and green-yellow) from the complex with POSS system were red-shifted, especially in solid state for a molecule with two complex moieties in a molecule due to the excimer/aggregate states. Nevertheless, these systems were used in the fabrication of electroluminescent devices that revealed quantum efficiency to be enhanced.

Yu et al. reported on the efficient formation of silsesquioxane-based phosphorescent materials consisting of an emissive Ir(III) complex and carbazole moieties (Figure 34) [114,115].

These systems were based on 3-(pyridin-2-yl)coumarinato N,C^4) as ancillary ligand and varied the additional ligand that was used to bind the POSS core with allyl group (i.e., 8-hydroxyquinolinolate and acetylacetone derivatives). The authors obtained compounds with one and two Ir moieties anchored to one POSS molecule with 12.5%, 23% and 55% yields, respectively. Their enhanced thermal stability was proven. The photoluminescence spectra of the POSS systems (in DCM and solid state) were obtained. Again, the absorption spectra were derived from carbazole units ($\lambda_{ab} = 360$ nm) as well as spin-allowed and spin-forbidden metal-to-ligand charge transfers (MLCT) ($\lambda_{ab} = 400$ nm). The emissions of these systems were similar and revealed the presence a weak green emission at λ_{em} = 535 nm, along with a dominant red emission at λ_{em} = 635 nm for 8-hydroxyquinolinolate, a strong green emission at λ_{em} = 530 nm and a dominant green emission λ_{em} = 567 nm. The quenching of the emission within the concentration rise was diminished due to the separation of the chromophores onto the POSS core. These compounds were also successfully applied to the fabrication of electroluminescence devices exhibiting high brightness and higher external quantum efficiencies.

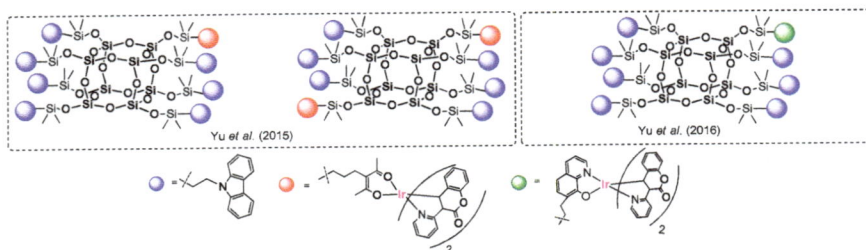

Figure 34. The Ir complexes of spherosilicate obtained through hydrosilylation.

In 2016, You and Su reported on a synthesis of POSS-based systems containing one, two and three moieties of Ir(III) complex that were also based on 3-(pyridin-2-yl)coumarinato N,C⁴) as an ancillary ligand, and allyl acetylacetone modified with n-Bu-carbazole molecule (Figure 35) [116].

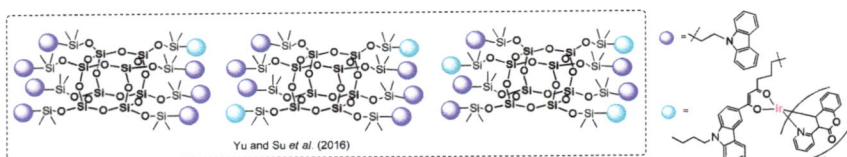

Figure 35. The spherosilicate-based complexes with one, two and three Ir complexes obtained by hydrosilylation.

These systems were obtained using the abovementioned general synthetic approach via hydrosilylation, with 26.4%, 14.7% and 6.4% yields, respectively, that may suggest the optimization of reaction conditions for better yields. Obtained compounds were verified in terms of their thermal resistance, which was proven. The absorption and emission spectra were, in general, analogous to the previously reported ones (i.e., absorption derived from the carbazole unit and spin-allowed and -forbidden metal-to-ligand charge-transfer (MLCT) transitions). The emission (λ_{ex} = 430 nm) spectra resembled each other and exhibited mainly green emission at λ_{em} = 530 nm and minor green-yellow at λ_{em} = 570 nm. The results also present a reduction of quenching from Ir-complex moiety interactions due to a bulky POSS core. The electroluminescent device fabricated using the abovementioned compounds revealed an enhanced quantum efficiency that proved its application utility.

3.6. Other Catalytic Reactions leading to the Formation of POSS-Based Systems of Interesting Photophysical Properties

A group of catalytic processes known as "click reactions" has gained a profound interest in (bio-) organic chemistry and also materials chemistry, a term that was successfully introduced into literature by Sharpless in 2001 [117]. In general, these are perceived as mainly one-pot, modular processes, with a wide tolerance to other functional groups that generate minimal and inoffensive byproducts. They are supposed to be stereospecific and are characterized by a high thermodynamic driving force to favor a single product formation with a high yield. Additionally, the process would preferably have mild and simple reaction conditions, along with commercially available reagents to obtain products easily isolated, mainly via crystallization or distillation. Many of these criteria are subjective and depend on a certain type of a process, but several reactions fit the concept better than others, and have been applied in the chemistry of silsesquioxanes [118]:

- [3+2] cycloadditions, such as Huisgen 1,3-dipolar cycloaddition and particularly Cu(I)-catalyzed azide-alkyne cycloaddition (CuAAC)
- thiol-ene coupling reaction (TEC)
- strain-promoted azide-alkyne cycloaddition (SPAAC)
- Diels–Alder reaction
- thiol-Michael addition
- oxime ligation

One of these processes, applied in the synthesis of silsesquioxanes, is the 1,3-dipolar Huisgen cycloaddition of azides to alkynes, to give form 1,2,3-triazoles (CuAAC) [119]. This is a catalytic reaction, mediated mostly by copper(I) salts (but also ruthenium or silver) and reducing agents (e.g., sodium ascorbate) [118,120].

A combination of the 1,3-dipolar Huisgen process followed by palladium Sonogashira coupling was used in the synthesis of oligophenyl-ethynylenes with heptaisobutylsilsesquioxane used as pendant groups by Ervithayasuporn et al. (Figure 36) [121].

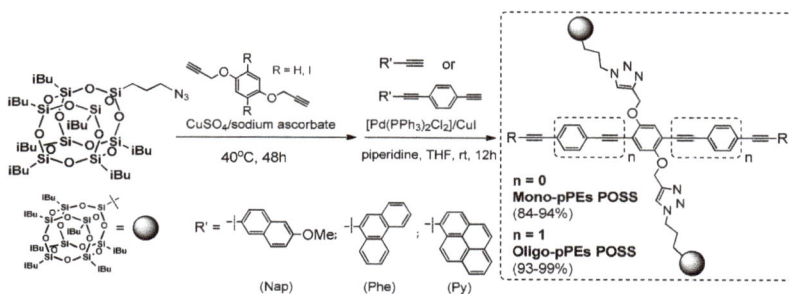

Figure 36. Synthesis of oligophenyl-ethynylenes with heptaisobutylsilsesquioxane as pendant groups via a 1,3-dipolar Huisgen reaction followed by palladium Sonogashira coupling.

These two consecutive processes turned out to be efficient methods (84–99% yields) for the synthesis of interesting, highly π-conjugated organic systems, with silsesquioxane units at two independent molecule fragment, of high thermal stability ($T_d^{5\%}$ up to 369 °C and exhibited T_g in a range of 80–197 °C for R' = Phe oligo-pPE POSS). Their photophysical properties were analyzed in solution (THF) and also in a solid state. While the maxima absorption in solution equaled λ_{ab} = 373–403 nm, the emission spectra (excitation at 350 nm) shift from deep blue to sky blue, and red-shifted (λ_{em} = 409–443 nm) as the amount of aromatic fragments of end-capping molecules increased, as well as with the extended conjugation of in the main chain. The emission spectra of photoluminescence measured in the solid state were strongly shifted towards longer wavelengths (λ_{em} = 470–533), which resulted in interchain orbital interactions referred to aggregation bands. Nevertheless, the extended π-conjugation of oligo-pPEs POSS maintained relatively high photoluminescence quantum efficiencies when compared to mono-pPEs POSS. The authors reported on the preliminary and successful use of mono- and oligo-pPEs POSS (R' = Py) as active dopants in the fabrication of dye-doped OLED devices.

An interesting example of octazido-functional POSS was reported by Xu et al. with two types of functionalities of different stoichiometry obtained via a 1,3-dipolar Huisgen reaction (Figure 37) [122]. The authors used octa(azidomethylene)spherosilicate obtained as an efficient azido-bering reagent [123]. The synthetic procedure was based on a different kind of copper complex, along with the reducing agent (i.e., [Cu(PPh₃)₃Br] and KF) that resulted in very high yields of products bearing two types of functionalities B and Y: W_{71}, W_{62} and W_{53}, which abbreviations derive from the stoichiometry of the acetylenes applied.

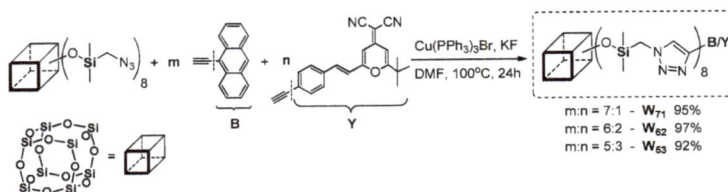

m:n = 7:1 - W_{71} 95%
m:n = 6:2 - W_{62} 97%
m:n = 5:3 - W_{53} 92%

Figure 37. Synthetic route to octa(azidomethylene)spherosilicate derivatives obtained by a 1,3-dipolar Huisgen reaction.

The obtained compounds were thermally more stable (T_g in a range of 192–273 °C) than their organic counterparts (B and Y). Their photophysical properties were analyzed in comparison to the acetylenes (B and Y) as reference molecules in solution and in a solid state. The absorption and emission spectra (shape, intensity and also λ_{em}) of the acetylenes were dependent on the solvent use, and were collected in THF and also THF/water (from a 0 to 90% water fraction). Additionally, the emission spectra of Y were strongly fixed on the amount of water content and were red-shifted ca. 60 nm. As a result, the emission spectra of W_{62} and W_{53} possessing two (W_{62}) and three (W_{53}) Y groups also exhibited the THF/water mixture emission dependence. The emission spectrum of W_{62} in THF presented two peaks at λ_{em} = 410 and 440 nm, but with the increasing content of water an enhancement of two new red-shifted peaks at ~530 and 570 nm were noted. As a result, the light band was fully covered from 400 to 700 nm, making the realization of a white light emission possible. Interestingly, this feature was preserved in the solid state and enabled its application as dopants in the synthesis of white-light-emitting materials.

López-Arbeloa, García-Moreno and Chiara et al. presented T_8 silsesquioxane-based dyes that were obtained by a copper(I)-catalyzed 1,3-dipolar cyclo-addition of octakis(3-azidopropyl)octasilsesquioxane with alkyne-substituted 4,4-di-fluoro-4-bora-3a,4a-diaza- *s*-indacene (BDP) chromophore (Figure 38) [124].

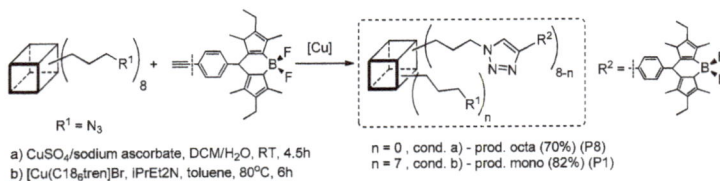

R^1 = N_3

a) CuSO$_4$/sodium ascorbate, DCM/H$_2$O, RT, 4.5h
b) [Cu(C18$_6$tren]Br, iPrEt2N, toluene, 80°C, 6h

n = 0 , cond. a) - prod. octa (70%) (P8)
n = 7 , cond. b) - prod. mono (82%) (P1)

Figure 38. Reaction path towards octa- and mono-BDP derivatives of T_8 silsesquioxane (BDP–POSS).

The reaction conditions (also reagent stoichiometry) as well as the type of Cu catalyst affected the number of BDP substituents attached to the T_8 core, and the process was carried out to obtain mono-BDP-substituted and fully (i.e., octa-BDP) substituted product. These two systems, along with their respective BDP analogues (as model compounds), were analyzed in terms of their photophysical properties. P1 and P8 exhibited similar absorption and fluorescence properties (P1: λ_{ab} = 526 nm, λ_{em} = 540 nm; P8: λ_{ab} = 523 nm, λ_{em} = 546 nm) but P8 revealed substantial fluorescence decay that was attributed to the strong π–π interaction of BDP units and formation of H-type or "sandwich" aggregates. Additionally, the fluorescence of P8 was strongly dependent on the type of solvent used (its poor solubility in non-polar media), and acetone was found to be the best. The compounds were analyzed for their lasing properties (in solution and in a solid state) in comparison to the BDP molecular analogues. Interestingly, compound P1 reached a laser efficiency of up to 56% (in the solid matrix) and up to 60% (in ethyl acetate solution) while maintaining a high photostability when compared to the parent BDP model (35%). This proves the high application potential of BDP–POSS-based systems in the formation of new photonic materials as alternative sources for optoelectronic devices.

An interesting work on the radical thiol–ene reaction (TEC) (mediated via DMPA = 2,2-Dimethoxy-2-phenylacetophenone) of propargyl thiol and octa(vinyl)POSS (OVS) functionalization, and its further ligation with rhodamine 6G, was presented by Mironenko et al. (Figure 39.) [125].

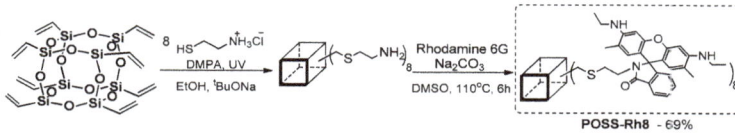

Figure 39. Reaction path towards octa-functionalized silsesquioxane with rhodamine 6G.

The authors reported on the photophysics of compound and related absorption (λ_{ab} = 532 nm)—emission spectra (λ_{em} = 554 nm) that were pH dependent and increased in intensity with a decrease in pH value. Due to their problems with solubility, an increase of intensity was noted for higher-content ethanol in the water/ethanol solution. This compound was tested as a fluorescent chemosensor for a variety of metals, and the results indicated a positive response for Au^{3+} ion (50:50 water/ethanol, pH = 1 and molar concentration of rhodamine units of 10^{-5} M). Under these conditions, there was a strong fluorescence emission enhancement. As a result, POSS–Rh8 may be used as selective fluorescent chemosensor for the selective determination of Au(3+) ions in aqueous media.

One of the reactions applied in the chemistry of silsesquioxanes was free radical copolymerization leading to hybrid materials that are more amenable to solution techniques (e.g., spin-coating that results in large-area device fabrication). On the other hand, the phosphorescence of metal complexes, especially iridium or platinum, and their use in specific types of OLED devices was reported [110]. However, these complexes may suffer from self-quenching in the solid due to interaction aggregation. This may be improved by anchoring them onto the Si–O–Si core (described in Section 3.4). The idea of grafting of non-conjugated polymers (as a transparent matrix) with the metal complex covalently attached to the polymer backbone was studied. In this scenario, solubility would be additionally improved in comparison to the conjugated counterparts, thus facilitating application.

An interesting paper of Ling and Lin et al. reported on the efficient synthesis of poly(methylmethacrylate) via the AIBN (2,2'-Azobis(2-methylpropionitrile))-mediated radical copolymerization of two types of methylmethacrylates (bearing mono(propyl)heptaphenyl(silsesquioxne) = M3, carbazole unit = M1, with styryl-based iridium complex = M2) used in different molar ratio (POSS from 2 to 8 mol %) and resulting in copolymeric systems (Figure 40) [126]. Depending on the specific ratio of comonomers, the molecular weights varied (M_n = 5100–19,400 g/mol, suggesting formation of other oligomeric systems), with PDI from 1.4 to 2.2.

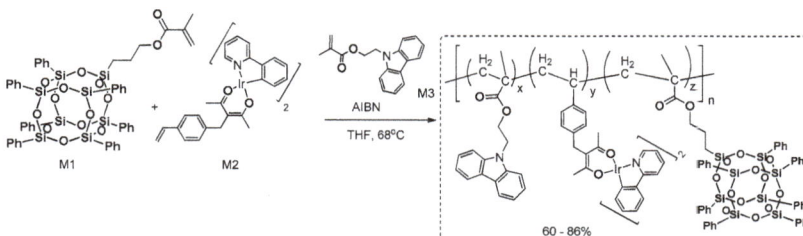

Figure 40. The synthetic route of the copolymer-based Ir complex.

The obtained materials were characterized in terms of their absorption emission properties, comparing the results with analyses for each monomer M2, M3 along with the copolymer without

POSS (i.e., M2 + M3 and final ternary copolymers M1 + M2 + M3 with 2–8 mol % POSS contribution). The absorption spectra of copolymers in DCM showed peaks at λ_{ab} = 237, 261, 294, 328 and 342 nm, which came from the carbazole segment and iridium complex moiety. The photoluminescence spectra (λ_{ex} = 365 nm) in solution exhibited blue and green emission bands λ_{em} ca. 420 and 435 nm (derived from the host carbazole unit M1), with additional green emission bands at λ_{em} ca. 530 nm (derived from the guest iridium complex fragment M2). The band at 530 nm was enhanced with the increase of iridium complex content from 0.5 to 5.0 mol % while the bands around 420 and 435 nm decreased in intensity. There was energy transference suggested by the host (donor) carbazole moiety to the guest (acceptor) iridium complex moiety, which is also dependent on the content of Ir units along with the POSS moiety. However, the ternary copolymer with 5 mol % or iridium units and 2 mol % of POSS exhibited a significant enhancement of the green emission that may have indicated the separation of Ir complex units, preventing their aggregation, so that the concentration-quenching was restrained in the solution. In a solid state, the emission spectra presented only one little red-shifted green emission band (up to ~10 nm) but there was an interesting enhancement of the spectra intensity for the copolymer with 0.5 mol % of Ir complex content, meaning that the POSS unit prevented the aggregation of Ir units more effectively—a phenomenon that was more intensive with an increase in POSS content. The quantum efficiencies for these systems in a solid state were also higher, up to 52% (when compared to binary copolymers M2 + M3, Φ = 7–24%), which suggests that incorporation of POSS units enhances luminescence profoundly. Thermal resistance was also improved. These aspects make the obtained materials definite candidates for OLED devices.

4. Conclusions

The continuous demand for novel hybrid materials of specific applications inspires researchers to develop new synthetic procedures in modular and efficient ways. In this review, we have concentrated on the last decade, most effective routes for synthesizing functionalized polysilsesquioxanes to be employed in POSS-based photoactive materials and/or their precursors.

- The first and essential step towards the synthesis of the abovementioned frameworks is the stoichiometric, hydrolytic condensation of prefunctionalized chloro- or alkoxysilanes, to constitute a basic silsesquioxane core with respective reactivity. These systems are, in turn, the basis for further modification (i.e., the introduction of highly functionalized organic moieties of complex architecture, usually and profoundly via catalytic methods).
- Heck coupling (HC) reactions catalyzed by palladium complex proceeding between C=C bond (especially SiCH=CH$_2$) and aryl halides is a reaction very commonly applied for the synthesis of π-conjugated arene systems and related optoelectronic compounds. The respective Sonogashira coupling (i.e., a reaction of aryl halides with terminal alkynes) is much less frequently used.
- Olefin cross-metathesis (CM), as well as silylative coupling (SC) reactions, enable the formation of silylalkenyl moieties, which may be obtained in the presence of Grubbs ruthenium carbene (Ru = CHPh) (CM) or Ru–H/Si (SC) catalyst. These are the other generally effective and highly selective methods. Also, Ge–HC=CH$_2$ group in germasilsesquioxane undergo effective cross-metathetic transformation.
- Hydrosilylation of olefins and/or acetylenes is performed predominantly with hydrospherosilicate and catalyzed by TM (usually Pt) complex, and is prevalent reaction that can be efficient step towards the synthesis of optoelectronic materials. Additionally, when dihydro-substituted silsesquioxane (especially DDSQ) is applied as a reagent, the formation of respective macromolecular (polymeric) material with a Si–O–Si core embedded can be also obtained.
- In the last part of this review, other catalytic processes leading to the functionalized silsesquioxane of potent optoelectronic materials was collected.

The methodology presented for the synthesis of POSS-based photoactive materials involves mono-, octa-functional cubic or double-decker silsesquioxanes. As was illustrated, the crucial aspect is to design a certain organic dye of respective photophysical properties (color of emitted light) and to anchor it on to the POSS core using the proper reaction procedure, which is dependent on the type of functional group at the Si–O–Si core. According to the perspective presented by experts in the chemistry of POSS-based photoactive compounds, this is now the next step required for the development of attractive and selective routes leading to multi-functional silsesquioxane derivatives with a markedly broadened application.

Author Contributions: B.D. and P.Ż.—contributed equally to Conceptualization, Supervision, Writing original draft and Writing-review & editing; B.M.—Conceptualization.

Funding: This research was funded by the National Centre for Research and Development of Poland [project No. PBS3/A1/16/2015] and the National Science Centre of Poland [project OPUS DEC-2016/23/B/ST5/00201].

Conflicts of Interest: The authors declare no conflict of interest.

References

1. POSS-Hybrid Plastics. Registered Trademark. Available online: https://hybridplastics.com/ (accessed on 15 March 2019).
2. Hartmann-Thompson, C. *Applications of Polyhedral Oligomeric Silsesquioxanes*; Springer: London, UK; New York, NY, USA, 2011; ISBN 978-90-481-3786-2.
3. Cordes, D.B.; Lickiss, P.D.; Rataboul, F. Recent developments in the chemistry of cubic polyhedral oligosilsesquioxanes. *Chem. Rev.* **2010**, *110*, 2081–2173. [CrossRef] [PubMed]
4. Yoshizawa, K.; Morimoto, Y.; Watanabe, K.; Ootake, N. Silsesquioxane derivative and process for producing the same. U.S. Patent 7319129 B2, 2008.
5. Morimoto, Y.; Watanabe, K.; Ootake, N.; Inagaki, J.; Yoshida, K.; Ohguma, K. Silsesquioxane derivative and production process for the same. U.S. Patent 7449539 B2, 2008.
6. Dudziec, B.; Marciniec, B. Double-decker Silsesquioxanes: Current Chemistry and Applications. *Curr. Org. Chem.* **2017**, *28*, 2794–2813. [CrossRef]
7. Li, G.; Wang, L.; Ni, H.; Pittman, C.U., Jr. Polyhedral Oligomeric Silsesquioxane (POSS) Polymers and Copolymers: A Review. *J. Inorg. Organomet. Polym.* **2001**, *11*, 123–154. [CrossRef]
8. Zhou, H.; Ye, Q.; Xu, J. Polyhedral oligomeric silsesquioxane-based hybrid materials and their applications. *Mater. Chem. Front.* **2017**, *1*, 212–230. [CrossRef]
9. Laine, R.M. Nanobuilding blocks based on the [OSiO1.5]x (x = 6, 8, 10) octasilsesquioxanes. *J. Mater. Chem.* **2005**, *15*, 3725–3744. [CrossRef]
10. Ayandele, E.; Sarkar, B.; Alexandridis, P. Polyhedral Oligomeric Silsesquioxane (POSS)-Containing Polymer Nanocomposites. *Nanomaterials* **2012**, *2*, 445–475. [CrossRef] [PubMed]
11. Ye, Q.; Zhou, H.; Xu, J. Cubic polyhedral oligomeric silsesquioxane based functional materials: Synthesis, assembly, and applications. *Chem. Asian J.* **2016**, *11*, 1322–1337. [CrossRef] [PubMed]
12. Zhang, W.; Müller, A. Architecture, self-assembly and properties of well-defined hybrid polymers based on polyhedral oligomeric silsequioxane (POSS). *Prog. Polym. Sci.* **2013**, *38*, 1121–1162. [CrossRef]
13. Tanaka, K.; Chujo, Y. Advanced functional materials based on polyhedral oligomeric silsesquioxane (POSS). *J. Mater. Chem.* **2012**, *22*, 1733–1746. [CrossRef]
14. Quadrelli, E.A.; Basset, J.M. On silsesquioxanes' accuracy as molecular models for silica-grafted complexes in heterogeneous catalysis. *Coord. Chem. Rev.* **2010**, *254*, 707–728. [CrossRef]
15. Haxton, K.J.; Cole-Hamilton, D.J.; Morris, R.E. The structure of phosphine-functionalised silsesquioxane-based dendrimers: A molecular dynamics study. *Dalt. Trans.* **2004**, 1665–1669. [CrossRef] [PubMed]
16. Crowley, C.; Klanrit, P.; Butler, C.R.; Varanou, A.; Hynds, R.E.; Chambers, R.C.; Seifalian, A.M.; Birchall, M.A.; Janes, S.M. Biomaterials Surface modi fi cation of a POSS-nanocomposite material to enhance cellular integration of a synthetic bioscaffold Manuela Plat e. *Biomaterials* **2016**, *83*, 283–293. [CrossRef]
17. John, Ł. Selected developments and medical applications of organic–inorganic hybrid biomaterials based on functionalized spherosilicates. *Mater. Sci. Eng. C* **2018**, *88*, 172–181. [CrossRef]

18. Chan, K.L.; Sonar, P.; Sellinger, A. Cubic silsesquioxanes for use in solution processable organic light emitting diodes (OLED). *J. Mater. Chem.* **2009**, *19*, 9103–9120. [CrossRef]

19. Yang, Z.; Gao, M.; Wu, W.; Yang, X.; Sun, X.W.; Zhang, J.; Wang, H.C.; Liu, R.S.; Han, C.Y.; Yang, H.; et al. Recent advances in quantum dot-based light-emitting devices: Challenges and possible solutions. *Mater. Today* **2018**. [CrossRef]

20. Liu, B.; Nie, H.; Zhou, X.; Hu, S.; Luo, D.; Gao, D.; Zou, J.; Xu, M.; Wang, L.; Zhao, Z.; et al. Manipulation of charge and exciton distribution based on blue aggregation-induced emission fluorophors: A novel concept to achieve high-performance hybrid white organic light-emitting diodes. *Adv. Funct. Mater.* **2016**, *26*, 776–783. [CrossRef]

21. Luo, D.; Yang, Y.; Xiao, Y.; Zhao, Y.; Yang, Y.; Liu, B. Regulating Charge and Exciton Distribution in High-Performance Hybrid White Organic Light-Emitting Diodes with n-Type Interlayer Switch. *Nano-Micro Lett.* **2017**, *9*, 37. [CrossRef] [PubMed]

22. Luo, D.; Chen, Q.; Gao, Y.; Zhang, M.; Liu, B. Extremely Simplified, High-Performance, and Doping-Free White Organic Light-Emitting Diodes Based on a Single Thermally Activated Delayed Fluorescent Emitter. *ACS Energy Lett.* **2018**, *3*, 1531–1538. [CrossRef]

23. Liu, B.-Q.; Wang, L.; Gao, D.-Y.; Zou, J.-H.; Ning, H.-L.; Peng, J.-B.; Cao, Y. Extremely high-efficiency and ultrasimplified hybrid white organic light-emitting diodes exploiting double multifunctional blue emitting layers. *Light Sci. Appl.* **2016**, *5*, e16137. [CrossRef] [PubMed]

24. Sellinger, A.; Laine, R.M. Organic-Inorganic Hybrid Light Emitting Devices (HLED). U.S. Patent 6517958 B1, 2003.

25. Lucenti, E.; Botta, C.; Cariati, E.; Righetto, S.; Scarpellini, M.; Tordin, E.; Ugo, R. New organic-inorganic hybrid materials based on perylene diimide-polyhedral oligomeric silsesquioxane dyes with reduced quenching of the emission in the solid state. *Dye. Pigment.* **2013**, *96*, 748–755. [CrossRef]

26. Clarke, D.; Mathew, S.; Matisons, J.; Simon, G.; Skelton, B.W. Synthesis and characterization of a range of POSS imides. *Dye. Pigment.* **2012**, *92*, 659–667. [CrossRef]

27. Du, F.; Bao, Y.; Liu, B.; Tian, J.; Li, Q.; Bai, R. POSS-containing red fluorescent nanoparticles for rapid detection of aqueous fluoride ions. *Chem. Commun.* **2013**, *49*, 4631–4633. [CrossRef] [PubMed]

28. Liu, Y.; Wang, K.-R.; Guo, D.-S.; Jiang, B.-P. Supramolecular Assembly of Perylene Bisimide with β-Cyclodextrin Grafts as a Solid-State Fluorescence Sensor for Vapor Detection. *Adv. Funct. Mater.* **2009**, *19*, 2230–2235. [CrossRef]

29. Du, F.; Tian, J.; Wang, H.; Liu, B.; Jin, B.; Bai, R. Synthesis and Luminescence of POSS-Containing Perylene Bisimide-Bridged Amphiphilic Polymers. *Macromolecules* **2012**, *45*, 3086–3093. [CrossRef]

30. Asuncion, M.Z.; Laine, R.M. Fluoride rearrangement reactions of polyphenyl- and polyvinylsilsesquioxanes as a facile route to mixed functional phenyl, vinyl T_{10} and T_{12} Silsesquioxanes. *J. Am. Chem. Soc.* **2010**, *132*, 3723–3736. [CrossRef]

31. Ronchi, M.; Sulaiman, S.; Boston, N.R.; Laine, R.M. Fluoride catalyzed rearrangements of polysilsesquioxanes, mixed Me, vinyl T_8, Me, vinyl T_{10} and T_{12} cages. *Appl. Organomet. Chem.* **2010**, *24*, 551–557. [CrossRef]

32. Boatz, J.A.; Rzsp, A.; Mabry, J.M.; Rzsm, A.; Mitchell, C. Structural Investigation of Fluoridated POSS Cages Using Ion Mobility Mass Spectrometry and Molecular Mechanics Preprint. *Chem. Mater.* **2008**, *20*, 4299–4309. [CrossRef]

33. Kunthom, R.; Piyanuch, P.; Wanichacheva, N.; Ervithayasuporn, V. Cage-like silsesquioxanes bearing rhodamines as fluorescence Hg2+ sensors. *J. Photochem. Photobiol. A Chem.* **2018**, *356*, 248–255. [CrossRef]

34. Zhou, H.; Ye, Q.; Wu, X.; Song, J.; Cho, C.M.; Zong, Y.; Tang, B.Z.; Hor, T.S.A.; Yeow, E.K.L.; Xu, J. A thermally stable and reversible microporous hydrogen-bonded organic framework: Aggregation induced emission and metal ion-sensing properties. *J. Mater. Chem. C* **2015**, *3*, 11874–11880. [CrossRef]

35. Çakal, D.; Ertan, S.; Cihaner, A.; Önal, A.M. Electrochemical and optical properties of substituted phthalimide based monomers and electrochemical polymerization of 3,4-ethylenedioxythiophene-polyhedral oligomeric silsesquioxane (POSS) analogue. *Dye. Pigment.* **2019**, *161*, 411–418. [CrossRef]

36. Ertan, S.; Kaynak, C.; Cihaner, A. A platform to synthesize a soluble poly(3,4-ethylenedioxythiophene) analogue. *J. Polym. Sci. Part A Polym. Chem.* **2017**, 1–7. [CrossRef]

37. Ertan, S.; Cihaner, A. Improvement of optical properties and redox stability of poly(3,4-ethylenedioxythiophene). *Dye. Pigment.* **2018**, *149*, 437–443. [CrossRef]

38. Huang, H.; Lin, H.; Kershaw, S.V.; Susha, A.S.; Choy, W.C.H.; Rogach, A.L. Polyhedral Oligomeric Silsesquioxane Enhances the Brightness of Perovskite Nanocrystal-Based Green Light-Emitting Devices. *J. Phys. Chem. Lett.* **2016**, *7*, 4398–4404. [CrossRef]

39. Huang, H.; Chen, B.; Wang, Z.; Hung, T.F.; Susha, A.S.; Zhong, H.; Rogach, A.L. Water resistant CsPbX3nanocrystals coated with polyhedral oligomeric silsesquioxane and their use as solid state luminophores in all-perovskite white light-emitting devices. *Chem. Sci.* **2016**, *7*, 5699–5703. [CrossRef]

40. Beletskaya, I.P.; Cheprakov, A.V. Heck reaction as a sharpening stone of palladium catalysis. *Chem. Rev.* **2000**, *100*, 3009–3066. [CrossRef]

41. Liras, M.; Pintado-Sierra, M.; Amat-Guerri, F.; Sastre, R. New BODIPY chromophores bound to polyhedral oligomeric silsesquioxanes (POSS) with improved thermo- and photostability. *J. Mater. Chem.* **2011**, *21*, 12803–12811. [CrossRef]

42. Liu, Y.; Yang, W.; Liu, H. Azobenzene-functionalized cage silsesquioxanes as inorganic-organic hybrid, photoresponsive, nanoscale, building blocks. *Chem. A Eur. J.* **2015**, *21*, 4731–4738. [CrossRef]

43. Sellinger, A.; Tamaki, R.; Laine, R.M.; Ueno, K.; Tanabe, H.; Williams, E.; Jabbour, G.E. Heck coupling of haloaromatics with octavinylsilsesquioxane: solution processable nanocomposites for application in electroluminescent devices. *Chem. Commun. (Camb)* **2005**, 3700–3702. [CrossRef]

44. Chanmungkalakul, S.; Ervithayasuporn, V.; Hanprasit, S.; Masik, M.; Prigyai, N.; Kiatkamjornwong, S. Silsesquioxane cages as fluoride sensors. *Chem. Commun.* **2017**, *53*, 12108–12111. [CrossRef]

45. Chanmungkalakul, S.; Ervithayasuporn, V.; Boonkitti, P.; Phuekphong, A.; Prigyai, N.; Kladsomboon, S.; Kiatkamjornwong, S. Anion identification using silsesquioxane cages. *Chem. Sci.* **2018**, *9*, 7753–7765. [CrossRef]

46. Lo, M.Y.; Zhen, C.; Lauters, M.; Ghassan, J.E.; Sellinger, A. Organic−Inorganic Hybrids Based on Pyrene Functionalized Octavinylsilsesquioxane Cores for Application in OLEDs. *J. Am. Chem. Soc.* **2007**, *18*, 5808–5809. [CrossRef]

47. Yang, X.H.; Giovenzana, T.; Feild, B.; Jabbour, G.E.; Sellinger, A. Solution processeable organic–inorganic hybrids based on pyrene functionalized mixed cubic silsesquioxanes as emitters in OLEDs. *J. Mater. Chem.* **2012**, *22*, 12689–12694. [CrossRef]

48. Wang, S.; Guang, S.; Xu, H.; Ke, F. Controllable preparation and properties of active functional hybrid materials with different chromophores. *RSC Adv.* **2015**, *5*, 1070–1078. [CrossRef]

49. Ke, F.; Wang, S.; Guang, S.; Liu, Q.; Xu, H. Synthesis and properties of broad-band absorption POSS-based hybrids. *Dye. Pigment.* **2015**, *121*, 199–203. [CrossRef]

50. Pu, K.Y.; Li, K.; Zhang, X.; Liu, B. Conjugated oligoelectrolyte harnessed polyhedral oligomeric silsesquioxane as light-up hybrid nanodot fortwo-photon fluorescence imaging of cellular nucleus. *Adv. Mater.* **2010**, *22*, 4186–4189. [CrossRef]

51. Ding, D.; Pu, K.-Y.; Li, K.; Liu, B. Conjugated oligoelectrolyte-polyhedral oligomeric silsesquioxane loaded pH-responsive nanoparticles for targeted fluorescence imaging of cancer cell nucleus. *Chem. Commun.* **2011**, *47*, 9837–9839. [CrossRef]

52. Sun, R.; Feng, S.; Wang, D.; Liu, H. Fluorescence-Tuned Silicone Elastomers for Multicolored Ultraviolet Light-Emitting Diodes: Realizing the Processability of Polyhedral Oligomeric Silsesquioxane-Based Hybrid Porous Polymers. *Chem. Mater.* **2018**, *30*, 6370–6376. [CrossRef]

53. Sun, R.; Huo, X.; Lu, H.; Feng, S.; Wang, D.; Liu, H. Recyclable fluorescent paper sensor for visual detection of nitroaromatic explosives. *Sens. Actuators B Chem.* **2018**, *265*, 476–487. [CrossRef]

54. Brick, C.M.; Tamaki, R.; Kim, S.; Asuncion, M.Z.; Roll, M.; Nemoto, T.; Ouchi, Y.; Chujo, Y.; Laine, R.M. Spherical, Polyfunctional Molecules Using Poly (bromophenylsilsesquioxane) s as Nanoconstruction Sites. *Macromolecules* **2005**, *38*, 4655–4660. [CrossRef]

55. Roll, M.F.; Asuncion, M.Z.; Kampf, J.; Laine, R.M. para-Octaiodophenylsilsesquioxane, [p-IC6H4SiO1.5]8, a Nearly Perfect Nano-Building Block. *ACS Nano* **2008**, *2*, 320–326. [CrossRef]

56. Roll, M.F.; Kampf, J.W.; Kim, Y.; Yi, E.; Laine, R.M. Nano Building Blocks via Iodination of [PhSiO 1.5] n, Forming High-Surface-Area, Thermally Stable, Microporous Materials via Thermal Elimination of I 2. *J. Am. Chem. Soc.* **2010**, *132*, 10171–10183. [CrossRef]

57. Brick, C.M.; Ouchi, Y.; Chujo, Y.; Laine, R.M. Robust Polyaromatic Octasilsesquioxanes from Polybromophenylsilsesquioxanes, Br x OPS, via Suzuki Coupling. *Macromolecules* **2005**, *38*, 4661–4665. [CrossRef]

58. Laine, R.M.; Sulaiman, S.; Brick, C.; Roll, M.; Tamaki, R.; Asuncion, M.Z.; Neurock, M.; Filhol, J.-S.; Lee, C.-Y.; Zhang, J.; et al. Synthesis and Photophysical Properties of Stilbeneoctasilsesquioxanes. Emission Behavior Coupled with Theoretical Modeling Studies Suggest a 3-D Excited State Involving the Silica Core. *J. Am. Chem. Soc.* **2010**, *132*, 3708–3722. [CrossRef] [PubMed]

59. Roll, M.F.; Mathur, P.; Takahashi, K.; Kampf, J.W.; Laine, R.M. [PhSiO1.5]8 promotes self-bromination to produce [o-BrPhSiO1.5]8: further bromination gives crystalline [2,5-Br2PhSiO1.5]8 with a density of 2.32 g cm−3 and a calculated refractive index of 1.7 or the tetraicosa bromo compound [Br3PhSiO1.5]8. *J. Mater. Chem.* **2011**, *21*, 11167–11176. [CrossRef]

60. Sulaiman, S.; Zhang, J.; Goodson, T., III; Laine, R.M. Synthesis, characterization and photophysical properties of polyfunctional phenylsilsesquioxanes: [o-RPhSiO1.5]8, [2,5-R2PhSiO1.5]8, and [R3PhSiO1.5]8. compounds with the highest number of functional units/unit volume. *J. Mater. Chem.* **2011**, *21*, 11177–11187. [CrossRef]

61. Jung, J.H.; Furgal, J.C.; Clark, S.; Schwartz, M.; Chou, K.; Laine, R.M. Beads on a Chain (BoC) Polymers with Model Dendronized Beads. Copolymerization of [(4-NH2C6H4SiO1.5)6(IPhSiO1.5)2] and [(4-CH3OC6H4SiO1.5)6(IPhSiO1.5)2] with 1,4-Diethynylbenzene (DEB) Gives Through-Chain, Extended 3-D Conjugation in the Excited State Tha. *Macromolecules* **2013**, *46*, 7580–7590. [CrossRef]

62. Hwan Jung, J.; Furgal, J.C.; Goodson, T.; Mizumo, T.; Schwartz, M.; Chou, K.; Vonet, J.F.; Laine, R.M. 3-D molecular mixtures of catalytically functionalized [vinylSiO 1.5] 10/[vinylSiO 1.5] 12. Photophysical characterization of second generation derivatives. *Chem. Mater.* **2012**, *24*, 1883–1895. [CrossRef]

63. Furgal, J.C.; Jung, J.H.; Clark, S.; Goodson, T.; Laine, R.M. Beads on a Chain (BoC) Phenylsilsesquioxane (SQ) Polymers via F− Catalyzed Rearrangements and ADMET or Reverse Heck Cross- coupling Reactions: Through Chain, Extended Conjugation in 3-D with Potential for Dendronization. *Macromolecules* **2013**, *46*, 7591–7604. [CrossRef]

64. Chinchilla, R.; Nájera, C. The Sonogashira reaction: A booming methodology in synthetic organic chemistry. *Chem. Rev.* **2007**, *107*, 874–922. [CrossRef]

65. Asuncion, M.Z.; Roll, M.F.; Laine, R.M. Octaalkynylsilsesquioxanes, Nano Sea Urchin Molecular Building Blocks for 3-D-Nanostructures. *Macromolecules* **2008**, *41*, 8047–8052. [CrossRef]

66. Gon, M.; Sato, K.; Tanaka, K.; Chujo, Y. Controllable intramolecular interaction of 3D arranged π-conjugated luminophores based on a POSS scaffold, leading to highly thermally-stable and emissive materials. *RSC Adv.* **2016**, *6*, 78652–78660. [CrossRef]

67. Pietraszuk, C.; Pawluć, P.; Marciniec, B. Handbook of Metathesis. In *Handbook on Metathesis. Vol. 2: Applications in Organic Synthesis*; Grubbs, R., O'Leary, D.J., Eds.; WILEY-VCH: Weinheim, Germany, 2015; pp. 583–631.

68. Feher, F.J.; Soulivong, D.; Eklund, A.G.; Wyndham, K.D. Cross-metathesis of alkenes with vinyl-substituted silsesquioxanes and spherosilicates: A new method for synthesizing highly-functionalized Si/O frameworks. *Chem. Commun.* **1997**, 1185–1186. [CrossRef]

69. Cheng, G.; Vautravers, N.R.; Morris, R.E.; Cole-hamilton, D.J. Synthesis of functional cubes from octavinylsilsesquioxane (OVS). *Org. Biomol. Chem.* **2008**, *6*, 4662–4667. [CrossRef]

70. Żak, P.; Dudziec, B.; Kubicki, M.; Marciniec, B. Silylative Coupling versus Metathesis-Efficient Methods for the Synthesis of Difunctionalized Double-Decker Silsesquioxane Derivatives. *Chem. A Eur. J.* **2014**, *20*, 9387–9393. [CrossRef]

71. Sulaiman, S.; Bhaskar, A.; Zhang, J.; Guda, R.; Goodson, T.; Laine, R.M. Molecules with perfect cubic symmetry as nanobuilding blocks for 3-D assemblies. Elaboration of octavinylsilsesquioxane. Unusual luminescence shifts may indicate extended conjugation involving the silsesquioxane core. *Chem. Mater.* **2008**, *20*, 5563–5573. [CrossRef]

72. Vautravers, N.R.; André, P.; Slawin, A.M.Z.; Cole-Hamilton, D.J. Synthesis and characterization of photoluminescent vinylbiphenyl decorated polyhedral oligomeric silsesquioxanes. *Org. Biomol. Chem.* **2009**, *7*, 717–724. [CrossRef]

73. Vautravers, N.R.; André, P.; Cole-Hamilton, D.J. Fluorescence activation of a polyhedral oligomeric silsesquioxane in the presence of reducing agents. *J. Mater. Chem.* **2009**, *19*, 4545–4550. [CrossRef]

74. Żak, P.; Marciniec, B.; Majchrzak, M.; Pietraszuk, C. Highly effective synthesis of vinylfunctionalised cubic silsesquioxanes. *J. Organomet. Chem.* **2011**, *696*, 887–891. [CrossRef]

75. Araki, H.; Naka, K. Syntheses and properties of dumbbell-shaped POSS derivatives linked by luminescent pi-conjugated units. *J. Polym. Sci. Part A Polym. Chem.* **2012**, *50*, 4170–4181. [CrossRef]

76. Żak, P.; Pietraszuk, C.; Marciniec, B.; Spólnik, B.; Danikiewicz, W. Efficient functionalisation of cubic monovinylsilsesquioxanes via cross-metathesis and silylative coupling with olefins in the presence of ruthenium complexes. *Adv. Synth. Catal.* **2009**, *351*, 2675–2682. [CrossRef]

77. Furgal, J.C.; Jung, J.H.; Goodson, T.; Laine, R.M. Analyzing Structure—Photophysical Property Relationships for Isolated T_8, T_{10}, and T_{12} Stilbenevinylsilsesquioxanes. *J. Am. Chem. Soc.* **2013**, *135*, 12259–12269. [CrossRef]

78. Cabrera-González, J.; Ferrer-Ugalde, A.; Bhattacharyya, S.; Chaari, M.; Teixidor, F.; Gierschner, J.; Núñez, R. Fluorescent carborane-vinylstilbene functionalised octasilsesquioxanes: Synthesis, structural, thermal and photophysical properties. *J. Mater. Chem. C* **2017**, *5*, 10211–10219. [CrossRef]

79. Marciniec, B.; Pietraszuk, C. Silylation of Styrene with Vinylsilanes Catalyzed by RuCl (SiR 3)(CO)(PPh 3) 2 and RuHCl (CO)(PPh 3) 3. *Organometallics* **1997**, *16*, 4320–4326. [CrossRef]

80. Marciniec, B.; Pietraszuk, C. Insertion of Vinylsilane into the Ruthenium-Silicon Bond-Direct Evidence. *J. Chem. Soc. Chem. Commun.* **1995**, 2003–2004. [CrossRef]

81. Wakatsuki, Y.; Yamazaki, H.; Nakano, M.; Yamamoto, Y. Ruthenium-catalysed disproportionation between vinylsilanes and mono-substituted alkenes via silyl group transfer. *J. Chem. Soc. Chem. Commun.* **1991**, *3*, 703–704. [CrossRef]

82. Marciniec, B. Catalytic Coupling of sp²- and sp-Hybridized Carbon–Hydrogen Bonds with Vinylmetalloid Compounds. *Acc. Chem. Res.* **2007**, *40*, 943–952. [CrossRef] [PubMed]

83. Frąckowiak, D.; Żak, P.; Spólnik, G.; Pyziak, M.; Marciniec, B. New Vinylgermanium Derivatives of Silsesquioxanes and Their Ruthenium Complexes - Synthesis, Structure, and Reactivity. *Organometallics* **2015**, *34*, 3950–3958. [CrossRef]

84. Żak, P.; Delaude, L.; Dudziec, B.; Marciniec, B. N-Heterocyclic carbene-based ruthenium-hydride catalysts for the synthesis of unsymmetrically functionalized double-decker silsesquioxanes. *Chem. Commun.* **2018**, *54*, 4306–4309. [CrossRef] [PubMed]

85. Żak, P.; Majchrzak, M.; Wilkowski, G.; Dudziec, B.; Dutkiewicz, M.; Marciniec, B. Synthesis and characterization of functionalized molecular and macromolecular double-decker silsesquioxane systems. *RSC Adv.* **2016**, *6*, 10054–10063. [CrossRef]

86. Żak, P.; Frąckowiak, D.; Grzelak, M.; Bołt, M.; Kubicki, M.; Marciniec, B. Olefin Metathesis of Vinylgermanium Derivatives as Method for the Synthesis of Functionalized Cubic and Double-Decker Germasilsesquioxanes. *Adv. Synth. Catal.* **2016**, *358*, 3265–3276. [CrossRef]

87. Marciniec, B.; Maciejewski, H.; Pietraszuk, C.; Pawluć, P. *Hydrosilylation: A Comprehensive Review on Recent Advances*; Marciniec, B., Ed.; Springer Science+Business Media B.V.: Dordrecht, The Netherlands, 2009; ISBN 978-1-4020-8171-2.

88. Troegel, D.; Stohrer, J. Recent advances and actual challenges in late transition metal catalyzed hydrosilylation of olefins from an industrial point of view. *Coord. Chem. Rev.* **2011**, *255*, 1440–1459. [CrossRef]

89. Imae, I.; Kawakami, Y.; Ooyama, Y.; Harima, Y. Solid state photoluminescence property of a novel poss-based material having carbazole. *Macromol. Symp.* **2007**, *249–250*, 50–55. [CrossRef]

90. Imae, I.; Kawakami, Y. Unique photoluminescence property of a novel perfectly carbazole-substituted POSS. *J. Mater. Chem.* **2005**, *15*, 4581–4583. [CrossRef]

91. Kohri, M.; Matsui, J.; Watanabe, A.; Miyashita, T. Synthesis and Optoelectronic Properties of Completely Carbazole-substituted Double-decker-shaped Silsesquioxane. *Chem. Lett.* **2010**, *39*, 1162–1163. [CrossRef]

92. Walczak, M.; Januszewski, R.; Majchrzak, M.; Kubicki, M.; Dudziec, B.; Marciniec, B. The unusual cis- and trans-architecture of dihydrofunctional double-decker shaped silsesquioxane–design and construction of its ethyl bridged π-conjugated arene derivatives. *New J. Chem.* **2017**, *41*, 3290–3296. [CrossRef]

93. Stefanowska, K.; Franczyk, A.; Szyling, J.; Pyziak, M.; Pawluć, P.; Walkowiak, J. Selective hydrosilylation of alkynes with octaspherosilicate (HSiMe2O)8Si8O12. *Chem. Asian J.* **2018**, *13*, 2101–2108. [CrossRef] [PubMed]

94. Walczak, M.; Stefanowska, K.; Franczyk, A.; Walkowiak, J.; Wawrzyńczak, A.; Marciniec, B. Hydrosilylation of alkenes and alkynes with silsesquioxane (HSiMe2O)(i-Bu)7Si8O12 catalyzed by Pt supported on a styrene-divinylbenzene copolymer. *J. Catal.* **2018**, *367*, 1–6. [CrossRef]

95. Dutkiewicz, M.; Maciejewski, H.; Marciniec, B.; Karasiewicz, J. New fluorocarbofunctional spherosilicates: Synthesis and characterization. *Organometallics* **2011**, *30*, 2149–2153. [CrossRef]

96. Walczak, M.; Januszewski, R.; Franczyk, A.; Marciniec, B. Synthesis of monofunctionalized POSS through hydrosilylation. *J. Organomet. Chem.* **2018**, *872*, 73–78. [CrossRef]

97. Duszczak, J.; Mituła, K.; Januszewski, R.; Żak, P.; Dudziec, B.; Marciniec, B. Highly efficient route for the synthesis of a novel generation of tetraorganofunctional double-decker type of silsesquioxanes. *ChemCatChem* **2019**, *11*, 1086–1091. [CrossRef]

98. Chi, H.; Lim, S.L.; Wang, F.; Wang, X.; He, C.; Chin, W.S. Pure blue-light emissive poly(oligofluorenes) with bifunctional POSS in the main chain. *Macromol. Rapid Commun.* **2014**, *35*, 801–806. [CrossRef]

99. Cho, H.J.; Hwang, D.H.; Lee, J.I.J.; Jung, Y.K.; Park, J.H.; Lee, J.I.J.; Lee, S.K.; Shim, H.K. Electroluminescent polyhedral oligomeric silsesquioxane-based nanoparticle. *Chem. Mater.* **2006**, *18*, 3780–3787. [CrossRef]

100. Froehlich, J.D.; Young, R.; Nakamura, T.; Ohmori, Y.; Li, S.; Mochizuki, A.; Lauters, M.; Jabbour, G.E. Synthesis of multi-functional POSS emitters for OLED applications. *Chem. Mater.* **2007**, *19*, 4991–4997. [CrossRef]

101. Eom, J.H.; Mi, D.; Park, M.J.; Cho, H.J.; Lee, J.; Lee, J.I.; Chu, H.Y.; Shim, H.K.; Hwang, D.H. Synthesis and properties of a polyhedral oligomeric silsesquioxane-based new light-emitting nanoparticle. *J. Nanosci. Nanotechnol.* **2009**, *9*, 7029–7033. [CrossRef]

102. Su, X.; Guang, S.; Li, C.; Xu, H.; Liu, X.; Wang, X.; Song, Y. Molecular hybrid optical limiting materials from polyhedral oligomer silsequioxane: Preparation and relationship between molecular structure and properties. *Macromolecules* **2010**, *43*, 2840–2845. [CrossRef]

103. Cheng, C.C.; Chu, Y.L.; Chu, C.W.; Lee, D.J. Highly efficient organic-inorganic electroluminescence materials for solution-processed blue organic light-emitting diodes. *J. Mater. Chem. C* **2016**, *4*, 6461–6465. [CrossRef]

104. Xiang, K.; Li, Y.; Xu, C.; Li, S. POSS-based organic-inorganic hybrid nanomaterials: Aggregation-enhanced emission, and highly sensitive and selective detection of nitroaromatic explosives in aqueous media. *J. Mater. Chem. C* **2016**, *4*, 5578–5583. [CrossRef]

105. Miniewicz, A.; Tomkowicz, M.; Karpinski, P.; Sznitko, L.; Mossety-Leszczak, B.; Dutkiewicz, M. Light sensitive polymer obtained by dispersion of azo-functionalized POSS nanoparticles. *Chem. Phys.* **2015**, *456*, 65–72. [CrossRef]

106. Miniewicz, A.; Girones, J.; Karpinski, P.; Mossety-Leszczak, B.; Galina, H.; Dutkiewicz, M. Photochromic and nonlinear optical properties of azo-functionalized POSS nanoparticles dispersed in nematic liquid crystals. *J. Mater. Chem. C* **2014**, *2*, 432–440. [CrossRef]

107. Tkachenko, I.M.; Kobzar, Y.L.; Korolovych, V.F.; Stryutsky, A.V.; Matkovska, L.K.; Shevchenko, V.V.; Tsukruk, V.V. Novel branched nanostructures based on polyhedral oligomeric silsesquioxanes and azobenzene dyes containing different spacers and isolation groups. *J. Mater. Chem. C* **2018**, *6*, 4065–4076. [CrossRef]

108. Ledin, P.A.; Tkachenko, I.M.; Xu, W.; Choi, I.; Shevchenko, V.V.; Tsukruk, V.V. Star-shaped molecules with polyhedral oligomeric silsesquioxane core and azobenzene dye arms. *Langmuir* **2014**, *30*, 8856–8865. [CrossRef]

109. Su, X.; Guang, S.; Xu, H.; Yang, J.; Song, Y. The preparation and optical limiting properties of POSS-based molecular hybrid functional materials. *Dye. Pigment.* **2010**, *87*, 69–75. [CrossRef]

110. Zhang, Q.C.; Xiao, H.; Zhang, X.; Xu, L.J.; Chen, Z.N. Luminescent oligonuclear metal complexes and the use in organic light-emitting diodes. *Coord. Chem. Rev.* **2019**, *378*, 121–133. [CrossRef]

111. Chen, K.-B.; Chang, Y.-P.; Yang, S.-H.; Hsu, C.-S. Novel dendritic light-emitting materials containing polyhedral oligomeric silsesquioxanes core. *Thin Solid Films* **2006**, *514*, 103–109. [CrossRef]

112. Yang, X.; Froehlich, J.D.; Chae, H.S.; Li, S.; Mochizuki, A.; Jabbour, G.E. Efficient Light-Emitting Devices Based on Phosphorescent Polyhedral Oligomeric Silsesquioxane Materials. *Adv. Funct. Mater.* **2009**, *19*, 2623–2629. [CrossRef]

113. Yang, X.; Froehlich, J.D.; Chae, H.S.; Harding, B.T.; Li, S.; Mochizuki, A.; Jabbour, G.E. Efficient light-emitting devices based on platinum-complexes-anchored polyhedral oligomeric silsesquioxane materials. *Chem. Mater.* **2010**, *22*, 4776–4782. [CrossRef]

114. Yu, T.; Wang, X.; Su, W.; Zhang, C.; Zhao, Y.; Zhang, H.; Xu, Z. Synthesis and photo- and electro-luminescent properties of Ir(III) complexes attached to polyhedral oligomeric silsesquioxane materials. *RSC Adv.* **2015**, *5*, 80572–80582. [CrossRef]

115. Zhao, Y.; Qiu, X.; Yu, T.; Shi, Y.; Zhang, H.; Xu, Z.; Li, J. Synthesis and characterization of 8-hydroxyquinolinolato-iridium(III) complex grafted on polyhedral oligomeric silsesquioxane core. *Inorganica Chim. Acta* **2016**, *445*, 134–139. [CrossRef]

116. Yu, T.; Xu, Z.; Su, W.; Zhao, Y.; Zhang, H.; Bao, Y. Highly efficient phosphorescent materials based on Ir(III) complexes-grafted on a polyhedral oligomeric silsesquioxane core. *Dalt. Trans.* **2016**, *45*, 13491–13502. [CrossRef]

117. Kolb, H.C.; Finn, M.G.; Sharpless, K.B. Click Chemistry: Diverse Chemical Function from a Few Good Reactions. *Angew. Chemie Int. Ed.* **2001**, *40*, 2004–2021. [CrossRef]

118. Li, Y.; Dong, X.H.; Zou, Y.; Wang, Z.; Yue, K.; Huang, M.; Liu, H.; Feng, X.; Lin, Z.; Zhang, W.; et al. Polyhedral oligomeric silsesquioxane meets "click" chemistry: Rational design and facile preparation of functional hybrid materials. *Polymer* **2017**, *125*, 303–329. [CrossRef]

119. Huisgen, R. 1,3-Dipolar Cycloadditions. Past and Future. *Angew. Chemie Int. Ed. Engl.* **1963**, *2*, 565–598. [CrossRef]

120. Ervithayasuporn, V.; Kwanplod, K.; Boonmak, J.; Youngme, S.; Sangtrirutnugul, P. Homogeneous and heterogeneous catalysts of organopalladium functionalized-polyhedral oligomeric silsesquioxanes for Suzuki–Miyaura reaction. *J. Catal.* **2015**, *332*, 62–69. [CrossRef]

121. Ervithayasuporn, V.; Abe, J.; Wang, X.; Matsushima, T.; Murata, H. Synthesis, characterization, and OLED application of oligo (p-phenylene ethynylene) s with polyhedral oligomeric silsesquioxanes (POSS) as pendant groups. *Tetrahedron* **2010**, *66*, 9348–9355. [CrossRef]

122. Zhao, G.; Zhu, Y.; Guang, S.; Ke, F.; Xu, H. Facile preparation and investigation of the properties of single molecular POSS-based white-light-emitting hybrid materials using click chemistry. *New J. Chem.* **2018**, *42*, 555–563. [CrossRef]

123. Zhu, Y.K.; Guang, S.Y.; Xu, H.Y. A versatile nanobuilding precursor for the effective architecture of well-defined organic/inorganic hybrid via click chemistry. *Chin. Chem. Lett.* **2012**, *23*, 1095–1098. [CrossRef]

124. Costela, Á.; López-Arbeloa, Í.; García-Moreno, I.; Trastoy, B.; Bañuelos, J.; Chiara, J.L.; Pérez-Ojeda, M.E. Click Assembly of Dye-Functionalized Octasilsesquioxanes for Highly Efficient and Photostable Photonic Systems. *Chem. A Eur. J.* **2011**, *17*, 13258–13268. [CrossRef]

125. Tutov, M.V.; Sergeev, A.A.; Zadorozhny, P.A.; Bratskaya, S.Y.; Mironenko, A.Y. Dendrimeric rhodamine based fluorescent probe for selective detection of Au. *Sens. Actuators B Chem.* **2018**, *273*, 916–920. [CrossRef]

126. Lin, M.; Luo, C.; Xing, G.; Chen, L.; Ling, Q. Influence of polyhedral oligomeric silsesquioxanes (POSS) on the luminescence properties of non-conjugated copolymers based on iridium complex and carbazole units. *RSC Adv.* **2017**, *7*, 39512–39522. [CrossRef]

polymers

Review

Functional Polyimide/Polyhedral Oligomeric Silsesquioxane Nanocomposites

Mohamed Gamal Mohamed [1,2] **and Shiao Wei Kuo** [1,3,*]

1   Department of Materials and Optoelectronic Science, Center of Crystal Research, National Sun Yat-Sen University, Kaohsiung 80424, Taiwan; mgamal.eldin12@yahoo.com
2   Chemistry Department, Faculty of Science, Assiut University, Assiut 71516, Egypt
3   Department of Medicinal and Applied Chemistry, Kaohsiung Medical University, Kaohsiung 807, Taiwan
*   Correspondence: kuosw@faculty.nsysu.edu.tw; Tel.: +886-7-525-4099

Received: 11 December 2018; Accepted: 23 December 2018; Published: 25 December 2018

Abstract: The preparation of hybrid nanocomposite materials derived from polyhedral oligomeric silsesquioxane (POSS) nanoparticles and polyimide (PI) has recently attracted much attention from both academia and industry, because such materials can display low water absorption, high thermal stability, good mechanical characteristics, low dielectric constant, flame retardance, chemical resistance, thermo-redox stability, surface hydrophobicity, and excellent electrical properties. Herein, we discussed the various methods that have been used to insert POSS nanoparticles into PI matrices, through covalent chemical bonding and physical blending, as well as the influence of the POSS units on the physical properties of the PIs.

Keywords: polyhedral oligomeric silsesquioxane (POSS); double-decker-shaped silsesquioxane (DDSQ); polyimide; thermal stability; dielectric constant

---

1. Introduction

Polyhedral oligomeric silsesquioxanes (POSS) form a class of nanostructured materials having diameters on the order of 1–3 nm; they can be considered as the smallest silica nanoparticles (NPs) [1–4]. Polyhedral oligomeric silsesquioxanes (POSS) moieties have the empirical formula $(RSiO_{1.5})_n$, where R may be an organic functional group (e.g., alkyl, alkylene, epoxide unit, acrylate, hydroxyl) or a hydrogen atom, controlled porosity and nanometer sized structure; the structures of POSS NPs can be divided into partial cage, ladder, and random structures [5–9]. The properties of POSS-containing polymers can improve (e.g., decreased flammability, viscosity, and heat discharge, and increased rigidity, strength, and modulus) via the degree of dispersion of the POSS NPs into the polymer matrix [10–26]. Two general approaches have been used to incorporate POSS units into polymer matrices: (i) physical blending [27–30] and (ii) covalent attachment [29–35]. In the physical blending approach, the POSS NPs are blended (through melt-mixing or solvent-casting) with the polymers without forming covalent bonds; the success of this approach depends on the compatibility of the POSS NPs with the polymers [36–38]. In the covalent approach, the POSS NPs are attached to the polymer chain through covalent bonds. Although many types of POSS NP architectures (non-functional, mono-functional, bifunctional, and multi-functional) have been incorporated into polymer matrices (Figure 1), the preparation of polymer/POSS nanocomposites remains challenging because it can be expensive to perform on large scales, can require long equilibrium times, and can result in aggregation of the POSS NPs [39–43].

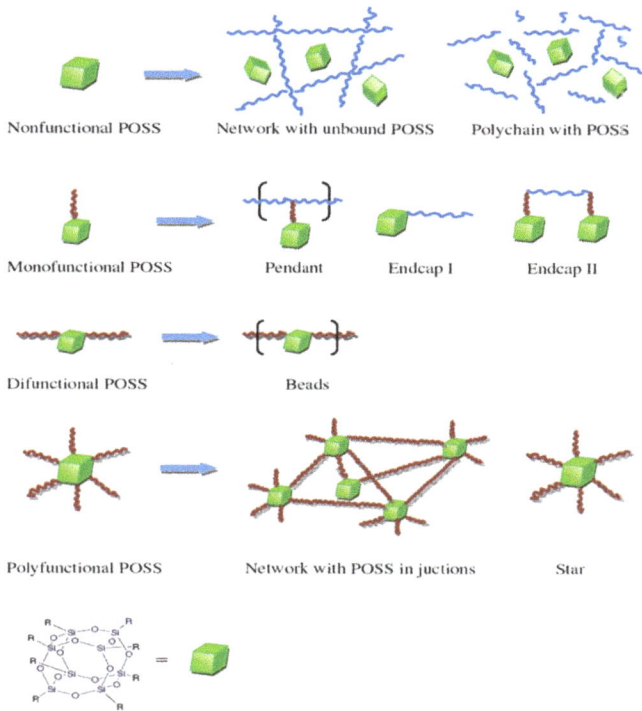

Figure 1. Schematic representation of possible architectures for incorporation of polyhedral oligomeric silsesquioxanes (POSS) nanoparticles (NPs) into polymer matrices [2]. Reproduced with permission from Elsevier.

Polyimide (PI) is an important high-performance material because of its outstanding thermo-oxidative stability, chemical resistance, and mechanical and electrical stability [44–46]. Aromatic polyimides have been applied as insulating materials in many areas, including aerospace and microelectronics [47–61]. Aromatic PIs are polymers featuring stiff aromatic backbones; they are synthesized in two steps: condensation polymerization of an aliphatic or aromatic dianhydride acid and a diamine under ambient conditions in a dipolar aprotic solvent [e.g., *N*,*N*-dimethylacetamide (DMAc), *N*-methylpyrrolidinone (NMP), dimethylsulfoxide (DMSO)] to afford a corresponding poly(amic acid) (PAA), and subsequent ring closure [62–65]. Most aromatic PIs are insoluble and infusible because of their heteroaromatic structures and planar aromatic units. The thermal and mechanical properties of PIs can be enhanced, and their dielectric constants and linear coefficients of thermal expansion decreased, after the incorporation of various inorganic materials (e.g., silica NPs and ceramics) [65–70].

In this Review, we focused on the various methods that have been used to prepare PI/POSS hybrid nanocomposites based on Figure 1, including the blending of non-functional POSS NPs with PIs, the covalent linking of mono-functional POSS NPs with PIs at the chain end or side chain, the covalent incorporation of bifunctional POSS NPs into the PI main chain, and the formation of thermally crosslinked PIs with multi-functional POSS NPs. Additionally, we discussed the physical properties of the resulting PI/POSS nanocomposites, including their dielectric constants and dynamic mechanical, thermal, electrical, and surface properties.

2. PI/POSS Nanocomposites

2.1. Non-Functional POSS NPs Blended with PIs

Chang et al. prepared PI/POSS nanocomposites exhibiting good thermal properties and low dielectric constants through the blending of fluorinated POSS precursors with PI [71]. The fluorinated POSS precursors was prepared by reacting octakis(dimethylsiloxyhexafluoropropyl) silsesquioxane (OF) with allyl 1,1,2,3,3,3-hexafluoropropyl ether (AHFPE) to form OF-POSS NPs, and with allyl propyl ether (APE) to form octakis(propyl ether) silsesquioxane (OP-POSS) NPs, using platinum 1,3-divinyl-1,1,3,3-tetramethyldisiloxane [Pt(dvs)] as the catalyst. Morphological studies revealed the possible occurrence of phase separation in these two PAA/OP and PAA/OF systems after a process of thermal imidization and solvent removal (Figure 2). Based on dynamic mechanical analysis (DMA), the glass transition temperature (T_g) of the PI (371 °C) increased after introduction 10 wt % of the OF-POSS NPs. In contrast, after the incorporation of 15 wt % of the OP-POSS NPs, the values of T_g of the PI (359 °C) decreased relative to that of the corresponding pure PI (366 °C), due to the bulky and rigid POSS units coming into close contact with the polymer chain matrix and increasing the free volume around them [72,73].

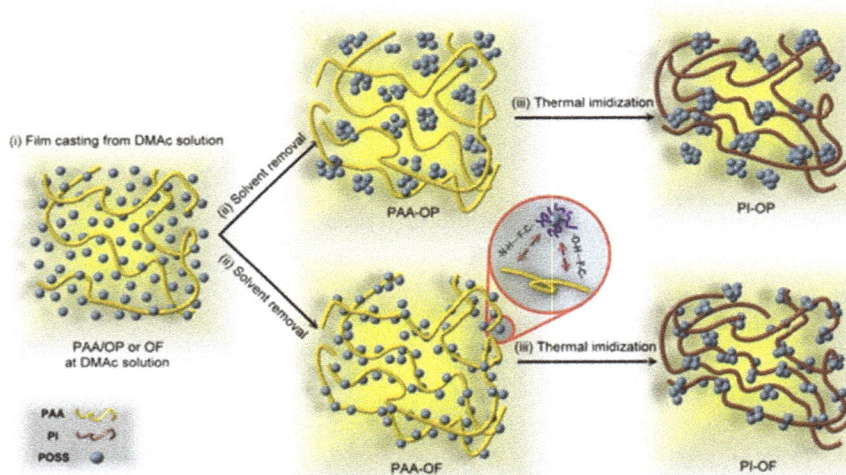

Figure 2. Cartoon representation of the deformation processes of Polyimide (PI)/octakis(dimethylsiloxyhexafluoropropyl) silsesquioxane (OF)-POSS and (OP)-POSS NPs during imidization [71]. Reproduced with permission from Wiley.

Liao et al. [74] prepared octa(aminophenyl)silsesquioxane (OAPS)-functionalized graphene oxide (GO)–reinforced PI nanocomposite films (Figure 3). The tensile strength of the OAPS-GO/PI composite increased 11.2-fold after adding 3 wt % OAPS-GO (Figure 4(a)). In addition, the dielectric constant of the OAPS-GO/PI composite film decreased upon increasing the concentration of OAPS-GO (Figure 4b). Furthermore, we have prepared nanoporous PI films of low dielectric constant (k = 2.25) through blending PI with a star poly(ethylene oxide) PEO-POSS hybrid as a template and then removing the PEO segments through oxidative thermolysis to produce voids inside the polymer matrix (pore sizes: 10–40 nm) [75]. Figure 5 displays the storage moduli and values of tan δ for the pure PI and the porous PI templated by the star PEO-POSS NPs. Through DMA, the pure PI displayed a single value of T_g of 370 °C, while the porous PI exhibited two T_g at 360 and 385 °C. The lower value of T_g corresponded to that typically observed for micro-phase separation from the porous PI, suggesting that the storage modulus decreased as a result of its foam structure. These foam structures would have collapsed, such that the corresponding porous structures would no longer exist, at higher temperatures, with the residual silica from the star PEO-POSS after thermal calcination enhancing the value of T_g (385 °C) of the porous PI. More importantly, the dielectric constant decreased significantly, from 3.25 to 2.25, for this porous PI matrix.

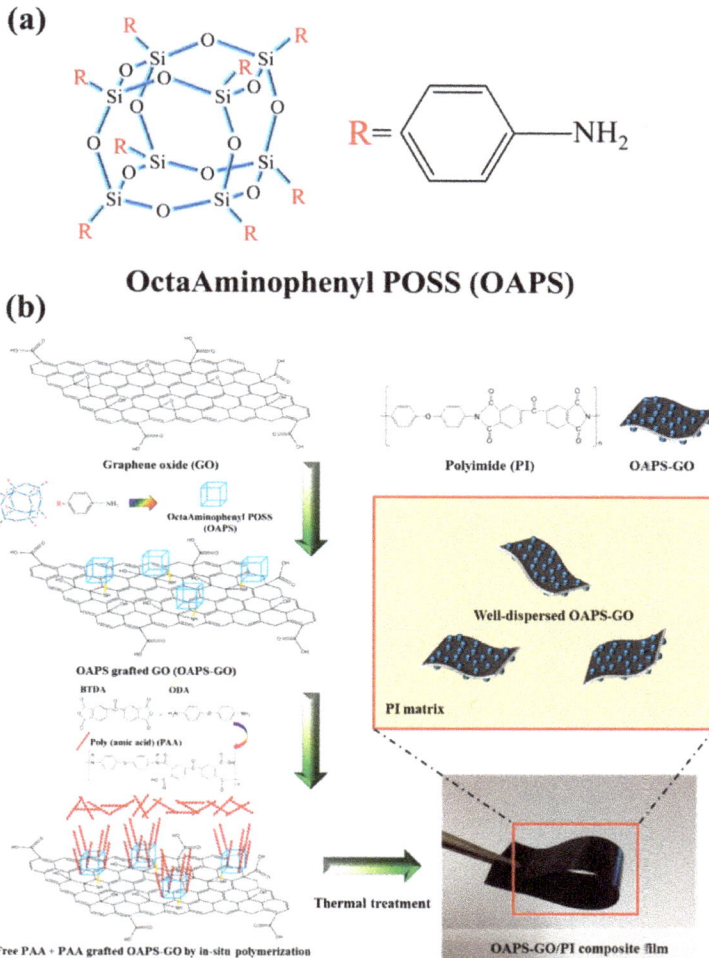

Figure 3. (a) Chemical structure of octa(aminophenyl)silsesquioxane (OAPS). (b) Schematic representation of the preparation of OAPS-graphene oxide (GO)/PI hybrid nanocomposites [74]. Reproduced with permission from the American Chemical Society.

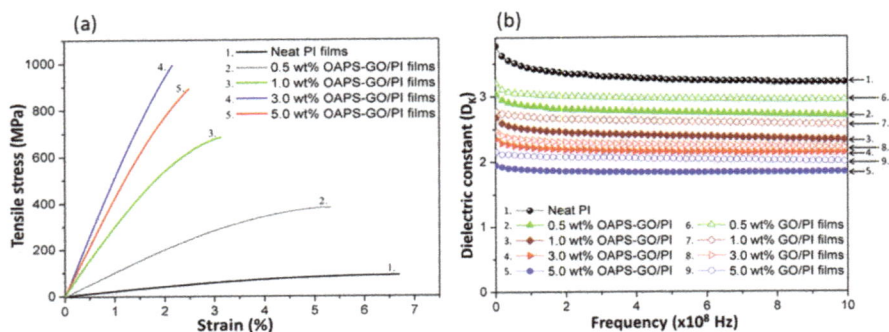

Figure 4. (**a**) Stress–strain curves of neat PI films and OAPS-GO/PI films with different contents of various amounts of OAPS-GO. (**b**) Profile dielectric constants of neat PI films, GO/PI films, and OAPS-GO/PI films [74]. Reproduced with permission from the American Chemical Society.

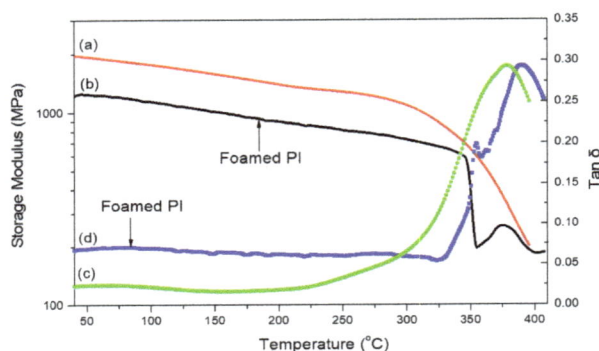

Figure 5. Dynamic mechanical analysis (DMA) curves for (**a,c**) pure PI and (**b,d**) porous PI (obtained from 10 wt % PEO-POSS), recorded at a heating rate of 2 °C/min [75]. Reproduced with permission from Elsevier.

2.2. Mono-Functional POSS NPs with PIs

2.2.1. PIs Containing POSS NPs at Polymer Chain End

Wei et al. [76] prepared nanoporous PI/POSS nanocomposites through a multistep process (Figure 6). First, pyromellitic dianhydride (PMDA) reacted with 4,4′-oxydianiline (ODA) in DMAc under a N_2 atmosphere at room temperature to afford PAA. Then, POSS-NH_2 reacted with PAA in DMAc to obtain PAA/POSS nanocomposites. The PI linked through its chain ends to POSS NPs was obtained after thermal imidization of the PAA/POSS nanocomposites at 300 °C. The dielectric constant of the resultant polyimide POSS nanocomposites decreased from 3.40 for the neat PI to 3.09 after adding 2.5 mol % of POSS units; this behavior was attributed to the phase-separated system and the uniformly porous structure (nanometer-scale) of the POSS molecules. In addition, the resultant PI linked through its chain ends to the POSS NPs formed lamellae or cylinders that were in the range of 60–70 nm long and 5 nm wide (through transmission electron microscope (TEM) image).

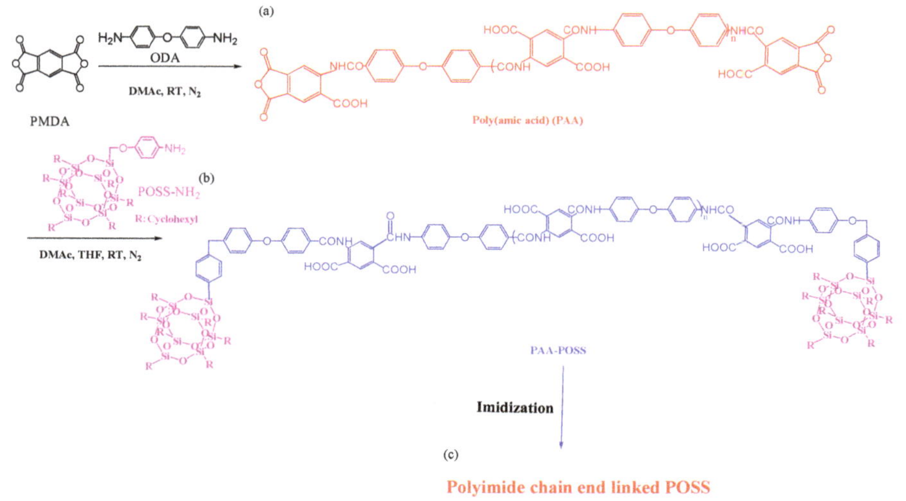

Figure 6. Preparation of (**a**) poly(amic acid) (PAA); (**b**) PAA/POSS nanocomposites; and (**c**) PI linked through its chain ends to POSS [76]. Reproduced with permission from the American Chemical Society.

Zhao et al. [77] synthesized a series of fluorinated PI/POSS hybrid materials through a simple route from 2,2′-bis(trifluoromethyl)benzidine, 4,4′-oxydiphthalic dianhydride, and a monofunctional POSS in m-cresol and isoquinoline as solvents (Figure 7). These hybrid polymers possessed low dielectric constants (in the range 2.47–2.92), high thermal stability, film formation ability, excellent solubility, and good hydrophobic and mechanical properties.

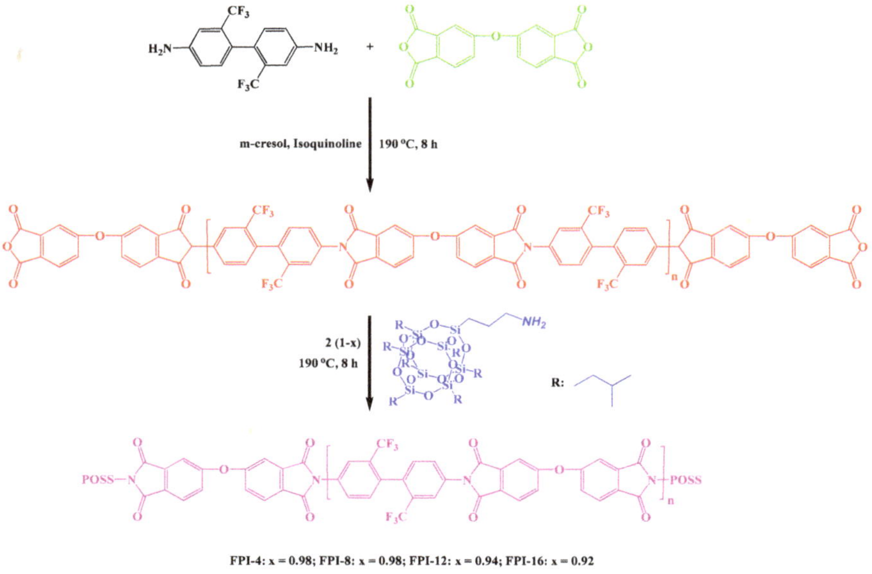

FPI-4: x = 0.98; FPI-8: x = 0.98; FPI-12: x = 0.94; FPI-16: x = 0.92

Figure 7. Synthesis of fluorinated PI/POSS hybrid nanocomposites [77]. Reproduced with permission from Springer.

2.2.2. PIs presenting POSS NPs Grafted to the Side Chains

Wei et al. [78] prepared PI-tethered POSS hybrid materials through copolymerization of a POSS-diamine, ODA, with PMDA (Figure 8). When the POSS content was 16 mol %, this PI/POSS composite displayed the large-scale self-assembled layer-by-layer structure of POSS. This layer-by-layer structure of POSS possessed (as elucidated by TEM observations) a layer length of greater than 100 nm and a layer spacing of 2–4 nm. Chen et al. [79] prepared hybrid film materials having tunable dielectric constants, lower than that of the neat PI, through thermally initiated free-radical graft polymerization of methacrylcyclopentyl-POSS (MA-POSS) with ozone-pretreated poly[N,N-(1,4-phenylene)-3,3′,4,4′-benzophenonetetracarboxylic amic acid] and subsequent thermal imidization to afford PI-g-PMA-POSS nanocomposite films. Nuclear magnetic resonance (NMR) spectroscopy, X-ray diffraction (XRD), and thermogravimetric analysis (TGA) confirmed the chemical composition and structure of these PI-g-PMA-POSS nanocomposite films.

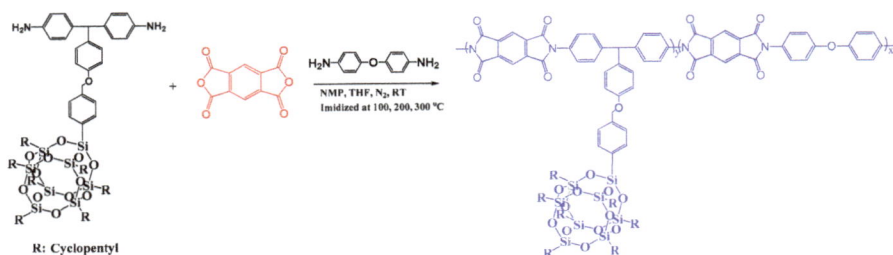

Figure 8. Schematic representation of side chain-tethered PI/POSS nanocomposites [78]. Reproduced with permission from the American Chemical Society.

2.3. Bifunctional POSS NPs Incorporated within the Main Chain of PIs

Kakimoto et al. [80] synthesized various semi-aromatic PIs containing double-decker-shaped silsesquioxane (DDSQ) in the main chain (POSS-PIs) through the reactions of DDSQ-diamine, which was prepared through hydrosilylation of DDSQ with cis-5-norbornene-endo-2,3-dicarboxylic anhydride and a subsequent reaction with ODA. The DDSQ-NH$_2$ reacted with various aromatic tetracarboxylic dianhydrides to obtain POSS-PIs nanocomposites (Figure 9). The POSS-PIs had low water absorption, low dielectric constants, and good thermal stabilities. In addition, the polymer hybrid materials films had good mechanical properties, with an elongation at breakage of 2.9–6.0%. Zheng et al. [81] synthesized a PI containing a tetrafunctional POSS through thermal imidization of 5,11,14,17-tetranilino DDSQ with 4,4′-diaminophenylether (ODA) and 3,3′,4,4′-benzo phenonetetracarboxylic dianhydride (DTDA) in DMAc, affording organic/inorganic PI nanocomposites incorporating variable contents of POSS units. Based on TGA and surface contact angle measurements (Figure 10a,b), these nanocomposite materials exhibited greater thermal stability and surface hydrophobicity, relative to those of the unmodified PI, upon increasing the POSS content. TEM imaging revealed (Figure 11) that the POSS molecules self-assembled into spherical microdomains having diameters in the range from 40 to 80 nm.

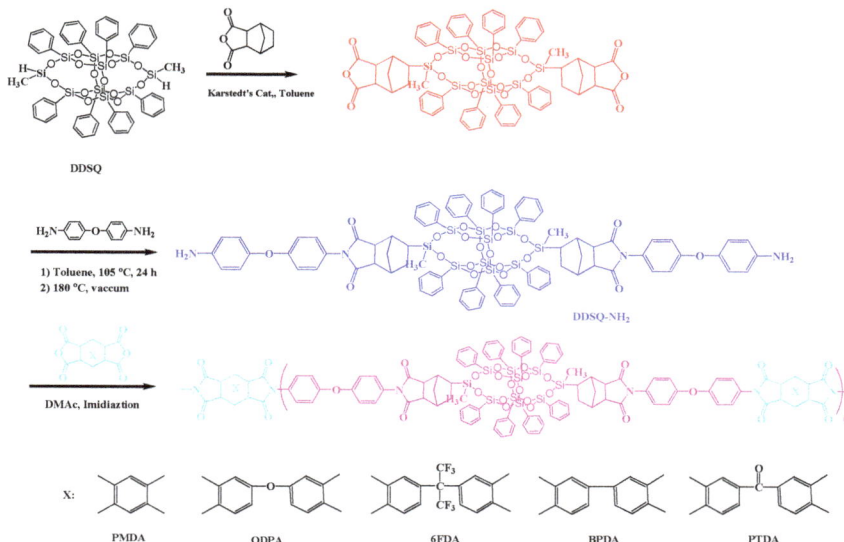

Figure 9. Preparation of semi-aromatic PIs containing a double-decker-shaped silsesquioxane (DDSQ) in the main chain [80]. Reproduced with permission from the American Chemical Society.

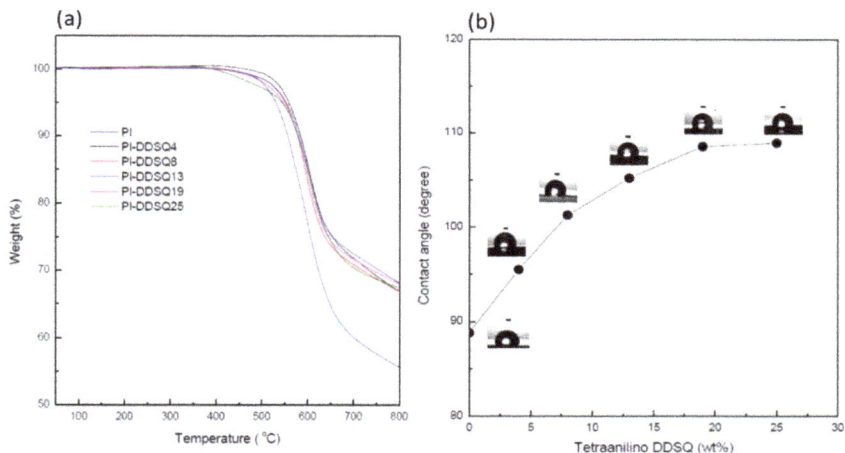

Figure 10. (a) Thermogravimetric analysis (TGA) thermogram and (b) water contact angles of PIs containing various contents of 5,11,14,17-tetranilino DDSQ [81]. Reproduced with permission from Wiley.

Zheng et al. [82] reported a facile synthetic method for the preparation of organic/inorganic PI with DDSQ in the main chain (Figure 12). First, a DDSQ-diamine was synthesized through a Heck reaction of 3,13-divinyloctaphenyl double-decker silsesquioxane (3,13-divinyl DDSQ) with 4-bromoaniline in the presence of a palladium catalyst; next, thermal imidization of the 3,13-dianilino DDSQ with ODA and 3,3′,4,4′-benzophenonetetracarboxylic in DMAc afforded PI/DDSQ nanocomposites.

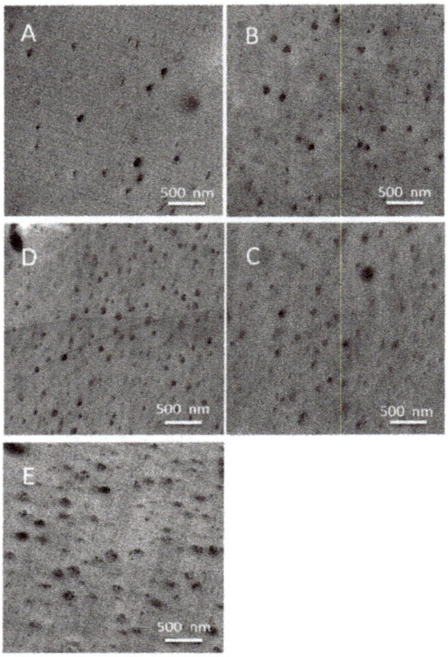

Figure 11. TEM micrographs of the synthesis of PIs containing various contents of 5,11,14,17 tetranilino DDSQ [81]. Reproduced with permission from Wiley.

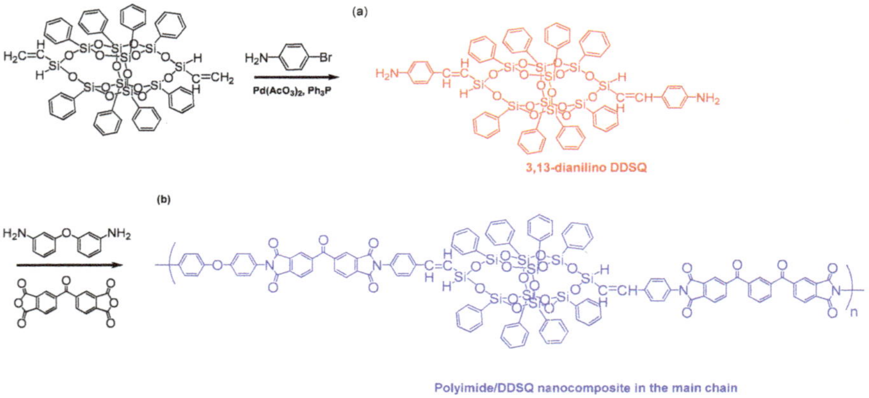

Figure 12. Preparation of organic/inorganic PIs with DDSQ in the main chain [82]. Reproduced with permission from the Royal Society of Chemistry.

According to the dielectric analysis, the dielectric constant of the PI/DDSQ nanocomposites decreased upon incorporating larger amounts of the DDSQ into the main chain (Figure 13). Furthermore, the thermal stability and surface hydrophobicity of the organic/inorganic nanocomposites were enhanced after the inclusion of the DDSQ in the main chain.

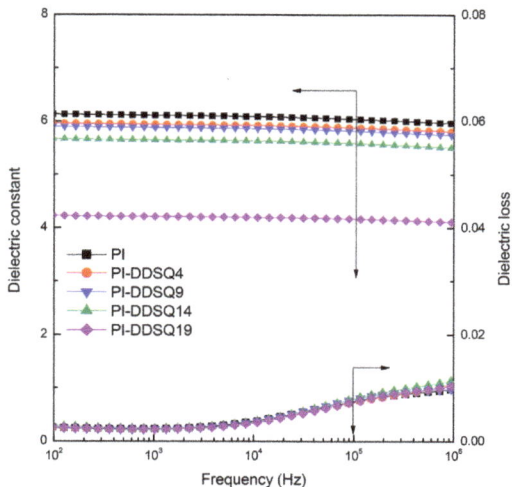

Figure 13. Plots of dielectric constant of organic/inorganic PIs with DDSQ in the main chain [82]. Reproduced with permission from the Royal Society of Chemistry.

2.4. Thermally Crosslinked PIs Incorporating Multi-Functional POSS NPs

We have prepared well-defined PI hybrid materials containing POSS NPs through the copolymerization of octakis(glycidyldimethylsiloxy)octasilsesquioxane (OG-POSS), 4,4′-carbonyldiphthalic anhydride (BTDA), and 4,4′-oxydianiline diamine (ODA) (Figure 14) [83]. The PI-POSS hybrid materials had low dielectric constants and thermal expansion coefficients that decreased from 66.23 to 63.28 to 58.25 ppm/°C when the POSS content increased from 0 to 10 wt %, presumably because of the increase in free volume caused by the presence of the POSS-tethered network.

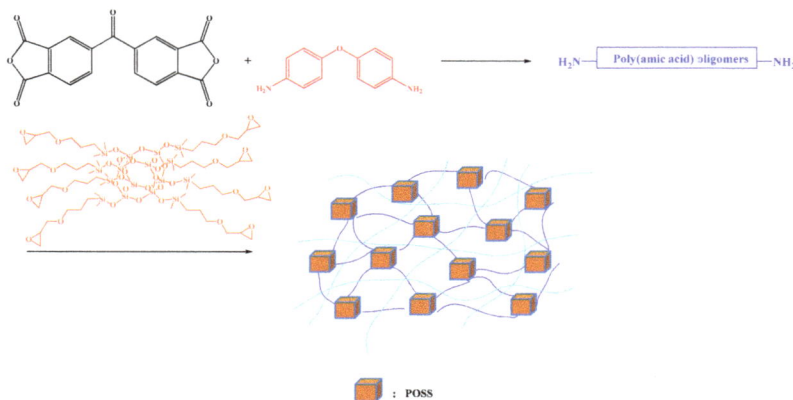

Figure 14. Preparation of OG-POSS cross-linked PIs [83]. Reproduced with permission from Elsevier.

Li et al. prepared a series of octa(aminophenyl)silsesquioxane (OAPS) covalently cross-linked sulfonated PIs (SPIs) (Figure 15) [84]. These OAPS cross-linked PI membranes possessed greater chemical and thermal stability, higher strength, and excellent solution processability when compared with those of the linear sulfonated PI membrane. Moreover, when the OAPS content was 0.4 mol %, the proton conductivity of the cross-linked SPI membrane reached its highest value (0.111 S/cm).

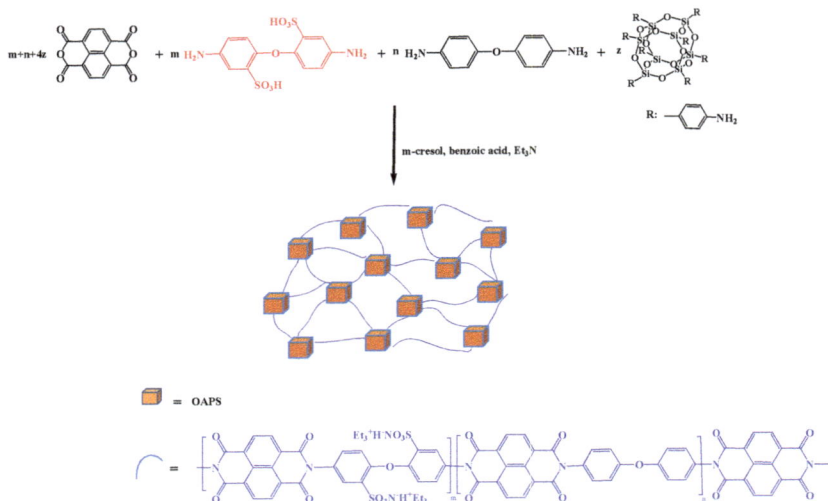

Figure 15. Preparation of POSS cross-linked cross-linked sulfonated PIs (SPIs) [84]. Reproduced with permission from Elsevier.

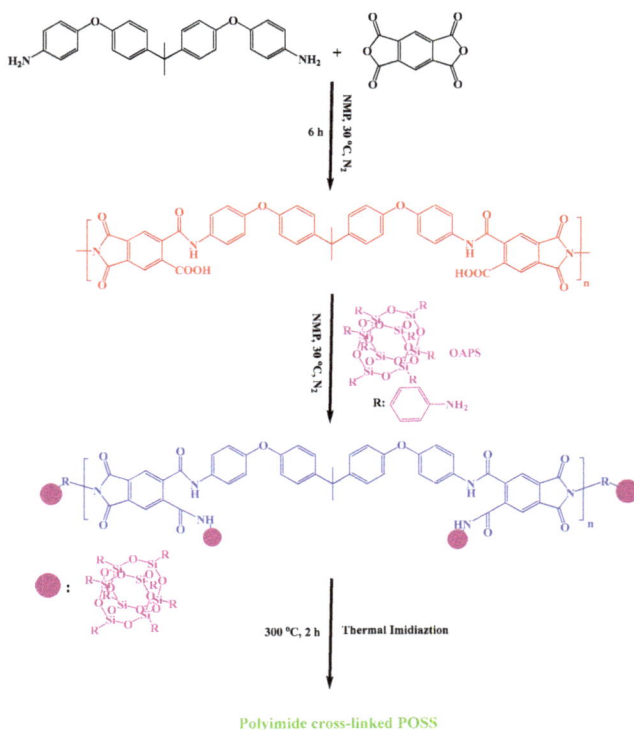

Figure 16. Schematic representation of PI cross-linked POSS [85]. Reproduced with permission from Elsevier.

Alagar et al. synthesized PI/POSS hybrid) materials through a two-step method (Figure 16) [85]. First, the formation of PAA from bisphenol-A ether diamine (BAED) and PMDA in NMP under N$_2$ at 30 °C, and its blending with various percentages of OAPS; subsequently, thermal imidization of the PAA/OAPS nanocomposites at 300 °C to obtain the POSS-PI nanocomposites. Thermal studies revealed that the glass transition temperature, char yield, thermal stability, and flame-retardant properties of the POSS-PI hybrid nanocomposite were all greater than those of the pristine PI. Adıguzel et al. [86] prepared star polyimides containing polyhedralsilsesquioxane as the core through in situ thermal curing of PAA with octa(aminopolyhedralsilsesquioxane) (Figure 17). The POSS-PI star nanocomposites exhibited unique characteristics including low water absorption, low dielectric constant, and good mechanical and thermal properties arising from the insertion of the rigid POSS units in the PI backbone.

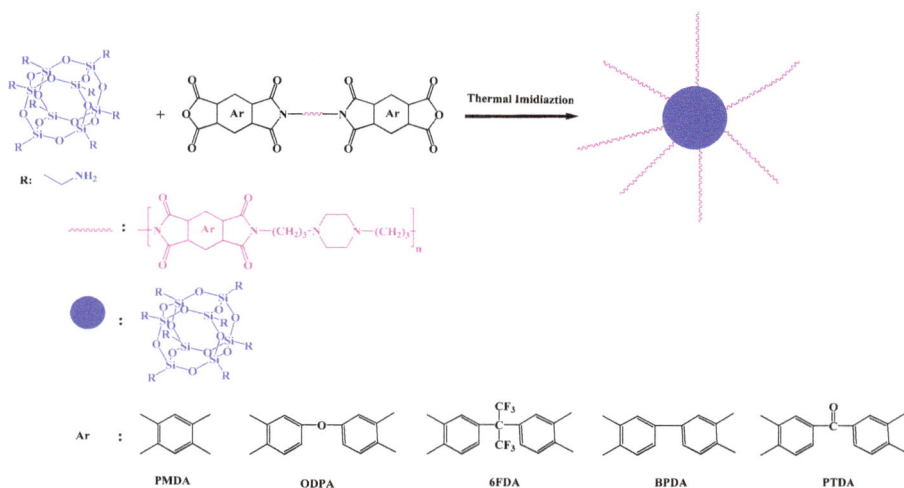

Figure 17. Preparation of star POSS/PI hybrid materials [86]. Reproduced with permission from Elsevier.

3. Conclusions

In this Review, we discussed various preparation types of PIs containing POSS NPs that displayed attractive characteristics, including low dielectric constants, low water absorption, and methanol permeability; excellent mechanical and electrical properties; and high thermal stability, surface hydrophobicity, flame retardance, chemical resistance, proton conductivity, and thermo-redox stability compared with those of their pristine PI precursors. The insertion of these POSS units into the PI precursors has been performed through either physical blending or covalent bonding. The preparation of organic/inorganic hybrid materials containing POSS and PI remains an interesting topic in academic and industrial science because of their potential applications in fuel cells and gas separation membranes, as insulator materials in microelectronics, and in the aerospace industry.

Author Contributions: M.G.M. and S.W.K. both contributed to the literature review and to the writing of this paper.

Conflicts of Interest: The authors declare no conflict of interest.

References

1. Shea, K.J.; Loy, D.A. Bridged polysilsesquioxanes. Molecular-engineered hybrid Organic-inorganic materials. *Chem. Mater.* **2001**, *13*, 3306–3319. [CrossRef]
2. Kuo, S.W.; Chang, F.C. POSS related polymer nanocomposites. *Prog. Polym. Sci.* **2011**, *36*, 1649–1696. [CrossRef]
3. Cordes, D.B.; Lickiss, P.D.; Rataboul, F. Recent developments in the chemistry of cubic polyhedral oligosilsesquioxanes. *Chem. Rev.* **2010**, *110*, 2081–2173. [CrossRef] [PubMed]
4. Wu, J.; Mather, P.T. POSS polymers: Physical properties and biomaterials applications. *Polym. Rev.* **2009**, *49*, 25–63. [CrossRef]
5. Li, Z.; Kong, J.; Wang, F.; He, C. Polyhedral oligomeric silsesquioxanes (POSSs): An important building block for organic optoelectronic materials. *J. Mater. Chem. C* **2017**, *5*, 5283–5298. [CrossRef]
6. Li, Z.; Tan, B.H.; Jin, G.; Li, K.; He, C. Design of polyhedral oligomeric silsesquioxane (POSS) based thermo-responsive amphiphilic hybrid copolymers for thermally denatured protein protection applications. *Polym. Chem.* **2014**, *5*, 6740–6753. [CrossRef]
7. Blanco, I.; Bottino, F.A.; Bottino, P. Influence of symmetry/asymmetry of the NPs structure on the thermal stability of polyhedral oligomeric silsesquioxane/polystyrene nanocomposites. *Polym. Compos.* **2012**, *33*, 1903–1910. [CrossRef]
8. Moore, B.M.; Ramirez, S.M.; Yandek, G.R.; Haddad, T.S.; Mabry, J.M. Asymmetric aryl polyhedral oligomeric silsesquioxanes (ArPOSS) with enhanced solubility. *J. Organomet. Chem.* **2011**, *696*, 2676–2680. [CrossRef]
9. Kuo, S.W. Building Blocks Precisely from Polyhedral Oligomeric Silsesquioxane Nanoparticles. *ACS Cent. Sci.* **2016**, *2*, 62–64. [CrossRef]
10. Mohamed, M.G.; Jheng, Y.R.; Yeh, S.L.; Chen, T.; Kuo, S.W. Unusual Emission of Polystyrene-Based Alternating Copolymers Incorporating Aminobutyl Maleimide Fluorophore-Containing Polyhedral Oligomeric Silsesquioxane NPs. *Polymers* **2017**, *9*, 103. [CrossRef]
11. Zhang, W.; Camino, G.; Yang, R. Polymer/polyhedral oligomeric silsesquioxane (POSS) nanocomposites: An overview of fire retardance. *Prog. Polym. Sci.* **2017**, *67*, 77–125. [CrossRef]
12. Zhou, Z.; Cui, L.; Zhang, Y.; Zhang, Y.; Yin, N. Preparation and properties of POSS grafted polypropylene by reactive blending. *Eur. Polym. J.* **2008**, *44*, 3057–3066. [CrossRef]
13. Gnanasekaran, D.; Madhaven, K.; Reddy, B. Developments of polyhedral oligomeric silsesquioxanes (POSS), POSS nanocomposites and their applications: A review. *J. Sci. Ind. Res.* **2009**, *68*, 437–464.
14. Chou, C.H.; Hsu, S.L.; Dinakaran, K.; Chiu, M.Y.; Wei, K.H. Synthesis and Characterization of Luminescent Polyfluorenes Incorporating Side-Chain-Tethered Polyhedral Oligomeric Silsesquioxane Units. *Macromolecules* **2005**, *38*, 745–751. [CrossRef]
15. Ak, M.; Gacal, B.; Kiskan, B.; Yagci, Y.; Toppare, L. Enhancing electrochromic properties of polypyrrole by silsesquioxane nanocages. *Polymer* **2008**, *49*, 2202–2210. [CrossRef]
16. Guenthner, A.J.; Lamison, K.R.; Lubin, L.M.; Haddad, T.S.; Mabry, J.M. Hansen Solubility Parameters for Octahedral Oligomeric Silsesquioxanes. *Ind. Eng. Chem. Res.* **2012**, *51*, 12282–12293. [CrossRef]
17. Zhang, W.; Müller, A.H.E. Synthesis of tadpole-shaped POSS-containing hybrid polymers via "click chemistry". *Polymer* **2010**, *51*, 2133–2139. [CrossRef]
18. Guo, H.; Meador, M.A.B.; McCorkle, L.; Quade, D.J.; Guo, J.; Hamilton, B.; Cakmak, M. Tailoring Properties of Cross-Linked Polyimide Aerogels for Better Moisture Resistance, Flexibility, and Strength. *ACS Appl. Mater. Interface* **2012**, *4*, 5422–5429. [CrossRef]
19. Maji, K.; Halder, D. POSS-Appended Diphenylalanine: Self-Cleaning, Pollution-Protective, and Fire-Retardant Hybrid Molecular Material. *ACS Omega* **2017**, *2*, 1938–1946. [CrossRef]
20. Pramudya, I.; Rico, C.G.; Lee, C.; Chung, H. POSS-Containing Bioinspired Adhesives with Enhanced Mechanical and Optical Properties for Biomedical Applications. *Biomacromolecules* **2016**, *17*, 3853–3861. [CrossRef]
21. Ueda, K.; Tanaka, K.; Chujo, Y. Fluoroalkyl POSS with Dual Functional Groups as a Molecular Filler for Lowering Refractive Indices and Improving Thermomechanical Properties of PMMA. *Polymers* **2018**, *10*, 1332. [CrossRef]
22. Liu, Y.; Wu, X.; Sun, Y.; Xie, W. POSS Dental Nanocomposite Resin: Synthesis, Shrinkage, Double Bond Conversion, Hardness, and Resistance Properties. *Polymers* **2018**, *10*, 369. [CrossRef]

23. Blanco, I. The Rediscovery of POSS: A Molecule Rather than a Filler. *Polymers* **2018**, *10*, 904. [CrossRef]
24. Huang, K.W.; Tsai, L.W.; Kuo, S.W. Influence of octakis-functionalized polyhedral oligomeric silsesquioxanes on the physical properties of their polymer nanocomposites. *Polymer* **2009**, *50*, 4876–4887. [CrossRef]
25. Blanco, I.; Abate, L.; Bottino, F.A. Mono substituted octaphenyl POSSs: The effects of substituents on thermal properties and solubility. *Thermochim. Acta* **2017**, *655*, 117–123. [CrossRef]
26. Blanco, I.; Bottino, F.A. The influence of the nature of POSSs cage's periphery on the thermal stability of a series of new bridged POSS/PS nanocomposites. *Polym. Degrad. Stab.* **2015**, *121*, 180–186. [CrossRef]
27. Hu, W.H.; Huang, K.W.; Chiou, C.W.; Kuo, S.W. Complementary Multiple Hydrogen Bonding Interactions Induce the Self-Assembly of Supramolecular Structures from Heteronucleobase-Functionalized Benzoxazine and Polyhedral Oligomeric Silsesquioxane Nanoparticles. *Macromolecules* **2012**, *45*, 9020–9028. [CrossRef]
28. Wu, Y.C.; Kuo, S.W. Self-assembly supramolecular structure through complementary multiple hydrogen bonding of heteronucleobase-multifunctionalized polyhedral oligomeric silsesquioxane (POSS) complexes. *J. Mater. Chem.* **2012**, *22*, 2982–2991. [CrossRef]
29. Chiou, C.W.; Lin, Y.C.; Wang, L.; Hirano, C.; Suzuki, Y.; Hayakawa, T.; Kuo, S.W. Strong Screening Effect of Polyhedral Oligomeric Silsesquioxanes (POSS) Nanoparticles on Hydrogen Bonded Polymer Blends. *Polymers* **2014**, *6*, 926–948. [CrossRef]
30. Chen, W.C.; Lin, R.C.; Tseng, S.M.; Kuo, S.W. Minimizing the Strong Screening Effect of Polyhedral Oligomeric Silsesquioxane Nanoparticles in Hydrogen-Bonded Random Copolymers. *Polymers* **2018**, *10*, 303. [CrossRef]
31. Mohamed, M.G.; Kuo, S.W. Polybenzoxazine/Polyhedral Oligomeric Silsesquioxane (POSS) Nanocomposites. *Polymers* **2016**, *6*, 225. [CrossRef]
32. Mohamed, M.G.; Hsu, K.C.; Hong, J.L.; Kuo, S.W. Unexpected fluorescence from maleimide-containing polyhedral oligomeric silsesquioxanes: Nanoparticle and sequence distribution analyses of polystyrene-based alternating copolymers. *Polym. Chem.* **2016**, *7*, 135–145. [CrossRef]
33. Laine, R.M.; Roll, M.F. Polyhedral Phenylsilsesquioxanes. *Macromolecules* **2011**, *44*, 1073–1109. [CrossRef]
34. Lin, Y.C.; Kuo, S.W. Hierarchical self-assembly structures of POSS-containing polypeptide block copolymers synthesized using a combination of ATRP, ROP and click chemistry. *Polym. Chem.* **2012**, *3*, 882–891. [CrossRef]
35. Lin, Y.C.; Kuo, S.W. Hierarchical self-assembly and secondary structures of linear polypeptides graft onto POSS in the side chain through click chemistry. *Polym. Chem.* **2012**, *3*, 162–171. [CrossRef]
36. Tomasz, P.; Anna, K.; Bozena, Z.; Marek, J.P. Structure, dynamics, and host-guest interactions in POSS functionalized cross-linked nanoporous hybrid organic-inorganic polymers. *J. Phys. Chem. C* **2015**, *119*, 26575–26587.
37. Liao, Y.T.; Lin, Y.C.; Kuo, S.W. Highly Thermally Stable, Transparent, and Flexible Polybenzoxazine Nanocomposites by Combination of Double-Decker-Shaped Polyhedral Silsesquioxanes and Polydimethylsiloxane. *Macromolecules* **2017**, *50*, 5739–5747. [CrossRef]
38. Hong, B.; Thoms, T.P.S.; Murfee, H.J.; Lebrun, M.L. Highly dendritic macromolecules with core polyhedral silsesquioxane functionalities. *Inorg. Chem.* **1997**, *36*, 6146–6147. [CrossRef]
39. Phillips, S.H.; Haddad, T.S.; Tomczak, S.J. Development in nanoscience: Polyherdral oligomeric silsesquioxane (POSS)-polymers. *Curr. Opin. Solid State Mater. Sci.* **2004**, *8*, 21–29. [CrossRef]
40. Ayandele, E.; Sarkar, B.; Alexandridis, P. Polyhedral Oligomeric Silsesquioxane (POSS)-Containing Polymer Nanocomposites. *Nanomaterials* **2012**, *2*, 445–475. [CrossRef]
41. Balazs, A.C.; Emrick, T.; Russell, T.P. Nanoparticle polymer composites: Where two small worlds meet. *Science* **2006**, *314*, 1107–1110. [CrossRef] [PubMed]
42. Anderson, J.A.; Sknepnek, R.; Travesset, A. Design of polymer nanocomposites in solution by polymer functionalization. *Phys. Rev. E* **2010**, *82*, 021803–0218011. [CrossRef] [PubMed]
43. Bockstaller, M.R.; Mickiewicz, R.A.; Thomas, E.L. Block copolymer nanocomposites: Perspectives for tailored functional materials. *Adv. Mater.* **2005**, *17*, 1331–1349. [CrossRef]
44. Ahmad, Z.; Mark, J.E. Polyimide-Ceramic Hybrid Composites by the Sol-Gel Route. *Chem. Mater.* **2001**, *13*, 3320–3330. [CrossRef]
45. Govindaraj, B.; Sundararajanb, P.; Muthusamy Sarojadevi, M. Synthesis and characterization of polyimide/polyhedral oligomeric silsesquioxane nanocomposites containing quinolyl moiety. *Polym. Int.* **2012**, *61*, 1344–1352. [CrossRef]

46. Ando, S.; Harada, M.; Okada, T.; Ishige, R. Effective Reduction of Volumetric Thermal Expansion of Aromatic Polyimide Films by Incorporating Interchain Crosslinking. *Polymers* **2018**, *10*, 761. [CrossRef]

47. Mathews, A.S.; Kim, I.; Ha, C.S. Fully aliphatic polyimides from adamantane-based diamines for enhanced thermal stability, solubility, transparency, and low dielectric constant. *J. Appl. Polym. Sci.* **2006**, *102*, 3316–3326. [CrossRef]

48. Liaw, D.-J.; Liaw, B.-Y.; Li, L.-J.; Sillion, B.; Mercier, R.; Thiria, R.; Sekiguchi, H. Synthesis and characterization of new soluble polyimides from 3,3′,4,4′-benzhydrol tetracarboxylic dianhydride and various diamines. *Chem. Mater.* **1998**, *10*, 734–739. [CrossRef]

49. Wachsman, E.D.; Frank, C.W. Effect of cure history on the morphology of polyimide: Fluorescence spectroscopy as a method for determining the degree of cure. *Polymer* **1988**, *29*, 1191–1197. [CrossRef]

50. Sroog, C.E. Polyimides. *J. Polym. Sci. Macromol. Rev.* **1976**, *11*, 161–208. [CrossRef]

51. Choi, J.Y.; Yu, H.C.; Lee, J.; Jeon, J.; Im, J.; Jan, J.; Jin, S.W.; Kim, K.K.; Cho, S.; Chung, C.M. Preparation of Polyimide/Graphene Oxide Nanocomposite and Its Application to Nonvolatile Resistive Memory Device. *Polymers* **2018**, *10*, 901. [CrossRef]

52. Chen, Y.; Lin, B.; Zhang, X.; Wang, J.; Lai, C.; Sun, Y.; Liu, Y.; Yang, H. Enhanced Dielectric Properties of Amino-Modified-CNT/Polyimide Composite Films with A Sandwich Structure. *J. Mater. Chem. A* **2014**, *2*, 14118–14126. [CrossRef]

53. Lei, X.; Chen, Y.; Qiao, M.; Tian, L.; Zhang, Q. Hyperbranched Polysiloxane (HBPSi)-Based Polyimide Films with Ultralow Dielectric Permittivity, Desirable Mechanical and Thermal Properties. *J. Mater. Chem. C* **2016**, *14*, 2134–2146. [CrossRef]

54. Simpson, J.O.; St. Clair, A.K. Fundamental Insight on Developing Low Dielectric Constant Polyimides. *Thin Solid Films* **1997**, *309*, 480–485. [CrossRef]

55. Lyulin, S.V.; Larin, S.V.; Gurtovenko, A.A.; Nazarychev, V.M.; Falkovich, S.G.; Yudin, V.E.; Svetlichnyi, V.M.; Gofman, I.V.; Lyulin, A.V. Thermal Properties of Bulk Polyimides: Insights from Computer Modeling Versus Experiment. *Soft Matter* **2014**, *10*, 1224–1232. [CrossRef] [PubMed]

56. Chi, Q.; Sun, J.; Zhang, C.; Liu, G.; Lin, J.; Wang, Y.; Wang, X.; Lei, Q. Enhanced Dielectric Performance of Amorphous Calcium Copper Titanate/Polyimide Hybrid Film. *J. Mater. Chem. C* **2014**, *2*, 172–177. [CrossRef]

57. Liao, W.H.; Yang, S.Y.; Hsiao, S.T.; Wang, Y.S.; Li, S.M.; Tien, H.W.; Ma, C.C.M.; Zeng, S.J. A Novel Approach to Prepare Graphene Oxide/Soluble Polyimide Composite Films with A Low Dielectric Constant and High Mechanical Properties. *RSC Adv.* **2014**, *4*, 51117–51125. [CrossRef]

58. Lei, X.F.; Chen, Y.; Zhang, H.P.; Li, X.J.; Yao, P.; Zhang, Q.Y. Space Survivable Polyimides with Excellent Optical Transparency and Self-Healing Properties Derived from Hyperbranched Polysiloxane. *ACS Appl. Mater. Interfaces* **2013**, *5*, 10207–10220. [CrossRef]

59. Kim, S.; Wang, X.; Ando, S.; Wang, X. Low Dielectric and Thermally Stable Hybrid Ternary Composites of Hyperbranched and Linear Polyimides with SiO$_2$. *RSC Adv.* **2014**, *4*, 27267–27276. [CrossRef]

60. Hasegawa, M.; Horiuchi, M.; Wada, Y. Polyimides Containing Trans-1,4-cyclohexane Unit (II). Low-K and Low-CTE Semi- and Wholly Cycloaliphatic Polyimides. *High Perform. Polym.* **2007**, *19*, 175–193. [CrossRef]

61. Fan, H.; Yang, R. Flame-Retardant Polyimide Cross-Linked with Polyhedral Oligomeric Octa(aminophenyl)silsesquioxane. *Ind. Eng. Chem. Res.* **2013**, *52*, 2493–2500. [CrossRef]

62. Zhang, L.; Tian, G.; Wang, X.; Qi, S.; Wu, Z.; Wu, D. Polyimide/ladder-like polysilsesquioxane hybrid films: Mechanical performance, microstructure and phase separation behaviors. *Compos. Part B* **2014**, *56*, 808–810. [CrossRef]

63. Wu, Y.W.; Zhang, W.C.; Yang, R.J. Ultralight and Low Thermal Conductivity Polyimide-Polyhedral Oligomeric Silsesquioxanes Aerogels. *Macromol. Mater. Eng.* **2017**, 1700403–1700415. [CrossRef]

64. Tyan, H.L.; Leu, C.M.; Wei, K.H. Effect of Reactivity of Organics-Modified Montmorillonite on the Thermal and Mechanical Properties of Montmorillonite/Polyimide Nanocomposites. *Chem. Mater.* **2001**, *13*, 222–226. [CrossRef]

65. Wu, S.; Hayakawa, T.; Kakimoto, M.; Oikawa, H. Synthesis and Characterization of Organosoluble Aromatic Polyimides Containing POSS in Main Chain Derived from Double-Decker-Shaped Silsesquioxane. *Macromolecules* **2008**, *41*, 3481–3487. [CrossRef]

66. Jiang, L.Y.; Wei, K.H. Bulk and surface properties of layered silicates/fluorinated polyimide nanocomposites. *J. Appl. Phys.* **2002**, *92*, 6219–6223. [CrossRef]

67. Huang, J.; He, C.; Xiao, Y.; Mya, K.Y.; Dai, J.; Siow, Y.P. Polyimide/POSS nanocomposites: Interfacial interaction, thermal properties and mechanical properties. *Polymer* **2003**, *44*, 4491–4499. [CrossRef]

68. Tamaki, R.; Choi, J.; Laine, R.A. Polyimide Nanocomposite from Octa(aminophenyl)silsesquioxane. *Chem. Mater.* **2003**, *15*, 793–797. [CrossRef]

69. Lin, C.H.; Chang, S.L.; Cheng, P.W. Soluble high-T_g polyetherimides with good flame retardancy based on an asymmetric phosphinated etherdiamine. *J. Polym. Sci. Part A Polym. Chem.* **2011**, *49*, 1331–1340. [CrossRef]

70. Ye, Y.S.; Chen, W.Y.; Wang, Y.Z. Synthesis and Properties of Low-Dielectric-Constant Polyimides with Introduced Reactive Fluorine Polyhedral Oligomeric Silsesquioxanes. *J. Polym. Sci. Part A Polym.* **2006**, *44*, 5391–5402. [CrossRef]

71. Ye, Y.S.; Yen, Y.C.; Chen, W.Y.; Cheng, C.C.; Chang, F.C. A simple approach toward low-dielectric polyimide nanocomposites: Blending the polyimide precursor with a fluorinated polyhedral oligomeric silsesquioxane. *J. Polym. Sci. Part A Polym. Chem.* **2008**, *46*, 6292–6304. [CrossRef]

72. Liu, H.; Zheng, S.; Nie, K. Morphology and Thermomechanical Properties of Organic-Inorganic Hybrid Composites Involving Epoxy Resin and an Incompletely Condensed Polyhedral Oligomeric Silsesquioxane. *Macromolecules* **2005**, *38*, 5088–5097. [CrossRef]

73. Xu, H.; Kuo, S.W.; Lee, J.S.; Chang, F.C. Glass transition temperatures of poly(hydroxystyrene-co-vinylpyrrolidone-co-isobutylstyrylpolyhedraloligosilsesquioxanes). *Polymer* **2003**, *43*, 5117–5124. [CrossRef]

74. Liao, W.H.; Yang, Y.S.; Hsiao, T.S.; Wang, Y.S.; Li, M.S.; Ma, C.C.; Tien, H.W.; Zeng, J.S. Effect of Octa(aminophenyl) Polyhedral Oligomeric Silsesquioxane Functionalized Graphene Oxide on the Mechanical and Dielectric Properties of Polyimide Composites. *ACS Appl. Mater. Interfaces* **2014**, *6*, 15802–15812. [CrossRef] [PubMed]

75. Lee, Y.J.; Huang, J.M.; Kuo, S.W.; Chang, F.C. Low-dielectric, nanoporous polyimide films prepared from PEO-POSS NPs. *Polymer* **2005**, *46*, 10056–10065. [CrossRef]

76. Leu, C.M.; Reddy, G.M.; Wei, K.H.; Shu, C.F. Synthesis and Dielectric Properties of Polyimide-Chain-End Tethered Polyhedral Oligomeric Silsesquioxane Nanocomposites. *Chem. Mater.* **2003**, *15*, 2261–2265. [CrossRef]

77. Wang, C.Y.; Chen, W.T.; Xu, C.; Zhao, X.Y.; Jian, L. Fluorinated Polyimide/POSS Hybrid Polymers with High Solubility and Low Dielectric Constant. *Chin. J. Polym. Sci.* **2016**, *34*, 1363–1372. [CrossRef]

78. Leu, C.M.; Chang, Y.T.; Wei, K.H. Synthesis and Dielectric Properties of Polyimide-Tethered Polyhedral Oligomeric Silsesquioxane (POSS) Nanocomposites via POSS-diamine. *Macromolecules* **2003**, *36*, 9122–9127. [CrossRef]

79. Chen, Y.; Chen, L.; Nie, H.; Kang, E.T. Low-k Nanocomposite Films Based on Polyimides with Grafted Polyhedral Oligomeric Silsesquioxane. *J. Appl. Polym. Sci.* **2006**, *99*, 2226–2232. [CrossRef]

80. Wu, S.; Hayakawa, T.; Kikuchi, R.; Grunzinger, S.J.; Kakimoto, M. Synthesis and Characterization of Semiaromatic Polyimides Containing POSS in Main Chain Derived from Double-Decker-Shaped Silsesquioxane. *Macromolecules* **2007**, *40*, 5698–5705. [CrossRef]

81. Qiu, J.; Xu, S.; Liu, N.; Wei, K.; Li, L.; Zheng, S. Organic–inorganic polyimide nanocomposites containing a tetrafunctional polyhedral oligomeric silsesquioxane amine: Synthesis, morphology and thermomechanical properties. *Polym. Int.* **2018**, *67*, 301–312. [CrossRef]

82. Zheng, S.; Wang, L.; Liu, N. Organic-inorganic polyimides with double decker silsesquioxane in the main chains. *Polym. Chem.* **2016**, *7*, 1158–1167.

83. Lee, Y.J.; Huang, J.M.; Kuo, S.W.; Lu, J.S.; Chang, F.C. Polyimide and polyhedral oligomeric silsesquioxane nanocomposites for low-dielectric applications. *Polymer* **2005**, *46*, 173–181. [CrossRef]

84. Gong, C.; Liang, Y.; Qi, Z.; Li, H.; Wu, Z.; Zhang, Z.; Zhang, S.; Zhang, X.; Li, Y. Solution processable octa(aminophenyl)silsesquioxane covalently cross-linked sulfonated polyimides for proton exchange membranes. *J. Membr. Sci.* **2015**, *476*, 364–372. [CrossRef]

Polymers **2019**, *11*, 26

85. Devaraju, S.; Vengatesan, M.R.; Selvi, M.; Kumar, A.A.; Alagar, M. Synthesis and characterization of bisphenol-A ether diamine-based polyimide POSS nanocomposites for low K dielectric and flame-retardant applications. *High Perform. Polym.* **2012**, *24*, 85–96. [CrossRef]

86. Seckin, T.; Köytepe, S.; Adıgüzel, H.I. Molecular design of POSS core star polyimides as a route to low-k dielectric materials. *Mater. Chem. Phys.* **2008**, *112*, 1040–1046. [CrossRef]

MDPI

St. Alban-Anlage 66

4052 Basel

Switzerland

Tel. +41 61 683 77 34

Fax +41 61 302 89 18

www.mdpi.com

Polymers Editorial Office

E-mail: polymers@mdpi.com

www.mdpi.com/journal/polymers

www.ingramcontent.com/pod-product-compliance
Lightning Source LLC
Chambersburg PA
CBHW051716210326
41597CB00032B/5499

* 9 7 8 3 0 3 9 2 1 9 9 4 0 *